Carbon Nanotube Enhanced Aerospace Composite Materials

SOLID MECHANICS AND ITS APPLICATIONS
Volume 188

Series Editor: G.M.L. GLADWELL
Department of Civil Engineering
University of Waterloo
Waterloo, Ontario, Canada N2L 3GI

Aims and Scope of the Series

The fundamental questions arising in mechanics are: *Why?*, *How?*, and *How much?* The aim of this series is to provide lucid accounts written by authoritative researchers giving vision and insight in answering these questions on the subject of mechanics as it relates to solids.

The scope of the series covers the entire spectrum of solid mechanics. Thus it includes the foundation of mechanics; variational formulations; computational mechanics; statics, kinematics and dynamics of rigid and elastic bodies: vibrations of solids and structures; dynamical systems and chaos; the theories of elasticity, plasticity and viscoelasticity; composite materials; rods, beams, shells and membranes; structural control and stability; soils, rocks and geomechanics; fracture; tribology; experimental mechanics; biomechanics and machine design.

The median level of presentation is the first year graduate student. Some texts are monographs defining the current state of the field; others are accessible to final year undergraduates; but essentially the emphasis is on readability and clarity.

For further volumes:
http://www.springer.com/series/6557

A.S. Paipetis • V. Kostopoulos
Editors

Carbon Nanotube Enhanced Aerospace Composite Materials

A New Generation of Multifunctional Hybrid Structural Composites

Springer

Editors
A.S. Paipetis
Materials Science and Engineering
University of Ioannina
Ioannina, Greece

V. Kostopoulos
Mechanical Engineering and Aeronautics
University of Patras
Patras, Greece

ISSN 0925-0042
ISBN 978-94-017-8468-9 ISBN 978-94-007-4246-8 (eBook)
DOI 10.1007/978-94-007-4246-8
Springer Dordrecht Heidelberg New York London

Printed on acid-free paper

Springer is part of Springer Science+Business Media (www.springer.com)

Preface

The well-documented increase in the use of high performance composites as structural materials in aerospace components is continuously raising demands on manufacturers in terms of dynamic performance, structural integrity, reliable life monitoring systems and adaptive actuating abilities. Current technologies are now addressing the above issues separately; material property tailoring and custom design practices are being aimed at enhancement of dynamic and damage tolerance characteristics; at the same time, life monitoring and actuation is being performed with embedded sensors/actuators that may prove to be detrimental to the structural integrity of components.

This contributed volume focuses on current research on the unique properties of carbon nanotubes (CNTs) as an additive in the matrix of Fibre-Reinforced Plastics (FRPs), for producing structural composites with improved mechanical performance as well as sensing/actuating capabilities. The development of new generation composites using CNTs as an additional phase within the matrix is expected to result in enhancement of the damping properties of materials, increased fracture toughness and extension of their individual fatigue life. This is expected to occur due to the multiplicity of energy dispersive mechanisms within materials. At the same time, the percolated CNT network within a composite is expected (1) to be strain sensitive and (2) closely related to internal damage mechanisms within the material, providing thus a sensing and life-assessment tool throughout the service life of the material. The electromechanical response of CNTs may also provide a field for the design of actuating systems comprised of CNT structures of varying degrees of anisotropy that will be incorporated in the composite. Additionally, dependence of the Raman shift on the local stress of CNTs can provide unique insights into stress fields at nanoscale level and their interaction with the macroscale.

The successful combination of CNT properties and existing sensing actuating technologies has led to realization of a multifunctional FRP structure. The current volume presents the state of the art research in the field. The contributions cover key aspects of novel composite systems, i.e. modeling from nanoscale to macroscale, enhancement of structural efficiency, dispersion and manufacturing, integral health monitoring abilities, Raman monitoring, and durability, as well as the capabilities that ordered carbon nanotube arrays offer in terms of sensing and/or actuating in aerospace composites.

June 2011 Alkis S. Paipetis and Vassilis Kostopoulos

Contents

Chapter 1
Carbon Nanotubes for Novel Hybrid Structural Composites with Enhanced Damage Tolerance and Self-Sensing/Actuating Abilities

A.S. Paipetis and V. Kostopoulos

Contents

Abstract Damage tolerance, reliability, and sensing/actuating abilities are within the forefront of research for aerospace composite materials and structures. The scope of this chapter is to identify the potential application of incorporating carbon nanotubes (CNTs) in novel hybrid material systems. CNTs may be employed as an additive in the matrix of Fibre Reinforced Plastics (FRP) for producing structural composites with improved mechanical performance as well as sensing/actuating capabilities. The novel multi-scale reinforced composite materials are by definition multifunctional as they combine better structural performance with smart features that may include strain monitoring, damage sensing and even actuation capabilities. This introductory chapter provides an overview of the concepts and technologies related to the hierarchical composite systems that will be elaborated in the following chapters, i.e. modelling, enhancement of structural efficiency, dispersion and manufacturing, integral health monitoring abilities, Raman monitoring, as well as the capabilities that ordered carbon nanotube arrays offer in terms of sensing and/or actuating in aerospace composites.

A.S. Paipetis (✉)
Department of Materials Engineering, University of Ioannina, 45110 Ioannina, Greece
e-mail: paipetis@cc.uoi.gr

V. Kostopoulos
Applied Mechanics Laboratory, Department of Mechanical Engineering and Aeronautics, University of Patras, 26500 Patras, Greece
e-mail: kostopoulos@mech.upatras.gr

A.S. Paipetis and V. Kostopoulos (eds.), *Carbon Nanotube Enhanced Aerospace Composite Materials*, Solid Mechanics and Its Applications 188,
DOI 10.1007/978-94-007-4246-8_1, © Springer Science+Business Media Dordrecht 2013

Keywords Aerospace composite materials • Multifunctional materials • Carbon nanotubes • Damage tolerance • Structural health monitoring

1.1 Introduction

Current aerospace technology is more than ever focusing on stretching the properties of advanced materials towards their limits. Advanced aerospace composite materials have reached excellent specific properties. A route towards further exploiting advanced structural material is by using enabling technologies for additional functionalities, without compromising structural integrity. In the past few years, novel materials such as carbon nanotubes (CNTs) and related technologies have posed a strong candidacy for providing an integrated approach towards enhanced structural integrity and multifunctionality.

CNTs possess excellent properties in terms of stiffness, strength, and conductivity, and they have exhibited promising properties in terms of actuation. In principle, CNTs may be employed for the realization of a new generation of nano-reinforced composite systems which could potentially replace "conventional composites" in aerospace and other applications. However, being a nano-scale reinforcement, CNTs lack the typical advantages of fibres or of reinforcement at the micron scale, in that they cannot be easily "tailored" to benefit most of their properties by inducing a controlled anisotropy in the structure.

To this end, the concept of "hybrid" or multi-scale composite has been developed (Fig. 1.1). Novel hybrid or hierarchical composite systems may benefit from the advantages of traditional structural composites and, at the same time, gain in properties and functionalities for the incorporation of CNTs as additives in their matrix (Baur and Silverman 2007). In order to benefit from the use of CNTs in conventional fibrous composites, three different levels of complexity may be applied.

1. Nano-Augmentation, meaning that by randomly and homogeneously dispersing CNTs into the matrix material, and following the already used manufacturing routes, improved multifunctional composites may be realised.
2. Nano-Engineering, meaning that by using organized CNT structures, such as 1D in fibre form, 2D in the form of bucky papers or aligned CNTs in plane form or 3D in the form of CNT forests or other special structures and introducing them in the composite laminate, improvement of some characteristics of their mechanical performance as well as additional functionalities can be introduced into conventional laminates.
3. Nano-Design, meaning that starting from the multifunctional performance envelope of the composite and having available the entire span of numerical tools from the molecular dynamic up to macro-scale multi-physics, we may design an appropriate multi-scale hybrid composite in order to serve the specific application needs.

The possibilities offered by the hierarchical approach may be summarized in the following two principles; (i) reinforcement at the nanoscale will enhance the structural

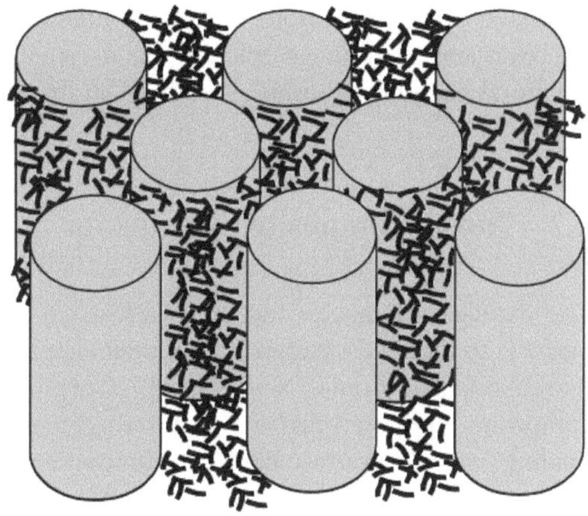

Fig. 1.1 The concept of multi-scale reinforcement in hybrid composites (Reprinted from Vlasveld et al. (2005). With permission from Elsevier)

properties of an otherwise conventional composite by triggering all the mechanisms that make structural composites so attractive at an additional scale, the nanoscale, and (ii) exploitation of the unique properties of CNTs will provide functionalities as real-time strain sensing, structural health monitoring or even actuation capabilities (Thostenson et al. 2001). The research route towards structural enhancement relates to inherent weaknesses of composite laminates such as interlaminar strength or toughness; through thickness reinforcement may be feasible at the nanoscale with mechanisms such as crack bridging at the nanoscale, and as a result increased toughness may be achieved via the energy dissipation mechanisms activated at the additional interface between the matrix and the nano reinforcement (Sun et al. 2009). Undoubtedly, the research in the aforementioned area has raised further issues which relate to dispersion of CNTs in the matrix and the matrix nanotube interface itself (Zhang 2010; Ma et al. 2010). It also raises the question whether the reinforcement at the nanoscale is governed by the same principles as reinforcement at the micro or macro-scale (Duncan et al. 2010).

The research towards additional functionalities was met with immense interest, particularly in the field of strain and damage sensing employing the real-time changes in the resistivity of the material. Reversible changes are due to strain and irreversible changes are due to damage (Li et al. 2008). The monitoring principle lies with the creation of a percolated network within the structure (Bauhofer and Kovacs 2009) that follows the far field applied strain field and is disrupted at any discontinuity induced due to damage initiation and accumulation (Deng and Zheng 2009). Additionally, other properties such as the stress induced changes of the Raman vibrational modes to monitor stress (De la Vega et al. 2011) or the actuating capabilities in electrolytic environments have also been extensively studied (Cooper et al. 2001).

In view of the above, the scope of this chapter is to provide an overview of the research work performed towards exploitation of the aforementioned properties for multi-scale reinforced composite materials, highlighting the problems and enabling technologies for the achievement of a new generation of advanced hybrid composite materials. More analytically, the tailored use of CNTs as nano-reinforcement in advanced aerospace fibrous composite materials will be explored towards (i) the improvement of damage tolerance and (ii) the provision of functionalities for structural health monitoring, stress and strain sensing and actuation.

1.2 Novel Composite Systems for Structural Enhancement

The damage tolerance concept in aerospace structures relates to their ability to perform to required standards within damage limits, which at the same time define its remaining life time (Nettles et al. 2011). This is the main design criterion for composite structures that are exposed to a number of events during in-service loading, which can cause damage initiation and structural degradation. The generally good fatigue resistance of composites aid in the durability and damage tolerance of their design (Lazzeri and Mariani 2009). As far as damage initiation and propagation is concerned, the design of composite structural components is the main challenge. As the reinforcing phase (mainly carbon fibres in the case of aerospace composites) is extremely brittle, the task of increasing the damage tolerance of the material lies with the matrix material. However, most matrix resins are also brittle and hence have limited resistance to damage, which manifests itself as matrix cracks and delaminations. These matrix damage mechanisms may occur as a result of an impact event, some form of environmental degradation or out-of-plane fatigue load. At the same time, as structural composite parts increase in size with a subsequent reduction of structural joints, the problem of passive damping in aerospace materials and structures has reemerged (Li and Crocker 2005). The designer's needs focus on control of unwanted vibrations as well as the need for improved resistance in the distribution of cracks and imperfections of the structure. This resistance will limit the extent of damage that is created in structures by composite materials due to impact with objects of relatively small mass with low speed (Raju Mantena et al. 2009).

Damping is also governed by matrix properties and consequently research has been focused on resin systems (matrix additives, interleaves etc.) (Sager et al. 2011). More analytically, the modification of matrix properties is a key mechanism in improving the damage tolerance of composite materials. Increased matrix toughness leads to improved delamination fracture toughness. In the past decade, research has been focusing on techniques that allow tailoring of the resin properties. These techniques target the maximization of dissipated energy through either a plastic deformation of the matrix (e.g. the inclusion of elastomers which increase the resin toughness (Lee et al. 2010)), or altering of the fracture process (e.g. ceramic modified polymers that inhibit interlaminar crack propagation (Brostow et al. 2011)). Hybrid resin systems such as thermoset/thermoplastic blends (Olmos et al. 2011) are also reported to improve the interlaminar fracture toughness of composite systems. However, brittle resin systems may exhibit high mode II delamination toughness which is attributed to the formation of microcracks ahead of the crack tip; these microcracks dissipate the energy and redistribute the load (Hojo et al. 1997). The inherent constraint of locally controlling the toughness of the matrix ahead of the crack tip is purely geometrical, as the high volume fraction of the reinforcing phase only allows formation of a space restricted plastic zone.

As a subsequent step to matrix properties tailoring, interleaves are also reported to improve the toughness of composites (Hojo et al. 2006). The interleaving technique consists of selective placement of soft and tough strips of resin (or composite) material in interlaminar interfaces that are most prone to delamination. This technique is particularly applied at or near free edges. Interleaving is promising as far as toughness improvement is concerned and its selective application reduces adverse effects on the structural integrity of the system. However, it is obvious that interleaves introduce additional sources of damage and degrade the mechanical properties of the load-bearing elements of the structure by decreasing their stiffness to weight ratio (Zhao et al. 2008b). At the same time, the technique poses limitations on design allowables and the reliability of aerospace structural parts. An obvious geometrical constraint is also present in this technique, as the structural integrity of the component limits the thickness of the interleave (Zhao et al. 2008a).

Last but not least, a method for improving the toughness of composite systems lies with the tailoring of the interface between fibres and matrix. A variety of energy dissipating mechanisms, such as interfacial debonding, post debonding friction and fibre pull-out are directly attributed to the fibre-matrix interface (Fu et al. 2008). The interface is also responsible for the stress magnification and redistribution around a discontinuity (such as a fibre crack) which is directly linked to crack propagation or arrest, the critical flaw size and the failure of the composite. The limits set regarding interfacial modification lie between a strong interface that will not allow crack deflection and lead to brittle failure of the composite and a tough interface that will allow the crack deflection up to the point where the created flaw size within the composite material will be critical to the structural integrity of the component (Krstic 1998).

An alternative approach to interfacial modification that combines the modification of the matrix properties as a macroscopically homogeneous material with the additional benefits of interfacial energy dissipation mechanisms is the inclusion of other phases in the matrix material which are not of the same order of magnitude of the reinforcing phase. This is a well-known technique ranging from carbon black modified rubbers to the use of other modifiers, such as piezoceramic materials (Tsantzalis et al. 2007a). These additives change the toughness as well as the dynamic properties of the material (e.g. both modulus and damping properties).

An interesting scenario is the use of CNTs as an additive (Cho et al. 2009). Due to their nanoscale size and huge aspect ratio and free surface, CNTs are expected to increase by several orders of magnitude the interfacial area in a composite system that employs as a matrix a resin with CNT addition (Fig. 1.2). Moreover, a minimum addition of the order of a few percent can dramatically modify the properties of the matrix material (Colbert 2003). The use of CNT in resin systems has been the basis of the development of new technologies, which explore the compatibility of matrices and CNT tubes and lead to spectacular improvement in structural material properties. As an example, CNT doped PBO fibres have been reported to exhibit twice the energy absorbing capability in relation to conventional PBO fibres (Shelley 2003; Kumar et al. 2002).

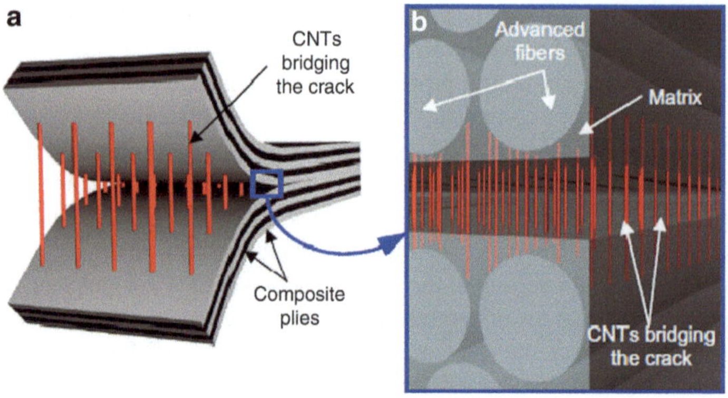

Fig. 1.2 Toughening in multi-scale reinforced composites (Reprinted from Garcia et al. (2008). With permission from Elsevier)

Finally, all matrix modifications do change the dynamic properties of the material (Gibson et al. 2007). Tougher matrices lead to higher damping properties which is a crucial issue in composite structures. The tailoring of the damping properties of the material, as structural joints are minimized and larger structures are feasible, is also a major issue that is currently being dealt with by the aforementioned techniques. As an irreversible process, damping is directly linked to the damage tolerance of the structure.

1.3 Novel Composite Systems for Structural Health Monitoring

The continuous assessment of remaining life of aerospace components at every stage of aircraft service life remains critical in order to ensure its structural integrity and service capacity. Therefore, it plays a major role in the design phase of aerospace components. This has led to the emergence of various structural health monitoring technologies, which by using the appropriate sensing technology aim to provide capabilities for monitoring structural integrity during the service life of an aircraft. Some of the more promising health monitoring concepts are based on smart materials and structures techniques, and incorporate embedded piezoelectric and/or fibre-optic sensors (Luyckx et al. 2011). These can provide continuous local strain field monitoring in real-time during service life, which can provide damage detection and assessment of remaining structural life. On the other hand, the incorporation of active elements, such as piezoceramics and shape memory alloy actuators, provide exciting new horizons in the near future realization of flight control surfaces, active vibration and noise control capabilities (Song et al. 2006). However, current smart technologies are limited by sensor and actuator size, their placement and distribution, and in some cases have detrimental effects on structural integrity of the host component (Yuan et al. 2010). Hence, the development of novel structural material systems combining advanced properties and sensing-actuating capabilities at the micro- and nano-scale is central to the composite design phase.

Fibrous composites provide an ideal medium for implementing smart material technologies as their internal structure and manufacturing methods enable the incorporation of various sensor and actuator forms that will provide health monitoring capability throughout the lifetime span of the component. In this aspect, smart composites are truly multifunctional materials, combining high properties and structural integrity with sensing and actuating capabilities (Akdogan et al. 2005). Yet, the development of smart composite materials remains an open research area, and many issues require consideration.

Nowadays, readily available embedded sensor technologies include fibre optic sensors, piezoelectric sensors and MEMS. Actuator technologies include ferroelectric and electrostrictor ceramics (Wheat et al. 1999), shape memory alloys (Bogue 2009) and magnetostrictive materials (Tuinstra and Koenig 1970; Wilson et al. 2007). Interferometric and – fibre Bragg Grating optic sensors are currently being used for real-time strain monitoring in aerospace structures, such as helicopter blades (Majumder et al. 2008). Fibre optic arrays are also used to assess local failure due to optical signal loss, whereas the change of the speckle pattern from multimode fibres due to mode scrambling has been correlated to a global strain field. Very recently, dynamic fibre Bragg Grating methodologies, accompanied by neural network techniques, have been proposed as a robust tool for SHM of aerospace components (Panopoulou et al. 2011). The main problems associated with fibre optic sensors are (i) the fibre diameter (approximately an order of magnitude bigger than the reinforcing fibre) which in many cases act as stress concentration site, (ii) their low strength at fibre-splicing locations, and (iii) their need for electro-optical signal conversion modules (Barton et al. 2002).

Piezoelectric (piezoceramic and piezopolymer) sensors and piezoceramic actuators are of major interest to the Aerospace industry. In piezoelectric sensors, local dynamic strain is converted to electrical signal, thus providing the ability for real-time monitoring systems (Akdogan et al. 2005). Using this direct piezoelectric effect, mostly surface attached piezoceramic sensors have been used for health monitoring and damage detection in composite structures. Moreover, using the converse piezoelectric effect, piezoceramic forms, such as patches, wafers and stack assemblies, are being used as electromechanical actuators. They have been applied to actively change the shape of aircraft wings, to provide active and passive damping (Horst and Kronig 2001) to avoid resonance phenomena, as in the case of tail buffet in High Performance Twin Tail Aircrafts, and to enhance aeroelastic performance in helicopter blades. The major advantages of piezoelectric materials are their high frequency and their direct electromechanical strain conversion. Disadvantages include low induced strain capability, high density, brittleness, and limited fatigue life.

Shape memory alloys are also used as actuators (Bogue 2009). They are actually quasi-static thermomechanical actuators which can induce very high strains due to martensitic phase transformation. Their major problem is their low frequency bandwidth, their complex thermomechanical behaviour and their limited fatigue

Table 1.1 Comparison of typical properties of sensor and actuator materials

	Piezo-eramic PZT	Piezo-polymer PVDF	Magneto-strictor Terfenol-D	Shape memory alloys	CNTs
Young's Modulus/GPa	70	2	40	20–80	270–1,800
Tensile strength/ MPa	80	180	28	1,000	3,600–63,000
Max. elastic strain/%	0.1	0.2	0.1	0	-
Max. temp./°C	160	80–120	280	400	2,800
Dyn. response bandwidth	<500 kHz	<500 kHz	<10 kHz	<2 Hz	<1 kHz

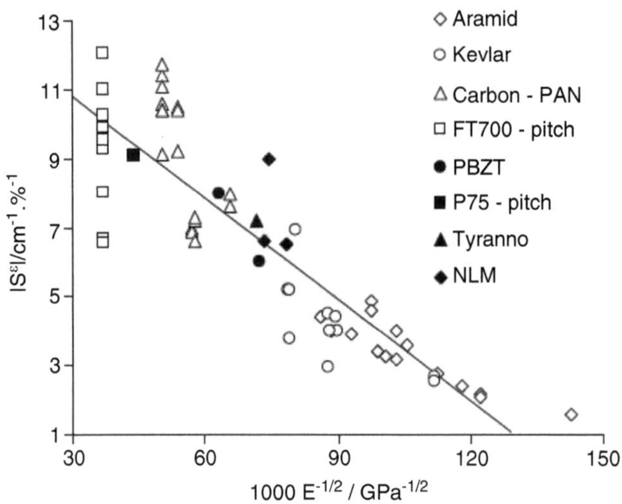

Fig. 1.3 Stress dependent shift of the G band vs. the inverse of Young's modulus square root (Reprinted from Gouadec and Colomban (2007). With permission from Elsevier)

life. The properties of materials used in current sensing/actuating technologies are shown in Table 1.1.

Apart from the aforementioned sensing/actuating techniques, a different approach is to consider the structural phases present in the composite as sensors themselves (Sureeyatanapas et al. 2010). The Raman technique is one of them (Fig. 1.3). The fundamental principle is that the change in the Raman shift frequency of a highly crystalline material – such as a carbon fibre – is directly related to the local stress (Frank et al. 2011). The technique has the resolving power of a focused laser beam that is on the order of a micrometer. Moreover, polarised Raman microscopy can provide preferential information in the case of a randomly dispersed reinforcement phase. Although this is not a competing technology for aerospace structures because of a number of drawbacks such as the complexity of the optical/acquisition system, the low penetration depth of laser light which allows only for surface information, and the long acquisition times, it is the only technique that directly relates to the stress field of structural components, and it is excellent in the characterisation of the interrogated material (Parthenios et al. 2002; Dassios et al. 2003; Young et al. 2004; Zhao et al. 2002).

As a different approach, the electric conductivity of the composite is monitored and related to the damage state (Bauhofer and Kovacs 2009; Vavouliotis et al. 2011). Monitoring changes in the electrical conductivity of carbon fibres may be a direct damage indicator (Thostenson et al. 2002). The technique is simple and requires no other embedded sensors; however, it is highly dependent on composite anisotropy and service induced damage, and does not directly relate to matrix properties which dominate the material toughness (Gibson 2010). Similarly, conductive polymer matrices loaded with conductive fillers (carbon blacks for example) are used as sensors (Böger et al. 2008). When stretched, some contacts between the conductive particles can be lost and the conductivity decreases. Conversely, when the material is compressed more contacts can be established and the conductivity increases. However, the composite has to be generally highly loaded, with often more than 20%wt, such that the conductive fillers form an electrically percolating network, consequently this technology can only be used for relatively soft polymer or elastomers which can exhibit large deformations. In a third approach carbon patches are embedded between ply interfaces to monitor changes in through-thickness electric conductivity (Gou et al. 2006). The technique appears to be sensitive to matrix damage; however it usually requires a large number of carbon patches and may adversely affect interlaminar strength.

In the past few years, there has been significant development regarding sensing technologies related to carbon nanotube (CNT) properties, primarily to their electric conductivity (Bauhofer and Kovacs 2009). When small volume fractions of CNTs are added into a polymer matrix, the electrical properties change significantly. In addition, the loading needed to render the polymer conductive is about an order of magnitude less than the respective loading required with carbon black conducting fillers (Sandler et al. 1999; Coleman et al. 1998). This is attributed to the fact that CNTs form a percolating network within the polymer, which due to their high surface aspect ratio is formed at low concentrations. This percolation network can serve to make conductive polymer blends or conductive polymer fibres that can be used to fabricate smart composite systems (Fig. 1.4). The conductivity of these textiles may vary when the material is loaded. More importantly, recent analytical and experimental studies show that the electronic structure and electric conductivity of CNTs can vary upon deformation (Rochefort et al. 1999). This has given significant boost to emerging nano-and micro-technologies (NMT) such as nanometer scale electromechanical sensors and switches. This effect, mainly investigated on a nanoscale, could be exploited to build new NMT strain sensors on micro- and macro-scale, embedded into the matrix, ply interfaces and composite plies of smart composite structures.

Additionally, it has been shown theoretically that the length of CNTs can change by changing their density of charge, acting thereby as new electromechanical

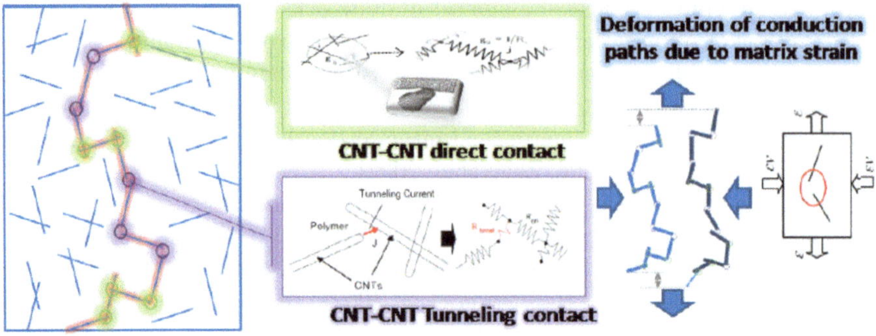

Fig. 1.4 Conductivity changes due to far field strain (Vavouliotis 2008)

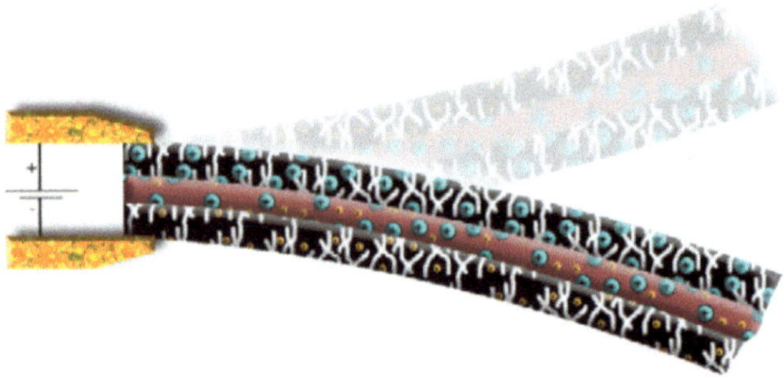

Fig. 1.5 CNT bimorph actuator (Reprinted from Biso et al. (2011). With permission from Elsevier)

actuators (Fig. 1.5). From a theoretical point of view, CNT actuators could exceed by far the properties of other available actuator technologies (Li et al. 2008). The stress and strain generated by CNTs is expected to be one or two orders of magnitude larger than that of piezoceramics, and their time response much faster than that of shape memory alloys. The main challenge to demonstrate and exploit these unique properties in the macroscale remains in fabricating optimized materials mostly comprised of organized nanotubes (Ahir et al. 2008). A first breakthrough was achieved in 1999, with the first experimental evidence of electrochemical actuation using a macroscopic piece of bucky paper comprised of CNTs in a liquid electrolyte. The stress generated by the bucky paper was about 0.8 MPa (twice as much the stress generated by a human muscle) when stimulated with a voltage of only 1 V (Baughman et al. 2002). In comparison, tens or even hundreds of volts are usually required by piezoceramics. Nevertheless, due to the absence of alignment of the CNTs in the bucky paper, the obtained macroscopic properties are still a small fraction of what can be expected to be the properties of individual CNTs. More recently, new processes have been developed to produce macroscopic assemblies of oriented CNTs, such as fibres (Terrones et al. 1997; Li et al. 2000; Cheng et al. 1998; Zhong et al. 2010; Liu et al. 2000). Even though the alignment in nanotubes fibres is not yet optimal, it has been experimentally shown that their properties can be significantly enhanced and the stress generated today by a typical nanotube fibre is about 15 MPa, which is about 20 times greater than the stress generated by isotropic bucky paper (Poulin 2005). Clearly, at this stage of technology, the properties of CNTs actuators start to become really competitive with competing technologies, yet they remain still far from their full potential.

In conclusion, CNTs offer the possibility to perform as nanosensors and microsensors, and at the same time demonstrate opportunities for the creation of new actuator systems embedded as structural elements in future aerospace structures. Compared to existing sensor and actuator technologies, which appear to have inherent limitations, CNTs appear to provide a unique opportunity to develop superior structural composite materials with their reinforcing elements acting as sensors and actuators. The latter provides unprecedented possibilities and applications in aerospace structures.

1.4 The Roadmap to Advanced Hybrid Composite Systems

The scope of this contributed book is to provide an overview of scientific and state of the art technologies that have been leading toward realization of novel composite materials and structural components, which on one hand can exhibit superior structural performance with emphasis in their damage tolerance, and on the other hand can possess inherent sensing capabilities. The enabling actuation technologies in future aerospace structural components via the presence of the nanoscale will also be addressed (Gibson 2010).

To this end, the second chapter of this book is dedicated to the ability of nanoreinforcement to provide sensing functionalities for strain and damage. The approach is based on the principle that CNTs doped within the matrix of a novel composite can form a percolating network at volume fractions much lower than that usually required with carbon blacks or other types of conducting fillers. The conductivity of such a composite has proven to be extremely sensitive to mechanical deformation. In the typical aerospace composite material where the epoxy matrix is an insulator, the conductivity directly depends on the "contacts" between the conductive phases (Li et al. 2008). However, an additional and unique sensing effect will come into play with CNTs, as applied stress and strains are expected to directly affect the electronic structure and electric conductivity of the individual nanotubes. This unique capability opens significant new possibilities because now hard and/or highly cross-linked polymers can be used as the mechanical stress and could be revealed via the change of the conductivity of the nanotubes alone. The change in conductivity is expected to be sensitive enough to provide real-time strain monitoring; it macroscopically remains an irreversible process, which is expected to be directly linked to the residual life of the structural component (Fig. 1.6). That is, the "ageing" of the percolation network manifested through link breakage events can be directly linked to the fatigue life of the system (Kostopoulos et al. 2009). Because of their aspect ratio, if nanotubes are incorporated in the

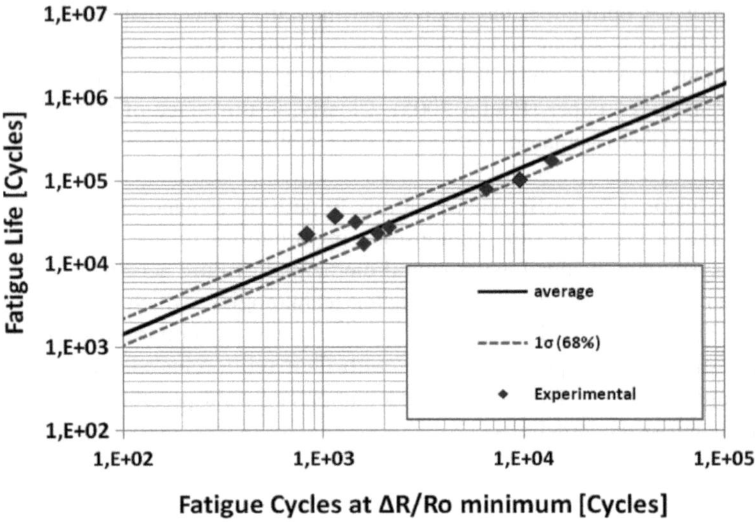

Fig. 1.6 Fatigue life prediction for a hybrid composite material based on its electrical response to fatigue loading (Reprinted from Vavouliotis et al. (2011). With permission from Elsevier)

composite matrix in an oriented manner, the conductive properties will also be anisotropic. This is a unique opportunity to fabricate new sensing composites with the possibility to detect not only the amplitude, but also the orientation of a mechanical load.

The third chapter is dedicated to the employment of ordered nanotube structures as sensors and actuators when embedded in typical aerospace composites. As has been shown, CNTs can be spun into fibres or ribbons of oriented CNTs. Nanotube fibres in particular have successfully been employed as embedded strain sensors in fibrous composites. As in textiles comprised of conductive polymer fibres, nanotubes fibres can serve as sensors. However, in contrast to classical conductive polymer fibres, nanotubes fibres are significantly more stable and thus more suitable for composite applications. CNTs can resist up to approximately 600°C in air (Triantafyllidis et al. 2008), and almost up to 2,000°C in an inert atmosphere (Purcell et al. 2002). Tight-binding (TB) molecular dynamics (MD) simulations revealed that this nanotube is mechanically stable at temperatures as high as 1,100°C (Peng et al. 2000). In addition, because of their great chemical stability, nanotubes are not degraded by UV or by other molecules like surfactants. Organised nanotube structures are also considered as materials for high performance actuators (Poulin 2005), and key aspects of such macroscopic devices are highlighted for their use in composite materials (Vigolo et al. 2000). By improving the manufacturing process of nanotube assemblies, the efficiency of energy conversion in nanotube fibres is further enhanced and thus these structures are among the most promising materials for actuator applications. Actuating abilities are demonstrated in liquid electrolytes although solid systems that allow diffusion and migration of ions are

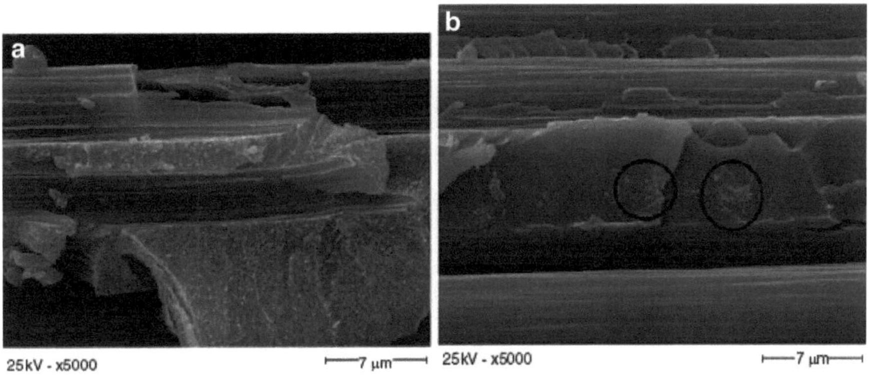

Fig. 1.7 SEM picture of the fracture surface of CNT enhanced composites under Mode I loading, (**a**) efficient dispersion and (**b**) inadequate dispersion as indicated by the presence of agglomerates

promising for rigid actuators which may be developed as model systems for future aerospace applications (Tsai et al. 2010).

The fourth and fifth chapters are dedicated to the dispersion technologies involved with inclusion of nano-reinforcement in the epoxy matrix. In particular, the fifth chapter deals with the technologies of mechanical dispersion of nanotubes in the matrix (Chow and Tan 2010). As has been extensively studied in the past few years, dispersion is probably the key parameter for exploitation of the enhanced properties of nano-reinforcement. Inadequate dispersion (Fig. 1.7) may lead to adverse effects, where the agglomerates of the nanophase are operating as defects rather than reinforcement (Fiedler et al. 2006). On the other hand, the dispersion process itself may damage the nanotubes – initially by reducing their aspect ratio – and consequentially reducing their reinforcing ability. The sixth chapter is devoted to the chemical compatibilisation of the nanophase. The routes towards achievement of this target are highlighted i.e., the use of organophilic CNTs, i.e. (a) nanotubes with attached organic moieties on their surface, and (b) nanotubes with increased interfacial bonding with epoxy matrix by attaching reactive functional groups (Ma et al. 2007).

Raman spectroscopy of CNTs and related structures has proven to be a unique tool for characterization of the structure of the nanotube and for study of the stresses developed within the nano-reinforcement due to stress transfer from the environment (Zhao and Wagner 2003). The latter is directly related (i) to the reinforcing ability of the nanophase and (ii) to employment of nanotubes as stress sensors within composite materials. To this end, the fourth chapter is dedicated to Raman Spectroscopy of CNTs (Dresselhaus et al. 2005) with emphasis on their response to stress fields. The Raman Spectrum of all graphitic structures is presented starting from graphite fibres (Melanitis and Galiotis 1990), to Single Wall CNTs to Multi Wall CNTs and finally to Single and Multi-layer Graphene (Frank et al. 2010) and distinct differences are highlighted. The induced changes in the Raman Spectrum

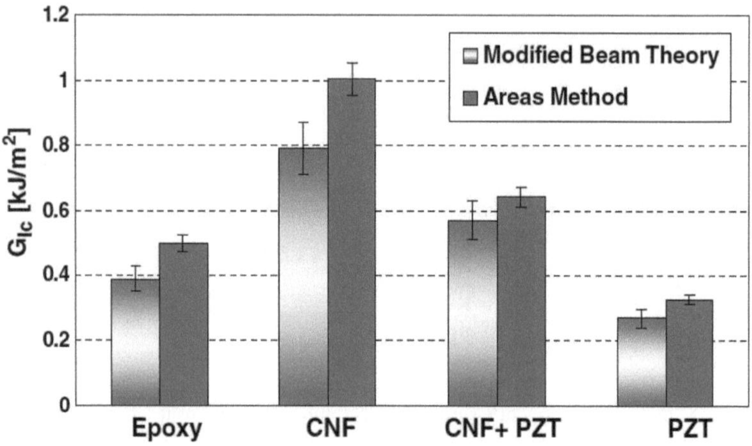

Fig. 1.8 Enhanced fracture properties for CNT modified composite materials

of Graphite fibres, Nanotubes and Graphene is presented, either via pressure (Papagelis et al. 2007) or direct stress application. Polarised Raman Spectroscopy has also been used in the study of structural characterization of the CNTs, monitoring of the stress field developed along any axis, and assessment of the induced anisotropic dispersion in candidate ordered CNT arrays for sensing/actuating applications (Zhao et al. 2002). Aspects relating to the reinforcing ability of the nanophase (Blighe et al. 2011), the stress sensing capability, as well as the stress transfer at multiple interfaces as studied with the technique are also highlighted (Cui et al. 2009).

The approach towards modeling of the behavior of hierarchical systems like those studied in this work should include multiple scales of reinforcement. Chapter 7 is dedicated to modeling of the Mechanical and Electrical Response of CNTs (Xiao et al. 2008). The coupling of electric and mechanical fields on nanotubes is studied via (i) an atomistic molecular mechanics approach for prediction of the mechanical response of CNTs (Arroyo and Belytschko 2002), (ii) a subatomic tight-binding approach for prediction of the pizeoresistive response of individual CNTs, and (iii) a homogenized microscale model for prediction of the pizeoresistive response of carbon nanotube doped insulating polymers (Fang and Wang 2010). The models are also compared to experimental results and good agreement is reported for small deformations.

As aforementioned, design for damage tolerance is the property of a material or a structure to sustain defects or cracks safely. In the eighth chapter, the addition of CNTs in small quantities as a means of improving damage tolerance properties of polymers, fibre reinforced polymer composites and their structures is presented. Novel composite systems have exhibited enhanced fracture toughness under mode I and mode II remote loading conditions, see Fig. 1.8 (Tsantzalis et al. 2007b), as well as fatigue life extension (Paipetis et al. 2009). This is in part attributed to the high surface aspect ratio of CNTs, leading to the creation of several orders of magnitude larger interface areas than those present in conventional composites. Thus enhanced energy dissipating mechanisms which will inhibit delaminations after impact, and at the same time provide the prerequisite for increased matrix toughness, are activated (Karapappas et al. 2009). The use of CNTs in aerospace composite structures has been proven to increase fracture toughness, impact strength, post-impact properties and the fatigue life of composites, making them less susceptible to damage. This is critical when designing both primary and secondary aircraft structures. Fewer joints would be used in a structure, reducing as a consequence the total weight of the structure and increasing the flexibility of a design concept.

Last, but not least Chap. 9 is dedicated to the environmental degradation of carbon nanotube hybrid aerospace composites. Although hybrid aerospace systems may exhibit improved mechanical properties, toughness and damage sensing abilities as discussed in detail in previous chapters, their environmental response was of key interest in order to be qualified for the aerospace industry. As these materials are newly developed, there is not extensive literature on their environmental exposure. However, if the hierarchical approach to reinforcement of new generation composite materials is to be widely accepted by the aerospace industry, the issue of environmental response will be of primary importance (Barkoula et al. 2009).

As contended above, novel hybrid composite systems are strong candidates towards the creation of structural components that will combine enhanced mechanical properties with sensing and life monitoring capacities. These structural components may consist of pin joints, adhesive joints with improved toughness properties and life monitoring abilities, and a smart composite shell panel with strain monitoring abilities and higher damping properties. The multi-scale multifunctional reinforcement may offer major advantages compared with existing technologies; CNTs are an integral part of the structural material system and improve the time dependent behaviour of the composite; they provide the possibility for strain and damage monitoring; only small weight fractions of CNTs are needed, which is a major advantage for processing and overall cost effectiveness of the materials. Typical applications in the field of Aeronautics and Space that will benefit from application of these novel hybrid composites include lightweight, multifunctional structural components for aerospace vehicles (with increased strength and longevity, improved energy efficiency, improved vehicle payload mass to lift-off mass ratios and having both sensing and actuating capabilities), structural components for high-value civilian transportation applications (for example, more extensive use of composites for airframes, helicopter rotors, and skins), multifunctional structural components for the space station (examples include skins, struts, and other structural members that combine strength, insulation, and shielding). Further applications may expand to advanced materials for fabrics and coatings used in space suits and other space applications, coatings and bonding agents for high-value components and equipment examples, including EMI shielding materials, ESD protection, ultra-strong adhesives, and conductive coatings for aerospace systems and components) or composites for satellite armor.

References

Ahir, S.V., Huang, Y.Y., Terentjev, E.M.: Polymers with aligned carbon nanotubes: active composite materials. Polymer **49**(18), 3841–3854 (2008)

Akdogan, E.K., Allahverdi, M., Safari, A.: Piezoelectric composites for sensor and actuator applications. IEEE Trans. Ultrason. Ferroelectr. Freq. Control. **52**(5), 746–775 (2005)

Arroyo, M., Belytschko, T.: An atomistic-based finite deformation membrane for single layer crystalline films. J Mech. Phys. Solid. **50**(9), 1941–1947 (2002)

Barkoula, N.M., Paipetis, A., Matikas, T., Vavouliotis, A., Karapappas, P., Kostopoulos, V.: Environmental degradation of carbon nanotube-modified composite laminates: a study of electrical resistivity. Mech. Compos. Mater. **45**(1), 21–32 (2009)

Barton, E.N., Ogin, S.L., Thorne, A.M., Reed, G.T.: Optimisation of the coating of a fibre optical sensor embedded in a cross-ply GFRP laminate. Compos Part A: Appl. Sci. Manuf **33**(1), 27–34 (2002)

Baughman, R.H., Zakhidov, A.A., De Heer, W.A.: Carbon nanotubes - the route toward applications. Science **297**(5582), 787–792 (2002)

Bauhofer, W., Kovacs, J.Z.: A review and analysis of electrical percolation in carbon nanotube polymer composites. Compos. Sci. Technol. **69**(10), 1486–1498 (2009)

Baur, J., Silverman, E.: Challenges and opportunities in multifunctional nanocomposite structures for aerospace applications. MRS Bull. **32**(4), 328–334 (2007)

Biso, M., Ansaldo, A., Futaba, D.N., Hata, K., Ricci, D.: Cross-linking super-growth carbon nanotubes to boost the performance of bucky gel actuators. Carbon **49**(7), 2253–2257 (2011). doi:10.1016/j.carbon.2011.01.057

Blighe, F.M., Young, K., Vilatela, J.J., Windle, A.H., Kinloch, I.A., Deng, L., Young, RJ., Coleman, J.N.: The effect of nanotube content and orientation on the mechanical properties of polymer-nanotube composite fibers: separating intrinsic reinforcement from orientational effects. Adv. Funct. Mater. **21**(2), 364–371 (2011)

Böger, L., Wichmann, M.H.G., Meyer, L.O., Schulte, K.: Load and health monitoring in glass fibre reinforced composites with an electrically conductive nanocomposite epoxy matrix. Compos. Sci. Technol. **68**(7–8), 1886–1894 (2008)

Bogue, R.: Shape-memory materials: a review of technology and applications. Assem. Autom. **29**(3), 214–219 (2009)

Brostow, W., Datashvili, T., Geodakyan, J., Lou, J.: Thermal and mechanical properties of EPDM/PP + thermal shock-resistant ceramic composites. J. Mater. Sci. **46**(8), 2445–2455 (2011)

Cheng, H.M., Li, F., Su, G., Pan, H.Y., He, L.L., Sun, X., Dresselhaus, M.S.: Large-scale and low-cost synthesis of single-walled carbon nanotubes by the catalytic pyrolysis of hydrocarbons. Appl. Phys. Lett. **72**(25), 3282–3284 (1998)

Cho, J., Boccaccini, A.R., Shaffer, M.S.P.: Ceramic matrix composites containing carbon nanotubes. J. Mater. Sci. **44**(8), 1934–1951 (2009)

Chow, W.S., Tan, P.L.: Epoxy/multiwall carbon nanotube nanocomposites prepared by sonication and planetary mixing technique. J. Reinf. Plast. Compos. **29**(15), 2331–2342 (2010)

Colbert, D.T.: Single-wall nanotubes: a new option for conductive plastics and engineering polymers. Plastics Addit Compound. **5**(1), 18–25 (2003)

Coleman, J.N., Curran, S., Dalton, A.B.., Davey, A.P., McCarthy, B., Blau, W., Barklie, R.C.: Percolation-dominated conductivity in a conjugated-polymer-carbon-nanotube composite. Phys. Rev. B Condens. Matter Mater. Phys. **58**(12), R7492–R7495 (1998)

Cooper, C.A., Young, R.J., Halsall, M.: Investigation into the deformation of carbon nanotubes and their composites through the use of Raman spectroscopy. Compos. Part A: Appl. Sci. Manuf. **32**(3-4), 401–411 (2001). doi:10.1016/s1359-835x(00)00107-x

Cui, S., Kinloch, I.A., Young, R.J., Noé, L., Monthioux, M.: The effect of stress transfer within double-walled carbon nanotubes upon their ability to reinforce composites. Adv. Mater. **21**(35), 3591–3595 (2009)

Dassios, K.G., Galiotis, C., Kostopoulos, V., Steen, M.: Direct in situ measurements of bridging stresses in CFCCs. Acta Materialia **51**(18), 5359–5373 (2003)

De la Vega, A., Kinloch, I.A., Young, R.J., Bauhofer, W., Schulte, K.: Simultaneous global and local strain sensing in SWCNT-epoxy composites by Raman and impedance spectroscopy. Compos. Sci. Technol. **71**(2), 160–166 (2011)

Deng, F., Zheng, Q.: Interaction models for effective thermal and electric conductivities of carbon nanotube composites. Acta Mechanica Solida Sinica **22**(1), 1–17 (2009). doi:10.1016/s0894-9166(09)60085-9

Dresselhaus, M.S., Dresselhaus, G., Saito, R., Jorio, A.: Raman spectroscopy of carbon nanotubes. Physics Reports **409**(2), 47–99 (2005)

Duncan, R.K., Chen, X.G., Bult, J.B., Brinson, L.C., Schadler, L.S.: Measurement of the critical aspect ratio and interfacial shear strength in MWNT/polymer composites. Compos. Sci. Technol. **70**(4), 599–605 (2010)

Fang, L., Wang, D.: On scale transition and homogenization in multiscale computational analysis of heterogeneous solids. Intel. Comput. Intel. **3**, 373–376 (2010)

Fiedler, B., Gojny, F.H., Wichmann, M.H.G., Nolte, M.C.M., Schulte, K.: Fundamental aspects of nano-reinforced composites. Compos. Sci. Technol. **66**(16), 3115–3125 (2006)

Frank, O., Tsoukleri, G., Parthenios, J., Papagelis, K., Riaz, I., Jalil, R., Novoselov, K.S., Galiotis, C.: Compression behavior of single-layer graphenes. ACS Nano **4**(6), 3131–3138 (2010)

Frank, O., Tsoukleri, G., Riaz, I., Papagelis, K., Parthenios, J., Ferrari, A.C., Geim, A.K., Novoselov, K.S., Galiotis, C.: Development of a universal stress sensor for graphene and carbon fibres. Nat. Commun. **2**(1) (2011). doi:10.1038/ncomms1247

Fu, S.Y., Feng, X.Q., Lauke, B., Mai, Y.W.: Effects of particle size, particle/matrix interface adhesion and particle loading on mechanical properties of particulate-polymer composites. Compos. Part B: Eng. **39**(6), 933–961 (2008)

Garcia, E.J., Wardle, B.L., John Hart, A.: Joining prepreg composite interfaces with aligned carbon nanotubes. Compos. Part A: Appl. Sci. Manuf. **39**(6), 1065–1070 (2008). doi:10.1016/j.compositesa.2008.03.011

Gibson, R.F.: A review of recent research on mechanics of multifunctional composite materials and structures. Compos. Struct. **92**(12), 2793–2810 (2010). doi:10.1016/j.compstruct.2010.05.003

Gibson, R.F., Ayorinde, E.O., Wen, Y.F.: Vibrations of carbon nanotubes and their composites: a review. Compos. Sci. Technol. **67**(1), 1–28 (2007)

Gou, J., Sumerlin, S., Gu, H.C., Song, G.: Damping enhancement of hybrid nanocomposites embedded with engineered carbon nanopaper. ASME Conf. Proc. **2006**(47659), 637–642 (2006)

Gouadec, G., Colomban, P.: Raman spectroscopy of nanomaterials: how spectra relate to disorder, particle size and mechanical properties. Prog. Cryst. Growth Charact. Mater. **53**(1), 1–56 (2007)

Hojo, M., Matsuda, S., Higuchi, T., Ochiai, S.: Fracture mechanism for mode II propagation of delamination fatigue cracks in CF/PEEK laminates. Zairyo/J. Soc. Mater. Sci. Jpn. **46**(4), 366–373 (1997)

Hojo, M., Ando, T., Tanaka, M., Adachi, T., Ochiai, S., Endo, Y.: Modes I and II interlaminar fracture toughness and fatigue delamination of CF/epoxy laminates with self-same epoxy interleaf. Int. J. Fatigue **28**(10 Spec. Iss), 1154–1165 (2006)

Horst, H.G., Kronig, K. Active vibration damping with piezoceramic actuators at an elastic beam structure [Aktive Schwingungsminderung mit piezokeramischen Aktuatoren an einer elastischen Balkenstruktur] (2001) VDI Berichte, (1606), pp. 143–162

Huang, M., Yan, H., Heinz, T.F., Hone, J.: Probing strain-induced electronic structure change in graphene by Raman spectroscopy. Nano Lett **10**, 4074–4079 (2010). doi:10.1021/nn103493g

Karapappas, P., Vavouliotis, A., Tsotra, P., Kostopoulos, V., Paipetis, A.: Enhanced fracture properties of carbon reinforced composites by the addition of multi-wall carbon nanotubes. J. Compos. Mater. **43**(9), 977–985 (2009)

Kostopoulos, V., Vavouliotis, A., Karapappas, P., Tsotra, P., Paipetis, A.: Damage monitoring of carbon fiber reinforced laminates using resistance measurements. Improving sensitivity using carbon nanotube doped epoxy matrix system. J. Intell. Mater. Syst. Struct. **20**(9), 1025–1034 (2009)

Krstic, V.D.: Ceramic matrix composites-present status and future trends. Mater. Sci. Forum **282–283**, 209–218 (1998)
Kumar, S., Dang, T.D., Arnold, F.E., Bhattacharyya, A.R., Min, B.G., Zhang, X., Vaia, R.A., Park, C., Wade Adams, W., Hauge, R.H., Smalley, R.E., Ramesh, S., Willis, P.A.: Synthesis, structure, and properties of PBO/SWNT composites. Macromolecules **35**(24), 9039–9043 (2002)
Lazzeri, L., Mariani, U.: Application of damage tolerance principles to the design of helicopters. Int. J. Fatigue **31**(6), 1039–1045 (2009)
Lee, V.A., Cardenas, H.L., Rawls, H.R.: Rubber-toughening of dimethacrylate dental composite resin. J. Biomed. Mater. Res. Part B Appl. Biomater **94**(2), 447–454 (2010)
Li, Z., Crocker, M.J.: A review on vibration damping in sandwich composite structures. Int. J. Acoust. Vib. **10**(4), 159–169 (2005)
Li, F., Cheng, H.M., Bai, S., Su, G., Dresselhaus, M.S.: Tensile strength of single-walled carbon nanotubes directly measured from their macroscopic ropes. Appl. Phys. Lett. **77**(20), 3161–3163 (2000)
Li, C., Thostenson, E.T., Chou, T.-W.: Sensors and actuators based on carbon nanotubes and their composites: a review. Compos. Sci. Technol. **68**(6), 1227–1249 (2008). doi:10.1016/j.compscitech.2008.01.006
Liu, C., Cheng, H.M., Cong, H.T., Li, F., Su, G., Zhou, H.L., Dresselhaus, M.S.: Synthesis of macroscopically long ropes of well-aligned single-walled carbon nanotubes. Adv. Mater. **12**(16), 1190–1192 (2000)
Lomax, R.: Statistical Concepts. Lawrence Erlbaum Associates, Mahwah (2007)
Luyckx, G., Voet, E., Lammens, N., Degrieck, J.: Strain measurements of composite laminates with embedded fibre bragg gratings: criticism and opportunities for research. Sensors **11**(1), 384–408 (2011)
Ma, P.C., Kim, J.K., Tang, B.Z.: Effects of silane functionalization on the properties of carbon nanotube/epoxy nanocomposites. Compos. Sci. Technol. **67**(14), 2965–2972 (2007)
Ma, P.-C., Siddiqui, N.A., Marom, G., Kim, J.-K.: Dispersion and functionalization of carbon nanotubes for polymer-based nanocomposites: a review. Compos. Part A: Appl. Sci. Manuf. **41**(10), 1345–1367 (2010). doi:10.1016/j.compositesa.2010.07.003
Majumder, M., Gangopadhyay, T.K., Chakraborty, A.K., Dasgupta, K., Bhattacharya, D.K.: Fibre Bragg gratings in structural health monitoring-present status and applications. Sensors Actuators A: Physical **147**(1), 150–164 (2008)
Melanitis, N., Galiotis, C.: Compressional behaviour of carbon fibres – Part 1: a Raman spectroscopic study. J. Mater. Sci. **25**(12), 5081–5090 (1990)
Nettles, A., Hodge, A., Jackson, J.: An examination of the compressive cyclic loading aspects of damage tolerance for polymer matrix launch vehicle hardware. J. Compos. Mater. **45**(4), 437–458 (2011)
Olmos, D., Bagdi, K., Mózcó, J., Pukánszky, B., González-Benito, J.: Morphology and interphase formation in epoxy/PMMA/glass fiber composites: effect of the molecular weight of the PMMA. J. Colloid Interface Sci. **360**, 289–299 (2011)
Paipetis, A., Matikas, T.E., Barkoula, N.M., Karapappas, P., Vavouliotis, A., Kostopoulos, V.: Nano-enhanced aerospace composites for increased damage tolerance and service life damage monitoring, Proc. SPIE **7292**, 72921O (2009); doi:10.1117/12.815210
Panopoulou, A., Loutas, T., Roulias, D., Fransen, S., Kostopoulos, V.: Dynamic fiber Bragg gratings based health monitoring system of composite aerospace structures. Acta Astronaut. **69**(7–8), 445–457 (2011)
Papagelis, K., Andrikopoulos, K.S., Arvanitidis, J., Christofilos, D., Galiotis, C., Takenobu, T., Iwasa, Y., Kataura, H., Ves, S., Kourouklis, G.A.: High pressure Raman study of the second-order vibrational modes of single- and double-walled carbon nanotubes. Physica Status Solidi (B) Basic Res. **244**(11), 4069–4073 (2007)
Parthenios, J., Katerelos, D.G., Psarras, G.C., Galiotis, C.: Aramid fibers; a multifunctional sensor for monitoring stress/strain fields and damage development in composite materials. Eng. Fract. Mech. **69**(9), 1067–1087 (2002)

Peng, L.M., Zhang, Z.L., Xue, Z.Q., Wu, Q.D., Gu, Z.N., Pettifor, D.G.: Stability of carbon nanotubes: how small can they be? Phys. Rev. Lett. **85**(15), 3249–3252 (2000)

Poulin, P.: Carbon nanotube fibers: smart, strong and sensitive. Fibres de nanotubes de carbone: Intelligentes, musclées et sensibles **282**, 13–17 (2005)

Purcell, S.T., Vincent, P., Journet, C., Binh, V.T.: Hot nanotubes: stable heating of individual multiwall carbon nanotubes to 2000 K induced by the field-emission current. Phys. Rev. Lett. **88**(10), 1055021–1055024 (2002)

Raju Mantena, P., Al-Ostaz, A., Cheng, A.H.D.: Dynamic response and simulations of nanoparticle-enhanced composites. Compos. Sci. Technol. **69**(6), 772–779 (2009)

Rochefort, A., Avouris, P., Lesage, F., Salahub, D.R.: Electrical and mechanical properties of distorted carbon nanotubes. Phys. Rev. B: Condense. Matter Mater Phys. **60**(19), 13824–13830 (1999)

Sager, R.J., Klein, P.J., Davis, D.C., Lagoudas, D.C., Warren, G.L., Sue, H.J.: Interlaminar fracture toughness of woven fabric composite laminates with carbon nanotube/epoxy interleaf films. J. Appl. Polym. Sci. **121**(4), 2394–2405 (2011)

Sandler, J., Shaffer, M.S.P., Prasse, T., Bauhofer, W., Schulte, K., Windle, A.H.: Development of a dispersion process for carbon nanotubes in an epoxy matrix and the resulting electrical properties. Polymer **40**(21), 5967–5971 (1999)

Shelley, S.: Newsfront: carbon nanotubes: a small-scale wonder. Chem. Eng. **110**(1), 27–29 (2003)

Song, G., Sethi, V., Li, H.N.: Vibration control of civil structures using piezoceramic smart materials: a review. Eng. Struct. **28**(11), 1513–1524 (2006)

Sun, L., Gibson, R.F., Gordaninejad, F., Suhr, J.: Energy absorption capability of nanocomposites: a review. Compos. Sci. Technol. **69**(14), 2392–2409 (2009)

Sureeyatanapas, P., Hejda, M., Eichhorn, S.J., Young, R.J.: Comparing single-walled carbon nanotubes and samarium oxide as strain sensors for model glass-fibre/epoxy composites. Compos. Sci. Technol. **70**(1), 88–93 (2010)

Terrones, M., Grobert, N., Olivares, J., Zhang, J.P., Terrones, H., Kordatos, K., Hsu, W.K., Hare, J. P., Townsend, P.D., Prassides, K., Cheetham, A.K., Kroto, H.W., Walton, D.R.M.: Controlled production of aligned-nanotube bundles. Nature **388**(6637), 52–55 (1997)

Thostenson, E.T., Ren, Z., Chou, T.W.: Advances in the science and technology of carbon nanotubes and their composites: a review. Compos. Sci. Technol. **61**(13), 1899–1912 (2001)

Thostenson, E.T., Li, W.Z., Wang, D.Z., Ren, Z.F., Chou, T.W.: Carbon nanotube/carbon fiber hybrid multiscale composites. J. Appl. Phys. **91**(9), 6034 (2002)

Triantafyllidis, K.S., Karakoulia, S.A., Gournis, D., Delimitis, A., Nalbandian, L., Maccallini, E., Rudolf, P.: Formation of carbon nanotubes on iron/cobalt oxides supported on zeolite-Y: effect of zeolite textural properties and particle morphology. Microporous Mesoporous Mater. **110** (1), 128–140 (2008)

Tsai, S.-Y., Chen, I.-W.P., Labrador, D., Porter, S., Liang, Z., Wang, B., Zhang, C.: Lightweight long nanotube-solid electrolyte composite actuators for morphing structure applications: fabrication and property characterization. Collection of Technical Papers - AIAA/ASME/ ASCE/AHS/ASC Structures, Structural Dynamics and Materials Conference, art. no. 2010–2658 (2010)

Tsantzalis, S., Karapappas, P., Vavouliotis, A., Tsotra, P., Kostopoulos, V., Tanimoto, T., Friedrich, K.: On the improvement of toughness of CFRPs with resin doped with CNF and PZT particles. Compos. Part A: Appl. Sci. Manuf. **38**(4), 1159–1162 (2007a)

Tsantzalis, S., Tsotra, P., Karapappas, P., Vavouliotis, A., Fanis, N., Kostopoulos, V., Friedrich, K.: Influence of carbon nanofibers and piezoelectric particles on the thermomechanical behavior of epoxy mixtures. **121–123,** 1419–1424 (2007b)

Tuinstra, F., Koenig, J.L.: Characterization of graphite fiber surfaces with Raman spectroscopy. J. Compos. Mater. **4**, 494–499 (1970)

Vavouliotis, A.: A new approach for the monitoring of the developing damage under mechanical loading in fibrous composite materials with carbon nanotube reinforced matrix PhD thesis. University of Patras, Greece (2008)

Vavouliotis, A., Paipetis, A., Kostopoulos, V.: On the fatigue life prediction of CFRP laminates using the electrical resistance change method. Compos. Sci. Technol. **71**(5), 630–642 (2011). doi:10.1016/j.compscitech.2011.01.003

Vigolo, B., Penicaud, A., Coulon, C., Sauder, C., Pailler, R., Journet, C., Bernier, P., Poulin, P.: Macroscopic fibers and ribbons of oriented carbon nanotubes. Science **290**(5495), 1331–1334 (2000)

Vlasveld, D.P.N., Bersee, H.E.N., Picken, S.J.: Nanocomposite matrix for increased fibre composite strength. Polymer **46**(23), 10269–10278 (2005). doi:10.1016/j.polymer.2005.08.003

Wheat, T.A., Ahmad, A., Minhas, R.K.: Ceramic actuators: Part 1 - materials. Can. Ceram. **68**(3), 56–62 (1999)

Wilson, S.A., Jourdain, R.P.J., Zhang, Q., Dorey, R.A., Bowen, C.R., Willander, M., Wahab, Q.U., Al-hilli, S.M., Nur, O., Quandt, E., Johansson, C., Pagounis, E., Kohl, M., Matovic, J., Samel, B., van der Wijngaart, W., Jager, E.W.H., Carlsson, D., Djinovic, Z., Wegener, M., Moldovan, C., Abad, E., Wendlandt, M., Rusu, C., Persson, K.: New materials for micro-scale sensors and actuators. An engineering review. Mater. Sci. Eng. R: Rep. **56**(1–6), 1–129 (2007)

Xiao, S., Ni, J., Wang, S.: The bridging domain multiscale method and its high performance computing implementation. J. Comput. Theor. Nanosci. **5**(7), 1220–1229 (2008)

Young, R.J., Halary, J., Stanford, J.L., Lovell, P.A.: Remote strain-sensor coatings based upon carbon nanotube-polymer composites (2004) Annual Technical Conference - ANTEC, Conference Proceedings, 2, pp. 1805–1809

Yuan, J., Wang, D.W., Lin, H.B., Zhao, Q.L., Zhang, D.Q., Cao, M.S.: Effect of ZnO whisker content on sinterability and fracture behaviour of PZT peizoelectric composites. J. Alloys Compd. **504**(1), 123–128 (2010)

Zhang, L.: Fabrication, characterisation and application of carbon nanotube polymer composites. **139–141**, 1–8 (2010)

Zhao, Q., Wagner, H.D.: Two-dimensional strain mapping in model fiber-polymer composites using nanotube Raman sensing. Compos. Part A: Appl. Sci Manuf. **34**(12), 1219–1225 (2003)

Zhao, Q., Frogley, M.D., Wagner, H.D.: Direction-sensitive strain-mapping with carbon nanotube sensors. Compos. Sci. Technol. **62**(1), 147–150 (2002)

Zhao, L., Zhang, F., Hu, X.Z., Huang, B.Z.: Study on interlaminar shear and damage behavior of carbon fiber composites with short fiber interleaves: 1. The comparative test. **41–42,** 335–340 (2008a)

Zhao, L., Zhang, F., Hu, X.Z., Huang, B.Z.: Study on interlaminar shear and damage behavior of carbon fiber composites with short fiber interleaves: 2. Micro-mechanical simulation, **41–42**, 341–347 (2008b)

Zhong, X.H., Li, Y.L., Liu, Y.K., Qiao, X.H., Feng, Y., Liang, J., Jin, J., Zhu, L., Hou, F., Li, J.Y.: Continuous multilayered carbon nanotube yarns. Adv. Mater. **22**(6), 692–696 (2010)

Chapter 2
On the Use of Electrical Conductivity for the Assessment of Damage in Carbon Nanotubes Enhanced Aerospace Composites

Antonios I. Vavouliotis and Vassilis Kostopoulos

Contents

Abstract In this chapter a review on the research of nano-enabled self-sensing structural composite materials is performed. The self-sensing concept is attained by exploiting the intrinsic electrical properties of a structural composite material. Recent research on self-sensing was stimulated by the introduction of nanotechnology in the field of composite materials. Nano-scale fillers such as carbon nanotubes (CNTs), due to their excellent electrical properties, may offer benefits of additional reinforcing phase acting at the nano-scale. The research may be divided into two distinctive categories depending on the type of fibre reinforcement. One category is the research that used electrically non-conductive glass fibre reinforced plastics (GFRP) where carbon nanotubes in various forms are incorporated into the composite to enable sensing. The other category is the research that used electrical conductive carbon fibre reinforced plastics (CFRP) where the carbon nanotubes in various forms are used to enhance the electrical sensing capabilities of the composite.

Keywords Electrical properties • Structural health monitoring • Structural composites • Nanotubes • Nano-composites

---

A.I. Vavouliotis • V. Kostopoulos (✉)
Applied Mechanics Laboratory, Mechanical Engineering and Aeronautics Department, University of Patras, Patras, Greece
e-mail: kostopoulos@mech.upatras.gr

A.S. Paipetis and V. Kostopoulos (eds.), *Carbon Nanotube Enhanced Aerospace Composite Materials*, Solid Mechanics and Its Applications 188,
DOI 10.1007/978-94-007-4246-8_2,

2.1 Introduction

Fibre reinforced polymer (FRPs) materials and especially Carbon Fibre Reinforced Polymers (CFRPs) are a widely accepted material choice either for primary structural applications or for secondary or tertiary structures for the aerospace industry. Due to their high specific strength and stiffness, CFRPs provide better design options over typical metallic materials. Moreover, the fact that composites may be optimized by proper selection of material and processing parameters renders them the undisputed material choice for applications that require high design flexibility. CFRPs have been widely used for critical components and structures, such as aircraft fuselages and wing structures, helicopter rotors and windmill blades, road and marine vehicle body structures, and, bridges and large civil infrastructures. Additionally, multifunctionality is an aspect that aerospace technology has been focusing on during the last few decades. Design parameters such as mass reduction with increased system efficiency demand multifunctional approaches. The technology concept of multifunctional materials with sensing capabilities combined with enhanced mechanical–electrical and/or thermal properties could prove useful for the demanding requirements of the aerospace sector. Enhancing operational reliability is an ongoing continuous objective for contemporary aerospace composite structures. At the same time, there is an emerging demand for advanced life-cycle management systems which will allow continuous monitoring of the structural integrity and assessment of the damage that develops during the operation for composite structures. During the last years, the so-called structural health monitoring systems tend to be an integral part of a wide range of structures from composite materials, where the maximum safety and low operating and maintenance costs are equally important, if not more important, parameters than performance, since control of damage in aerospace structures requires regular costly inspections of aircraft systems to avoid catastrophic failures.

In general, structural health monitoring systems utilize various non-destructive damage detection techniques that on one hand supply dedicated analysis tools for damage diagnosis and on the other hand predictive tools of the remaining operational life. Sensors are the key elements of every non-destructive technique. Sensors are the devices that measure and record the physical property related directly or indirectly to the damage. Depending on the technique used, the sensors can be optical (FBG), acoustic emission (AE), strain gauges etc. An important problem in almost all existing sensors is the various restrictions of their use either because of non-conformity of their specifications to the operating conditions in demanding operational environments (e.g. fatigue, impact, humidity etc) or because they affect the structural performance of the composite material, especially when they are integrated inside the material. According to Abry et al. (2001) acceptable health monitoring sensors should meet specific requirements such as small weight and size, high sensitivity, structural compatibility in the case of built-in sensors, lifelong operation capacity, ability for health monitoring of large critical areas of the structures and possibility to transmit information to a central processor in real time.

An optimal way to proceed for such types of structural health monitoring is to use the material itself as sensor. This concept is referred to as self-sensing and is attained by exploiting the intrinsic behavior of a structural material (Kemp 1994). Towards this goal, measuring the electrical properties of composite materials is proposed. Electrical resistivity is an inherent material property defined by its material state. Since damage alters the material state, also electrical resistivity changes. Consequently, the material itself can act as a sensor of its own damage. Although electrical contacts and a resistance meter are typically needed for electrical property measurement, the composite is the sensor; neither the fibres nor the electrical contacts are the sensors. This function surpasses the specific distinction between electrical self-sensing material concepts and embedded or attached electrical sensors (e.g. strain gauges, optical fibres, piezoelectric sensors etc). The material self-sensing concept uses only constituent phases of the material, is by definition non-intrusive and does not deteriorate the structural performance of the composite material. Additionally it is directly applicable to existing composite structures, having the ability to cover large areas/volumes at low operational cost, since it usually requires typical electronic devices (e.g. multi-meters). A key advantage of the electrical self-sensing concept is that it has negligible effects on the structural weight from the sensor point of view.

In composites, this electrical self-sensing concept is applied mainly on carbon fibre reinforced polymer (CFRP) composites and was first reported by Schulte and Baron (1989). It has been studied for the last 20 years by various researchers worldwide (Chung 1998; Wang and Chung 1997, 1998a, b, 2000; Wang et al. 1998, 1999; Chung and Wang 2003; Khemiri et al. 2005; Angelidis et al. 2004, 2006; Chung et al. 2006; Todoroki and Yoshida 2004, 2005; Todoroki et al. 2002; Sirong and Chung 2007; Irving and Thiagarajan 1998; Dae-Cheol et al. 1999; Ceysson et al. 1996; Prasse et al. 2001; Abry et al. 1999; Kupke et al. 2001; Weber and Schwartz 2001; Park et al. 2001, 2002, 2003; Xia and Curtin 2007, 2008). The measurement of electrical resistance is most reliable for intermediate levels of resistance, such as resistance in the range from 0.1Ω to $1M\Omega$. A large resistance exceeding $1M\Omega$ is relatively difficult to measure, due to the need for a high voltage in order to pass a current through the large resistor. Conventional meters are incapable of measuring resistances exceeding $1M\Omega$, due to their voltage limitation. Due to the conductive nature of the carbon fibres the electrical properties of CFRPs are in the range of the most conventional measuring devices (multi-meters). This is not the case for other FRP materials with non-conductive fibre reinforcements (e.g. Glass-FRPs, Kevlar-FRPs). In order to overcome this problem, hybrid-FRP materials were proposed, by Nanni et al. (2006), that combined the use of non-conductive (glass fibres) and conductive fibre (carbon fibres). Furthermore Shin (2002) proposed the use of (carbon black) particle-filled electrical conductive polymer as matrix for GFRPs in the conductive phase for self-sensing.

Despite its advantages, the self-sensing concept has received less attention than the use of embedded or attached sensors, due to the scientific challenge of developing self-sensing structural materials. Although much attention has been given to their mechanical properties and durability, relatively little attention has been directed to their sensing behavior, which relates to their electrical behavior. Lately, research on self-sensing was encouraged and stimulated by the introduction of nanotechnology into the field of composite materials. Nano-scale fillers such as carbon nanotubes (CNTs) have been placed recently in the epicenter of composite research. Taking into consideration their high aspect ratio, large surface area and excellent electrical properties, they offer benefits of an additional reinforcing phase acting at the nano-scale. This evolution provided the necessary momentum for the development of advanced self-sensing structural materials and created a promising technological path towards the ultimate material engineering goal of providing multi-functional materials. In this chapter we review the aforementioned research of nano-enabled self-sensing concepts.

The main routes identified are connected with the various developments of nano-material research through the years that provided different forms of nano-engineered materials solutions (e.g. CNT-loaded polymers, CNT-bucky papers, CNT-fibres, CNT-sized fabrics etc.). The carbon nanotube (CNT) loaded polymers was the first route that was developed, inheriting the long industrial processing experience on particle filled engineered plastics. Fiedler et al. (2004) proposed the use of CNT loaded polymer matrices in glass fibre reinforced (GFRP) composites instead of micro-sized carbon particle (e.g. flakes etc.) filled matrices used in past works (Indada et al. 2005; Okuhara et al. 2000, 2001) in order to utilize the electrically conductive network of nanotubes formed in the polymer matrix surrounding the fibres. Kostopoulos et al. (2009a) proposed the use of carbon nanotubes (CNTs) as additives in the epoxy matrix of carbon fibre reinforced laminates (CFRPs) aiming to enhance the real-time damage monitoring via the electrical resistance change (ERC) method. In parallel, new very promising nano-engineered structures were developed providing new tools for constructing self-sensing composite materials. Such structures are the so-called carbon nanotube buckypapers, the nanotube-fibres and nanotube sized fabrics.

2.2 CNT-GFRPs

Glass fibres are electrical insulating materials and are globally the most widely used reinforcement in composites. The development of novel glass-fibre-reinforced plastics (GFRPs) with electrical conductivity has opened up new opportunities for damage sensing. As a pilot approach, Fiedler et al. (2004) proposed that adding a small amount of carbon nanotubes to form an electrically conductive network is a promising approach to monitor damage initiation and propagation for glass fibre-reinforced composites. Thostenson et al. (Thostenson and Chou 2006; Li et al. 2008) fabricated nanotube–epoxy–fibre composites where 0.5 wt.% of multiwalled CNTs were dispersed into the epoxy matrix, and they designed specific electro-mechanical tests in order to assess the damage monitoring capabilities for specific distinct failure modes. Five plies of unidirectional CNT-GFRPs with a discontinuity at the center ply of the laminate were used under tension in order to evaluate

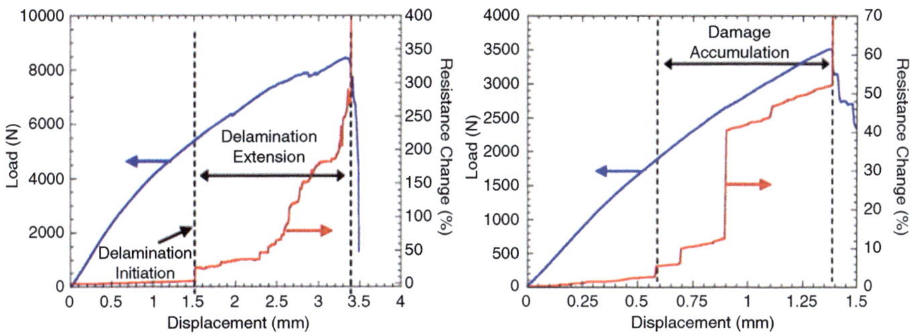

Fig. 2.1 (*left*) Load/displacement and resistance response of a five-ply unidirectional composite with the center ply cut to initiate delamination, (*right*) load/displacement and resistance response of a $(0/90)_s$ cross-ply composite showing accumulation of damage due to micro-cracks (Reprinted from Li et al. 2008, with permission from Elsevier)

inter-laminar delamination while cross-ply $[0/90]_s$ laminates were used also under tension to assess the influence of transverse micro-crack development. Moreover unidirectional specimens were tested in three-point bending at varying spans, in order to assess the capability of nanotubes to sense through-thickness inter-laminar fracture. Results (Fig. 2.1) from a five-ply unidirectional composite showed that at low strain there is a linear increase in the specimen resistance with deformation and a sharp increase in resistance occurs when the ply delamination is initiated. Results from the cross-ply symmetric laminates (Fig. 2.1) showed that during initial loading there is a linear increase in resistance with strain. Upon the initiation of micro-cracking in the 90° plies there is a sharp change in the electrical resistance. From the first initiation of cracking to the ultimate fracture of the laminate resistance, changes are marked by sharp step increases likely corresponding to the accumulation of micro-cracks and linear increases in resistance with deformation between the step increases. For both cases, there is a linear increase in resistance with deformation prior to damage initiation, indicating strong potential for both strain and damage detection. After the onset of damage and subsequent re-loading of damaged structures there is a remarkable shift in the sensing curve, indicating irreversible damage.

More recently the same group (2008) (Thostenson and Chou 2008) utilized electrically conductive networks of carbon nanotubes as *in situ* sensors for detecting damage accumulation during cyclic loading of glass fibre composites. They produced cross-ply laminates $[0/90]_2$ with 0.5 wt.% of nanotubes in the polymer matrix and they recorded simultaneously in real-time the electrical resistance, load, strain and crosshead displacement data during tensile deformation with increasing peak load followed by continuous cyclic loading. The electrical resistance versus strain curve showed substantial hysteresis due to the formation and opening/closing of cracks during cyclic loading that may be utilized as a quantitative measure of damage as depicted in the following figure (Fig. 2.2).

Sotiriadis et al. (2007) modified a plain vinylester resin with 1.0 wt.% multiwall CNTs and produced a twelve ply nanotube/glass-fibre/vinylester composite using a plain weave [0/90] E-Glass fabric. Two types of tests were carried out with on-line

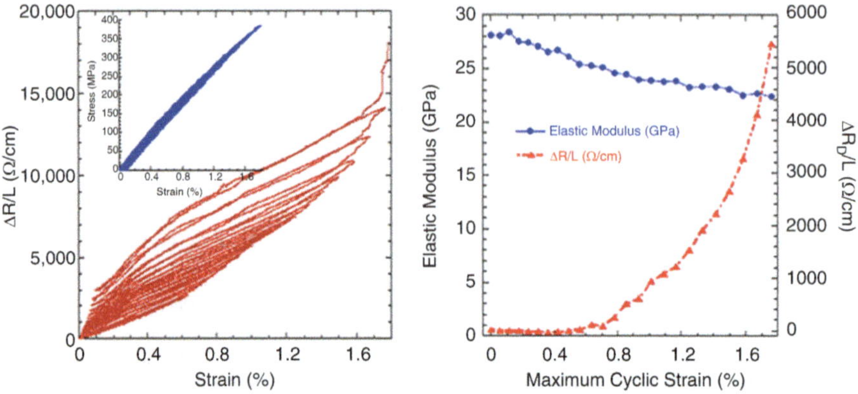

Fig. 2.2 (*left*) Resistance–strain response showing substantial hysteresis and (inset) stress–strain response, (*right*) change in elastic modulus and damaged resistance, ΔR_D, with maximum cyclic strain (Reprinted from Thostenson and Chou 2008)

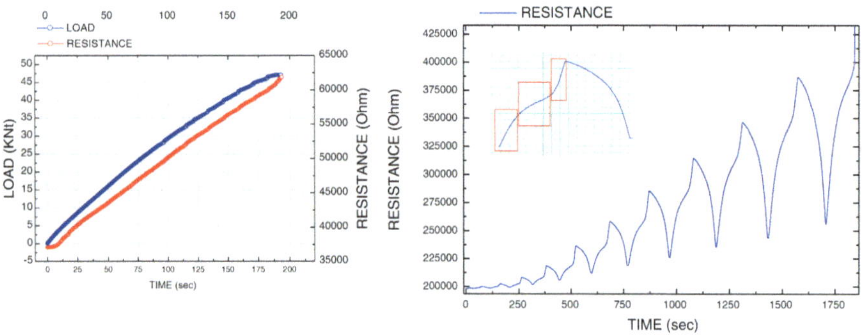

Fig. 2.3 (*left*) Resistance monitoring during the tensile test for the nano-composite GFRP specimen, (*right*) resistance monitoring throughout the tensile incremental loading and Magnified view of the resistance changes for a single loading cycle: three phases of resistance increase (Reprinted from Sotiriadis et al. 2007)

monitoring of the electrical resistance. During monotonic tensile tests a monotonic increase of the resistance with increasing strain took place until the fracture of the specimen. The monitored resistance changes stemmed from the deformation of the CNT network that was formed within the modified matrix of the composite. After percolation, the CNTs provided a 3D conductive network within the matrix, which followed the volumetric changes of the composite material. The resistance changes related directly to the microscopic strain, while at the same time were indicative of damage at the nano-scale; this damage was reflected in the breaches of the CNT network at the defect sites which, in their turn, led to an increase of macroscopic resistance. It is significant to note that the monotonic and almost linear increase was present until failure of the specimen. In the case of monotonic loading up to specimen failure (Fig. 2.3 left), this resistance change of the percolated CNT

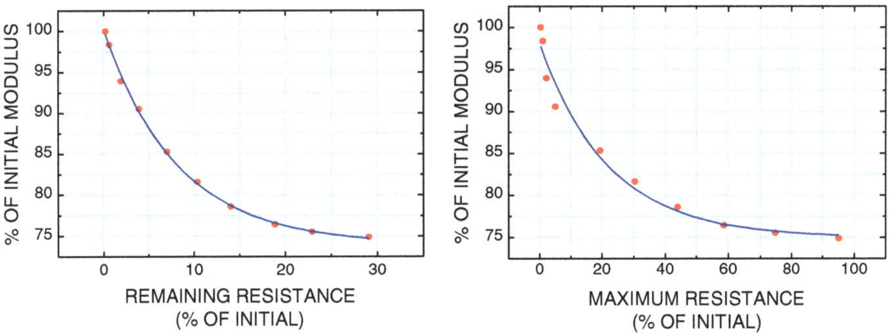

Fig. 2.4 Modulus deterioration for incremental loading of doped GFRPs as a function of irreversible resistance changes (Reprinted from Sotiriadis et al. 2007)

network reflected all macroscopic changes which were due to active mechanisms, that is, on one hand macroscopic deformation of the CNT network and on the other hand successive cumulative damage mechanisms such as microscopic damage initiation in the composite, transverse cracking up to saturation and cumulative 0° fibre fractures up to macroscopic failure of the composite.

Resistance measurements throughout step-wise cyclic loading tests with gradual increase of the maximum load for each cycle until the fracture of the specimen (Fig. 2.3 right) revealed a monotonic increase of resistance at the loading phase of the test followed by a decrease through the unloading phase. A detailed study of the resistance changes during all loading cycles indicated that a resistance increase consistently followed distinct phases. The first phase is dominated by a high increase rate of the resistance. During the second phase a lower increase rate was observed. The end of the second phase occurs when the load reaches the maximum load level of the previous cycle. The final phase features a steep increase of the resistance until the maximum load is reached. Additionally, the rate of the resistance changes was indicative of (i) the loading history of the materials and (ii) the onset of more cumulative damage: when the load level exceeded the previous maximum load, the change of the resistance increase rate was readily identified. Moreover, the resistance value for each cycle at zero load level was higher than the one at the start of the previous cycle, indicating that there is a residual resistance for each consecutive cycle. This irreversible increase of the resistance revealed that resistance changes were not only related to stress or strain but were also dependent on the accumulated irreversible damage within the material. The maximum resistance for each loading cycle was observed at the peak load of the respective cycle. The minimum or remaining resistance recorded for each cycle was recorded at the onset of the respective cycle. The maximum resistance was found to be more sensitive with regard to the stiffness degradation of the material compared to the initial resistance. The change of maximum resistance reached approximately 100% compared to the remaining resistance which reached approximately 30%. However, the remaining resistance was more indicative of the residual damage in the composite as it corresponded to the unloaded stage. For both cases an exponential decrease of the stiffness E versus the resistance increase was observed (Fig. 2.4).

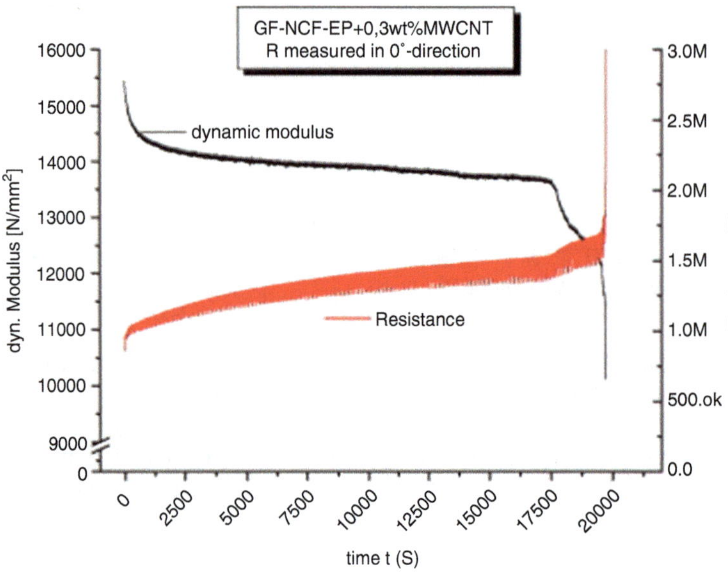

Fig. 2.5 Stiffness and resistance change for a specimen under dynamic tensile load. Resistance measured in longitudinal 0°-direction (Reprinted from Böger et al. 2008, with permission from Elsevier)

In addition, the resistance correlated better to the stiffness loss than the applied load. Results of the resistance decay are proposed to be used as a direct index of the stiffness degradation of the material for this type of mechanical loading.

Böger et al. (2008) modified an epoxy resin by the addition of 0.3 wt.% of nano-scaled carbon particles. Three different types of nano-particles were used in this study: double wall carbon nanotubes (DWCNT), multi-wall carbon nanotubes (MWCNT) and carbon black. The electrical conductive matrices were used to produce glass fibre reinforced composites (GFRP) by resin transfer moulding (RTM). The electromechanical characterization included simultaneous monitoring of the electrical resistance during incremental tensile tests, fatigue tests and inter-laminar shear strength (ILSS). During the incremental tensile tests and fatigue tests it was possible to clearly measure resistance changes in the materials that were related to micro-scale damage, such as inter-fibre failure. This kind of damage cannot be detected by other damage sensing methods. Later in the fatigue tests, when macroscopic damage (delamination or rupture of fibre bundles) occurred, these events were leading to explicit signals in the z-direction (through the thickness) electrical conductivity measurement of the materials (Figs. 2.5 and 2.6). Furthermore, also a dependency of the electrical resistance on the load applied to the composite was found. Therefore, authors suggest that by these measurements not only the accumulation of damage can be detected but also the strain state of a composite structure.

Nofar et al. (2009) obtained also valuable conclusions by measuring the electrical resistance change in the 1.0 wt.% nanotube-glass fibre–epoxy composites during

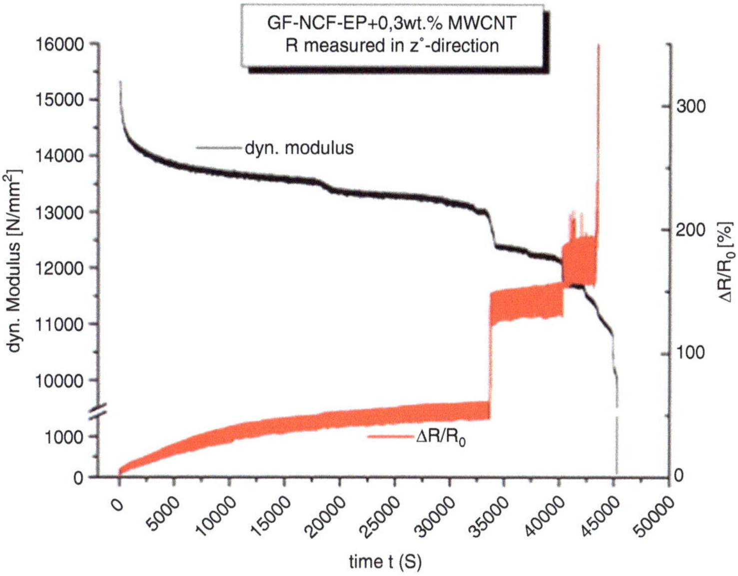

Fig. 2.6 Stiffness and resistance change for a specimen under dynamic tensile load. Resistance measured in z-direction (Reprinted from Böger et al. 2008, with permission from Elsevier)

tensile and fatigue tests. By partitioning the tensile and fatigue samples with electrically conductive probes, it is shown that with both increasing tensile load and number of cycles, different resistance changes are detected in different regions and failure happens in the part in which higher resistance change was detected. Moreover the change in slope of the electrical resistance versus strain curve enabled the detection of an elastic limit (Fig. 2.7). The absence of residual change in resistance for fatigue loading up to maximum loads that are lower than the elastic limit supported the aforementioned conclusion.

The more sensitive residual change in resistance observed (Fig. 2.8 left) coupled with the presence of matrix cracks for loads where there is large change in residual resistance and little change in residual strain (Fig. 2.8 right), proposed that the carbon nanotube network created has better sensitivity in detecting damage versus conventional strain gauges. This is further supported by fatigue results done on samples with and without cracks. The better sensitivity of the carbon nanotube network as compared to strain gages can be explained by the fact that carbon nanotubes are spread throughout the matrix in the composites, and most of the initial cracks and delaminations take place within the matrix material. The stress path around a crack may go around the gauge if the strain gauge is located too close to the crack. On the other hand, since the nanotube networks are connected all over the sample, the occurrence of any defect or damage can cause disconfiguration of the nanotube network. This in turn produces an increase in resistance along the sample, regardless of the location of failure.

Fernberg et al. (2009) focused on an experimental investigation of the relation between resistivity changes as measure of transverse cracking damage

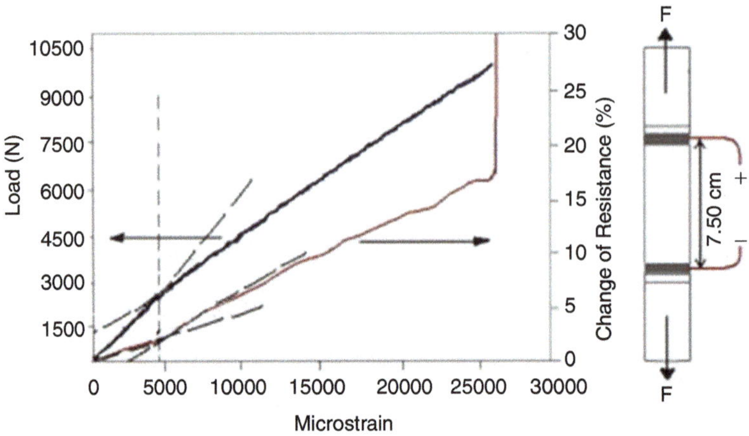

Fig. 2.7 Change of resistance in tension for 1.0 wt.% nanotube-glass fibre–epoxy composite (Reprinted from Nofar et al. 2009, with permission from Elsevier)

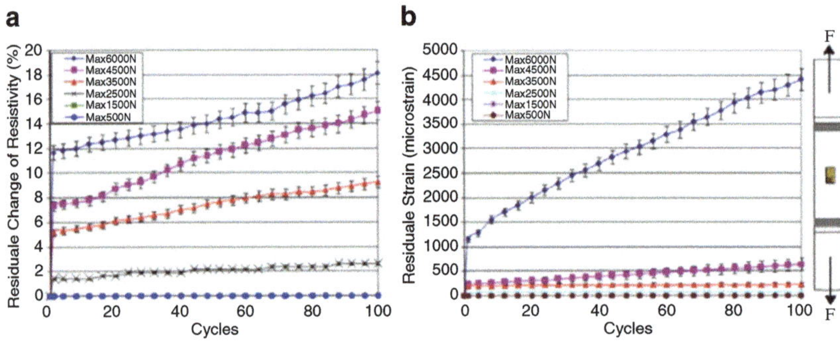

Fig. 2.8 (*left*) Residual change of resistance and (*right*) residual strain using strain gauge measured for six maximum loads in the first 100 cycles (Reprinted from Nofar et al. 2009, with permission from Elsevier)

accumulation in glass fibre cross-ply composites. These tests were performed as loading–unloading experiments under a controlled and constant displacement rate. A maximum tensile strain of 0.1% was attained during the first cycle. The maximum strain was stepwise increased by 0.1% for each subsequent cycle until a strain of 1.1% was attained in the last cycle. The electrical resistance was continuously measured and monitored by a multi-meter during the loading and unloading sequence. A value of the specimen electrical resistance R was registered after completion of each loading cycle i.e., in relaxed and unloaded mode. The specimen was thereafter also removed from the testing machine and the number of transverse cracks in the 90° layer was counted and registered using an optical microscope. Applied stresses and strains were continuously registered throughout the tests and hence facilitated monitoring of the stiffness degradation during a test sequence. Additional testing on

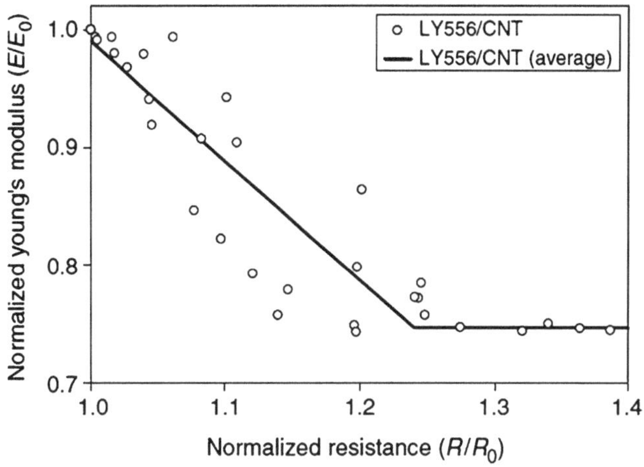

Fig. 2.9 Stiffness changes vs resistance change for CNT-doped cross-ply laminates (Reprinted from Fernberg et al. 2009)

cross-ply composites involved stepwise tensile loading–unloading experiments. In these tests the sample electrical resistance was measured both in loaded and unloaded state. The loading cycle involved loading to a maximum tensile strain of 0.06% during the first cycle followed by unloading to 0.04%. In the following cycles the maximum strain was increased by 0.02% for each cycle until a strain of 0.8% was attained. Unloading to 0.04% was performed between each loading cycle. A multimeter was used to manually register the electrical resistance between the contact areas of the sample both in the stressed and the relaxed state of a loading cycle.

The normalized modulus plotted vs. normalized electrical resistance changes (for electrical resistances measured once the materials are in an unloaded state) clearly demonstrated (Fig. 2.9) the potential to use CNT-doped resin for indirect determination of damage state of polymer composite. Although there is some scatter in the curve, there is sufficient correlation to state dependence between the two measured quantities. A possible interpretation of these results according to the authors is that the initial stiffness decrease is a consequence of damage in the form of transverse cracks. These cracks contribute to breaking of electrical pathways within the resistive percolated network. The consequence of broken pathways is increased electrical resistance. With increasing damage in the form of transverse cracks the electrical resistance hence increases. Apart from transverse cracking there is also other type of damage that may occur during loading. Such damage is e.g., intra-laminar delamination between 90° and 0° plies. In comparison, this damage only influences the stiffness marginally, whereas they are likely to have severe impact on electrical resistance. Therefore, the last part – where electrical resistance increases and no stiffness reduction occurs – can be attributed to the delamination growth.

Yesil et al. (2010) demonstrated that diamine and surfactant modifications on CNTs cause improvements of damage sensing capability under fatigue and impact,

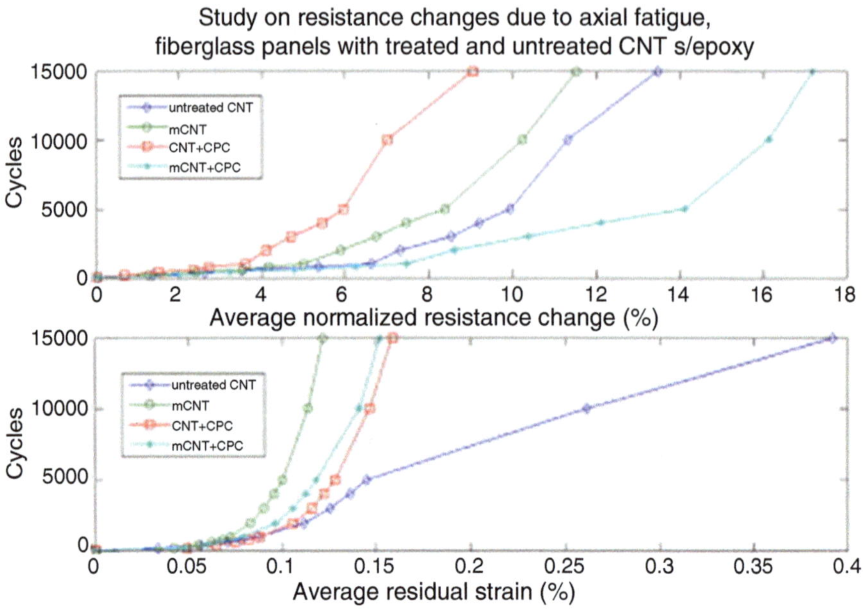

Fig. 2.10 Normalized resistance changes, $(R - R_0)/R_0 \times 100$, versus cycles (*top*) and average residual strain versus cycles (*bottom*) for fibreglass panel configurations, under tensile fatigue. The data are averages of three specimens for each group, except for the mCNT-CPC and mCNT cases, where one specimen failed prematurely (at approximately 4,000 cycles) (Reprinted from Yesil et al. 2010)

in addition to an increase in axial strength and stiffness. This was demonstrated on fibre glass-reinforced panels prepared with treated CNT/epoxy through hand lay-up. Four different composite panels were manufactured with fibreglass reinforcement and different matrix types: (a) as-received CNT/epoxy; (b) as-received CNTs modified with CPC, mixed with epoxy (CNT-CPC/epoxy); (c) CNTs treated with diamine, mixed with epoxy, without CPC (mCNT/epoxy) and (d) with CPC (mCNT-CPC/epoxy). Baseline panels with fibreglass and neat resin were also prepared. The CNT loading was 0.5 wt.% for all the panels containing CNTs. The electrical resistance changes of the composite panels during the static, fatigue and impact tests were also measured by the two-point probe method, with the help of metal electrodes, which were placed on the specimens using silver paste for decreasing the contact resistance. By correlating normalized resistance changes with numbers of cycles and residual strain (for the fatigue tests, Fig. 2.10), and transferred impact energy and inelastic energy curves (for the impact tests, Fig. 2.11) authors showed that the mCNT/epoxy panels with and without CPC exhibit larger resistance changes than the respective non-diamine treated panels.

Nanni et al. (2009, 2011) concluded that conductive filler dispersion and its intrinsic properties are very important features when aimed at prepared self-monitoring materials. In particular, the use of nano-fillers with high surface areas, high OAN and low particle dimensions are recommended to achieve a reliable self-monitoring system.

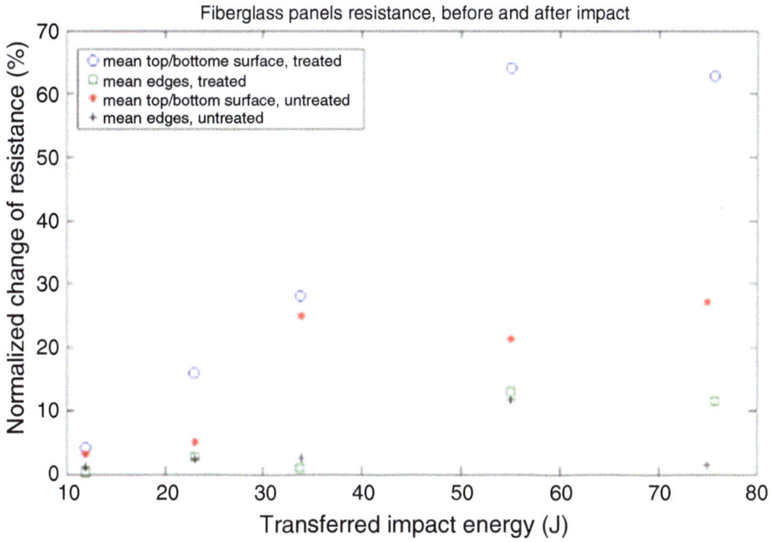

Fig. 2.11 Normalized resistance changes, $(R - R_0)/R_0 \times 100$, versus transferred energy, for surface measurements and edge measurements, of fibreglass panels with epoxy and treated CNT (mCNT-CPC) and as-received untreated CNT/epoxy (Reprinted from Yesil et al. 2010)

They developed two types of hybrid self-monitoring composite rods, made of an internal conductive core surrounded by an external structural part (Fig. 2.12). Both the internal core and the external part were made of glass fibre-epoxy; nevertheless, electrical conductivity was achieved in the inner core by incorporating carbon nano-particles within the resin. In particular, the manufactured self-monitoring composite materials contained, as an alternative, two types of carbon black nano-particles with different surface areas, OAN and particle size. The aim was to correlate the composite self-monitoring performance to the conductive filler properties, by characterizing the filler interaction with the epoxy matrix. The results showed that only samples containing high surface area (HSA) nano-particles (Fig. 2.13) show true self-monitoring behavior, while low surface area (LSA) nano-particles (Fig. 2.14) are not suitable for such applications, since electrical resistance recovery was found at high loads.

Rheological measurements demonstrated that samples with high surface area nano-particles show a more uniform filler dispersion, while large aggregates are present in the case of LSA ones. This occurrence could be responsible for the electrical resistance recovery, due to aggregates breakage at high loads, with consequent release of a number of carbon nano-particles in the matrix that increase electrical conductivity. SEM observations confirm the different microstructures of the two types of specimens, validating this theory.

Rausch and Mäder (2010a, b) presented a novel approach for interphase sensing by modifying glass fibre coatings with CNTs towards health monitoring in continuous glass fibre reinforced thermoplastics. Cyclic tensile loading of the model

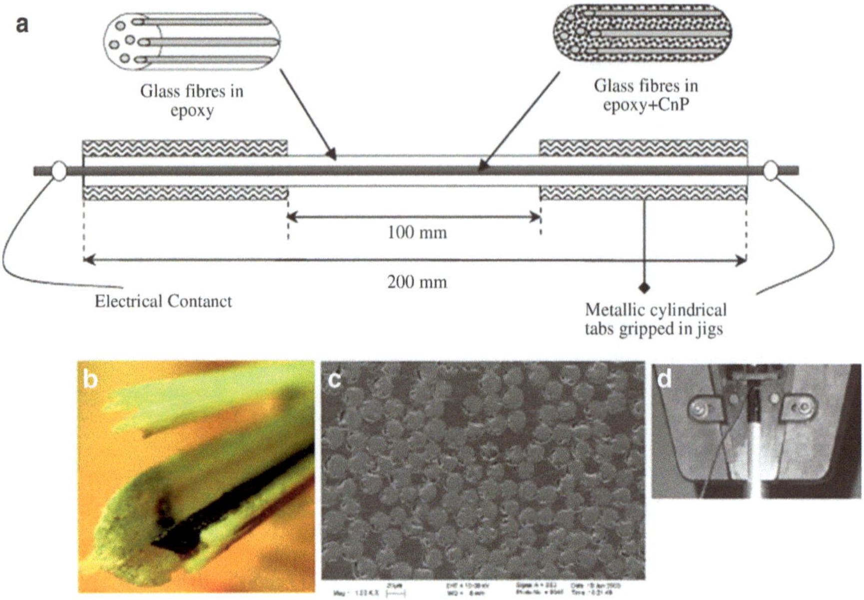

Fig. 2.12 (**a**) Sketch of hybrid CnP-GFRP, (**b**) sample cut and open to show inner part, (**c**) SEM micrograph of the internal conductive core cross section (CnP in epoxy resin + glass fibres and (**d**) particular of sample gripped in tensile machine) (Reprinted from Nanni et al. 2011)

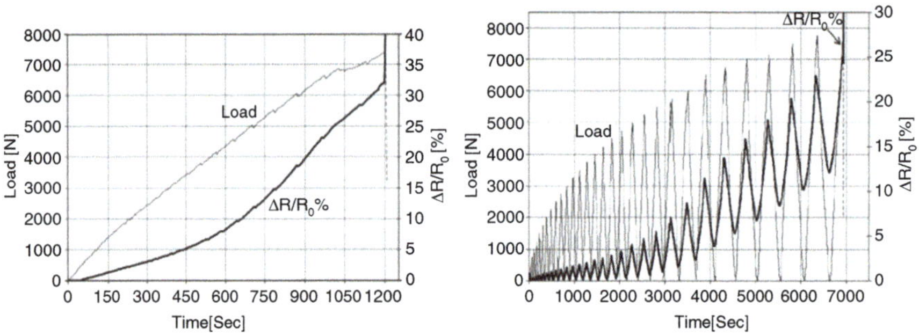

Fig. 2.13 Self-monitoring results for HSA_3 sample. Load/time curve (*light gray line*) and $\Delta R/R_0\%$/time curve (*dark gray line*) both under continuous and cyclic loading (Reprinted from Nanni et al. 2011)

composites is performed highlighting the potential of the sensor for detection of interphase failure. Based on the resistance change curve during cyclic loading, they introduced new parameters allowing the quantification of the accumulated interphase damage. They report on different approaches for tailoring the resistance as well as the sensitivity of interphase sensors based on carbon nanotubes (CNTs). The two main aspects in affecting their initial resistance as well as the sensitivity of the systems during mechanical loading are the yarn coating content and the

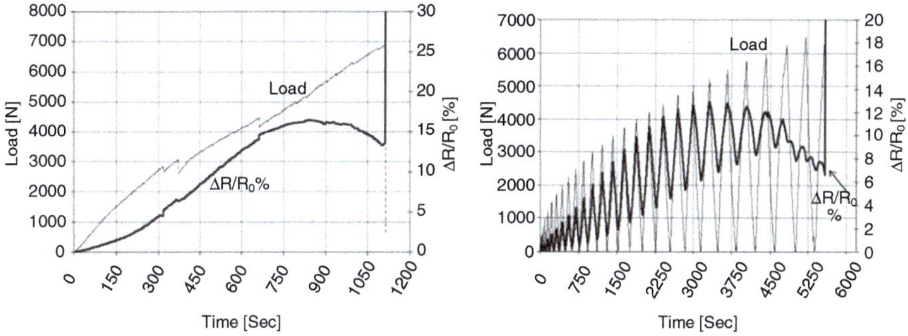

Fig. 2.14 Self-monitoring results for LSA_4 specimens: Load/time curve (*light gray line*) and $\Delta R/R_0$%/time curve (*dark gray line*) both under continuous and cyclic loading (Reprinted from Nanni et al. 2011)

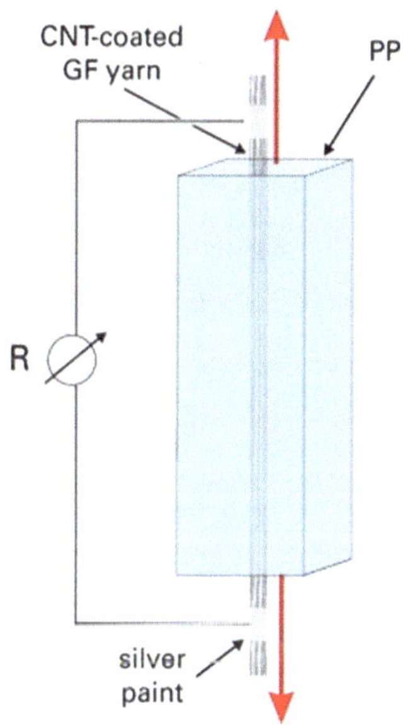

Fig. 2.15 Scheme of specimen for tensile testing and simultaneous recording of resistance change of the CNT-coated GF yarn embedded in a PP matrix (Reprinted from Rausch and Mäder 2010a)

CNT-weight fraction of the coating (Fig. 2.15). Varying those factors, the conducted tensile tests showed that the initial resistance as well as the sensitivity of the interphase sensors can be adjusted within a certain range.

Additionally, it is shown that glass fibre (GF)-yarns with low coating contents allow identifying critical loads for the interphase, which are found to be below the ones for GF failure (Fig. 2.16). Performing cyclic tensile loading above and below this critical stress value has a significant effect on the interphase life-time. In order to assess the interphase damage quantitatively, new parameters based on the resistance change are introduced. Those parameters allow for direct comparison and characterization of different GF modifications, i.e. inter-phases, during mechanical testing by cyclic loading of the interphase sensors.

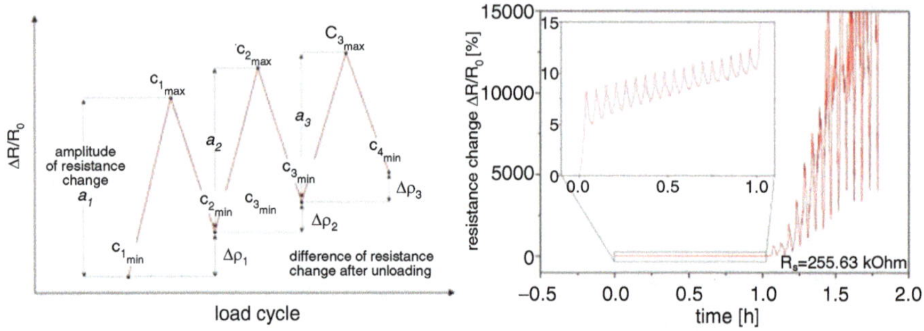

Fig. 2.16 Definition of characteristic points within a schematic resistance change curve during cyclic loading. Resistance change of interphase sensor during stress controlled cyclic loading between 0 and 22 MPa. The inset figure shows the amplitude of the resistance change before the occurrence of severe interphase damage, which causes a distinct change of the amplitude pattern (Reprinted from Rausch and Mäder 2010a)

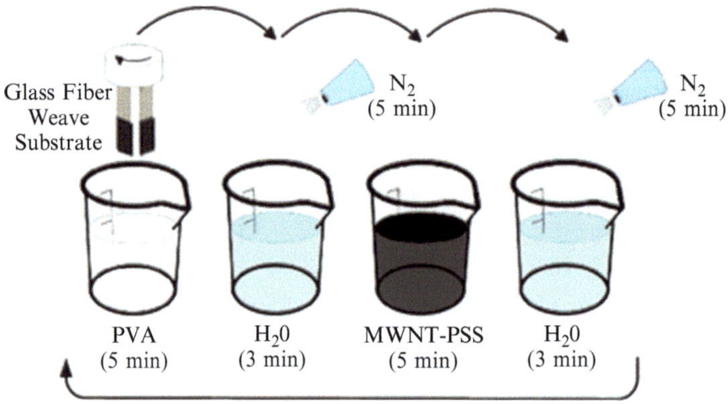

Fig. 2.17 Schematic illustrating the LbL deposition technique employed for fabricating (MWNT–PSS/PVA) $_n$ thin films on glass fibre weaves (Reprinted from Loyola et al. 2010b)

Loyola et al. (2010a, b) proposed utilization of in-situ strain monitoring of FRP composites via layer-by-layer multiwalled carbon nanotube-polyelectrolyte thin films deposited directly upon glass fibre weaves (Fig. 2.17). The nano-composite-coated fibreglass is embedded in GFRP during composite fabrication for creating a self-sensing composite structure. A layer-by-layer (LbL) thin film fabrication methodology is employed for depositing piezo-resistive MWNT–polyelectrolyte (PE) thin films onto the fibreglass weave.

Upon embedding this strain-sensitive fibreglass layer within GFRP samples, their strain-free electrical properties are characterized. Then, electro-mechanical testing is conducted for characterizing the strain-sensing performance of nanocomposite-enhanced GFRPs.

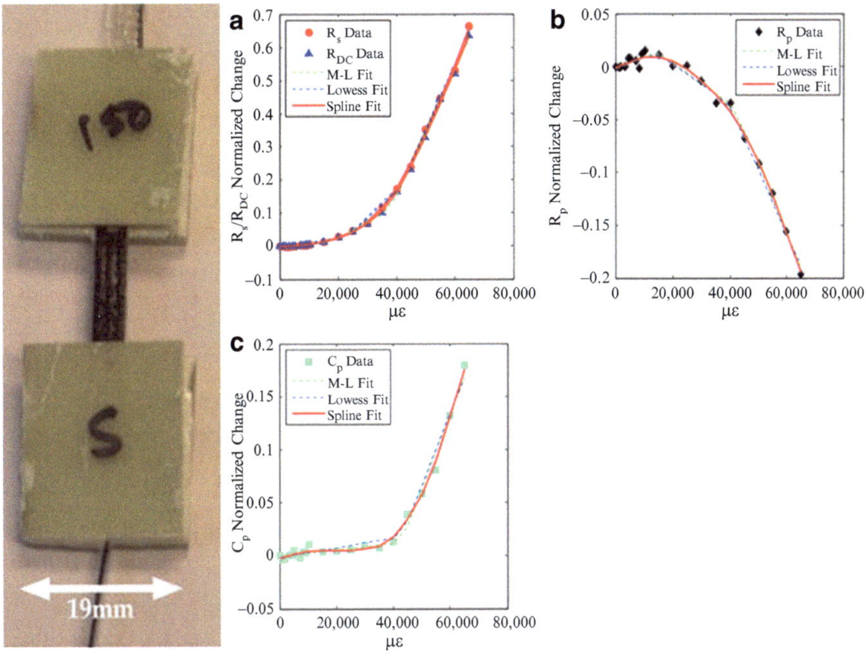

Fig. 2.18 The final manufactured nanocomposite-embedded GFRP specimen. (**a**) The (MWNT–PSS/PVA)$_{29}$ thin film's a bulk resistance (i.e., R_{DC} and R_s), (**b**) inter-nanotube resistance (R_p), and (**c**) inter-nanotube capacitance (C_p) response to applied strain (Reprinted from Loyola et al. 2010b)

The GFRP samples (Fig. 2.18) are loaded in uni-axial tension while (1) the time-domain surface resistivity is measured, and (2) electrical impedance spectroscopy (EIS) is conducted to characterize the complex impedance response of the self-sensing GFRP system. Using the experimental EIS measurements, a simple equivalent circuit model is proposed for modeling the thin film impedance response to applied strains. Using the equivalent circuit model, individual circuit parameters are examined for their sensitivity to strain. Finally, this study concludes by comparing the experimental results and model fits from the time – and frequency-domain strain sensing results. The nano-composites' piezo-resistive responses are well-captured by cubic smoothing spline fitting, and all the responses demonstrated two distinct sensitivities, depending on whether the film is strained at low (0–10,000 με) or high strain (>10,000 με). Within the lower strain regime of less than 10,000 με, the bulk piezo-resistivity exhibits a typical elastic response, while the inter-nanotube behavior is hypothesized to be due to carbon nanotube reorientation. In the higher strain regime, the bulk response is believed to suggest evidence for micro-cracking of the matrix and film. The inter-nanotube response, dependent on thin film thickness, exhibits a behavior that indicates that the thin film within the glass fibre bundles is subjected to compressive forces due to Poisson's effect, as evident from the negative and positive gage factors for R_p and C_p, respectively.

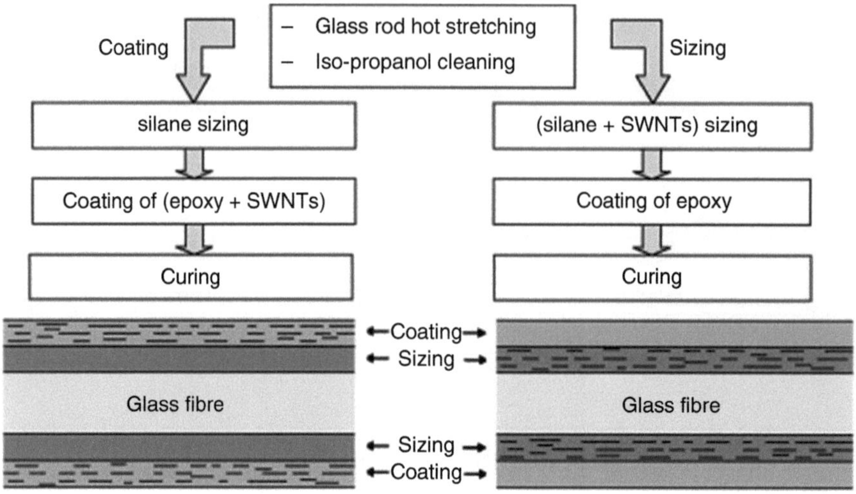

Fig. 2.19 Fibre preparation methods and schematic of finished samples (not to scale) (Reprinted from Sureeyatanapas and Young 2009)

Sureeyatanapas et al. (Sureeyatanapas and Young 2009; Sureeyatanapas et al. 2010) demonstrated the use of both a luminescent dopant material (Sm_2O_3) and single-walled carbon nanotubes (SWNTs) as strain sensors for glass fibres through the use of combined luminescence and Raman spectroscopy. Single-walled nanotubes can be combined with a silane coating on the surface of the doped fibres, and local strain can be simultaneously monitored using both techniques, despite the presence of this coating (Fig. 2.19).

It has been shown that good agreement with shear-lag theory can be obtained using both techniques, during fragmentation of the glass fibre. A maximum shear stress of 25 MPa was obtained for these samples, which is lower than for pure glass fibres with Sm^{3+} dopant ions present but without the presence of SWNTs, but higher than for a non-doped glass with SWNTs on the surface. This suggests that the SWNTs do affect the interface by reducing the interfacial shear stress between the resin and the fibre. A comparison between the two strain sensing techniques has validated the approach of using SWNTs for the monitoring of local strain, even though the material is only coated on the fibre's surface. A consistent relationship between the strain from the luminescence spectroscopy of the Sm^{3+} ions and the Raman spectroscopy from the SWNTs has demonstrated that the techniques are consistently correlated, even during complex events such as fragmentation of a glass fibre. It is clear that these techniques could be used for health monitoring of glass fibre reinforced plastics in a variety of applications.

Gao et al. (2010) described a method that introduces electrical conductivity to glass fibre surfaces by depositing MWCNT networks, and in turn, specifically forming an interconnected MWCNT-rich interphase within glass-fibre-reinforced epoxy composites. They used commercially carboxyl-functionalized MWCNTs produced via the catalytic carbon vapor deposition (CCVD) process, with average diameter of 9.5 nm, and average length of 1.5 μm. various aqueous dispersions were utilized and compared, including dispersion aids of non-ionic, cationic, or anionic surfactants. Based on optimum conditions for efficient dispersion, all samples had a constant MWCNTs : surfactant weight ratio of 2:3 and underwent equal batch-wise sonication employing a tip sonicator at constant output power of 180 W for 180 min. Then, the glass fibres were dipped into a MWCNT dispersion with the pH value of 5 $\sim$ 6 and 0.5 wt.% MWNTs for 15 min, withdrawn with their axes perpendicular to the solution surface, and dried in a vacuum oven at 40 °C for 8 h. Finally, the concentration of nanotubes on the glass fibre surface measured with an electronic balance was 2.3 wt.%. A commercial DGEBA-based epoxy with amine hardener in a weight ratio of 100:34 was used as the matrix, and the composites were cured at identical conditions (80 °C, 6 h). To avoid possible complex effects of coupling agents on the electrical properties of the nanotubes, no additional coupling agent was used. With their experimental work, authors demonstrated that single MWCNT–glass fibre and corresponding epoxy matrix composites show stress/strain, temperature, and relative humidity dependence in their electrical conductivity; as in situ multifunctional sensors, they are capable of detecting piezo-resistive effects as well as the local glass transition temperature. Moreover they reported that unidirectional composites fabricated via the MWCNT–glass fibres exhibit ultrahigh anisotropic semiconducting electrical properties and an ultralow electrical percolation threshold.

The same group in another work Zhang et al. (2010) developed the electrophoretic deposition (EPD) method and compared it with the dip coating method to deposit MWCNTs onto non-conductive glass fibres. By both techniques, they introduced new functional interphases inspired by the nano-scale interphase in biological bone and improved the interfacial strength. Through the EPD method or dip coating, a conductive pathway is created by the randomly oriented carbon nanotube networks on the curved fibre surface, and the electrical resistance value of coated fibres reached the semi-conductive range. The EPD coating was more homogeneous and continuous and the electrical resistance values of single fibres scattered much smaller than that by dip coating. Therefore, EPD proved to be a more efficient procedure to deposit MWCNTs onto insulative fibre surface. The EPD fibres also gained higher interfacial shear strength without degradation of the fibre strength compared with the control fibre and DIP fibre. The interfacial shear strength of single EPD fibre composites exhibited more than 30% improvement, irrespective of whether the coating includes a silane coupling agent or not. Related to the differently treated glass fibres, three interphase structures were proposed, which were consistent with fragment length results of Weibull distribution analysis. The EPD method produced mid-homogeneous coating, resulting in heterogeneous interphase coexistence similar with the structure of biological bone. The deposition of MWCNTs by dip coating created inhomogeneous interphases, led to a decrease of the single fibre tensile strength and inhomogeneous interphase stress distribution. The electrical resistance measurement of single fibre/epoxy composites under

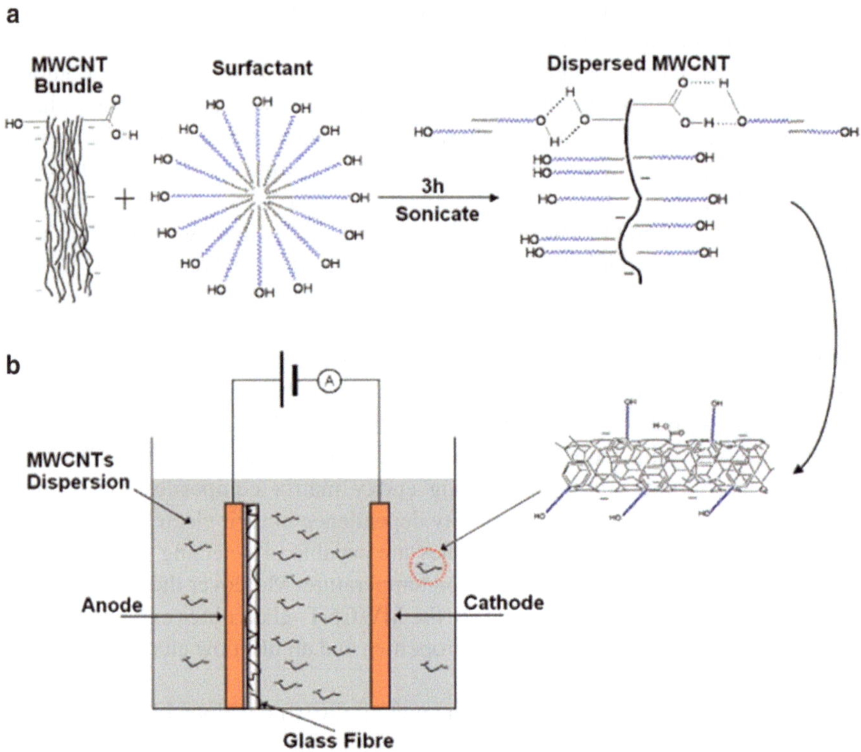

Fig. 2.20 Schematic illustrations of (**a**) MWCNTs dispersion process in water with surfactant and (**b**) deposition of MWCNTs onto insulative glass fibre surface by the electrophoretic deposition cell (Reprinted from Zhang et al. 2010, with permission from Elsevier)

tensile loading indicated this semi-conductive glass fibre composite are capable of early warning before composite fracture, and the inherent damage can be monitored simultaneously. This effect can be used for in situ sensor development for composite damage process instead of external sensors (Figs. 2.20 and 2.21).

Lim et al. (2011) aimed to apply resistance-based health monitoring towards the measurement of damage in composites during dynamic compression loading. Specifically, the effectiveness of an embedded carbon nanotube network in sensing damage arising from dynamic compression loading in a thick-section composite was evaluated. Experiments are performed using a split Hopkinson pressure bar experimental apparatus and the electrical response of a composite specimen is measured in parallel. The composite panel was produced using twenty layers of plain woven E-glass fabric, 5 × 5 yarns/in. The carbon nanotube sizing agent (SIZICYL™ XC R2G, NanoCyl) was first diluted with three volumetric parts distilled water prior to infusion via vacuum-assisted resin transfer molding (VARTM). After oven drying at 150 °C overnight, the sizing-treated fabric was infused with an epoxy cycloaliphatic amine, SC-15 from Applied Poleramic Inc.

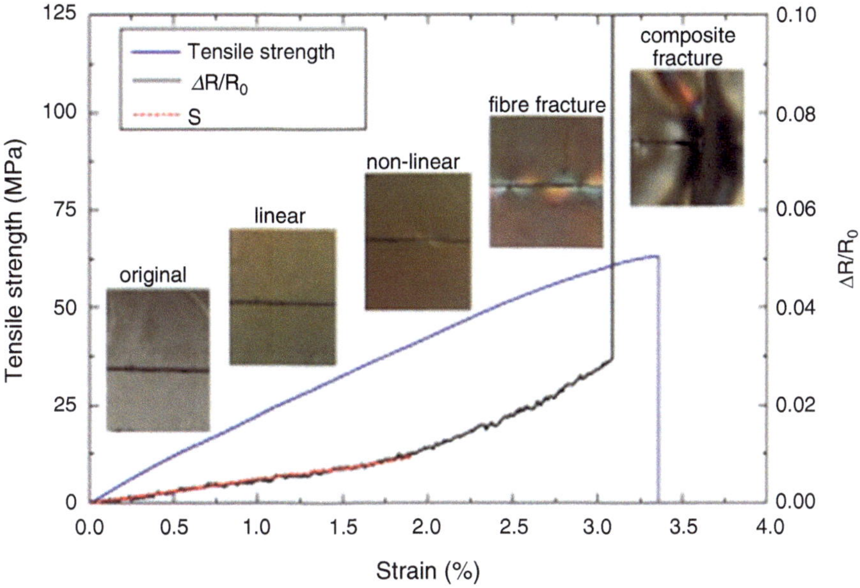

Fig. 2.21 Simultaneous change of electrical resistance and stress as a function of strain for single coated fibre/epoxy composite, the dashed *S* is the straight line simulation of $\Delta R/R_0$ at the linear increasing stage. Inserted figures are the photoelastic profiles during tensile process corresponding to the $\Delta R/R_0$ value at the stages of original, linear, non-linear, fibre fracture and composite fracture (Reprinted from Zhang et al. 2010, with permission from Elsevier)

(API), using VARTM and the part was cured at room temperature (22 °C) for 2 days and post-cured at 150 °C for 2 h. After curing, the panel was machined at a 45° angle to the longitudinal axis, resulting in 45° composite strips which were 0.35 in. (8.9 mm) thick. These strips were core-drilled to yield cylindrical specimens (3/4 in., 19.1 mm diameter) with parallel flat edges. These specimens (Fig. 2.22), with a cross-sectional area of 2.61 cm^2 (0.405 $in.^2$), provided a compromise in which images of the specimen surface could be taken easily with minimal interference of reflections of the free end of the Hopkinson bar.

Initially, quasi-static compression tests were performed to identify the stress level at which failure of the composite occurred as well as the resistance behavior associated with compressive failure of a 45° off-axis composite specimen. Following up, split Hopkinson pressure bar experiments were performed on the same geometry 45° specimens (Fig. 2.23). A single specimen was impacted multiple times, each time at an increased gas gun pressure until failure occurred. Evidence of damage is seen in the mechanical response of a 45° carbon nanotube/E-glass/SC-15 composite specimen under dynamic compression loading; this is correlated with increases in resistance, which occurs only after impacts that result in a decrease in specimen stiffness (Fig. 2.24).

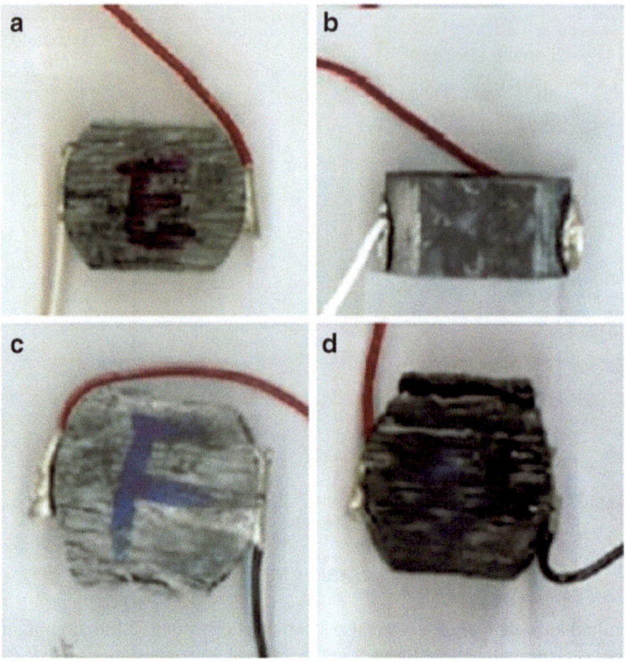

Fig. 2.22 Composite specimens with attached electrodes (*white* and *red wires*) used in split Hopkinson pressure bar evaluation: (**a**) untested impact face, (**b**) untested edge-on view, (**c**) quasi-static compression tested face and (**d**) SHPB tested face (Reprinted from Lim et al. 2011)

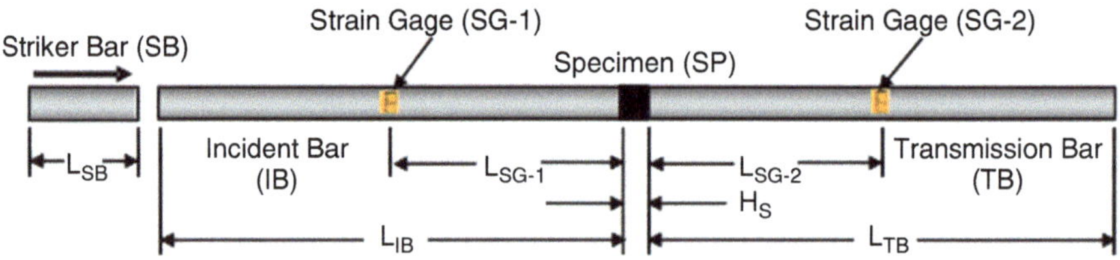

Fig. 2.23 Split Hopkinson pressure bar experimental apparatus (Reprinted from Lim et al. 2011)

2.3 CNT-CFRP

Kostopoulos et al. initially proposed early from 2005 (Kostopoulos et al. 2005, 2006, 2007a) the idea of incorporating conductive particles such as Carbon nanofibres (CNFs) as dopants into the matrix material of CFRPs for sensitivity enhancement of the electrical resistance monitoring technique. Their main goal was to use carbon nano-particles as a nano-sensor for damage detection within the matrix material of the CFRPs. For this reason, CNF-doped CFRPs and neat CFRPs were subjected to loading-unloading tension tests and the electrical resistance was measured at each maximum loading and unloading state. Significant changes

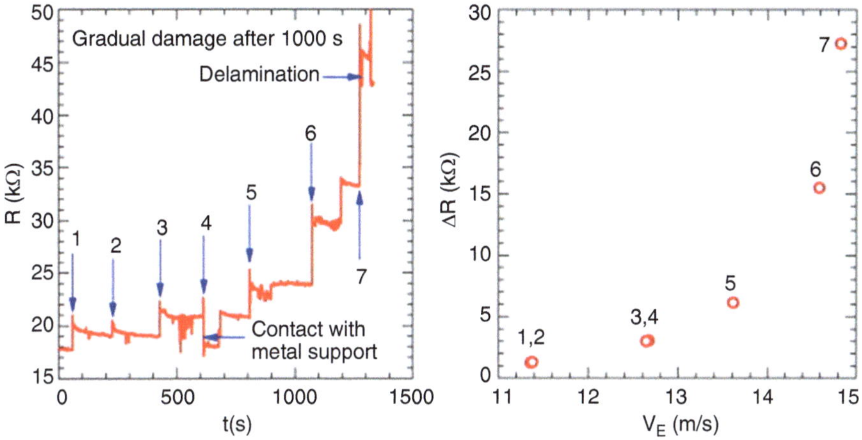

Fig. 2.24 (*left*) *R–t* of a carbon nanotube/E-glass/SC-15 composite specimen during SHPB experiments. (*right*) Change in baseline resistance vs. striker bar exit velocity, V_E; numbers indicate impact sequence (Reprinted from Lim et al. 2011)

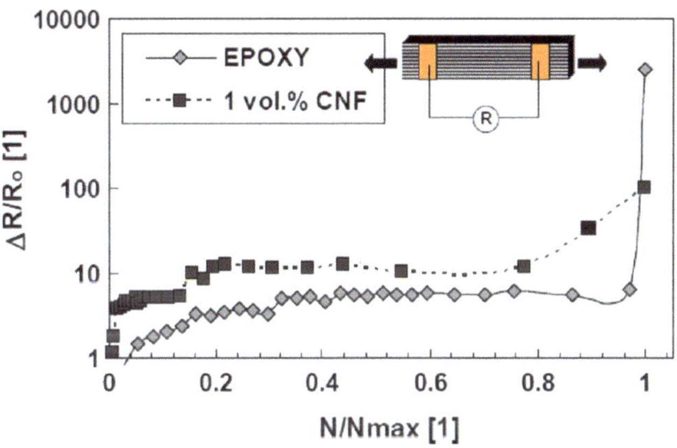

Fig. 2.25 Changes in the resistance versus the normalized number of cycles for laminates with neat and CNF-doped EP matrix

were noted in the electrical resistance of both types of materials. With increasing applied load the resistance increased due to the damage of the fibres and therefore diminishment of the percolating network. During the unloading of the specimens the broken fibres are forced to come in contact and consequently the resistance increased. For the doped sample smaller steps of resistance increase were noted. Furthermore they monitored the changes in the resistance under fatigue loading of laminates with neat and CNF-doped EP matrix and they reported (Fig. 2.25) that the doped sample was more sensitive to resistance changes, speculating that the presence of conductive CNFs can give evidence of the matrix cracking which takes place in the earlier cycles.

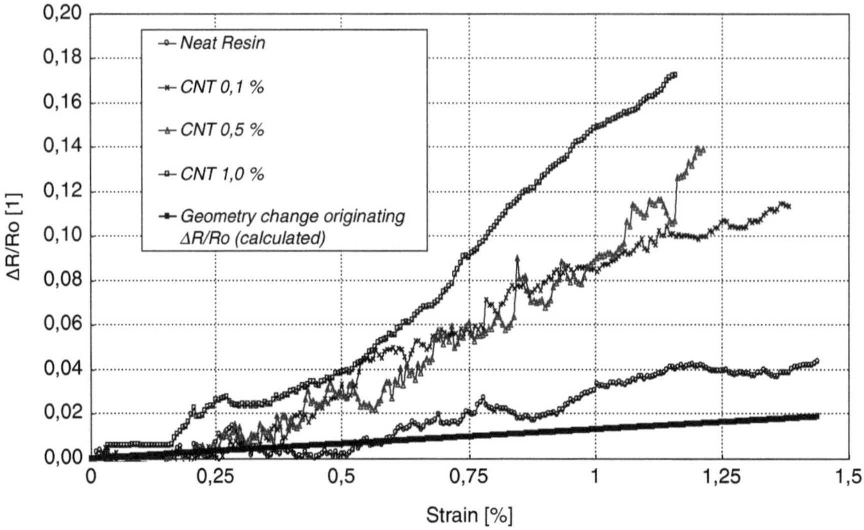

Fig. 2.26 Normalized Resistance change (ΔR/Ro) vs. strain of CFRP laminates with neat and CNT doped epoxy matrix

The same group from 2006 (Vavouliotis et al. 2006; Kostopoulos et al. 2007b, 2008, 2009b) explored the idea for MWCNT-modified fibre reinforced composites with nano-sensing capabilities. They reported that the presence of the CNTs in the epoxy matrix of continuous carbon fibre composites enhanced the real-time damage monitoring via electrical resistance change (ERC) method. This was established via direct comparison of the electromechanical behaviour of the CNT doped CFRP laminates with conventional laminates. The higher resistance changes ($\Delta R/R_o$) that were recorded for the modified CFRPs are directly related to the increase in sensitivity of the ERC technique for damage sensing, since all other parameters (piezo-resistivity, electrical contact degradation and geometrical deformation) are considered to be constant for all studied material configurations. These assumptions even have a reasonable logical basis, but they need further experimental investigations to be confirmed or rejected. The influence of the CNT on the Poisson ratio and the piezo-resistance of the CFRPs are two major topics suggested for future work. Moreover the selected electrode configuration (two surface probe/four wires) for the detection of damage even at low strains was feasible. This was attributed to the initial statistical fibre breakage and fibre-matrix debonding which affected the local fibre-matrix strain field induced local damage which was mirrored by the CNT percolation network. This effect was achieved for CNT concentrations above 0.5 wt.% (Fig. 2.26).

During the cyclic tensile loading-unloading-reloading experiments (Fig. 2.27), $\Delta R/R_o$ followed the pattern of loading. The laminates with CNT-doped matrix exhibited enhanced sensitivity and capability to track the loading variations. For all materials tested, the peak of $\Delta R/R_o$ increased with the increase of maximum load.

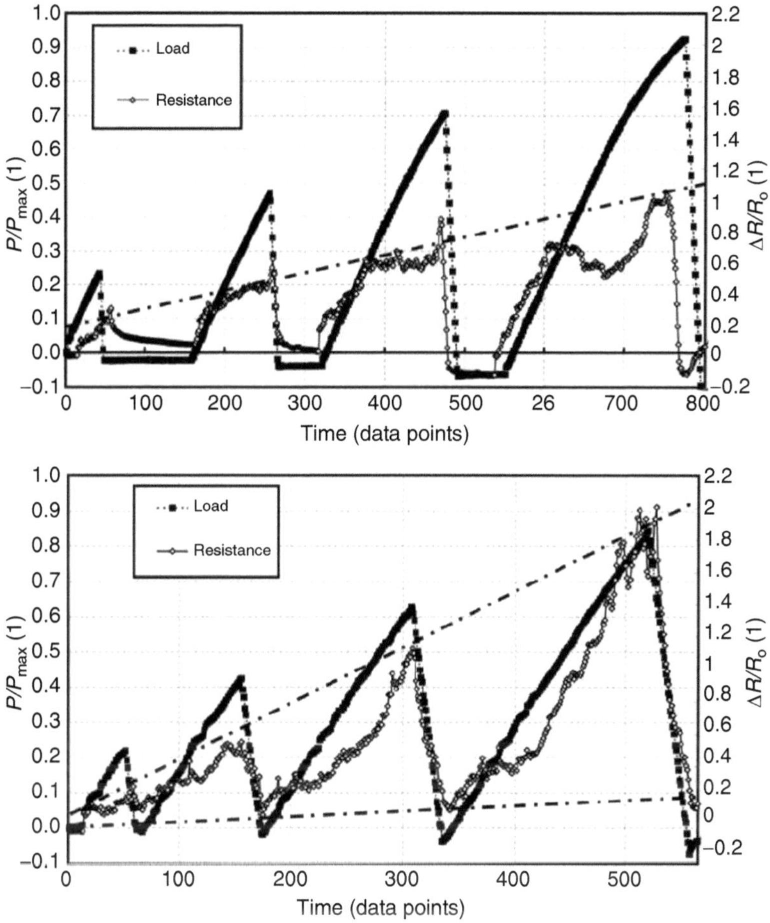

Fig. 2.27 Normalized applied load (P/P_{max}) and Normalized Resistance change ($\Delta R/R_o$) vs. experimental time during the four cycles of quasi-static loading-unloading-reloading test of CFRP laminates with 1 % CNT modified EP matrix

The relative resistance change was higher with increasing CNT content in the epoxy matrix of the laminates. Upon unloading, the nano-modified CFRPs exhibited a residual resistance change which increased at the end of each consecutive cycle. This residual change was more pronounced with increasing CNT content. The residual resistance change at the end of each loading cycle was attributed to irreversible damage phenomena related to the matrix. This damage can be directly quantified via monitoring of the monotonically increasing "zero load line". As was argued, this residual resistance is only related to matrix related damage that follows the primary fibre damage and is mirrored in the percolated CNT network. This is further supported by the finding that the residual resistance could not provide conclusive information about the induced damage in the case of plain CFRP laminates.

Zhang et al. (2007) investigated the ability of the nanotube additives to detect delamination growth. They developed a hierarchical (hybrid) composite wherein a modified Epoxy-2000 resin with 0.5% weight of MWCNT additives was used to produce an 8-ply twill-weaved graphite fibre composite laminate. Initial delamination of the hybrid composite is introduced by inserting a Teflon film between the central plies during the layup. Mode I delamination tests were performed on the hybrid MWNT/graphite-fibre/epoxy composite by application of load normal to the defect plane. While the graphite fibres in the twill-weave composite are conductive in a plane, the out-of-plane (or through-thickness) conductivity is negligible due to the insulating epoxy binder that interconnects the individual lamina. However they observed that with the addition of ~0.5% weight of MWNT in the resin, the through-thickness resistance is reduced by over three orders of magnitude. This is because the dispersed MWNT bridges the spaces between the graphite fibre layers and provides a continuous electrical conduction pathway. They observed that this through-thickness resistance is very sensitive to the delamination length, as shown in Fig. 2.28, indicating that the MWNT additives can detect in real-time mode the size of the delamination and its growth rate.

Barkoula et al. (2009) while studying the environmental durability of carbon nanotube (CNT)-modified carbon-fibre-reinforced polymers (CFRPs), also explored the moisture-caused changes in the resistivity of CFRPs and CNT-modified CFRPs. To examine this problem, CNT-modified CFRPs were exposed to hydrothermal loadings using a water bath with temperature control. At specified intervals, the composites were weighted, and the water uptake vs. time was recorded for both the modified and a reference system. The electrical conductivity of the composites was registered at the same time intervals. In the case of the doped CFRP laminates the resistance was monotonically increasing with weight gain. The inclusion of a small weight fraction of a conductive phase (CNTs) to an otherwise conductive material (due to the presence of carbon fibres), although it was hardly affecting the initial resistance of the system, was totally altering its electrical behavior. While in the neat CFRPs, the conductivity reached a plateau at approximately 0.2% relative weight gain and rapidly decreased thereafter, in the case of CNT-modified CFRPs, there was a clear monotonic increase in the resistance (Fig. 2.29). This can only be attributed to a synergistic effect between the main carbon fibre reinforcement and the CNTs that were included in the epoxy matrix. Last but not least, the monitoring of the hydrothermally induced damage via the electrical resistance technique for composite laminates may be made feasible with the CNT inclusion. This is not directly applicable for the conventional composite systems.

Vavouliotis et al. (2009) continued the effort on investigating of the capacity of the CNTs to be used as inherent sensors utilizing an improved and more sensitive electric resistance change method for common CFRP composites. At the beginning unidirectional composites with various CNT contents and a reference polymer resin matrix were used for quasi-static tensile and cyclic loading-unloading-reloading tests, to show that matrix damage at relatively low strain level causes detectable variation in the composite's resistance and to investigate systematically the electromechanical behavior versus the CNTs content. Moreover quasi-isotropic composites were used in order to quantify the CNT doping effect during tension-tension

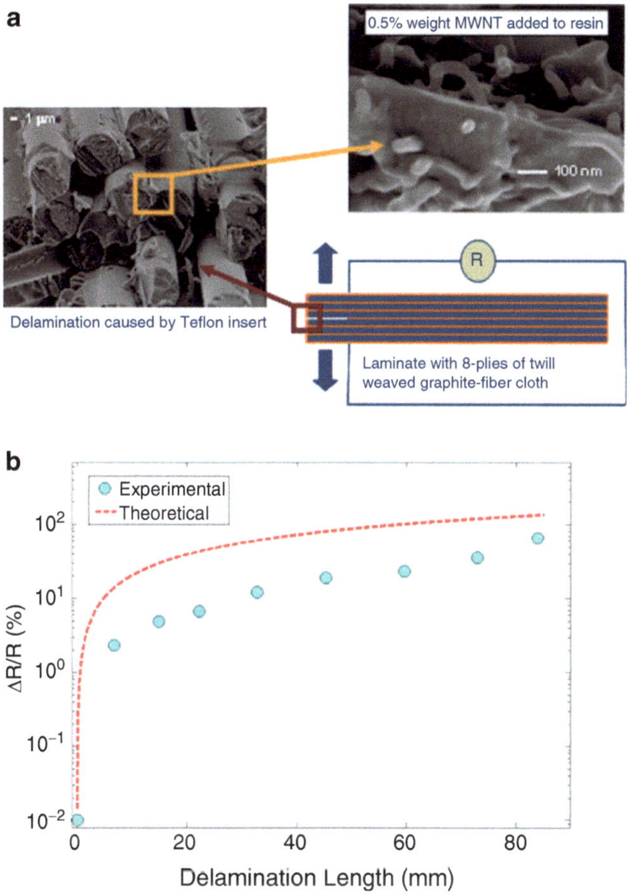

Fig. 2.28 (*Color online*) Detection of real-time delamination growth in a graphite-fibre/carbon-nanotube epoxy laminate. (**a**) A modified resin comprised of MWNT additives in an Epoxy-2000 system is used to lay up, vacuum bag, and cure an 8-ply twill-weaved graphite-fibre composite laminate. A Teflon insert is used to generate a delamination defect as shown in the schematic. (**b**) Changes in through-thickness electrical resistance across the delamination are plotted as a function of the delamination length. Predicted results for the resistance change are also shown for comparison (Reprinted with permission from Zhang et al. 2007, Copyright 2011, American Institute of Physics)

fatigue tests while in parallel the longitudinal resistance was monitored. In addition, Acoustic Emission and Acousto-Ultrasonic techniques were used for monitoring the fatigue process of the laminates. The real-time sensing of load variations via electrical resistance measurements is verified for quasi-isotropic composites in both cases, with 0.5% doped epoxy matrix and with reference matrix. It is also confirmed that the mean resistance changes during the tension-tension fatigue test could reflect the damage accumulation of both materials. Two main stages are distinguished (Fig. 2.30). During the initial stages of the fatigue, less than 10% of the fatigue life, the resistance suddenly drops, mainly due to the self-alignment of the conducting

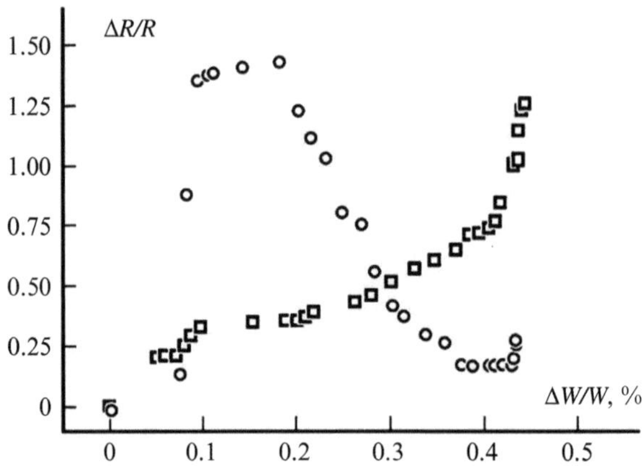

Fig. 2.29 Relative change in the electrical resistance DR/R versus weight gain DW/W for the neat (*cycle*) and 0.5 % CNT-modified (*cube*) CFRP specimens (Reprinted from Barkoula et al. 2009)

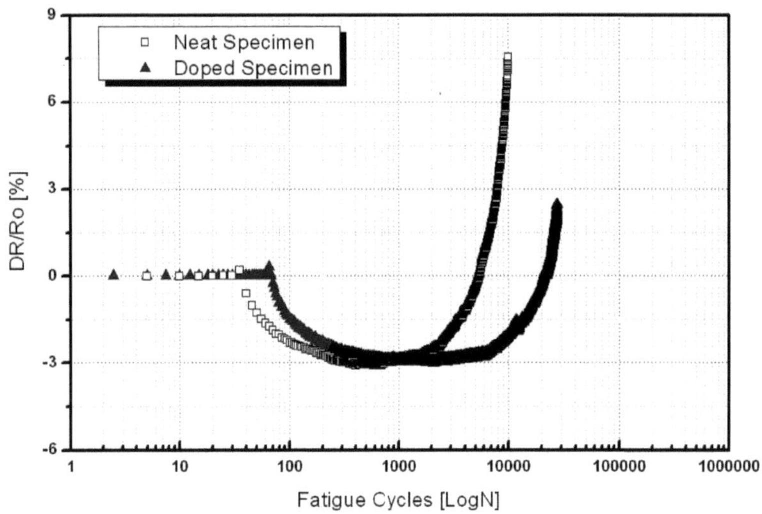

Fig. 2.30 Normalized Resistance Change versus fatigue cycles in logarithmic scale

network of the material with new electrical fibre contacts created after initial inter-laminar matrix cracking. Having the conducting network reached a new equilibrium, resistance is mainly affected by the more intense damage accumulation and is increased continuously up to the final breakage. Acoustic emission analysis proved very helpful in identifying the characteristic fatigue damage states though it is not clear how sensitive is the method to discriminate between the doped and the un-doped specimens.

In a more recent work (2011) (Vavouliotis et al. 2011) the same group evaluated the effect of dispersed Multiwall Carbon Nanotubes (MWCNT) into the epoxy matrix while studying for the first time the electromechanical response (Electrical Resistance Change method) as a damage index of quasi-isotropic Carbon Fibre

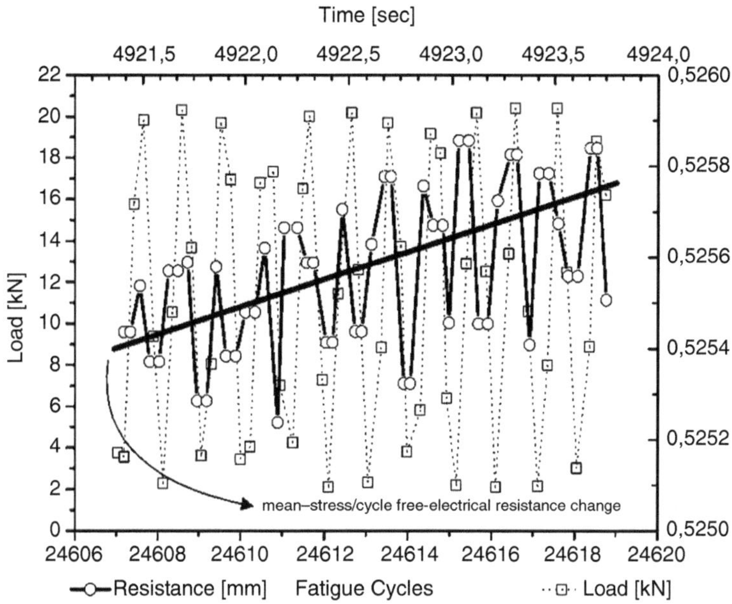

Fig. 2.31 Graph of specimen's electrical resistance and applied load versus time and fatigue cycles

Reinforced (CFRPs) laminates under fatigue loading (Fig. 2.31). The longitudinal resistance change of the specimens was monitored throughout the fatigue experiment. Three different stress levels were tested. The frequency and the ratio (R) of the minimum applied load (stress) to the maximum applied load (stress) were kept constant for the different stress levels. The temperature of the specimen was also monitored throughout the process in order to deduce its effect on the electrical resistance of the specimen. The electrical behavior of the quasi-isotropic CFRP deviated from the commonly observed electrical response of unidirectional or cross-ply CFRPs due to the presence of the 45° layers. During initial stages of loading the resistance drops and afterwards it follows a positive slope up to final fracture. This repeatable pattern was observed for both the neat and the CNT-doped specimens, with the latter having smoother electrical recordings. The effect of temperature was calculated to be limited for the specific material and test/measurement configuration. The electromechanical response was correlated to stiffness degradation (Fig. 2.32) and acoustic emission findings (Fig. 2.33) enabling identification of specific regions during the fatigue life referring to specific mechanisms of damage accumulation. More specifically the experimental results revealed that the occurrence of the initial drop of the electrical resistance is linked with the occurrence of the Characteristic Damage State (CDS), associated with a specific percentage of stiffness reduction. This finding was used in order to predict the remaining life independently from the applied stress level with a high degree of confidence, assuming a constant stress level throughout the whole lifetime. The presence of the

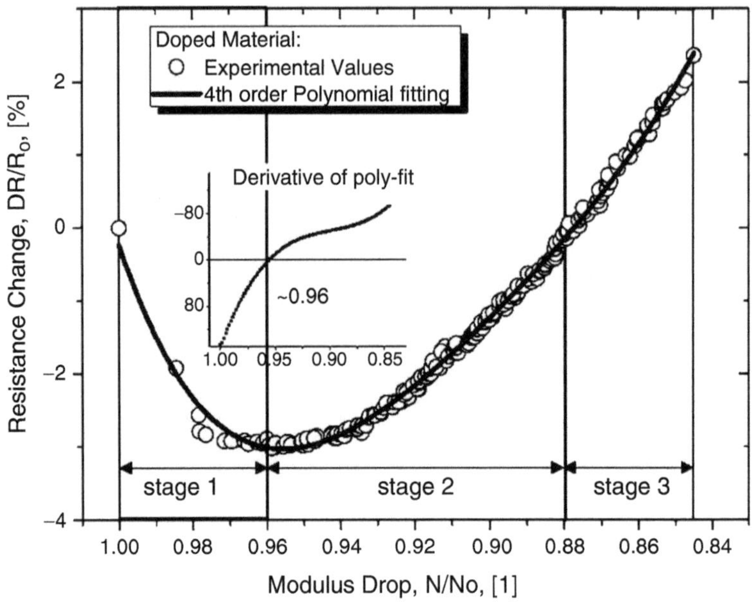

Fig. 2.32 Normalized Electrical resistance change versus modulus drop for CNT-doped quasi-CFRP material. Experimental values and polynomial fitting

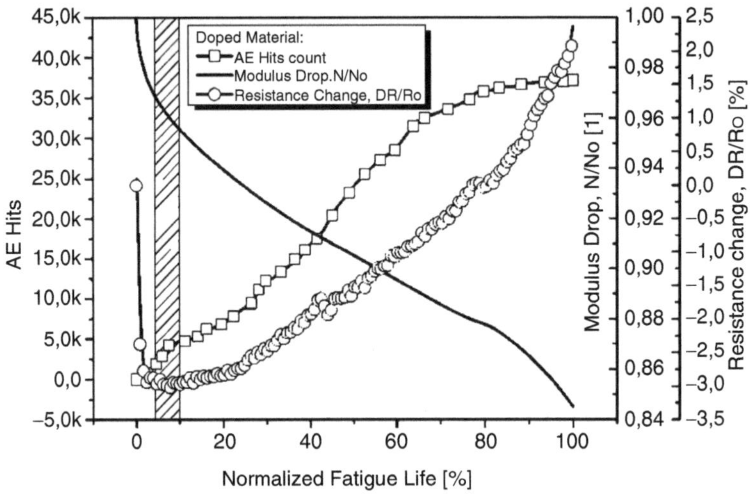

Fig. 2.33 Cumulative number of Acoustic Emission hits versus normalized fatigue life for CNT-doped quasi-CFRP material. Also in double axis format normalized electrical resistance and modulus drop

MWCNT in the epoxy matrix for carbon fibre based composites did not alter drastically the electrical response. Despite the fact that the CNT addition increases the electrical conductivity of the matrix by many orders of magnitude, the effect is masked by the presence of conductive carbon fibres. As a result, the nano-doped matrix did not contribute directly to the electrical conduction mechanism. Nevertheless in microscopic level, the presence of the nanotube influenced positively the mechanism of fibre to fibre electrical contact. During fatigue the range of the electrical resistance change ($\Delta R/R_o$) for the nano-doped CFRP specimens exhibited a decreasing trend. Additionally, the fatigue life prediction for the nano-composites had a higher coefficient of confidence (R^2)

2.4 Concluding Remarks

Based on the aforementioned review, it is evident that nano-enabled self-sensing structural composite materials with tailored electrical properties provide new momentum towards the development of multifunctional materials for advanced applications. The incorporation of carbon nanotubes in the otherwise electrical insulating epoxy matrix of either carbon fibre or glass fibre reinforced composites is suggested by some researchers as the most feasible methodology and experimental results showed improved online damage monitoring capabilities via the electrical resistance change (ERC) method. Nevertheless other methodologies aiming at the nanotube incorporation on the reinforcing fibres especially in the case of glass fibres (deposition, sizing etc) have been investigated proving interesting results. Various configurations and tailoring of nanotube concentration were explored in order to help shed light on the progression and characteristic of damage states. It has been shown that in the case of nanotube-enhanced GFRPs the conductive percolating nanotube networks in traditional fibre composites can accurately detect the onset, nature, and progression of damage. In parallel the CNT modified CFRPs demonstrated enhanced intrinsic damage sensing capabilities due to the conductive nature of their matrix. The concept is proven for various matrix and fibre dominated damage mechanisms stimulated by different experimental campaigns (tensile, bending, fatigue, hydrothermal etc.) while carbon nanotubes enabled the detection of damage accumulation at the nano-scale. The resulting enhanced sensitivity allowed the identification of early damage stages, requested for a potential structural health monitoring application towards the development of a tool for detecting and quantifying structural deterioration and therefore assessing the remaining lifetime of the composite structures. Furthermore potential self-sensing tools will reduce unscheduled and scheduled inspection times and will allow a rapid quality assurance and enhanced manufacturing process control, revolutionizing composite manufacture. Despite the promising results and the emerging need for such self-sensing materials by the aerospace industry, further research shall be made mainly towards the unification of the distributed experimental results around the world that will raise the confidence level required to allow the utilization in real structural health monitoring applications. Moreover the use of new nano-materials products such as graphene and/or concepts of hybrid use of nano-materials are believed to be the next steps for further enhancement of their damage monitoring capabilities.

References

Abry, J.C., Bochard, S., Chateauminois, A., Salvia, M., Giraud, G.: In situ detection of damage in CFRP laminates by electrical resistance measurements. Compos. Sci. Technol. **59**(6), 925–935 (1999)

Abry, J.C., et al.: In-situ monitoring of damage in CFRP laminates by means of AC and DC measurements. Compos. Sci. Technol. **61**, 855–864 (2001)

Angelidis, N., Wei, C.Y., Irving, P.E.: The electrical resistance response of continuous carbon fibre composite laminates to mechanical strain. Compos. A **35**, 1135–1147 (2004)

Angelidis, N., Wei, C.Y., Irving, P.E.: Response to discussion of paper: The electrical resistance response of continuous carbon fibre composite laminates to mechanical strain. Compos. A Appl. Sci. Manuf. **37**, 1495–1499 (2006)

Barkoula, N.M., Paipetis, A., Matikas, T., Vavouliotis, A., Karapappas, P., Kostopoulos, V.: Environmental degradation of carbon nanotube-modified composite laminates: A study of electrical resistivity. Mech. Compos. Mater. **45**(1), 21–32 (2009)

Böger, L., Wichmann, M.H.G., Meyer, L.O., Schulte, K.: Load and health monitoring in glass fibre reinforced composites with an electrically conductive nanocomposite epoxy matrix. Compos. Sci. Technol. **68**(7–8), 1886–1894 (2008)

Ceysson, O., Salvia, M., Vincent, L.: Damage mechanisms characterization of carbon fibre/epoxy laminates by both electrical resistance measurements and acoustic emission analysis. Scr. Mater. **34**, 1273–1280 (1996)

Chung, D.D.L.: Self-monitoring structural materials. Mater. Sci. Eng. Rep. **22**(2), 57–78 (1998)

Chung, D.D.L., Wang, S.: Self-sensing of damage and strain in carbon-fiber polymer-matrix structural composites by electrical resistance measurement. Polym. Polym. Compos. **11**, 515–525 (2003)

Chung, D.D.L., Wang, S.: Discussion on paper 'The electrical resistance response of continuous carbon fibre composite laminates to mechanical strain' by N. Angelidis, C.Y. Wei and P.E. Irving, Composites Part A. **35**, 1135–1147 (2004). Composites Part A: Appl. Sci. Manuf. **37**, 1490–1494 (2006)

Dae-Cheol Seo, Jung-Ju Lee: Damage detection of CFRP laminates using electrical resistance measurement and neural network. Composite Structures. In: Tenth International Conference on Composite Structures, vol. 47, pp. 525–530, 1–4 December 1999

Fernberg, P., Nilsson, G., Joffe, R.: Piezoresistive performance of long-fiber composites with carbon nanotube doped matrix. J. Intell. Mater. Syst. Struct. **20**(9), 1017–1023 (2009)

Fiedler, B., Gojny, F.H., Wichmann, M.H.G., Bauhofer, W., Schulte, K.: Can carbon nanotubes be used to sense damage in composites? Ann. Chim. Sci. Mat. **29**(6), 81–94 (2004)

Gao, S., Zhuang, R., Zhang, J., Liu, J., Mäder, E.: Class fibers with carbon nanotube networks as multifunctional sensors. Adv. Funct. Mater. **20**(12), 1885–1893 (2010)

Indada, H., Okuhara, Y., Kumagai, H.: Health monitoring of concrete structure using self-diagnosis materials. In: Asari, F. (ed.) Sensing issues in civil structural health monitoring, pp. 239–248. Springer, Berlin (2005)

Irving, P.E., Thiagarajan, C.: Fatigue damage characterization in carbon fibre composite materials using an electrical potential technique. Smart Mater. Struct. **7**, 456–466 (1998)

Kemp, M.: Self-sensing composites for smart damage detection using electrical properties. Proc. SPIE Int. Soc. Opt. Eng. **2361**, 136–139 (1994)

Khemiri, N., Angelidis, N., Irving, P.E.: Experimental and finite element study of the electrical potential technique for damage detection in CFRP laminates. Experimental and finite element. Smart Mater. Struct. **14**, 147–154 (2005)

Kostopoulos, V., Tsotra, P., Vavouliotis, A., Karapappas, P., Tsantzalis, S., Loutas, T.: Nano-modified fiber reinforced composites: A way towards the development of new materials for space applications. In: Proceeding of 5th ESA Symposium on Micro/Nano Technologies for Space, ESTEC, Noordwijk (ESA-WPP-255) (2005)

Kostopoulos, V., Tsotra, P., Karapappas, P., Vavouliotis, A., Tsantzalis, S., Friedrich, K.: Enhancement of the mechanical performance of carbon fiber reinforced epoxy resin composites by the introduction of carbon nano-fibers (CNF). Collection of Technical Papers - AIAA/ASME/ASCE/AHS/ASC Structures, Structural Dynamics and Materials Conference, p. 3273 (2006)

Kostopoulos, V., Tsotra, P., Vavouliotis, A., Karappapas, P., Tsantzalis, S., Loutas, T.: Damage detection during fatigue loading of CNF doped CFRPs via resistance measurements and AE. Diffus. Defect Data B. Solid State Phenom. **121–123**(Part 2), 1399–1402 (2007a)

Kostopoulos, V., Tsotra, P., Vavouliotis, A., Karapappas, P., Fiamegou, E. Tsantzalis, S.:, Improved sensing properties of epoxy resin polymers and carbon fiber reinforced epoxy resin composites by the introduction of multiwall carbon nanotubes (MWCNT). Collection of Technical Papers - AIAA/ASME/ASCE/AHS/ASC Structures, Structural Dynamics and Materials Conference, p. 4727 (2007b)

Kostopoulos, V., Vavouliotis, A., Karapappas, P. Carbon nanotube epoxy modified CFRPs: Towards improved mechanical and sensing for multifunctional aerostructures. In: Proceedings of SPIE – The International Society for Optical Engineering, 6929, art. no. 69292M (2008)

Kostopoulos, V., Vavouliotis, A., Karapappas, P., Tsotra, P., Paipetis, A.: Damage monitoring of carbon fiber reinforced laminates using resistance measurements. Improving sensitivity using carbon nanotube doped epoxy matrix system. J. Intell. Mater. Syst. Struct. **20**(9), 1025–1034 (2009a)

Kostopoulos, V., Vavouliotis, A., Karapappas, P., Tsotra, P., Paipetis, A.: Damage monitoring of carbon fiber reinforced laminates using resistance measurements. Improving sensitivity using carbon nanotube doped epoxy matrix system. J. Intell. Mater. Syst. Struct. **20**(9), 1025–1034 (2009b)

Kupke, M., Schulte, K., Schuler, R.: Non-destructive testing of FRP by d.c. and a.c. electrical methods. Compos. Sci. Technol. **61**(6), 837–847 (2001)

Li, C., Thostenson, E.T., Chou, T.-W.: Sensors and actuators based on carbon nanotubes and their composites: A review. Compos. Sci. Technol. **68**(6), 0266–3538 (2008). doi:10.1016/j.compscitech.2008.01.006. ISSN 0266-3538

Lim, A.S., An, Q., Chou, T., Thostenson, E.T.: Mechanical and electrical response of carbon nanotube-based fabric composites to Hopkinson bar loading. Compos. Sci. Technol. **71**(5), 616–621 (2011)

Loyola, B.R., La Saponara, V., Loh, K.J.: Characterizing the self-sensing performance of carbon nanotube-enhanced fiber-reinforced polymers. In: Proceeding of SPIE, vol. 7649, 76490F, doi:10.1117/12.848712 (2010a)

Loyola, B.R., La Saponara, V., Loh, K.J.: In situ strain monitoring of fiber-reinforced polymers using embedded piezoresistive nanocomposites. J. Mater. Sci. **45**(24), 6786–6798 (2010b)

Nanni, F., Auricchio, F., Sarchi, F., Forte, G., Gusmano, G.: Self-sensing CF-FRP rods as mechanical reinforcement and sensors of concrete beams. Smart Mater. Struct. **15**, 182–186 (2006)

Nanni, F., Ruscito, G., Nad, L., Gusmano, G.: Self-sensing nanocomposite CNP-GFRP rods as reinforcement and sensors of concrete beams. J. Intell. Mater. Syst. Struct. **20**(13), 1615–1623 (2009)

Nanni, F., Ruscito, G., Puglia, D., Terenzi, A., Kenny, J.M., Gusmano, G.: Effect of carbon black nanoparticle intrinsic properties on the self-monitoring performance of glass fibre reinforced composite rods. Compos. Sci. Technol. **71**(1), 1–8 (2011)

Nofar, M., Hoa, S.V., Pugh, M.D.: Failure detection and monitoring in polymer matrix composites subjected to static and dynamic loads using carbon nanotube networks. Compos. Sci. Technol. **69**(10), 1599–1606 (2009)

Okuhara, Y., Shin, S.G., Matsubara, H., Yanagida, H., Takeda, T.: Self-diagnosis function of FRP containing electrically conductive phase. In: Proceeding of SPIE's Sensory Phenomena and Measurements Instrumentation for Smart Structures and Materials, San Diego, CA, USA, pp. 191–198 (2000)

Okuhara, Y., Shin, S.G., Matsubara, H., Yanagida, H., Takeda, T.: Development of conductive FRP containing carbon phase for self-diagnosis structures. In: Proceeding of SPIE's Smart Structures and Materials. New Port Beach, CA, USA, pp. 314–322 (2001)

Park, J.B., Lee, S., Kim, K.W., Yoon, D.J.: Interfacial Aspect of Electrodeposited Conductive Fibers/Epoxy Composites Using Electro-Micromechanical Technique and Nondestructive Evaluation. J. Colloid Interface Sci. **237**(1), 80–90 (2001)

Park, J.B., Okabe, T., Takeda, N., Curtin, W.A.: Electromechanical modeling of unidirectional CFRP composites under tensile loading condition. Compos. A. Appl. Sci. Manuf. **33**, 267–275 (2002)

Park, J.B., Okabe, T., Takeda, N.: New concept for modeling the electromechanical behavior of unidirectional carbon-fibre-reinforced plastic under tensile loading. Smart Mater. Struct. **12**, 105–114 (2003)

Prasse, T., Michel, F., Mook, G., Schulte, K., Bauhofer, W.: A comparative investigation of electrical resistance and acoustic emission during cyclic loading of CFRP laminates. Compos. Sci. Technol. **61**(6), 831–835 (2001)

Rausch, J., Mäder, E.: Health monitoring in continuous glass fiber reinforced thermoplastics: Manufacturing and application of interphase sensors based on carbon nanotubes. Compos. Sci. Technol. **70**(11), 1589–1596 (2010a)

Rausch, J., Mäder, E.: Health monitoring in continuous glass fiber reinforced thermoplastics: Tailored sensitivity and cyclic loading of CNT-based interphase sensors. Compos. Sci. Technol. **70**(13), 2023–2030 (2010b)

Schulte, K., Baron, Ch: Load and failure analyses of CFRP laminates by means of electrical resistivity measurements. Compos. Sci. Technol. **36**, 63–76 (1989)

Shin, S.: Evolution of the health of concrete structures by electrically conductive GFRP (glass fiber reinforced plastic) composites. Met. Mater. Int. **8**(1), 63–68 (2002)

Sirong, Z., Chung, D.D.L.: Analytical model of piezoresistivity for strain sensing in carbon fiber polymer–matrix structural composite under flexure. Carbon **45**, 1606–1613 (2007)

Sotiriadis, G., Tsotra, P., Paipetis, A., Kostopoulos, V.: Stiffness degradation monitoring of carbon nanotube doped Glass/Vinylester composites via resistance measurements. J. Nanostruct. Polym. Nanocompos. **3**(3), 90–95 (2007)

Sureeyatanapas, P., Young, R.J.: SWNT composite coatings as a strain sensor on glass fibres in model epoxy composites. Compos. Sci. Technol. **69**(10), 1547–1552 (2009)

Sureeyatanapas, P., Hejda, M., Eichhorn, S.J., Young, R.J.: Comparing single-walled carbon nanotubes and samarium oxide as strain sensors for model glass-fibre/epoxy composites. Compos. Sci. Technol. **70**(1), 88–93 (2010)

Thostenson, E., Chou, T.-W.: Carbon nanotube networks: Sensing of distributed strain and damage for life prediction and self healing. Adv. Mater. **18**, 2837–2841 (2006). doi:10.1002/adma.200600977

Thostenson, E.T., Chou, T.: Real-time in situ sensing of damage evolution in advanced fiber composites using carbon nanotube networks. Nanotechnology **19**(21), 215713 (2008)

Todoroki, A., Yoshida, J.: Electrical resistance change of unidirectional CFRP due to applied load. JSME. Int. J. Ser. A. **47**(Part 3), 357–364 (2004)

Todoroki, A., Yoshida, J.: Apparent negative piezoresistivity of single-ply CFRP due to poor electrical contact of four-probe method. Key Eng. Mater. 297–300 I, 610–615 (2005)

Todoroki, A., Tanaka, M., Shimanura, Y.: Measurement of orthotropic electric conductance of cfrp laminates and analysis of the effect on delamination monitoring with an electric resistance change method. Compos. Sci. Technol. **62**, 619–628 (2002)

Vavouliotis, A., Tsotra, P., Kostopoulos, V., Karapappas, P., Miaris, A. Nikolaou, N.: MWCNT-modified fiber reinforced composites with nano-sensing capabilities: A way towards the development of the new functional materials for space applications. In: AIAA 57th International Astronautical Congress, IAC 2006, p. 5523 (2006)

Vavouliotis, A., Karapappas, P., Loutas, T., Voyatzi, T., Paipetis, A., Kostopoulos, V.: Multistage fatigue life monitoring on carbon fibre reinforced polymers enhanced with multiwall carbon nanotubes. Plast. Rubber Compos. **38**(2–4), 124–130 (2009)

Vavouliotis, A., Paipetis, A., Kostopoulos, V.: On the fatigue life prediction of CFRP laminates using the electrical resistance change method. Compos. Sci. Technol. **71**(5), 630–642 (2011)

Wang, X., Chung, D.D.L.: Real-time monitoring of fatigue damage and dynamic strain in carbon fiber polymer-matrix composite by electrical resistance measurement. Smart Mater. Struct. **6**, 504–508 (1997)

Wang, X., Chung, D.D.L.: Self-monitoring of fatigue damage and dynamic strain in carbon fiber polymer-matrix composite. Compos. B. Eng **29**, 63–73 (1998a)

Wang, X., Chung, D.D.L.: Residual stress in carbon fiber embedded in epoxy, studied by simultaneous measurement of applied stress and electrical resistance. Compos. Interfaces **5**(3), 277–281 (1998b)

Wang, S., Chung, D.D.L.: Piezoresistance in continuous carbon fiber polymer-matrix composite. Polym. Compos. **21**, 13–19 (2000)

Wang, S., Shui, X., Fu, X., Chung, D.D.L.: Early fatigue damage in carbon-fibre composites observed by electrical resistance measurement. J. Mater. Sci. **33**, 3875–3884 (1998)

Wang, X., Wang, S., Chung, D.D.L.: Sensing damage in carbon fiber and its polymer-matrix and carbon-matrix composites by electrical resistance measurement. J. Mater. Sci. **34**, 2703–2713 (1999)

Weber, I., Schwartz, P.: Monitoring bending fatigue in carbon-fibre/epoxy composite strands: a comparison between mechanical and resistance techniques. Compos. Sci. Technol. **61**(6), 849–853 (2001)

Xia, Z.H., Curtin, W.A.: Modeling of mechanical damage detection in CFRPs via electrical resistance. Compos. Sci. Technol. **67**(7–8), 1518–1529 (2007)

Xia, Z.H., Curtin, W.A.: Damage detection via electrical resistance in CFRP composites under cyclic loading. Compos. Sci. Technol. **68**(12), 2526–2534 (2008)

Yesil, S., Winkelmann, C., Bayram, G., La Saponara, V.: Surfactant-modified multiscale composites for improved tensile fatigue and impact damage sensing. Mater. Sci. Eng., A **527** (27–28), 7340–7352 (2010)

Zhang, W., Sakalkar, V., Koratkar, N.: In situ health monitoring and repair in composites using carbon nanotube additives. Appl. Phys. Lett. **91**(13), 133102 (2007)

Zhang, J., Zhuang, R., Liu, J., Mäder, E., Heinrich, G., Gao, S.: Functional interphases with multiwalled carbon nanotubes in glass fibre/epoxy composites. Carbon **48**(8), 2273–2281 (2010)

Chapter 3
Carbon Nanotube Structures with Sensing and Actuating Capabilities

C. Jaillet, N.D. Alexopoulos, and P. Poulin

Contents

Abstract We describe carbon nanotube (CNT) structures which are used as mechanical sensors, electromechanical actuators and shape memory materials. These structures include CNT mats and fibres of aligned CNTs. Mechanical sensors are based on the piezo-resistivity of the investigated CNT structures. They can be used as embedded sensors for sensing and damage monitoring of composites. CNT can also be used for novel actuator technologies. Indeed CNTs deform in response to charge injection and electrostatic phenomena. They can be stimulated under the form of electrodes in a given electrolyte. CNT structures can generate a large stress

C. Jaillet • P. Poulin (✉)
Centre de Recherche Paul Pascal, Université de Bordeaux, CNRS, Avenue Schweitzer, 33600 Pessac, France
e-mail: poulin@crpp-bordeaux.cnrs.fr

N.D. Alexopoulos
Department of Financial Engineering, University of the Aegean, 821 00 Chios, Greece

A.S. Paipetis and V. Kostopoulos (eds.), *Carbon Nanotube Enhanced Aerospace Composite Materials*, Solid Mechanics and Its Applications 188,
DOI 10.1007/978-94-007-4246-8_3, © Springer Science+Business Media Dordrecht 2013

because of their stiffness. In other classes of actuating materials, carbon nanotubes can be used as fillers of shape memory polymers (SMPs). SMPs have applications in packaging, biomedical devices, heat shrink tubing, deployable structures, etc. CNTs are ideal materials to improve the stiffness of shape memory polymers, which is critical for achieving large stress recovery. Their electrical conductivity is of particular interest in the engineering of SMPs which can be heated via Joule's heating and directly stimulated by an electrical current. We review in this chapter the properties of these new functional materials and highlight their potential for future applications.

Keywords Sensor • Actuator • Piezo-resistivity • Composite • Electrode • Shape memory material • Polymer

3.1 Introduction

3.1.1 General Concepts

Carbon nanotubes (CNTs) are currently considered as particularly promising particles in the fields of sensors and actively moving materials (Baughman et al. 2002; Li et al. 2008). Their properties can be exploited in several ways and for different classes of sensors and actuators. In this chapter we focus on CNT structures which contain a large fraction of nanotubes, or which are solely comprised of carbon nanotubes. The processes for making such structures are still under development and not as mature as processes used for making composites with a low fraction of carbon nanotubes. Nevertheless CNT structures with a large fraction of nanotubes, typically above 10 wt% and up to 100 wt%, are expected to exhibit unique properties with enhanced manifestations of the specific features of carbon nanotubes. We hope that research on such structures will lead in the future to novel technologies for sensors and actuators potentially useful in aircraft industries. We describe in particular in this chapter the properties of CNT mats, often called "buckypapers" and of CNT fibres. The latter structures are particularly interesting because they allow carbon nanotubes to be aligned on macroscopic scale. In addition they can be embedded in composites or weaved as textile structures. Carbon nanotube mats are ideal structures to design bimorph devices and novel classes of actuators. We review in this chapter the properties of such materials and highlight their potential for future applications. The chapter is divided into the following sections:

1. The first section describes the potential of the produced composite polyvinyl alcohol – carbon nanotube fibres to be used as embedded stress or strain sensors in glass fibre reinforced composites. Composite materials and structures offer great advantages in light-weight applications, e.g. very high specific mechanical properties, when compared to their competitive materials. However, their main disadvantage for widespread use in civil aircraft structures is their non-destructive inspection (NDI) or their *in-situ* identification of developed

non-visible damage under real loading conditions. Research in such a direction is of imperative importance for their wide use in aircraft structures. In such conditions, inspection and maintenance periods are often very critical and a methodology is needed to minimize the time intervals that the components are out-of service. An *in-situ* structural health monitoring system would primarily give on-line information regarding the structural safety of the structure and secondarily would significantly lower the inspection/maintenance costs.

An intelligent structural health monitoring system could provide firstly on-line information on the developed damage to a specific location of the composite structure and secondarily its extent. Typical state-of-the-art damage and sensing techniques are the active piezoelectric sensors, fibre optical sensors and acoustic emission sensors, e.g. (Boller et al. 2009; Balageas et al. 2006; Giurgiutiu 2008). These techniques had been applied to composite materials and structures by using embedded sensors. Nevertheless, each technique presents specific advantages and disadvantages, the latter being the limitations concerning resolution or clarity of the measured data. Furthermore, cost intensive external hardware is needed and for fibre optical and piezoelectric sensors, the sensor hardware has to be embedded into the composite structure. This has been proven to be detrimental to the composite properties. Furthermore, the introduction of health monitoring systems should be compatible with existing composite manufacturing processes. This is especially difficult in the case of embedded piezoelectric sensors or MEMS, as these devices are sensitive to high temperatures and pressures.

In the present chapter, the PVA-CNT fibre will be used as a new type of embedded sensor for sensing and damage monitoring of composites. Using the electrical conductivity of embedded nano-fibres into non-conductive composites, the structural health monitoring can be assessed by the *in-situ* measurements of the electrical resistance change of the nano-fibre. These fibres have extra small size (diameter of the fibre ranges from 10 to 20 μm) and do not impose any artificial defects on the composite material while embedded. In addition, they have proved to be compatible to existing composite manufacturing techniques, producing high-quality advanced composite materials.

2. We address in the second section of the present chapter the use of carbon nanotubes as electro-mechanical actuators. This section is divided into three sub-sections. The first one deals with actuators made of neat nanotubes. The CNT charge density in a given electrolyte can be varied by applying a low voltage with respect to a reference electrode. A double layer forms at the nanotube interface and the material expands or contracts in response to quantum mechanical and electrostatic effects (Baughman et al. 1999; Fraysse et al. 2002; Sun et al. 2002; Ghosh et al. 2005; Gupta et al. 2004; Hughes and Spinks 2005; Riemenschneider et al. 2009a, b; Bartholome et al. 2008; Barisci et al. 2003; Madden et al. 2006; Yun et al. 2006). CNTs operate at low voltage and can generate a large stress because of their stiffness. The first macroscopic manifestation of this phenomenon was reported by Baughman et al. (1999). These first actuators were made of so-called "buckypapers" which are mats of randomly orientated carbon nanotubes.

Several groups have theoretically and experimentally investigated the involved mechanisms and attempted to improve the performances of such nanotube-based actuators (Fraysse et al. 2002; Sun et al. 2002; Ghosh et al. 2005; Gupta et al. 2004; Hughes and Spinks 2005; Riemenschneider et al. 2009a, b; Bartholome et al. 2008; Barisci et al. 2003; Madden et al. 2006; Yun et al. 2006). Nevertheless optimization of nanotube structures for such applications remains challenging. Indeed an optimal actuator has to combine high mechanical strength, good electrical conductivity and a large surface specific area to maximize the interface exposed towards the electrolyte. In addition, alignment of the nanotubes is expected to be critical since it could promote macroscopic dimensional changes along a given direction. The shown actuators have been studied in liquid electrolytes. They are in such conditions not suitable for aircraft applications. Nevertheless, systems in liquid electrolytes are model systems which are particularly well suited to study the fundamental properties of CNT electro-mechanical actuators. Transferring these properties in practical devices will require the use of solid electrolytes. This point is addressed later in this chapter.

Alternatively, other groups (Raguse et al. 2003) have used thin films of conductive gold nanoparticles to make bimorph devices with actuating capabilities. In this case the films exhibited high electrical conductivity and were still sufficiently strong and porous to make bimorph actuators that can bend when they are electrically stimulated in a liquid electrolyte. The authors deduced from deflections of their device that gold nano-particle films could generate a stress up to 0.6 MPa (Raguse et al. 2003). Carbon nanotubes because of their intrinsic structural and physical properties (large surface area, high mechanical properties) are of particular interest for developing similar bimorph actuators with improved capabilities. The second subsection deals with actuators made of CNT bucky papers and bilayer structures. We will see further in the chapter that carbon nanotubes bilayer structures allow large amplification of the strain generated by the expansion or contraction of a nanotube mat stimulated in a liquid electrolyte.

Lastly, the third sub-section of this part deals with recent actuators made with solid electrolytes. The work in this field is still very recent but particularly important for the future development of actual applications in the field of aircraft applications.

3. In other classes of actuating materials, carbon nanotubes can be used as fillers of shape memory polymers (SMPs). SMPs have applications in packaging, biomedical devices, heat shrink tubing, deployable structures, micro-devices, etc. Shape memory polymers are usually deformed at high temperature (T_d) and then cooled down under fixed strain to trap the deformed polymer chains, thus storing mechanical energy. Upon reheating, typically in the vicinity of the glass transition temperature (T_g), the polymer chains become mobile and the material can relax by reverting towards its original and more stable shape. While this is a common mechanism several other phenomena can be exploited for generating shape memory effects in polymer materials (Lendlein and Kelch 2002; Liu et al. 2007). The efficiency of shape memory polymers is controlled by the

composition of the polymer, as defined by its chemical structure, molecular weight, degree of cross-linking and fraction of amorphous and crystalline domains (Lendlein and Kelch 2002; Liu et al. 2007; Kim et al. 1996; Ohki et al. 2004; Hu et al. 2005; Morshedian et al. 2003; Qin and Mather 2009; Chung et al. 2008). The energy which is restored upon shape recovery is a growing function of the energy supplied during the deformation at high temperature (Kim et al. 1996; Gall et al. 2005). Shape memory polymers can exhibit large strain when they revert towards their initial shape. Unfortunately, this large strain is usually associated with a low stress recovery from a few tenths of MPa to a few tens of MPa (Lendlein and Kelch 2002; Liu et al. 2007; Gall et al. 2005; Kornbluh et al. 2002; Lendlein and Langer 2002; Gupta et al. 1994). Consequently, the energy density, which results from a combination of stress and strain, is rather low. Combining large stress and large strain recovery as well as finding more controlled programming procedures remain critical challenges for the development of smarter and stronger shape memory materials. The inclusion of nano-particles has been shown to improve the behavior of shape memory polymers. These efforts include an increase in their mechanical properties (Gall et al. 2002, 2004; Liu et al. 2004; Meng et al. 2007; Gunes and Jana 2008; Luo and Mather 2009), addition of conductive nano-particles to achieve shape memory effects which can be triggered by Joule's heating (Koerner et al. 2004) or inclusion of magnetic nano-particles which can cause heating in the presence of an alternating magnetic field (Mohr et al. 2006). CNTs combine several interesting properties: they are stiff, rod-like in shape and electrically conductive. They are thus ideal materials to improve the stiffness of shape memory polymers, which is critical for achieving large stress recovery. Their electrical conductivity is of particular interest to engineering of materials which can be heated via Joule's heating and directly stimulated by an electrical current. Nevertheless the manifestation of these properties is here again expected to strongly depend on the fraction of nanotube and on their ordering. With the aim of improving nanotube-based shape memory materials we describe in the fourth section recent results on thermo-mechanical properties of fibres made of aligned CNTs.

3.1.2 CNT Structures

3.1.2.1 CNT Fibres

In all the sections of this chapter, the fibres are obtained by a coagulation spinning process which consists in injecting a nanotube dispersion in the co-flowing stream of a coagulating polymer solution (Vigolo et al. 2000). This process leads to the formation of polymer-nanotube composite fibres with a large fraction of embedded nanotubes. The nanotube fraction can greatly exceed 10 wt%. And even, as shown further, these materials can be used to achieve fibres solely comprised of nanotubes.

The polymer used in this process is the polyvinyl-alcohol (PVA). The fibre spinning process is presented in more details in the following sections of this chapter. As produced, PVA-nanotube composite fibres exhibit particularly interesting shape memory phenomena (Miaudet et al. 2007) which are described in the fourth section of the chapter. Those fibres can indeed absorb a large amount of mechanical energy when they are hot stretched (Miaudet et al. 2005; Dalton et al. 2003). As a result they generate a very large stress which exceeds 100 MPa when they are reheated after they have been cooled under tensile load. In addition composite PVA-nanotube fibres display a temperature memory with a peak of recovery stress at $T_{d,}$ the temperature of their initial deformation. The microscopic origin of this temperature is still unclear. Some possible mechanisms based on gradients of the glass transition temperature of amorphous PVA fractions at the interface of nanotubes or PVA crystallites will be discussed.

Fibres solely comprised of nanotubes are obtained by removing the PVA from the nanotube-PVA composite fibres which are described above. The polymer is removed by thermal degradation. This leads to porous fibres solely comprised of entangled CNTs which remain stuck because of strong van der Waals interactions. It is observed that fibre drawing of the initial composite fibres allows a substantial improvement of the nanotube alignment. The latter is reflected by a greater mechanical strength. Such fibres are used for making electromechanical actuators described in the third section of the present chapter. It is also observed that drawing leads to greater electrochemical capacitance of the fibres. This improvement is ascribed to the debundling of the nanotubes as the fibres are stretched. Electromechanical properties are characterized by measuring the isometric stress generated when the fibres are electrically stimulated in an aqueous electrolyte. The maximal observed stress is about 10 MPa which is an order of magnitude greater than the stress reported for random assemblies of nanotubes in buckypaper in similar conditions (Baughman et al. 1999).

3.1.2.2 CNT Buckypapers

Like fibres, mats are obtained from homogeneous liquid dispersions of carbon nanotubes. Several methods have been proposed over the last few years to make films of carbon nanotubes. In the present work, buckypapers have been prepared using oxidized carbon nanotubes which can be well dispersed in aqueous solution without surfactants and tip sonication. Moreover, the presence of carboxylic acid groups resultant from the oxidation process at the surface of the nanotubes allows strong interactions with a polymer such as PVA. Typically functionalized nanotubes are obtained by adding single wall nanotubes (SWNT) or multi-wall nanotubes (MWNT) to a solution of nitric acid under reflux. After a few days of acid treatment (1–3 days), the suspension is rinsed with distilled water up to neutralization and re-dispersed in water to obtain homogeneous oxidized-CNT dispersion. Then the dispersion is filtered on a membrane under vacuum to obtain pure nanotube assemblies or structures with a controlled fraction of nanotubes depending on the products added to the dispersion (polymers, ionic liquids, …). It is also possible to achieve carbon nanotube films by simply evaporating the solvent of a deposited dispersion layer. The preparation of nanotube mats and bimorph devices used for actuator applications is described in more detail further in the chapter.

3.2 PVA-CNT Fibres as Mechanical Sensors in Composites

In this section, the PVA-CNT fibres were used as embedded mechanical strain sensors in composite materials. The PVA-CNT fibres have an inherent electrical conductivity; they can be embedded in non-conductive media (e.g. glass fibre reinforced composites), in order to monitor their stress/strain field of the composite material via the electrical resistance change of the embedded PVA-CNT fibre during mechanical loading.

3.2.1 General Concepts

The aerospace industry focuses its research on producing multi-functional materials, driving design parameters being (a) weight reduction with increased mechanical properties as well as (b) monitoring their structural health by means of sensing capability. This means that the new generation of advanced composites should be manufactured with embedded sensors or they should act simultaneously as sensors (hybrid composites).

The electric resistance change method had been firstly used by Schulte and Baron (1989) for sensing of structural health monitoring, by means of identifying internal damage of carbon fibre reinforced (CFRP) laminates. Many research studies have employed the electrical resistance change for such purposes to composite materials, e.g. (Kaddour et al. 1994; Irving and Thiagarajan 1998; Seo and Lee 1999; Arby et al. 1999, 2001; Kupke et al. 2001). The main advantage of this method is that it does not require expensive equipment for instrumentation. The electrical conductivity of the carbon fibres was first used to monitor damage in carbon fibre reinforced polymers (CFRPs), which could be related to fibre breakage. The electrical methods have been extensively studied and have been used to study a variety of damage mechanisms, e.g. delamination, matrix cracking, under various loading conditions, e.g. (Todoroki and Tanaka 2002; Todoroki et al. 1995, 2004, 2006). Therefore, by exploiting the inherent conductivity of the carbon fibre, the monitoring of the structural health of CFRP materials by means of electrical conductivity is feasible.

Use of PVA-CNT fibre in composites that present inherent electrical resistance such as the carbon fibre reinforced plastics may not give realistic results as there is no possible way to solely measure the resistance of the nano-fibre. This type of fibre can be successfully embedded in non-conductive composites, such as the glass reinforced composites, and measure the electrical resistance change of the PVA-CNT fibre with variations in the mechanical stress/strain state of the composite. Notice that monitoring of carbon fibre composites that have inherent conductivity has been performed over the last two decades, e.g. (Parvisi and Bailey 1978; Highsmith and Reifsnider 1982; Gagel et al. 2006). The idea of monitoring a composite using a unique fibre has been made in (Muto et al. 2001), where a carbon fibre was embedded into GFRP. Mainly due to the difference in modulus of elasticity between the two media, the sensor 'carbon fibre' did not monitor the progressive damage of the composite but actually promoted its final fracture.

For a successful synergistic function with composite material, the conductive fibre for monitoring damage accumulation, should present the same or lower modulus of elasticity and higher ductility. A very promising, conductive material is the PVA-CNT fibre; they are thinner than a human hair and offer a promise for high strength and ductility, light weight, thermally and electrically conducting structural elements at a lower cost than other nanotube forms.

3.2.2 Manufacturing and Testing of Hybrid Composites

Embedding of the PVA-CNT fibre to composites has been performed by researchers of the University of the Aegean, Greece in collaboration with the Laboratory of Advanced Composites of the Research and Development Department of Hellenic Aerospace Industry. PVA-CNT fibres have been produced by researchers at the Centre de Recherche Paul Pascal, University of Bordeaux, France by injecting a multi-wall carbon nanotube dispersion into the co-flowing stream of a coagulating polyvinyl alcohol solution (Vigolo et al. 2000). The PVA-CNT fibre can be embedded between the plies of the composite before the resin infusion (or transfer). As it presents an inherent electrical conductivity, this should be measured with adequate equipment in the surface of the composite. The logical route of thinking to accomplish this task is to produce a conductive path from the place where the fibre was laid inside the composite and through the plies can measure the electrical conductivity of the fibre in the outer surface of the composite.

Typical configuration for manufacturing of the hybrid composites can be seen in Fig. 3.1. For manufacturing of the plate with the PVA-CNT fibres the following process was followed: ten plies of fabric, oriented at 0/90° had been cut and used for manufacturing. The first nine plies were laid and the PVA-CNT fibres were placed between the 9th and last ply. More details regarding manufacturing can be found in (Alexopoulos et al. 2010a, b). At a distance of 50 mm (the measuring distance for the PVA-CNT fibres) two marks were covered with silver paste (conductive epoxy) and finally the tenth ply was placed on top to complete the lay-up as sketched in Fig. 3.1. The marks covered with silver paste served to create a means of "connector" to the material's surface, where the cables will be placed for recording of resistance measurements during testing. Small quantities of silver paste had been also used in the outer surface of the tenth layer of the fabrics and above the marks

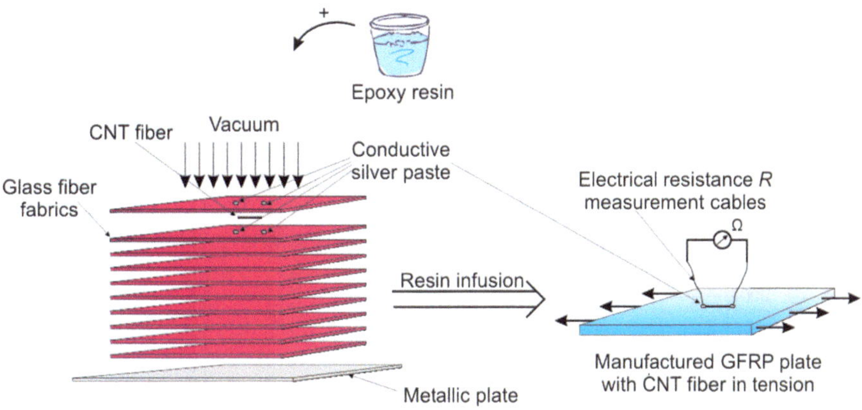

Fig. 3.1 Manufactured GFRP plate with embedded PVA-CNT fibre and wiring for resistance measurements

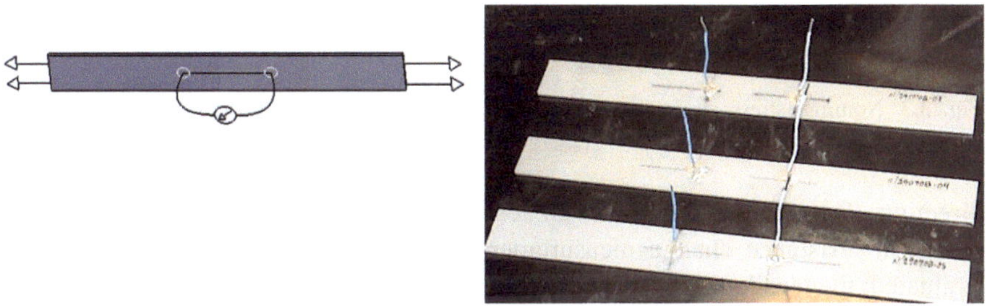

Fig. 3.2 Sketch and macrophotograph of three manufactured GFRP coupons

made in the previous layer. Small quantities were used such as not to impregnate the fabric and produce any large 'artificial defects' on the material that would decrease its mechanical performance.

The specimens with the embedded PVA-CNT fibre had been cut from the material plates according to the ASTM D3039 specification and edge-polished. The dimensions of the testing specimens were width × length = 25 × 250 mm. At the two marks of each specimen covered with silver paste, two cable connectors had been added again with silver paste in order to attach the multimeter for the resistance measurements (Fig. 3.2).

A servo-hydraulic Instron 100 kN testing machine had been used to record the force and displacement data, while a 50 mm extensometer was attached to record axial strain data of the coupons. An Agilent multimeter was used to record in situ the electrical resistance data of the specimen's embedded CNT fibre during mechanical loading. A DC voltage of 10 V was applied to cables connected to the PVA-CNT fibre of the specimens (Fig. 3.2), the current was measured and the resistance was calculated from these values. The resistance measurements were performed in a two-point, 50 mm distance measurement set-up in the longitudinal

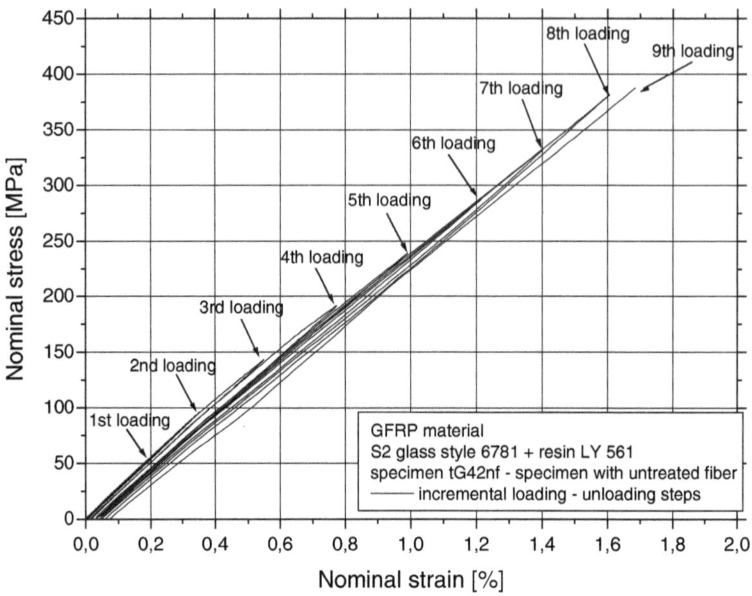

Fig. 3.3 Nominal stress–strain curves for nine different loading–unloading steps of reference GFRP coupon

direction. Two different mechanical tests were conducted and the potential for electrical resistance change measurements for structural health monitoring was evaluated: the tensile test and three-point bending test. Different quasi-static incremental loading – unloading steps or progressive damage accumulation (PDA) tests in different specimens had been made to seek the PVA-CNT fibre's electrical response to mechanical loading (Alexopoulos et al. 2010b). As the incremental loading steps had been made to specific levels of tensile fracture stress of the composite material, the testing machine was load-controlled.

3.2.3 Hybrid Composites in Service

3.2.3.1 Tension

Typical axial tensile nominal stress–strain diagrams for 11 incremental loadings for coupons without the PVA–CNT fibre can be seen in Fig. 3.3. As the tests were load controlled the specimen returned to its zero load (stress) condition, after every unloading. The incremental tensile loading steps of additional 50 MPa each, induced damage to the material that can be noticed as residual strain measurements after every unloading step.

Typical results of the various incremental loading–unloading steps of a coupon with embedded untreated PVA–CNT fibre can be graphically seen in Fig. 3.4.

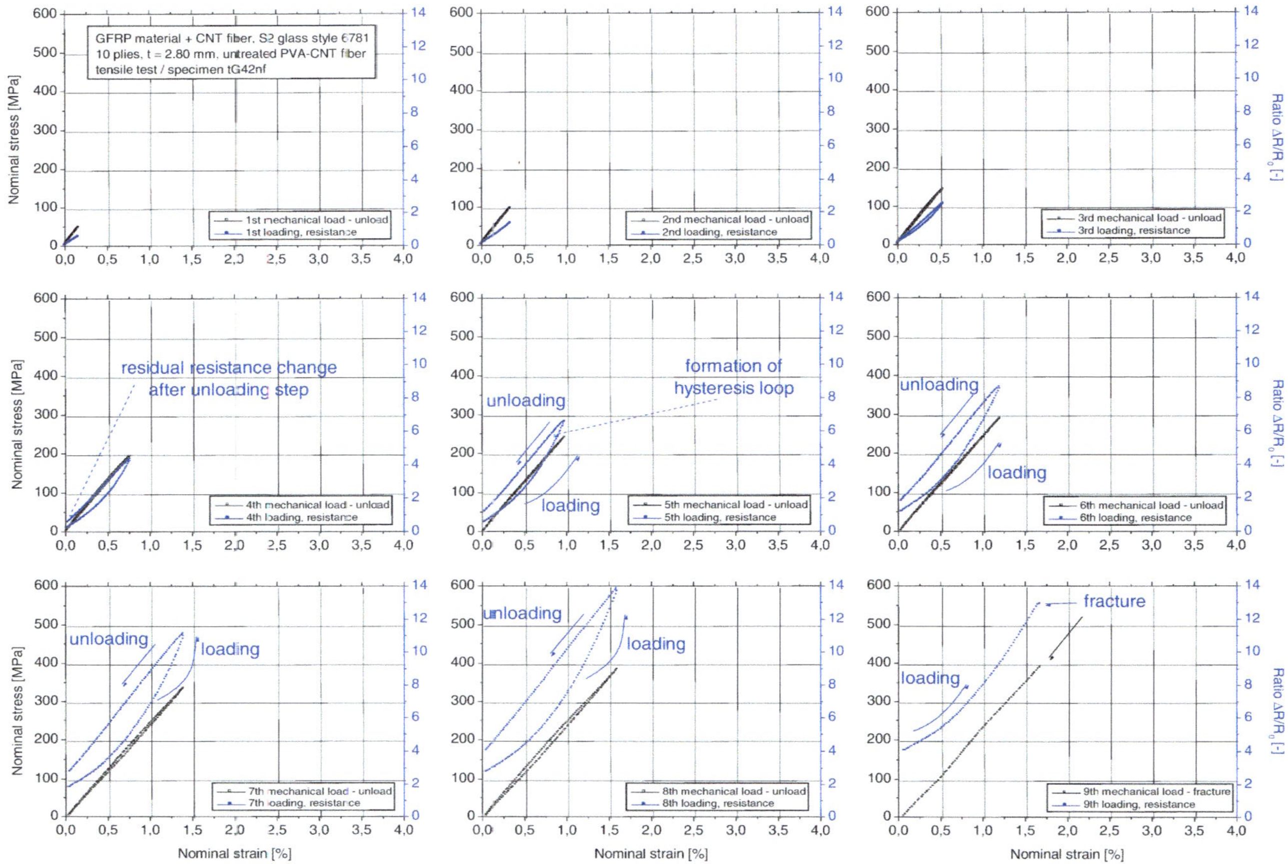

Fig. 3.4 Typical tensile mechanical and resistance results of GFRP specimen with embedded PVA-CNT fibre for six different incremental loading–unloading steps

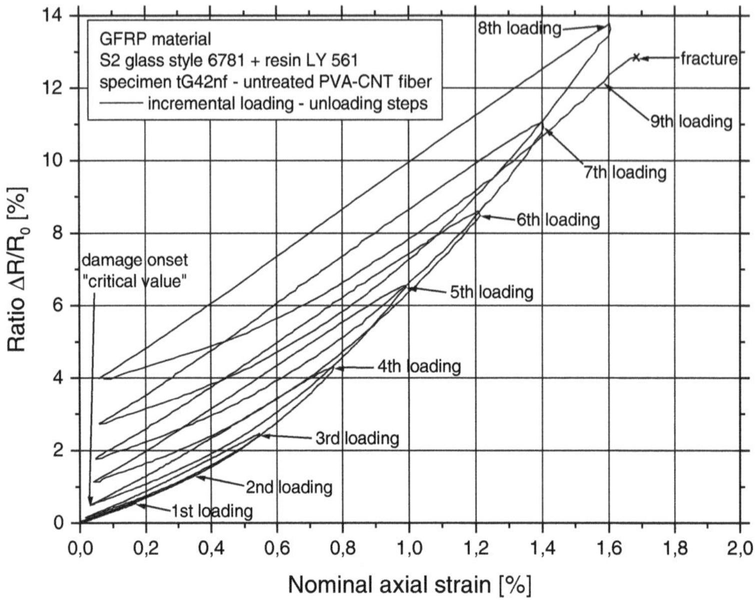

Fig. 3.5 Tensile mechanical strain and electrical resistance change $\Delta R/R_0$ measurements for the GFRP material with embedded untreated PVA–CNT fibre

An example of nine loading–unloading steps is shown to seek simultaneously the hybrid material's mechanical/electrical resistance response. The specific loading levels were 11, 22, 33, 44, 55, 66, 78, 89 and 100% of the fracture stress, respectively. The PVA–CNT fibre's electrical resistance change ($\Delta R/R_0$) follows the mechanical response (stress–strain) of the coupon; it increases when loaded and decreases when unloaded.

As shown graphically in Fig. 3.4 for the case of untreated PVA– CNT fibre, the PDA tests resulted in residual resistance change values (blue dots) of the PVA–CNT fibre after unloading. Larger residual resistance change measurements of the order of 2–4% were noticed (in right blue-colored Y-axis) after a higher level of incremental loading–unloading step. In addition, the loading– unloading branch of the resistance change of the PVA–CNT fibre in the same figure does follows the same pattern; the two branches are recognizable and are indicated via arrows. A hysteresis loop is formed that is clearly distinguishable for higher loading conditions of the hybrid composite. Noticeable is that a loading branch is always exponential/parabolic, while an unloading branch seems to be linear for all cases.

A typical tensile mechanical strain–electrical resistance change diagram for the untreated PVA–CNT fibre can be seen in Fig. 3.5 for nine different steps till fracture. Each loading maxima is marked in the figure, while – as noticed – a hysteresis loop is formed for all cases after unloading. This behavior is attributed to

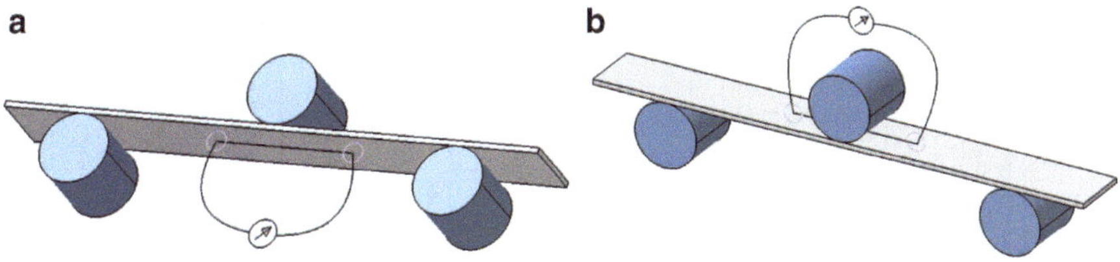

Fig. 3.6 (**a**) Three-point bending tests in GFRP material with embedded PVA-CNT fibre: (**a**) fibre tested in the tensile region and (**b**) fibre tested in the compressive region

possible plastic deformation of the PVA material of the fibre. Besides the expected residual strain measurements after every unloading step, residual resistance change measurements of the untreated PVA–CNT fibre are also noticeable. Residual resistance measurement is recorded, with a maximum value of 4% at the last loading step. The critical value of residual resistance is the value of almost 0.5% presented after the fourth unloading. It actually represents the threshold value for damage detection of the composite.

3.2.3.2 Bending

Specimens with embedded PVA-CNT fibres were tested in three-point bending tests for two different cases: (a) as shown in Fig. 3.6a, the fibre was placed at the 'bottom' of the specimen such as tensile stresses are developed in the region of the fibre, (b) while in Fig. 3.6b it was placed at the 'top' of the specimen, such as compressive stresses are developed in the fibre's region. The mechanical load had been converted to mechanical stress by taking into account the material's geometrical dimensions as well as its moment of inertia. The nominal stress of the fibre σ_{fib} has been calculated from the equation:

$$\sigma_{fib} = \frac{M_b}{I_Z} \cdot y_{fib}, \tag{3.1}$$

where M_b is the maximum bending moment at the specimen, I_z the moment of inertia and y_{fib} the distance of the CNT fibre from the middle thickness of the specimen.

For the case of fibre working in tension, monotonic loading up to fracture of the specimen gave results very close to the respective tensile results. The material and PVA-CNT fibre's response had been also tested in other specimens and for various incremental loading–unloading steps. Typical test results can be seen in Fig. 3.7 for a total of six incremental loading–unloading steps. For the first three (up to 17, 33 and 50 of the fracture stress, respectively), very small and noisy values of the ratio $\Delta R/R_0$ were calculated.

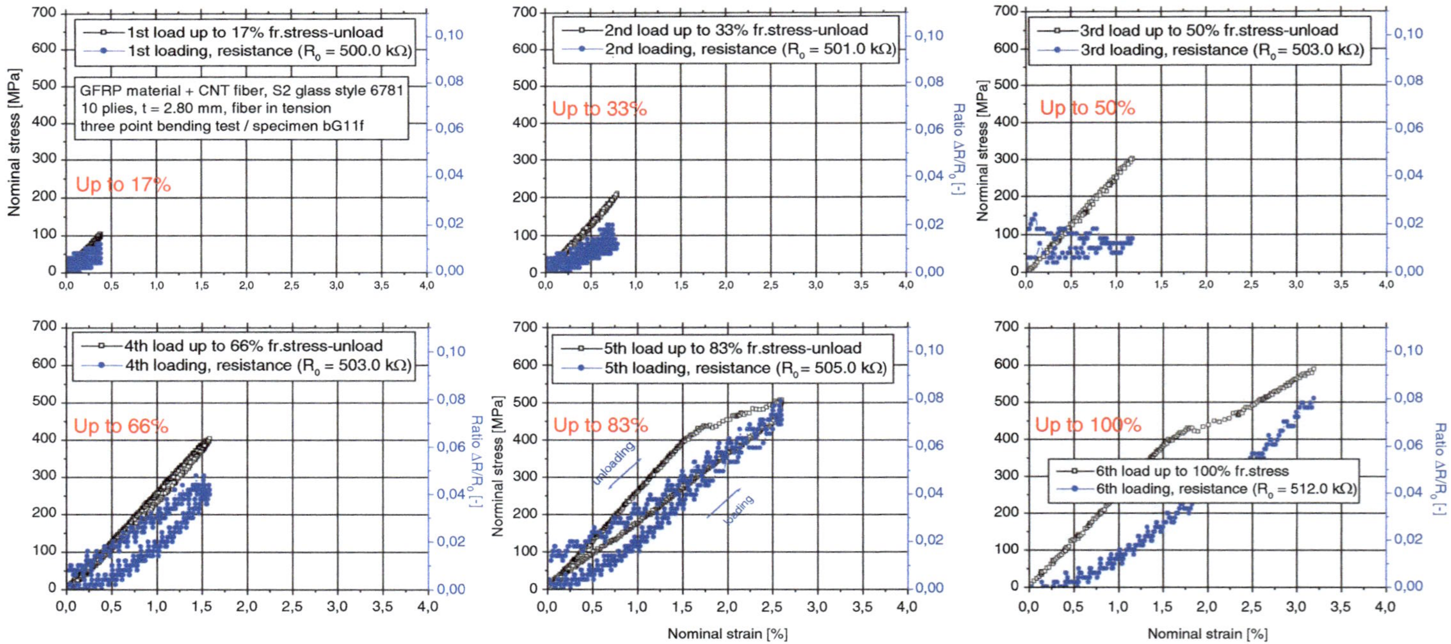

Fig. 3.7 Typical three point bending mechanical and resistance results of GFRP specimen with embedded PVA-CNT fibre in tension for six different incremental loading–unloading steps

The fourth loading–unloading step has been made up to the 400 MPa (66% of fracture stress). The fibre's response was very distinctive and a hysteresis loop of $\Delta R/R_0$ measurements after unloading was observed. It is also clear that after unloading, the resistance change does not return to the zero value; a residual resistance was noticed. It is eminent that in the fifth loading–unloading step (83%), essential damage to the material will occur. The next loading and unloading step of the CNT fibre (lower and upper branch of the hysteresis loop, respectively) resulted in a residual ratio $\Delta R/R_0$ of almost 0.01 or approximately 6 kΩ (from 505 to 512 kΩ). Finally, for the last loading up to fracture, a simultaneous increase in the CNT fibre's response can be observed.

The direct correlation between mechanical stress and ratio $\Delta R/R_0$ showed exactly the same quantitative results with the tensile test results. When essential damage occurred due to incremental loadings, the curve correlating mechanical stress and $\Delta R/R_0$ is shifted and becomes a function of the degree of internal damage that occurred in the previous loading.

The results of the monotonic loading up to fracture (mechanical stress–strain as well as the resistance change measurements) for the GFRP specimen with embedded PVA-CNT fibre in compression is shown in Fig. 3.8a. It is clear that during the continuously increasing mechanical load of the specimen's region with the PVA-CNT fibre, the readings of the electrical response of the fibre are negative. A magnification of the region of the plot for low applied strains can be seen in Fig. 3.8b. A local, negative peak is noticed for the $\Delta R/R_0$ measurements at approximately 0.85% applied strain that corresponds to almost 220 MPa compressive stresses. Due to the large deflection of the material's cross section (total deflection was approximately 16 mm), the section's loading alters to tension in this specific region and this is the reason for this negative peak.

With increasing mechanical load, the region of the fibre is tensile tested and begins to increase its electrical resistance change. Of course, when the readings of the fibre equal zero, the local computed stress level is zero and not compressive as noted in Fig. 3.8a. The resistance of the fibre increases till fracture since the fibre's region is in tension.

3.2.4 Damage Assessment

Damage develops in a composite with the incremental, quasi-static loading–unloading steps. Depending on the magnitude of the peak load value, a different kind of damage is developed, e.g. for the low loading values, mainly matrix cracking and debonding between matrix and fibres happens; for medium loading values delamination occurs, while for loadings close to the ultimate tensile load, the main damage mechanism is fibre breakage. Location of the development of damage is strictly linked with the main damage mechanism of the composite. Briefly, with

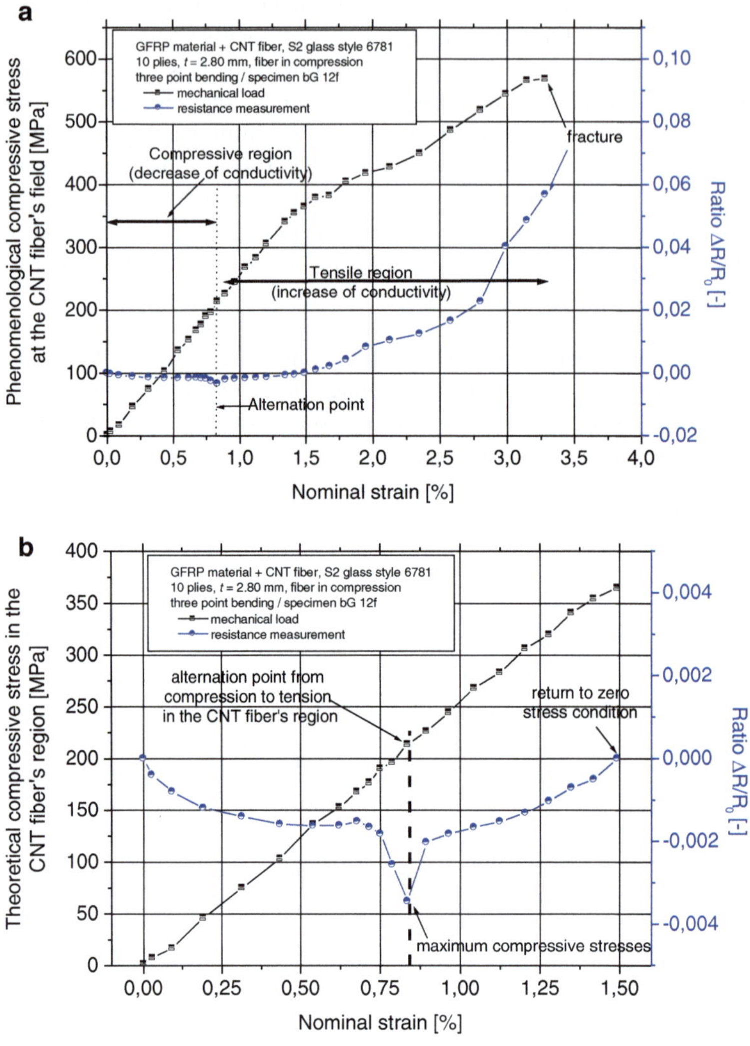

Fig. 3.8 (**a**) Typical three-point bending mechanical and electrical resistance results of the GFRP material with embedded PVA-CNT fibre in the compressive region and (**b**) enlargement of the true loading region where the fibre is in compression

increasing loading, location of the damage occurs firstly in the matrix, then to the interface between the plies of the composite and the final step is the failure of the fibres.

Despite that damage mechanisms in composites are well known, there is no absolute measure to quantify damage. Different researchers have used ultrasonics, acousto-ultrasonics and acoustic emission techniques or even advanced acoustic emission indices in order to characterize each individual damage mechanism and correlate their findings with residual mechanical properties of the composite material, e.g. (Pantelakis et al. 2001; Loutas and Kostopoulos 2009; Philippidis and Assimakopoulou 2008; Aggelis et al. 2010). Nevertheless, from the mechanical point of view, developed damage in the composite can be calculated by reduction of the modulus of elasticity or by the normalized values of modulus of elasticity E/E_0.

Figure 3.9a shows the decrease of the normalized modulus of elasticity for a number of coupons with incremental tensile loading steps. The test results for the coupons without and with embedded PVA–CNT fibres can be seen in the figure, as well as the main damage mechanisms of the investigated composite. Stiffness decrease is almost the same for specimens with and without embedded PVA–CNT fibre. It is also absolutely dependent on the number of loadings up to fracture and therefore by the introduced damage to the coupon; for the cases of low (Seo and Lee 1999) and many (Todoroki et al. 2006) loadings, the stiffness degradation follows the same pattern regardless of the presence or the type of PVA–CNT fibre. Additionally, fracture of the specimens always initiated and occurred within the gauge length of the coupon and not in the wiring connections of the PVA–CNT fibre and given the scatter in composites, it can be concluded that the addition of the fibre did not decrease the material's mechanical properties.

Available experimental data for two different types of PVA–CNT fibres can be seen in Fig. 3.9a by means of the residual resistance measurements as a function of the percentage of fracture stress of the composite. Coupons with untreated fibres are marked as rectangular while the pre-stretched fibres as pyramidal. For all cases, the residual resistance measurements of the PVA–CNT fibre are dependent on the number of loading–unloading steps. It is eminent that for a specimen that suffered a high number of loading–unloading steps till fracture, greater accumulative damage will be induced to the material.

The two different fibres exhibit completcly different trends in Fig. 3.9b; untreated fibre exhibits an exponential increase behavior while the pre-stretched fibre a fairly linear trend. Marked in the figure are also the mean trend lines as well as the upper and lower limits for the two fibres. Untreated fibre gives almost identical values for low-level loadings up to 50% with the pre-stretched fibre. Beyond this critical value, untreated fibre exhibits an essential increase that can be used for damage monitoring.

The most popular damage indication of a composite by means of mechanical testing is the stiffness decrease. For this cause, the residual resistance measurements of the fibres were plotted in Fig. 3.10 against their respective values of normalized modulus of elasticity E_i/E_0. Notice that both damage stages of matrix cracking and delamination are also marked in the figure. For the case of the untreated fibre, an exponential curve fit can be used, while for the case of calibration of the investigated pre-stretched PVA–CNT fibre, the mean induced damage for the investigated system of material and fibre can be fitted by a linear regression. This linear correlation can be used in all stages of damage accumulation in composites, as graphically noticed in the respective figure.

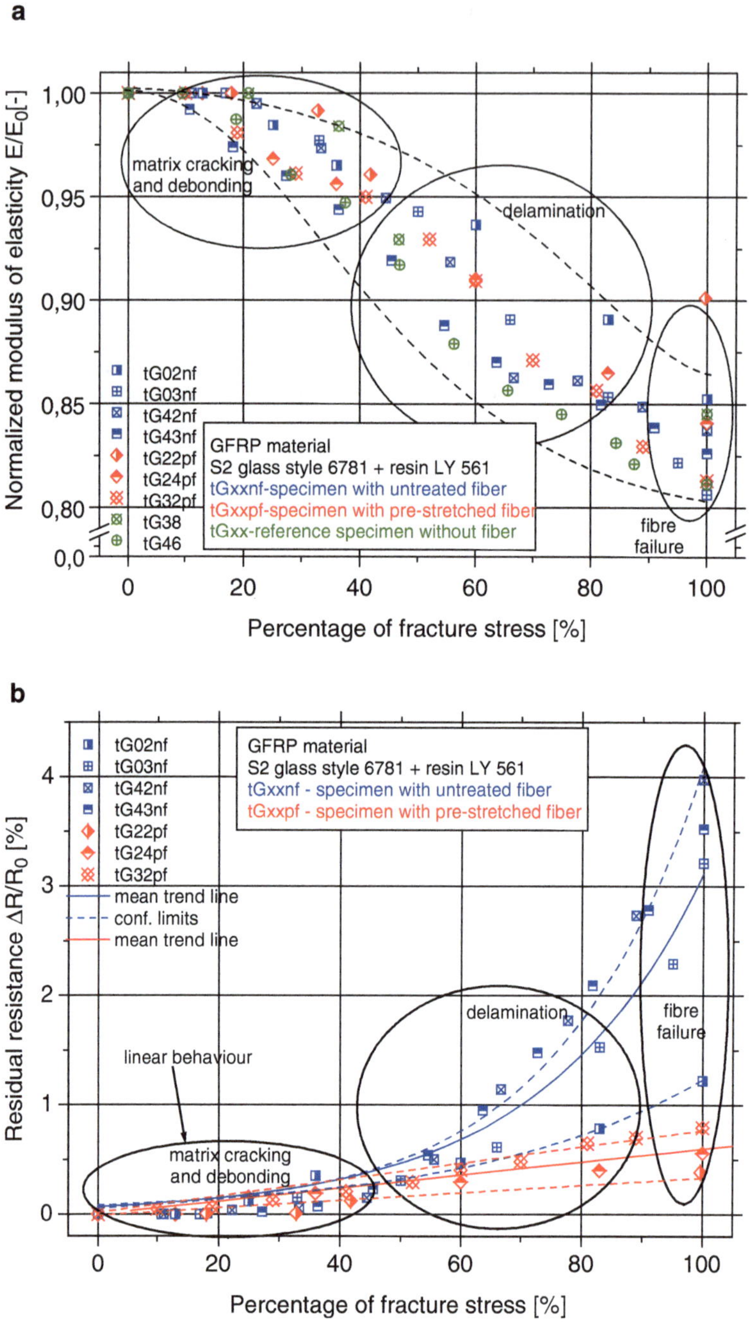

Fig. 3.9 (**a**) Modulus of elasticity degradation due to progressive damage accumulation tests of GFRP coupons with and without PVA-CNT fibre and (**b**) correlation of the levels of incremental loading steps with residual resistance measurements of the PVA–CNT fibre

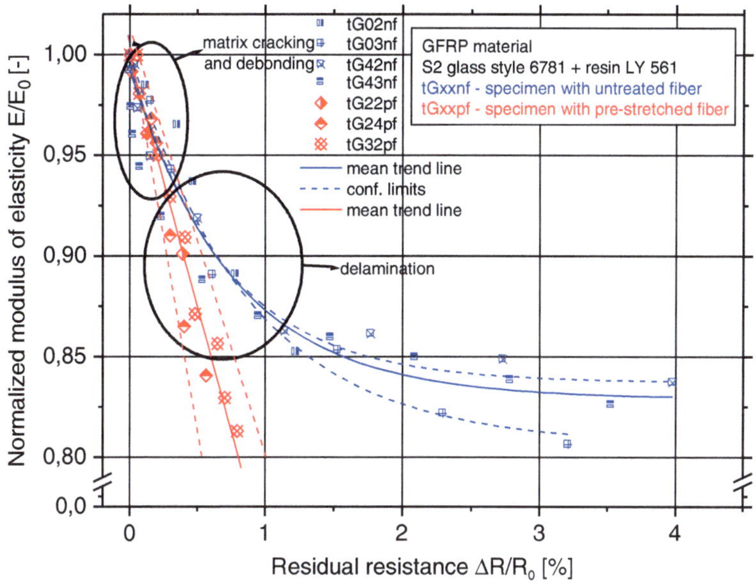

Fig. 3.10 Correlation of the residual resistance readings with induced damage to the composite via normalized modulus of elasticity

3.3 Electromechanical Actuators Made of CNT Structures

3.3.1 CNT Fibre Electromechanical Actuators

3.3.1.1 Nanotubes Fibres and Experimental Procedures

The nanotube fibres used for electromechanical actuators have been prepared by using a so-called coagulation process (Vigolo et al. 2000). This process consists in injecting a single wall nanotube dispersion into the co-flowing stream of a coagulating polyvinyl alcohol (PVA) solution. Sonication partially unbundles the nanotubes (Badaire et al. 2004a) but not sufficiently to yield a dispersion fully comprised of individual single wall nanotubes. The nanotubes coagulate when they meet the PVA solution and form a gel fibre. These fibres are extracted from the coagulating bath and dried vertically. The resultant systems consist of composite PVA-nanotube fibres with a fraction of carbon nanotubes of about 15 wt%. They exhibit a hierarchical structure with the formation of microfibrils (Neimark et al. 2003) resultant from the coagulation of the bundles. Fibres used for electro-chemical actuators are not directly used after their preparation. Specific treatments are performed. Indeed, the composite fibres used for electrochemical actuators are stretched in a mixture of 50 wt% water and 50 wt% acetone. This process was shown to be efficient at increasing the nanotube alignment and improving the stiffness of the fibres (Vigolo et al. 2002). The draw ratio varied from 0 to 500%.

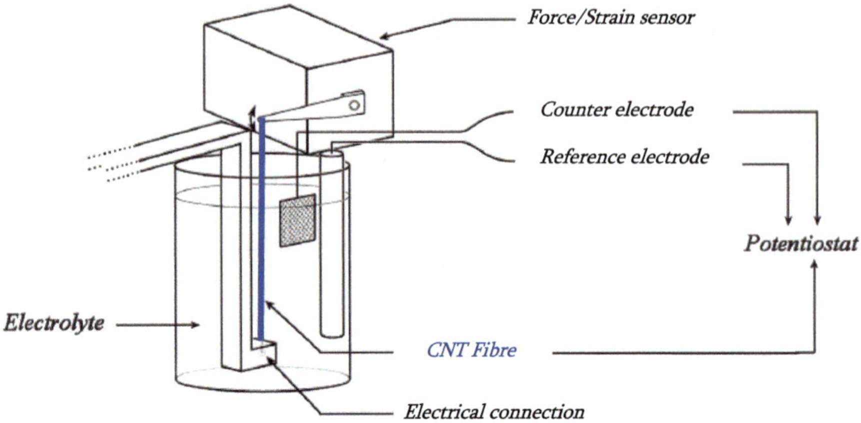

Fig. 3.11 A carbon nanotube (*CNT*) fibre is fixed on top to the lever arm of a force sensor. The fibre is fixed at its *bottom* onto a plastic holder. A conductive copper wire (not shown for sake of clarity) is attached to the sample holder and allows the electrical connection of the fibre. The fibre is immersed in a liquid electrolyte in a classical three electrodes device. Each electrode is described in the text

The un-stretched or stretched fibres are then baked at 600 °C under a nitrogen atmosphere to fully remove the PVA and achieve fibres solely comprised of carbon nanotubes. The mechanical, electrochemical and electromechanical properties have been characterized using a set-up shown in Fig. 3.11. This set-up includes a reference Ag/AgCl (3 M) electrode and a sheet of nanotube paper enclosed in a platinum mesh which serves as a counter electrode. The nanotube fibre acts as the working electrode. The liquid electrolyte in which the fibre is immersed is a 1 M NaCl aqueous solution. The immersed part of the fibre is 1.5 cm long. The fibre is stimulated with a potentiostat. It is fixed to the lever arm of a force sensor. This instrument allows mechanical characterizations in different modes. In particular it allows a given tensile load to be applied to the fibre. Elongation of the fibre in response to this load allows its Young's modulus to be measured. The fibre is electrically connected and stuck at its bottom onto the sample holder as shown in Fig. 3.11. The electrical contact is ensured by silver paint. The contacts are coated with an insulating resin to avoid exposure to the electrolyte and undesired electrochemical artifacts. For electromechanical characterizations, the fibre is mechanically loaded by the force sensor and its length is kept fixed. The stress generated in such isometric conditions is followed as a function of the applied potential. The electrochemical capacitance of the fibres is deduced from cyclic voltammetry experiments at different scan rates.

3.3.1.2 Electrochemical and Electromechanical Properties of CNT Fibres

As indicated in the introduction, carbon nanotubes can deform in response to charge injection and because of the presence of ions adsorbed at their interface

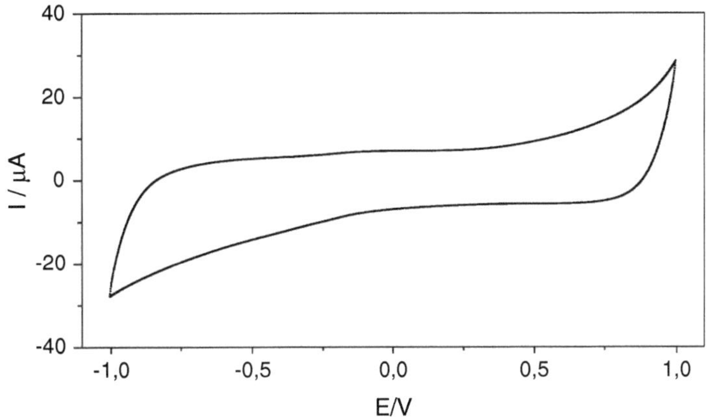

Fig. 3.12 Typical cyclic voltammogram of a carbon nanotube fibre immersed in a NaCl (1 M) solution. The current is shown as a function of the voltage vs a Ag/AgCl reference electrode. The scan rate is 50 mv/s

(Baughman et al. 1999). Varying the charge density and the surface potential of the nanotubes can be achieved electrochemically by stimulating an electrode comprised of nanotubes in a given electrolyte. This was first achieved in 1999 with mats of randomly oriented nanotubes by Baughman et al. (Baughman et al. 1999). The so-called isometric stress (stress at fixed strain) generated by such actuators was about 0.75 MPa. This value is still far from the potential that could be theoretically generated by individual and defect free nanotubes. The challenge for achieving large stress generation consists in optimizing nanotubes assemblies to make materials which combine strength electrical conductivity and porosity. This is not straightforward, but we show in the present chapter that assembling nanotubes under the forms of aligned fibres allows substantial improvements.

A typical cyclic voltammogram (CV) of a nanotube achieved in a NaCl (1 M) solution fibre is shown in Fig. 3.12. This CV was performed at a scan rate of 50 mV/s. The absence of peaks in the investigated potential window reflects the absence of faradic processes. This means that the presence of impurities or of surface functional groups can be considered as negligible for the present investigations. Series of CVs at different scan rates allowed the electrochemical capacitance of the fibres to be deduced (Bard and Faulkner 2001).

The properties of fibres which have been wet-stretched in a bath of acetone and water, as described in (Vigolo et al. 2002), are listed in Table 3.1.

It is observed that wet stretching improves the Young's modulus, electrical conductivity and electrochemical capacitance of the fibres. The improvement of Young's modulus and electrical conductivity can be understood by considering that stretching increases the nanotube alignment. The improvement of electrochemical capacitance is believed to arise from some unbundling of the nanotubes under mechanical load. Unbundling the nanotubes can increase the area of nanotubes exposed towards the electrolyte and thereby the electrochemical capacitance of

Table 3.1 Young's modulus, electrical conductivity and electrochemical capacitance of nanotube fibres for different stretching ratio

Stretching ratio (%)	Young's modulus (GPa)	Conductivity ($S.cm^{-1}$)	Capacitance ($F.cm^{-3}$)
0	4	35	16
100	7	53	23
200	12	107	40
500	16	160	52

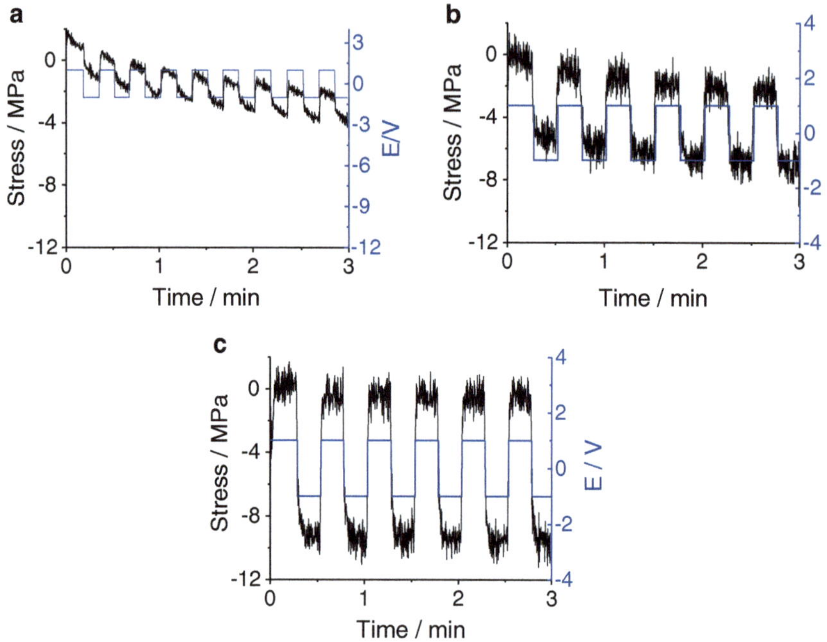

Fig. 3.13 Stress generated by nanotube fibres in isometric conditions for a stimulation of ± 1 V (vs Ag/AgCl) in aqueous solutions of NaCl (1 M). (**a**) Signal of an unstretched fibre tested *right* after preparation, (**b**) signal of a stretched fibre with a draw ratio of 500% tested right after preparation, (**c**) Signal of a stretched fibre with a draw ratio of 500% after 1 h of operation which correspond to 180 cycles. A load of 20 MPa is applied to the fibres

the fibres. Improvements of these properties are expected to yield better actuating properties. This is actually checked by measuring the isometric stress generated by fibres in aqueous solution of NaCl (1 M). The fibre is stimulated with a square wave potential of ± 1 V with respect to an Ag/AgCl reference electrode. A mechanical load is applied to the fibre during the experiments. The stimulation frequency is 30 mHz (three cycles per minute). It is observed that the generated stress increases with the stretching ratio, the applied load and also the operating time. Nevertheless, as shown in Fig. 3.3a, b, the envelope of the electrochemical response decreases in the first cycles of operation. This reflects some mechanical relaxation of the fibre after the load has been applied. After some tens of cycles the signal envelope becomes more stable (Fig. 3.13c). At the same time the amplitude of the generated

Table 3.2 Stress generated by nanotube fibres in isometric conditions for a stimulation of ±1 V (vs Ag/AgCl) in aqueous solutions of NaCl (1 M) for different stretching ratio

Stretching ratio (%)	Generated stress (MPa)
0	3
100	3.5
200	8
500	10

The fibres are tested after operating for 1 h. A mechanical load of 20 MPa is applied to the fibres

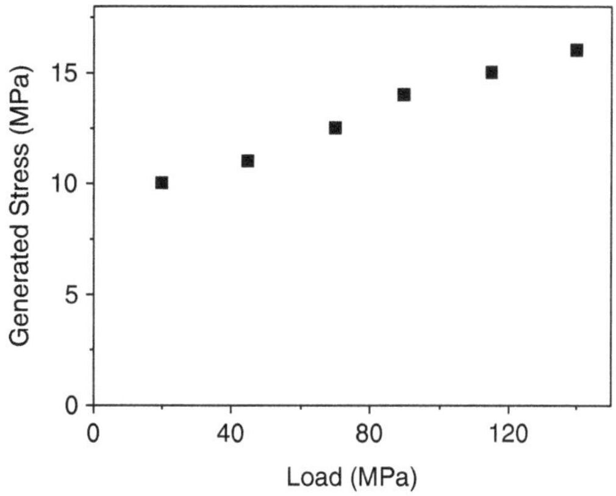

Fig. 3.14 Stress generated by stretched nanotube fibres in isometric conditions for a stimulation of ±1 V (vs Ag/AgCl) in aqueous solutions of NaCl (1 M) as a function of the applied mechanical load. The stretching ratio of the fibres is of 500%

stress increases. Typically no evolution of the generated stress is observed after 4–5 h of operation. The signal can be considered as optimal after that time. It is believed that the porosity and full wetting of the fibres is not yet achieved in the first operating cycles. After a few hundreds of operating cycles the electrolyte can fully penetrate within the porous structure of the fibre and yield a better electromechanical response.

The generated stress of fibres with different stretching ratio is shown in Table 3.2. The applied mechanical load is about 20 MPa.

As expected, it is observed that the generated stress increases with the stretching ratio. We can already note that the generated stress clearly exceeds 0.75 MPa, the stress generated by non-oriented buckypapers (Baughman et al. 1999). The influence of the applied load is shown in Fig. 3.14. The stress generated in isometric conditions after 1 h of operation is plotted as a function of the applied load. The fibres have been wet-stretched by 500%. The maximal generated stress is here achieved for a mechanical load of 140 MPa. This stress is about 17 MPa. This value is particularly high and confirms the excellent potential of carbon nanotubes for electrochemical actuators.

Even though the present results can be viewed as already promising, several challenges are still faced for further improvements. Indeed nanotube fibres solely comprised of nanotubes are brittle and their handling is still rather delicate. In addition even if progress has been achieved in terms of generated stress, we should keep in mind that we are still far from the most optimistic predictions

(Baughman et al. 1999, 2002; Sun et al. 2002). A strain deformation of about 0.1% of an individual nanotube that exhibits a Young's modulus of 1 TPa corresponds to a generated stress of about 1 GPa, which is two orders of magnitude greater than the stress achieved with nanotube fibres. This means that these materials are far from fully manifesting the potential of carbon nanotubes for electromechanical actuators technologies. It is therefore still necessary to achieve stiffer and stronger structures. Improvements would also be expected if nanotubes could be sorted as a function of their chirality and electronic properties. Indeed, the electromechanical response of carbon nanotubes should depend on their chirality. Therefore optimal dimensional changes should be achieved with nanotubes of a well-defined chirality. Finally, it is also critical to study the electromechanical properties of nanotube assemblies in solid electrolytes and check if interesting performances can also be obtained. This would enlarge the spectrum of potential applications of nanotube actuators that could operate in dry environments and in particular for aircraft applications.

3.3.2 CNT Buckypaper for Bilayer Electromechanical Actuators

3.3.2.1 Nanotubes Mats and Bimorph Device

We present an investigation of nanotube films reinforced with a polymer binder. We have used oxidized-multiwall nanotubes and polyvinyl alcohol (PVA) as binder. Several studies have shown that PVA is a good candidate to form interesting CNT/polymer composites (Vigolo et al. 2000; Shaffer and Windle 1999; Cadek et al. 2002; Zhang et al. 2003, 2004; Lin et al. 2003; Coleman et al. 2004, 2006; Badaire et al. 2004b; Chen et al. 2005; Bin et al. 2006; Minus et al. 2006; Bhattacharyya et al. 2006; Liu et al. 2005; Wang et al. 2006; Mazzoldi et al. 2008; Ryan et al. 2007; Cadek et al. 2004), including actuators (Mazzoldi et al. 2008). Indeed this semi-crystalline polymer has excellent film forming properties, high tensile strength and flexibility. Moreover, the hydroxyl groups present in the PVA chains can promote strong interactions with oxidized-CNT which have carboxyl and hydroxyl groups at their surface. It has also already been shown that crystallization of the polymer is enhanced by the presence of CNTs. This leads to an increase of the mechanical properties of the composite (Shaffer and Windle 1999; Coleman et al. 2004; Minus et al. 2006; Ryan et al. 2007; Cadek et al. 2004). Nevertheless, PVA is an insulating polymer. The addition of too much polymer will result in low porosity and low electrical conductivity. This is why a compromise has to be found to maintain a relatively good conductivity combined with sufficient porosity to allow migration of ions and good mechanical properties.

Different types of carbon nanotube mats and devices have been prepared. The fabrication processes are presented below.

Fig. 3.15 Picture of an oxidized-MWNT mat. The mat is of 4 cm in diameter and of 20 μm in thickness

Table 3.3 Mechanical and electrical properties of oxidized-MWNT bucky papers with varied weight fractions of PVA

Papers	PVA wt fraction (%)	Strain to failure (%)	Stress to failure (MPa)	Young modulus (GPa)	Conductivity ($S.cm^{-1}$)
Oxidized-MWNT	0	0.8	12	1.5	33
Oxidized-MWNT/PVA	18	1.7	21	2.0	17
	30	**5.5**	**51**	**3.4**	**9**
	60	44.6	44	1.4	0.2

Neat Oxidized-MWNT Mats

Oxidized carbon nanotubes are obtained by adding 300 mg of MWNT in 50 ml of nitric acid 3.6 M under reflux. After 3 days of acid treatment, the suspension is rinsed with distilled water and redispersed in water to obtain a homogeneous oxidized-MWNT dispersion. The MWNTs weight ratio is then adjusted to 0.7 and 10 g of the dispersion are filtered under vacuum on a regenerated cellulose membrane (pore size = 0.45 μm, diameter = 4.7 cm, thickness = 0.155 mm) to obtain an oxidized-MWNT paper sheet as shown in Fig. 3.15.

Composite Oxidized-MWNT/PVA Mats

Adding the desired quantity of aqueous solution of PVA (5 wt%, M_w = 198 Kg. mol $^{-1}$, hydrolysis 98%) to the oxidized-MWNT dispersion is sufficient to prepare oxidized-MWNT/PVA mats. After addition of the PVA solution the dispersions are ultrasonicated 5 min in a water bath and then filtered as mentioned above.

Table 3.3 presents the mechanical and electrical properties of carbon nanotubes mats with different weight fractions of PVA.

We can see in Table 3.3 that 30 wt% of PVA in the mat is useful to obtain a good compromise of mechanical and electrical properties. Considering these results, devices using oxidized-MWNT/PVA paper mats containing 30% of polyvinyl alcohol (Mw 195 Kg/mol) have been prepared and characterized in more detail regarding actuating capabilities.

Fig. 3.16 Sketch of a device of bimorph actuators

Bimorph Device Made of Oxidized-MWNT/PVA (30 wt%) Buckypaper

To make the device we have coated one side of the buckypaper with a thin layer of gold. This layer was deposited by an evaporation process. The gold layer deposition was done to increase the conductivity and to promote actuation all along the device. Then, on the same side, an inert layer of PVA has been deposited to achieve a bimorph as sketched in Fig. 3.16.

This device is made of three layers but can still be considered as a bimorph device. Indeed, the system can be viewed as a two-layer structure with an active layer made of nanotube paper and an inert layer made of two sheets: gold and pure PVA. The gold and inert PVA layers strongly adhere to each other. This is why we consider in the following that the gold and inert PVA layers as a single effective layer with a thickness that corresponds to the sum of the two thicknesses and a Young's modulus directly derived from a simple rule of mixture.

3.3.2.2 Bimorph Electromechanical Actuator in Organic Liquid Electrolyte

Actuator Device in Liquid Electrolyte

The pseudo-bimorph devices described in the previous sub-section are connected to a power generator through a metallic wire connected to one side of the device as schematically shown in Fig. 3.17. The voltage is applied between the device and a platinum mesh which serves as a counter electrode in the electrolyte.

The section of the device, which highlights the different layers, has been observed by scanning electron microscopy (SEM). A typical SEM picture is shown in Fig. 3.18.

The SEM micrograph of the section of the device revealed the different parts of the actuator and also allowed us to determine precisely the thicknesses of the three layers. At the top of the image we can see the oxidized-MWNT/PVA paper (thickness, $t_p = 26.3\ \mu m$), then the layer of gold ($t_{Au} = 1.5\ \mu m$) and lastly the PVA layer ($t_{PVA} = 6.8\ \mu m$) at the bottom.

The generated stress, using different applied voltage, has been determined from measurements of the bimorph deflections.

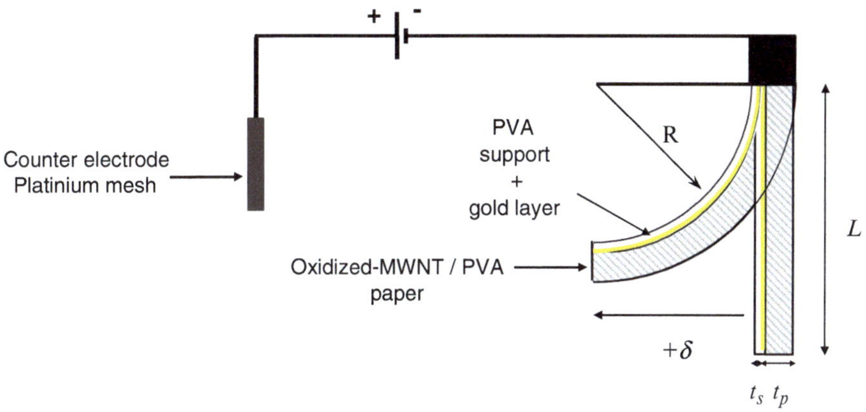

Fig. 3.17 Schematic of the actuator device

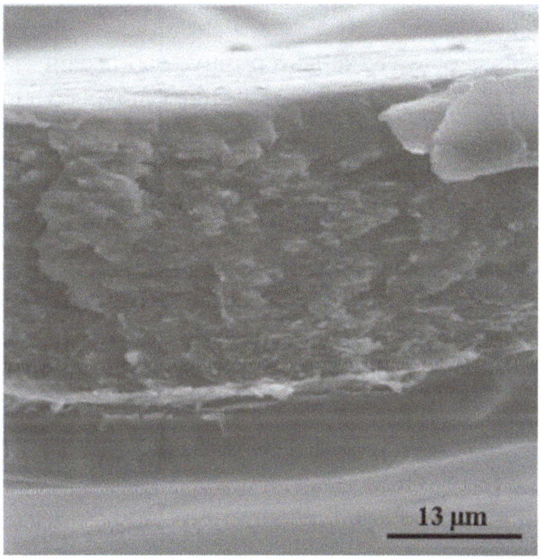

Fig. 3.18 SEM micrograph of the bimorph section. At the *top* of the image the oxidized-MWNT/PVA paper (thickness, $t_p = 26.3$ µm), then the layer of gold ($t_{Au} = 1.5$ µm) and finally the inert PVA layer ($t_{PVA} = 6.8$ µm)

Generated Stress

In analogy with other bilayer actuators (Raguse et al. 2003; Stoney 1909), the observed deflection allows us to evaluate the stress generated (σ) by the swelling of the oxidized-MWNT/PVA paper. The Stoney's equation (Stoney 1909) presented below is used for such a determination (Eq. 3.2):

$$\sigma \sim \frac{E_s.t_s.\delta}{3(1-\nu_s)} t_p.L^2 \tag{3.2}$$

E_s is the Young modulus of the inert PVA and gold layers calculated using the rule of mixture presented below (Eq. 3.3):

$$E_s = \frac{t_{Au}E_{Au} + t_{PVA}E_{PVA}}{t_{Au} + t_{PVA}} \tag{3.3}$$

E_{Au} and E_{PVA} are respectively the Young modulus of the gold layer (*77 GPa*) and of the PVA layer swollen with acetonitrile (*3.77 GPa*). t_s is the sum of the thickness of the gold ($t_{Au} = 1.5\ \mu m$) and PVA layer ($t_{PVA} = 6.8\ \mu m$). t_p is the thickness of the carbon nanotubes composite paper ($t_p = 26.3\ \mu m$). v_s is the Poisson ratio of the polyvinyl alcohol which is *0.49*. δ is the deflection of the device and L the length of the immersed part of the actuator. As long as $\delta < L$ and the Young modulus of the paper sheets $\leq E_s$, Eq. 3.3 can be considered as valid.

The device has been dipped in a solution of TBA/TFB (TetraButylAmmonium/TetraFluoroBorate) in acetonitrile (0.5 M/acetonitrile) and tested electromechanically at constant frequency (0.03 Hz) and different amplitude of applied voltages with respect to the platinum mesh counter electrode (2, 5, 8 and 10 V). Macroscopic deflections of the device were observed and measured. Expansion of the CNT mats is observed at both positive and negative voltages with respect to the counter electrode. Nevertheless, deflections are much more pronounced at negative voltages.

The bimorph device is bent because of the elongation of the PVA-nanotube sheet. Considering the voltage amplitude applied, the electro-mechanical response is presumably not only due to a charge injection phenomenon. Indeed, generally the bending behavior by the charge and the discharge of carbon nanotubes in a liquid electrolyte occurs at voltages lower than 2 V with respect to a reference electrode. Here the potential is controlled with respect to the counter electrode and probably lowered because of resistance losses of the resistive nanotube sheet. But actuation can involve electro-osmotic effects (Shahinpoor and Kim 2000, 2001; Paquette et al. 2003; Asaka et al. 1995; De Gennes et al. 2000; Kim and Shahinpoor 2002; Nemat-Nasser 2002; Tadokoro et al. 2001; Fukushima et al. 2005; Mukai et al. 2008) caused by the migration of the electrolyte ions within the carbon nanotubes structure. This induces subsequent swelling and de-swelling of the material.

The obtained results of the generated stress using Eq. 3.3 for applied voltages varying between 2 and 10 V are shown in Fig. 3.19.

The generated stress increases strongly with an increase in the amplitude of the applied voltage. It can reach an optimal value of 1.8 MPa. This value compares well and even exceeds the stress generated by recent bimorphs made of gold nanoparticles. A value of 1.8 MPa also compares well with actuation phenomena generated by charge injection. It can thus be concluded that actuation properties of CNT composites are interesting. Nevertheless we can not claim at this stage that this approach is substantially better than related technologies based on ionic swelling. In particular migration of ions is also involved in other classes of actuators such as ionic polymer metal composites (IPMCs). Such materials are made of a polymer layer sandwiched in between metal particles films. IPMCs are capable of generating a stress that exceeds 10 MPa (Tadokoro et al. 2001).

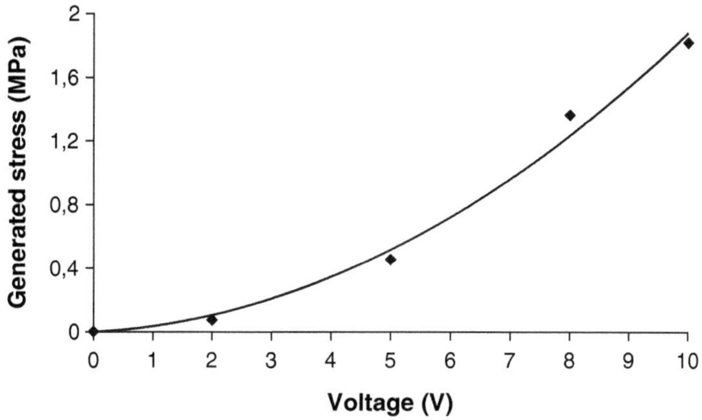

Fig. 3.19 Evolution of the generated stress as a function of the amplitude of the applied voltage to an oxidized-MWNT/PVA bimorph device dipped in TBA/TFB 0.5 M/acetonitrile. The voltage is negative with respect to a counter electrode

In addition, we should recall that the present actuators operate in liquid electrolytes and even if these studies are very useful for a better understanding they are not directly suitable for aircraft applications. Dried electrolytes are expected to be better candidates for this field of application.

3.3.3 Dry State Actuators

3.3.3.1 Preparation of Devices and Characterization of Deflections

Systems made of gels based on mixtures of CNT and ionic liquids have been investigated. These systems exhibit several advantages: indeed, they do not flow as aqueous electrolytes or acetonitrile solutions and do not evaporate at room temperature and pressure.

The new devices presented in this sub-section are inspired by the work by Aida et al. on carbon nanotube actuators (Fukushima et al. 2005; Mukai et al. 2008; Takeuchi et al. 2009). Dry state actuators consist of a three-layered film which includes a layer of polymer and ionic liquid (IL) sandwiched by two identical layers composed of a mixture of nanotubes, ionic liquid and polymer (Fukushima et al. 2005). Here the structures are made of MWNT, $BMIBF_4$ and PVdF(HFP). Devices are elaborated following the steps listed below.

Preparation of a Device for Dry State Actuation

The first 0.5 g of PVdF(HFP) are dispersed in 20 mL of acetone. The dissolution of the polymer was facilitated by heating the dispersion during 30 min at 60 °C.

Fig. 3.20 Device of a dry state actuator. Oxidized-MWNT/PVdF(HFP) + $BMIBF_4$/oxidized-MWNT

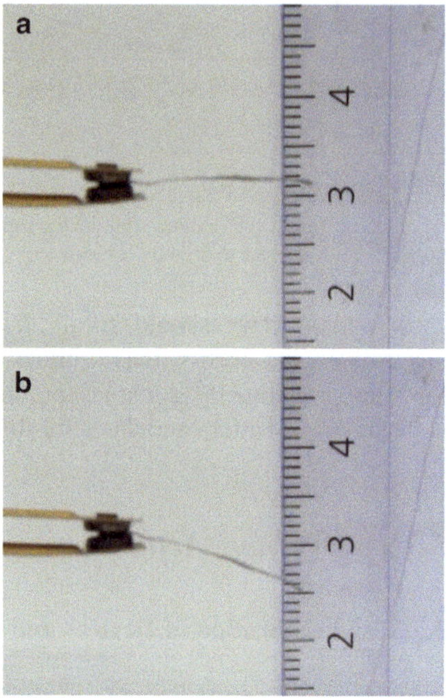

Fig. 3.21 Bending behavior of the dry state actuator oxidized-MWNT/PVdF (HFP) + $BMIBF_4$/oxidized-MWNT (dimension: 2.2 cm [length] × 1,9 mm [wide] × 80 µm [thick]) at constant frequency (0.1 Hz) and (**a**) *top* picture no applied voltage (**b**) *bottom* picture applied +15 V.

Then, 0.25 g of the ionic liquid $BMIBF_4$ in the PVdF(HFP) were added to the solution. After 5 min of mixing the solution is placed in a crystallizer, and the film formed upon solvent evaporation at room temperature.

When the PVdF(HFP) and $BMIBF_4$ film is formed we deposited 3 mL of an aqueous oxidized-MWNT dispersion with a solid content of 0.7 wt%. Once the oxidized-MWNT layer was formed and dried on one side, the same was done on the other face of the polymer-IL film.

The obtained three-layered film has a thickness of about 80 µm. A schematic representation of this device is shown in Fig. 3.20.

Images of Fig. 3.21 show the bending of the dry state carbon nanotube actuator.

The stress generated by this actuator was deduced from mechanical and electromechanical characterizations. A stress *vs* strain tensile curve is shown in Fig. 3.22. The strain to failure is 1.2% and the Young's modulus is of 385 MPa.

In analogy with other multilayered actuators (Shahinpoor and Kim 2004; Kim and Shahinpoor 2003), the observed deflections allow us to evaluate the stress

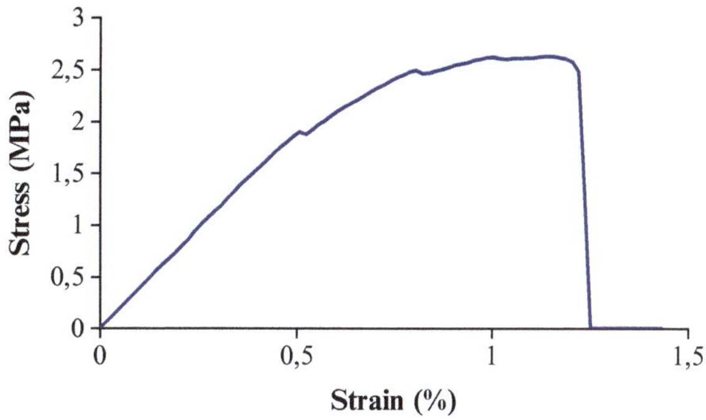

Fig. 3.22 Stress versus strain for the oxidized-MWNT/PVdF(HFP) + $BMIBF_4$/oxidized-MWNT device

generated (σ) by the present device. As discussed previously, the Stoney's equation (Stoney 1909) can be used for such a determination. In the present case, E_s and t_s respectively are the Young's modulus (*385 MPa*) and the thickness (*80 μm*) of the device. t_p is the thickness of the carbon nanotube paper ($t_p = 5\ \mu m$). v_s is the Poisson's ratio of the PVdF(HFP) which is *0.33*. δ is the deflection of the device and L the length of the actuator.

Considering the observation of the macroscopic deflections of the device and its mechanical properties, it is deduced that the maximal generated stress is about 0.3 MPa. This stress is achieved for a relatively high voltage of about 15 V.

3.4 CNT Fibre with Shape Memory Properties

For this application, composite PVA-nanotube fibres are tested in a temperature controlled chamber shown in Fig. 3.23. They are glued onto two Invar bars which are clamped by holders outside the temperature controlled chamber. The stress and strain are measured with a mechanical testing instrument. Invar is chosen to hold the fibres because of its very small thermal expansion coefficient, of about 10^{-6} K. Strain variations due to thermal expansion are estimated to be lower than 0.1%. The length of the initial samples, before stretching at high temperature, is about 1 cm. The fibres are stretched at a deformation temperature T_d and then cooled down to room temperature under fixed strain. Their length does not change when the load is released at room temperature, thus showing good "shape fixity". The fibres, however, shrink substantially when they are re-heated.

A qualitative evidence of the shape memory behavior of CNT-PVA composite fibres is shown in Fig. 3.24. A knot of a stretched fibre tightens when the fibre is heated. This reflects the shrinking and the large strain recovery of the deformed fibre.

Figure 3.25 shows more quantitatively the stress needed to stretch CNT-PVA composite fibres up to 800% at different temperatures. A greater stress is needed to deform the fibres at low T_d. At high T_d, the fibres become softer and can be more

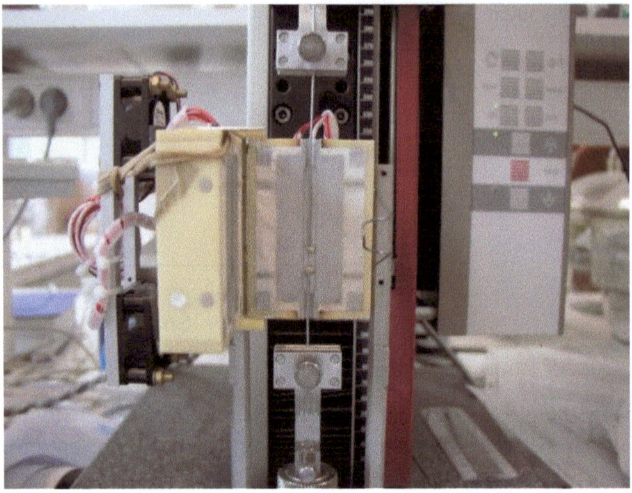

Fig. 3.23 Temperature controlled chamber used to characterize shape memory effects of CNT-PVA fibres

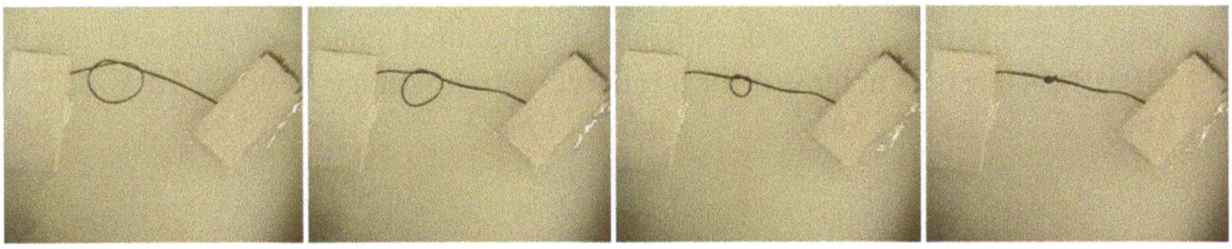

Fig. 3.24 Qualitative evidence of the shape memory behavior of a CNT-PVA composite fibre. The shown fibre has been stretched at 150 °C. It has then been cooled down to room temperature under tensile load. A knot was made with the stretched fibres. The fibre shrinks and the knot tightens when the fibre is reheated. Reheating is here simply achieved by blowing hot air toward the fibres. The series of picture shows the fibre shrinking as a function of time. The time interval between each picture is 3 s

easily deformed. This softness is associated with a lower supply of mechanical energy. This can be estimated in Fig. 3.25 where the area under each curve corresponds to the energy supplied to the fibres stretched at different T_d, from 70 to 180 °C.

As shown in Fig. 3.26, when reheated at fixed strain the fibres generate a strong stress with a maximum at a well-defined temperature (T_s). The occurrence of a peak recovery stress in conditions of fixed strain has already been observed for other shape memory polymers and nano-composites (Hu et al. 2005; Miyamoto et al. 2002), but with no direct link between T_s and T_d. In conventional materials the peak of recovery stress occurs in the vicinity of the glass transition of the neat polymer. In fact, this is interpreted in the literature as a direct manifestation of the glass transition of a pure polymer (Miyamoto et al. 2002). When the materials are initially deformed above the glass temperature transition, the peak disappears and

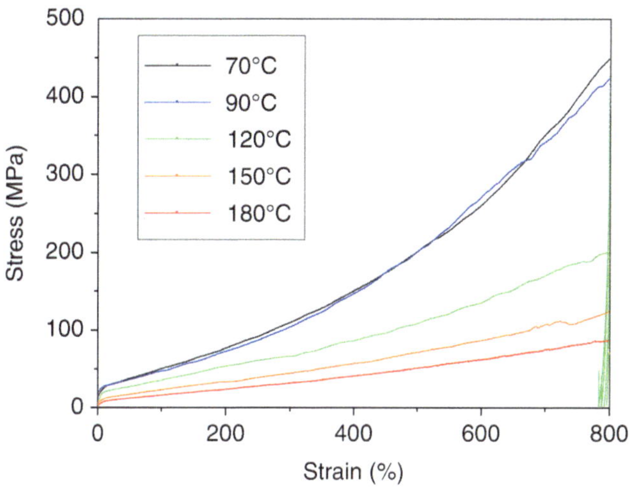

Fig. 3.25 Stress vs strain curves of nanotube-composite fibres. The fibres are stretched up to 800% at different temperatures (T_d)

Fig. 3.26 Stress generated by a nanocomposite fibre when it is re-heated. The strain is kept fixed and the temperature is increased from room temperature to 230 °C at a rate of 5 °C per minute. The different colours correspond to the temperatures T_d at which the fibres have been initially deformed. A peak is observed in each case for a temperature T_s roughly equal to T_d.

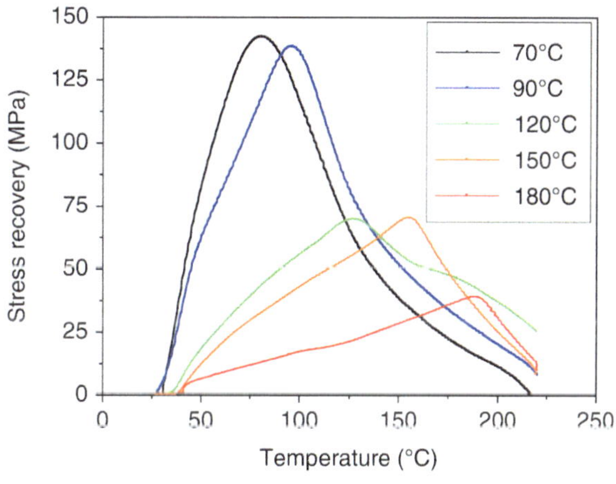

the stress generated upon shape recovery substantially decreases. This is due to the fact that polymer chains can relax when deformed at temperatures well above T_g, thus decreasing the potential for stored mechanical energy.

Here the peak is preserved well above the T_g of the neat PVA. Neat PVA can exhibit several thermo-mechanical relaxations depending on its degree of cross-linking and humidity (Park et al. 2001). The glass transition temperature of the material presently used is about 80 °C in its dry state. The samples were prepared several days before testing and not stored in a dried atmosphere. This is why they contained some undetermined fraction of moisture. As already reported (Park et al. 2001) and presently observed, the effect of humidity is seen in DMA experiments

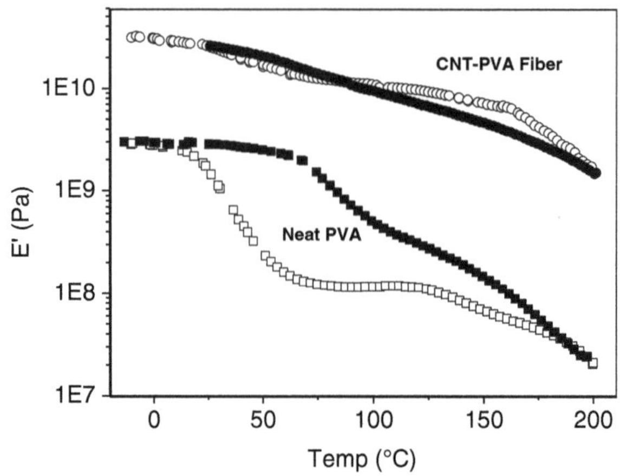

Fig. 3.27 Storage modulus E′ of neat PVA (*squares*) and CNT-PVA fibres (*circles*), upon heating (*open symbols*) and upon cooling (*black symbols*). The curves upon heating provide examples of the behavior of non-dried materials. By contrast, the curves upon cooling correspond to the behavior of materials in their dried states

with a shift of the T_g from 80 to 40 °C. Characterizations to determine the storage moduli of PVA and CNT-PVA composite fibres measured by DMA are shown in Fig. 3.27. After the heating stage the samples are dried and the mechanical properties measured upon cooling. Those correspond to the properties of materials in their dried states. The curves are reproducible after a single heating stage as long as the samples are not kept for a long period of time in a humid atmosphere. The effect of humidity was negligible in the shape memory experiments because the samples were tested right after hot-stretching.

Strikingly, when compared to previously investigated shape memory polymers, T_s and T_d are actually roughly equal. This near-equality means that the fibres memorize the temperature at which they have been deformed. The peak of stress generated can be observed up to 180 °C, which is about 100 °C above the T_g of the neat PVA. This distinctive feature provides an opportunity to rationally control T_s, without varying the chemical structure of the material. In addition, it is observed that the maximal stress generated by the fibre is close to 150 MPa. This value is from one to two orders of magnitude greater than the stress generated by conventional shape memory polymers. It is obtained for fibres which have been deformed at 70 and 90 °C; temperatures which are in the vicinity of the T_g of the neat PVA. They correspond to the conditions for which the greatest energy is supplied during initial deformation.

Additionally, because carbon nanotube fibres are electrically conductive, we note that the thermal shape memory effects can be triggered by Joule's heating when an electrical current is passed through the fibre. This can be useful for the direct use in micro-devices where heating via an external source can be difficult.

Some of the present results can be understood on the basis of previous knowledge of the structure of CNT-PVA fibres and on the main known features of nano-composites and shape memory polymers. Shape memory polymers usually involve two "phases" (Lendlein and Kelch 2002; Liu et al. 2007; Kim et al. 1996; Ohki et al. 2004; Meng et al. 2007): a fixed one, which can be made of crystallites, rigid segments or chemical cross-links, and a mobile one which is made of amorphous polymer. The latter drives shape memory effects through elongation and contraction of the polymer chains respectively during programming and shape recovery, but the fixed phase is necessary to lock deformations in the material and achieve good shape fixity. Shape memory effects are more pronounced in the vicinity of T_g because this temperature corresponds to the relaxation of the amorphous fractions of the polymer. CNTs substantially alter the properties of the composite fibres in several ways. First, and as shown in Fig. 3.27, they act as reinforcements characterized by an increase of one order of magnitude of the storage modulus. Second and as already reported (Miaudet et al. 2005; Cadek et al. 2002) they promote the stabilization of crystalline domains. This can contribute to the locking of mechanical constraints. Therefore carbon nanotubes by increasing the stiffness of the polymer can allow more energy to be absorbed and restored. This can explain the very large stress measured in the present experiments.

However, the origin of the temperature memory is still less clear. It could likely be arising from a broad glass transition with the contribution of confined polymers at the interface of nanotubes or crystalline domains. It has been shown that significant gradients of T_g can develop at the interface of nano-particles (Berriot et al. 2003). Amorphous polymer shells around the CNTs or around crystalline domains largely overlap and percolate, such as the CNTs themselves; meaning that there is a distribution of polymer-CNT or amorphous polymer-crystallites distances which ranges from molecular contact to several nanometers. This distribution of confinement results in a wide broadening of the relaxation time spectrum and specifically the glass transition through a distribution of polymer fractions which exhibit different T_g. This property could be responsible for peaks of stress recovery well above the T_g of the neat polymer. Indeed, when the material is stretched at T_d, the polymer fractions that have lower glass transition temperatures (far from the interface) can quickly relax and don't efficiently contribute to the storage of mechanical energy. In contrast, polymer fractions with glass transition temperatures close to T_d dominate the behavior by storing and restoring mechanical energy. We also note that composites treated in the vicinity of T_g still exhibit higher toughness and generate greater stress recovery. This indicates that the fractions of amorphous polymer with un-shifted or slightly shifted glass transition temperatures remain the major components of the composite. This scenario is still speculative and further research is needed to clarify the microscopic origin of the temperature memory. In particular, if it would actually be arising from a broadening of the glass transition and to confined polymers at the interface of nanotubes or crystallites, this temperature memory should take place in other nano-composites and even in neat semi-crystalline polymers which exhibit a sufficient fraction of crystalline domains. Research is currently underway to validate such predictions. Preliminary experiments suggest that neat PVA can also exhibit temperature memory. The recovery stress of neat PVA is not as high as that of PVA-CNT composite fibres but neat semi-crystalline PVA seems to also exhibit peaks of recovery stress at the temperatures of its initial deformation T_d. A full and systematic confirmation of these preliminary observations would be of great interest to advance the basic knowledge of shape memory phenomena in polymers. It would also be particularly interesting for applications since temperature memory allows tuning shape memory phenomena via material treatments without varying chemical composition.

3.5 Conclusion

Manifesting the properties of individual nanotubes for making functional materials potentially useful in aircraft applications such as sensors and actively moving materials is particularly appealing. We have seen in this chapter that carbon nanotubes can be used as fillers in shape memory polymer fibres and bring novel properties such as large stress generation and enhancement of temperature memory phenomena. Carbon nanotube fibres can also be used as stress and strain sensors and be embedded in composites for non-destructive health monitoring applications. Finally carbon nanotubes can be used to make neat or composite films that expand or contract upon electrical stimulations. But, regardless of the exact type of nanotube materials, it is critical to order and assemble nanotubes on macroscopic scale to optimize the materials properties. The development of fibres or compact films is a promising approach towards these objectives. Fibres can be easily processed into textile, cable, composite or electrode structures (Viry et al. 2010). More importantly they allow nanotubes to be easily oriented on macroscopic scale. Nevertheless, it is also clear that improvements are still needed because nanotube based functional materials are far from fully manifesting the potential of individual carbon nanotubes. Indeed the levels of reinforcements of polymer composite are still below the best expectations. The same holds for the electromechanical properties of fibres or mats comprised of nanotubes. Future research will be in particular valuable to achieve stiffer and stronger structures that will be capable of generating strong stresses combined with large strain deformations; it should thereby lead to novel and efficient technologies of functional materials potentially useful in aircraft applications.

References

Aggelis, D.G., Barkoula, N.M., Matikas, T.E., Paipetis, A.S.: Acoustic emission monitoring of degradation of cross ply laminates. J. Acoust. Soc. Am. **127**(6), 246–251 (2010)

Alexopoulos, N.D., Bartholome, C., Poulin, P., Marioli-Riga, Z.: Structural health monitoring of glass fiber reinforced composites using embedded carbon nanotube (CNT) fibers. Compos. Sci. Technol. **70**, 260–271 (2010a)

Alexopoulos, N.D., Bartholome, C., Poulin, P., Marioli-Riga, Z.: Damage detection of glass fiber reinforced composites using embedded PVA-carbon nanotube (CNT) fibers. Compos. Sci. Technol. **70**, 1733–1741 (2010b)

Arby, J.C., Bochard, S., Chateauminois, A., Salvia, M., Giraud, G.: In situ detection of damage in CFRP laminates by electrical resistance measurements. Compos. Sci. Technol. **59**, 925–935 (1999)

Arby, J.C., Choi, Y.K., Chateauminois, A., Dalloz, B., Giraud, G.: In-situ monitoring of damage in CFRP laminates by means of AC and DC measurements. Compos. Sci. Technol. **61**, 855–864 (2001)

Asaka, A., Oguro, K., Nishimura, N., Misuhata, M., Takenaka, H.: Bending of polyelectrolyte membrane – platinum composites by electric stimuli I. Response characteristics to various waveforms. Polym. J. **27**(4), 436–440 (1995)

Badaire, S., Poulin, P., Maugey, M., Zakri, C.: In situ measurements of nanotube dimensions in suspensions by depolarized dynamic light scattering. Langmuir **20**(24), 10367–10370 (2004a)

Badaire, S., Pichot, V., Zakri, C., Poulin, P., Launois, P., Vavro, J., Guthy, C., Chen, M., Fischer, J.E.: Correlation of properties with preferred orientation in coagulated and stretch-aligned single-wall carbon nanotubes. J. Appl. Phys. **96**(12), 7509–7513 (2004b)

Balageas, D., Fritzen, C., Guemes, A. (eds.): Structural Health Monitoring. ISTE, London/ Newport Beach (2006)

Bard, A.J., Faulkner, L.R.: Electrochemical Methods. Fundamentals and Applications. Wiley, New York (2001)

Barisci, J.N., Spinks, G.M., Wallace, G.G., Madden, J.D., Baughman, R.H.: Increased actuation rate of electromechanical carbon nanotube actuators using potential pulses with resistance compensation. Smart Mater. Struct. **12**(4), 549–555 (2003)

Bartholome, C., Derre, A., Roubeau, O., Zakri, C., Poulin, P.: Electromechanical properties of nanotube-PVA composite actuator bimorphs. Nanotechnology **19**((32) (2008)

Baughman, R.H., Cui, C.X., Zakhidov, A.A., Iqbal, Z., Barisci, J.N., Spinks, G.M., Wallace, G.G., Mazzoldi, A., De Rossi, D., Rinzler, A.G., Jaschinski, O., Roth, S., Kertesz, M.: Carbon nanotube actuators. Science **284**(5418), 1340–1344 (1999)

Baughman, R.H., Zakhidov, A.A., de Heer, W.A.: Carbon nanotubes – the route toward applications. Science **297**(5582), 787–792 (2002)

Berriot, J., Montes, H., Lequeux, F., Long, D., Sotta, P.: Gradient of glass transition temperature in filled elastomers. Europhys. Lett. **64**(1), 50–56 (2003)

Bhattacharyya, S., Salvetat, J.P., Saboungi, M.L.: Reinforcement of semicrystalline polymers with Collagen-modified single-walled carbon nanotubes. Appl. Phys. Lett. **88**, 2331191–2331193 (2006)

Bin, Y., Mine, M., Koganemaru, A., Jiang, X., Matsuo, M.: Morphology and mechanical and electrical properties of oriented PVA-VGCF and PVA-MWNT composites. Polymer **47**(4), 1308–1317 (2006)

Boller, C., Chang, F.-K., Fujino, Y.: Encyclopaedia of Structural Health Monitoring. Wiley, New York (2009)

Cadek, M., Coleman, J.N., Barron, V., Hedicke, K., Blau, W.: Morphological and mechanical properties of carbon-nanotube-reinforced semicrystalline and amorphous polymer composites. J. Appl. Phys. Lett. **81**(27), 5123 5125 (2002)

Cadek, M., Coleman, J.N., Ryan, K.P., Nicolosi, V., Bister, G., Fonseca, A., Nagy, J.B., Szostak, K., Beguin, F., Blau, W.J.: Reinforcement of polymers with carbon nanotubes: the role of nanotube surface area. Nano Lett. **4**(2), 353–356 (2004)

Chen, W., Tao, X., Xue, P., Cheng, X.: Enhanced mechanical properties and morphological characterizations of poly(vinyl alcohol)–carbon nanotube composite films. Appl. Surf. Sci. **252**(5), 1404–1409 (2005)

Chung, T., Rorno-Uribe, A., Mather, P.T.: Two-way reversible shape memory in a semicrystalline network. Macromolecules **41**(1), 184–192 (2008)

Coleman, J.N., Cadek, M., Blake, R., Nicolosi, V., Ryan, K.P., Belton, C., Fonseca, A., Nagy, J.B., Gun'ko, Y.K., Blau, W.J.: High performance nanotube-reinforced plastics: understanding the mechanism of strength increase. Adv. Funct. Mater. **14**(8), 791–798 (2004)

Coleman, J.N., Cadek, M., Ryan, K.P., Fonseca, A., Nagy, J.B., Blau, W.J., Ferreira, M.S.: Reinforcement of polymers with carbon nanotubes. The role of an ordered polymer interfacial region. Experiment and modeling. Polymer **47**(26), 8556–8561 (2006)

Dalton, A.B.., Collins, S., Munoz, E., Razal, J.M., Ebron, V.H., Ferraris, J.P., Coleman, J.N., Kim, B.G., Baughman, R.H.: Super-tough carbon-nanotube fibres – these extraordinary composite fibres can be woven into electronic textiles. Nature **423**(6941), 703–703 (2003)

De Gennes, P.G., Okumura, K., Shahinpoor, M., Kim, K.J.: Mechanoelectric effects in ionic gels. Europhys. Lett **50**, 513–518 (2000)

Fraysse, J., Minett, A.I., Jaschinski, O., Duesberg, G.S., Roth, S.: Carbon nanotubes acting like actuators. Carbon **40**(10), 1735–1739 (2002)

Fukushima, T., Azaka, K., Kosaka, A., Aida, T.: Fully plastic actuator through layer-by layer casting with ionic-liquid-based bucky gel. Angew. Chem. Int. Ed. **44**, 2410–2413 (2005)

Gagel, A., Lange, D., Schulte, K.: On the relation between crack densities, stiffness degradation, and surface temperature distribution of tensile fatigue loaded glass-fibre non-crimp-fabric reinforced epoxy. Composites Part A **37**, 222–228 (2006)

Gall, K., Dunn, M.L., Liu, Y.P., Finch, D., Lake, M., Munshi, N.A.: Shape memory polymer nanocomposites. Acta Mater. **50**(20), 5115–5126 (2002)

Gall, K., Dunn, M.L., Liu, Y.P., Stefanic, G., Balzar, D.: Internal stress storage in shape memory polymer nanocomposites. Appl. Phys. Lett. **85**(2), 290–292 (2004)

Gall, K., Yakacki, C.M., Liu, Y.P., Shandas, R., Willett, N., Anseth, K.S.: Thermomechanics of the shape memory effect in polymers for biomedical applications. J. Biomed. Mater. Res. A **73A**(3), 339–348 (2005)

Ghosh, S., Gadagkar, V., Sood, A.K.: Strains induced in carbon nanotubes due to the presence of ions: Ab initio restricted Hatree-Fock calculations. Chem. Phys. Lett. **406**(1–3), 10–14 (2005)

Giurgiutiu, V.: Structural Health Monitoring with Piezoelectric Wafer Active Sensors. Elsevier, Oxford (2008)

Gunes, I.S., Jana, S.C.: Shape memory polymers and their nanocomposites: a review of science and technology of new multifunctional materials. J. Nanosci. Nanotechnol. **8**(4), 1616–1637 (2008)

Gupta, V.B., Radhakrishnan, J., Sett, S.K.: Effect of processing history on shrinkage stress in axially oriented poly(ethylene-terephtathalate) fibres and films. Polymer **35**(12), 2560–2567 (1994)

Gupta, S., Hughes, M., Windle, A.H., Robertson, J.: Charge transfer in carbon nanotube actuators investigated using in situ Raman spectroscopy. J. Appl. Phys. **95**(4), 2038–2048 (2004)

Highsmith, A.L., Reifsnider, K.L.: Damage in composite materials. ASTM STP **775**, 103–117 (1982)

Hu, J.L., Yang, Z.H., Yeung, L.Y., Ji, F.L., Liu, Y.Q.: Crosslinked polyurethanes with shape memory properties. Polym. Int. **54**(5), 854–859 (2005)

Hughes, M., Spinks, G.M.: Multiwalled carbon-nanotube actuators. Adv. Mater. **17**(4), 443 (2005)

Irving, P.E., Thiagarajan, C.: Fatigue damage characterization in carbon fibre composite materials using an electrical potential technique. Smart Mater. Struct. **7**(4), 456–466 (1998)

Kaddour, A.S., Al-Salehi, F.A.R., Al-Hassani, S.T.S., Hinton, M.J.: Electrical resistance measurement technique for detecting failure in CFRP materials at high strain rates. Compos. Sci. Technol. **51**(3), 377–385 (1994)

Kim, K.J., Shahinpoor, M.: A novel method of manufacturing three dimensional Ionic polymer-metal composites (IPMCs) biomimetic sensors, actuators and artificial muscles. Polymer **43**, 797–802 (2002)

Kim, K.J., Shahinpoor, M.: Ionic polymer–metal composites: II. Manufacturing techniques. Smart Mater. Struct. **12**, 65–79 (2003)

Kim, B.K., Lee, S.Y., Xu, M.: Polyurethanes having shape memory effects. Polymer **37**(26), 5781–5793 (1996)

Koerner, H., Price, G., Pearce, N.A., Alexander, M., Vaia, R.A.: Remotely actuated polymer nanocomposites – stress-recovery of carbon-nanotube-filled thermoplastic elastomers. Nat. Mater. **3**(2), 115–120 (2004)

Kornbluh, R. Pelrine, R., Pei Q., Heydt R., Stanford S., Oh S., Eckerle J.: In: Proceedings of SPIE, Smart Structures and Materials 2002: Industrial and Commercial Applications of Smart Structures Technologies, San-Diego (CA) USA, p. 4698 (2002)

Kupke, M., Schulte, K., Schuler, R.: Non-destructive testing of FRP by d.c. and a.c. electrical methods. Compos. Sci. Technol. **61**, 837–847 (2001)

Lendlein, A., Kelch, S.: Shape-memory polymers. Angew. Chem. Int. Ed. **41**(12), 2034–2057 (2002)

Lendlein, A., Langer, R.: Biodegradable, elastic shape-memory polymers for potential biomedical applications. Science **296**(5573), 1673–1676 (2002)

Li, C., Thostenson, E.T., Chou, T.W.: Sensors and actuators based on carbon nanotubes and their composites: a review. Compos. Sci. Technol. **68**(6), 1227–1249 (2008)

Lin, Y., Zhou, B., Fernando, K.A., Liu, P., Allard, L.F., Sun, Y.: Polymeric carbon nanocomposites from carbon nanotubes functionalized with matrix polymer. Macromolecules **36**(19), 7199–7204 (2003)
Liu, Y., Gall, K., Dunn, M.L., McCluskey, P.: Thermomechanics of shape memory polymer nanocomposites. Mech. Mater. **36**, 929–940 (2004)
Liu, L., Barber, A.H., Nuriel, S., Wagner, H.D.: Mechanical properties of functionalized single-walled carbon-nanotube/poly(vinyl alcohol) nanocomposites. Adv. Funct. Mater. **15**, 975–980 (2005)
Liu, C., Qin, H., Mather, P.T.: Review of progress in shape-memory polymers. J. Mater. Chem. **17**(16), 1543–1558 (2007)
Loutas, T.H., Kostopoulos, V.: Health monitoring of carbon/carbon, woven reinforced composites. Damage assessment by using advanced signal processing techniques. Part I: Acoustic emission monitoring and damage mechanisms evolution. Compos. Sci. Technol. **69**(2), 265–272 (2009)
Luo, X.F., Mather, P.T.: Preparation and characterization of shape memory elastomeric composites. Macromolecules **42**(19), 7251–7253 (2009)
Madden, J.D.W., Barisci, J.N., Anquetil, P.A., Spinks, G.M., Wallace, G.G., Baughman, R.H., Hunter, I.W.: Fast carbon nanotube charging and actuation. Adv. Mater. **18**(7), 870 (2006)
Mazzoldi, A., Tesconi, M., Tognetti, A., Rocchia, W., Vozzi, G., Pioggia, G., Ahluwalia, A., De Rossi, D.: Carbon nanotube actuators: soft-lithographic fabrication and electro-chemical modelling. Mater. Sci. Eng. C **28**(7), 1057–1064 (2008)
Meng, Q.H., Hu, J.L., Zhu, Y.: Shape-memory polyurethane/multiwalled carbon nanotube fibres. J. Appl. Polymer Sci. **106**(2), 837–848 (2007)
Miaudet, P., Badaire, S., Maugey, M., Derre, A., Pichot, V., Launois, P., Poulin, P., Zakri, C.: Hot-drawing of single and multiwall carbon nanotube fibres for high toughness and alignment. Nano Lett. **5**(11), 2212–2215 (2005)
Miaudet, P., Derre, A., Maugey, M., Zakri, C., Piccione, P.M., Inoubli, R., Poulin, P.: Shape and temperature memory of nanocomposites with broadened glass transition. Science **318**(5854), 1294–1296 (2007)
Minus, M.L., Chae, H.G., Kumar, S.: Crystallization of poly(vinyl alcohol). Polymer **47**(11), 3705–3710 (2006)
Miyamoto, Y., Fukao, K., Yamao, H., Sekimoto, K.: Memory effect on the glass transition in vulcanized rubber. Phys. Rev. Lett. **88**(22) (2002)
Mohr, R., Kratz, K., Weigel, T., Lucka-Gabor, M., Moneke, M., Lendlein, A.: Initiation of shape memory effect by inductive heating of magnetic nanoparticles in thermoplastic polymers. Proc. Natl. Acad. Sci. USA **103**(10), 3540–3545 (2006)
Morshedian, J., Khonakdar, H.A., Mehrabzadeh, M., Eslami, H.: Preparation and properties of heat-shrinkable cross-linked low-density polyethylene. Adv. Polym. Technol. **22**(2), 112–119 (2003)
Mukai, K., Asaka, K., Kiyohara, K., Sugino, T., Takeuchi, I., Fukushima, T., Aida, T.: High performance fully plastic actuator based on ionic-liquid-based bucky gel. Electrochim. Acta **53**, 5555–5562 (2008)
Muto, N., Arai, Y., Shin, S.G., Matsubara, H., Yanagida, H., Sugita, M., et al.: Hybrid composites with self-diagnosing function for preventing fatal fracture. Compos. Sci. Technol. **61**, 875–883 (2001)
Neimark, A.V., Ruetsch, S., Kornev, K.G., Ravikovitch, P.I.: Hierarchical pore structure and wetting properties of single-wall carbon nanotube fibres. Nano Lett. **3**(3), 419–423 (2003)
Nemat-Nasser, S.: Micromechanics of actuation of ionic polymer-metal composites. J. Appl. Phys. **92**, 2899–2915 (2002)
Ohki, T., Ni, Q.Q., Ohsako, N., Iwamoto, M.: Mechanical and shape memory behavior of composites with shape memory polymer. Compos. Part A-Appl. Sci. Manuf. **35**(9), 1065–1073 (2004)

Pantelakis, S.G., Kyriakakis, E.C.H., Papanikos, P.: Non-destructive fatigue damage characterization of laminated thermosetting fibrous composites. Fatigue Fract. Eng. Mater. Struct. **24**(10), 651–662 (2001)

Paquette, J.W., Kim, K.J., Nam, J., Tak, Y.S.: An equivalent circuit model for ionic polymer-metal composites and their performance improvement by a clay-based polymer nano-composite technique. J. Intell. Mater. Syst. Struct. **14**, 633–642 (2003)

Park, J.S., Park, J.W., Ruckenstein, E.: On the viscoelastic properties of poly(vinyl alcohol) and chemically crosslinked poly(vinyl alcohol). J. Appl. Polymer Sci. **82**(7), 1816–1823 (2001)

Parvizi, A., Bailey, J.E.: On multiple transverse cracking in glass fibre epoxy cross-ply laminates. J. Mater. Sci. **13**, 2131–2136 (1978)

Philippidis, T.P., Assimakopoulou, T.T.: Strength degradation due to fatigue-induced matrix cracking in FRP composites: an acoustic emission predictive model. Compos. Sci. Technol. **68**(15–16), 3272–3277 (2008)

Qin, H.H., Mather, P.T.: Combined one-way and two-way shape memory in a glass-forming nematic network. Macromolecules **42**(1), 273–280 (2009)

Raguse, B., Müller, K.-H., Wieczorek, L.: Nanoparticles actuators. Adv. Mater. **15**(11), 922–926 (2003)

Riemenschneider, J., Opitz, S., Sinapius, M., Monner, H.P.: Modeling of carbon nanotube actuators: Part I – Modeling and electrical properties. J. Intell. Mater. Syst. Struct. **20**(2), 245–250 (2009a)

Riemenschneider, J., Opitz, S., Sinapius, M., Monner, H.P.: Modeling of carbon nanotube actuators: Part II – Mechanical properties, electro mechanical coupling and validation of the model. J. Intell. Mater. Syst. Struct. **20**(3), 253–263 (2009b)

Ryan, K.P., et al.: Carbon nanotubes for reinforcement of plastics. A case study with poly(vinyl alcohol). Compos. Sci. Technol. **67**(7–8), 1640–1649 (2007)

Schulte, K., Baron, C.: Load and failure analyses of CFRP laminates by means of electrical resistivity measurements. Compos. Sci. Technol. **36**(1), 63–76 (1989)

Seo, D.C., Lee, J.J.: Damage detection of CFRP laminates using electrical resistance measurement and neural network. Compos. Struct. **47**, 525–530 (1999)

Shaffer, M., Windle, A.H.: Fabrication and characterization of carbon nanotube/poly(vinyl alcohol) composites. Adv. Mater. **11**(11), 937–941 (1999)

Shahinpoor, M., Kim, K.J.: The effect of surface-electrode resistance on the performance of ionic polymer–metal composite (IPMC) artificial muscles. Smart Mater. Struct. **9**, 543–551 (2000)

Shahinpoor, M., Kim, K.J.: Ionic polymer–metal composites: I. Fundamentals. Smart Mater. Struct. **10**, 819–833 (2001)

Shahinpoor, M., Kim, K.J.: Ionic polymer–metal composites: III. Modeling and simulation as biomimetic sensors, actuators, transducers, and artificial muscles. Smart Mater. Struct. **13**, 1362–1388 (2004)

Stoney, G.G.: The tension of metallic films deposited by electrolysis. Proc. R. Soc. Lond. **A82**, 172 (1909)

Sun, G.Y., Kurti, J., Kertesz, M., Baughman, R.H.: Dimensional changes as a function of charge injection in single-walled carbon nanotubes. J. Am. Chem. Soc. **124**(50), 15076–15080 (2002)

Tadokoro, S., Takamori, T., Oguro, K.: Modeling IPMC for design of actuation mechanisms. In: Bar-Cohen, Y. (ed.) Electroactive Polymer (EAP). SPIE Press, Washington, DC (2001)

Takeuchi, I., Asaka, K., Kiyohara, K., Sugino, T., Terasawa, N., Mukai, K., Fukushima, T., Aida, T.: Electromechanical behavior of fully plastic actuators based on bucky gel containing various internal ionic liquids. Electrochim. Acta **54**, 1762–1768 (2009)

Todoroki, A., Tanaka, M.: Delamination identification of cross-ply graphite/epoxy composite beams using electric resistance change method. Compos. Sci. Technol. **62**, 629–639 (2002)

Todoroki, A., Kobayashi, H., Matuura, K.: Application of electric potential method to smart composite structures for detecting delamination. JSME Int. J. **38**(4), 524–530 (1995)

Todoroki, A., Tanaka, M., Shimamura, Y., Kobayashi, H.: Effects with a matrix crack on monitoring by electrical resistance method. Adv. Compos. Mater. **13**(2), 107–120 (2004)

Todoroki, A., Omagari, K., Shimamura, Y., Kobayashi, H.: Matrix crack detection of CFRP using electrical resistance change with integrated surface probes. Compos. Sci. Technol. **66**, 1539–1545 (2006)

Vigolo, B., Penicaud, A., Coulon, C., Sauder, C., Pailler, R., Journet, C., Bernier, P., Poulin, P.: Macroscopic fibres and ribbons of oriented carbon nanotubes. Science **290**(5495), 1331–1334 (2000)

Vigolo, B., Poulin, P., Lucas, M., Launois, P., Bernier, P.: Improved structure and properties of single-wall carbon nanotube spun fibres. Appl. Phys. Lett. **81**(7), 1210–1212 (2002)

Viry, L., Mercader, C., Miaudet, P., Zakri, C., Derre, A., Kuhn, A., Maugey, M., Poulin, P.: Nanotube fibres for electromechanical and shape memory actuators. J. Mater. Chem. **20**(17), 3487–3495 (2010)

Wang, X., Park, S.Y., Yoon, K.H., Lyoo, W.S., Min, B.G.: The effect of multi-walled carbon nanotubes on the molecular orientation of poly(vinyl alcohol) in drawn composite films. Fibres Polym. **7**(4), 323–327 (2006)

Yun, Y., Shanov, V., Tu, Y., Schulz, M.J., Yarmolenko, S., Neralla, S., Sankar, J., Subramaniam, S.: A multi-wall carbon nanotube tower electrochemical actuator. Nano Lett. **6**(4), 689–693 (2006)

Zhang, X., Liu, T., Sreekumar, T.V., Kumar, S., Moore, V.C., Hauge, R.H., Smalley, R.E.: Poly (vinyl alcohol)/SWNT composite film. Nano Lett. **3**(9), 1285–1288 (2003)

Zhang, X., Liu, T., Sreekumar, T.V., Kumar, S., Hu, X., Smith, K.: Gel spinning of PVA/SWNT composite fibre. Polymer **45**(26), 8801–8807 (2004)

Chapter 4
Mechanical Dispersion Methods for Carbon Nanotubes in Aerospace Composite Matrix Systems

Sergiy Grishchuk and Ralf Schledjewski

Contents

Abstract Utilizing the reinforcing effects CNTs might bring requires techniques resulting in separated and uniformly dispersed CNTs in the matrix resin system. Mechanical dispersion methods are available in various types. A review of the literature of these dispersion techniques and the accompanying dispersion mechanisms is presented. Starting with a general overview of problems that occur by modifying matrix materials with CNTs and a short description of pre-dispersion and additive assisted dispersion, the main focus is on mixing and milling techniques. Furthermore, the rapid expansion of supercritical suspensions and ultrasonication are discussed. Finally, possible ways of combining dispersive methods and controlling the dispersion quality are presented.

S. Grishchuk
Institut für Verbundwerkstoffe GmbH, Kaiserslautern, Germany

R. Schledjewski (✉)
Chair of Processing of Composites, Montanuniversität Leoben, Otto Gloeckel Str. 2, A-8700 Leoben, Austria
e-mail: Ralf.Schledjewski@unileoben.ac.at

A.S. Paipetis and V. Kostopoulos (eds.), *Carbon Nanotube Enhanced Aerospace Composite Materials*, Solid Mechanics and Its Applications 188,
DOI 10.1007/978-94-007-4246-8_4,

Keywords Carbon nanotubes • Dispersion • Mixing • Milling • Ultrasonication • Rapid Expansion of Supercritical Suspension (RESS)

4.1 Introduction

Polymer matrix composites (PMCs) are commonly composed of rigid reinforcements (e.g., fibres, fillers, etc.) embedded in a polymer matrix. The polymer can range from low molecular weight monomers (oligomers) used in thermosetting materials to high molecular weight polymers for thermoplastic applications. Different monomers (vinyl, acrylic etc.), thermosetting resins (epoxy, polyester, phenolics, bismaleimide, polyimide resins etc.), elastomers (different kinds of rubbers, polyurethanes, etc.) and amorphous and semi-crystalline thermoplastic materials are used for PMCs (U.S. Department of Defense 2002). Consequently, composite processing is dependent upon the polymer matrix. The chemical composition and physical properties of matrix materials may affect fundamentally the processing, fabrication and ultimate properties of composite materials. The broad range of polymer properties gives rise to various forms of composite materials, including prepreg (neat-resin-impregnated fabrics to form a tacky solid fabric), neat-resins (that can infuse a fibre-fabric during moulding), and filled resins (Gillham 1983). PMCs are widely used in different load-bearing structures in both commercial and military applications such as packaging, house furnishing, sporting goods, recreational products, high-strength high-durability adhesives, transportation (including automotive and aerospace applications), construction of ship hulls and surface ship superstructures, wind turbines, helicopter rotor blades, high-performance airframes, multiservice munitions etc. due to their good static mechanical properties (e.g. high specific strength and stiffness), corrosion resistance, heat resistance, solvent resistance etc. (U.S. Department of Defense 2002; Gillham 1983; Sands et al. 2001). The majority of engineering composite materials consist of a thermosetting epoxy matrix reinforced by continuous fibres. Different kinds of fibres (carbon, aramid, glass, boron, alumina, silicon carbide, quartz, natural fibres etc.) in the form of individual fibres (chopped or continuous), mats or 2D-3D fabrics can be used as reinforcement in polymer composites. However, the glass and carbon fibres are mostly widely used. In general, epoxies are known for their excellent adhesion, chemical and heat resistance, good-to-excellent mechanical properties (high modulus and failure strength, low creep etc.) and very good electrical insulating properties (U.S. Department of Defense 2002). However, similar to other thermosetting resins they are relatively brittle and have poor crack resistance (Garg and Mai 1988; Hwang et al. 1989; Salamone 1996). Safe operation of structural composite materials requires that, in addition to their good static mechanical and other properties, they need to have high fracture toughness and good fatigue-resistance. Additionally, great interest in industrial applications is concerned also with development of novel fire (Lu and Hamerton 2002; Kandola et al. 2003; Hshieh and Beeson 1997; Sorathia et al. 2001; Bourbigot and Le Bras 1996; Perez et al. 2006) and blast (Hebert et al. 2008;

Tekalur et al. 2008a, b, c; O'Toole et al. 2006) resistant composite materials. Therefore, most thermosetting resins are often used with reinforcing fillers and/or fire retardants to produce composite materials for different applications ranging from swimming pool liners and automotive components to corrosion resistant tanks and aircraft fuselages. Unlike traditional filled polymer systems, nanocomposites require relatively low nanofillers loadings for achievement of significant property improvements (Le Bras and Bourbigot 1996; Gilman et al. 1999; Si et al. 2007; Manjunatha et al. 2010; Thostenson et al. 2005; Hussain et al. 2006; Tjong 2006; Vlasveld et al. 2007; Gao et al. 2007; Mahmoodian et al. 2010; Sandler et al. 2003). The unique properties of nanoparticles and the possibility of combining them with conventional fibre reinforcements have a high potential to improve the material properties of polymers (Sandler et al. 2003; Njuguna et al. 2008; Bauer et al. 2008; Chen and Tolle 2004; Becker and Simon 2005; Gupta et al. 2007; Yasmin et al. 2006; Young and Eichhorn 2007; Gojny et al. 2005a). Nanocomposites usually exhibit light-weight, good dimensional stability, enhanced heat and flame resistance, improvements in strength and modulus as well as barrier properties with far less loading than conventional composite counterparts, however these properties depend on several factors such as type of nanoparticle, surface treatments, polymer matrix, synthesis methods, and polymer nanocomposites morphology. The nanofillers such as ceramic nanofillers, nanoclays, carbon nanotubes (CNTs), etc. are promising for a variety of new nanocomposites, adhesives, coatings, and other materials with specific improved properties. This is a reason why development of nano-reinforced polymer composites is one of the most promising approaches in the field of future engineering applications (Fiedler et al. 2006; Manocha et al. 2006; Hanemann and Szabo 2010; Gacitua et al. 2005). Nowadays composite materials are intended to be widely used as an alternative of aluminium structure in aircraft and aerospace applications. Previously, the composite materials were mostly used in secondary structures of aircraft such as fairings, small doors and control surfaces. However, with growing up of the technology, the use of composite materials for primary structures such as wings and fuselage has increased (Nurhaniza et al. 2010). The most extensively used fibres in aerospace application are glass, carbon and aramid (Mangalgiri 1999; Krishnadas Nair 1994; Njuguna and Pielichowski 2004a, b; Njuguna and Pielichowski 2003). Many structural applications (especially aerospace structural composites) require a significant reduction in weight for energy and/or environmental reasons. Carbon fibres, having low density but high mechanical performances, are the best candidate for this purpose and are widely used in structural composites for aerospace applications (Njuguna and Pielichowski 2003, 2004a, b; Loidl et al. 2005; Huang 2009; Budnitskii et al. 1993). Carbon fibre reinforced polymer (CFRP) composites are characterized by a combination of important material properties such as high specific strength and stiffness, etc. with light weight, which make their use especially attractive for aircraft and aerospace applications (Morioka et al. 2001; Williams et al. 2007; Firouzmanesh and Azar 2003; Abusafieh et al. 2001; Quilter 2001). From this point of view, development of light-weight CFRP nanocomposites has a high potential by use of low-density nanofillers as the matrix modifiers. Therefore, nanofillers such as carbon nanotubes

(CNTs) are promising candidates to produce a new class of nanocomposite materials (especially, CFRP composites) that are stronger than conventional composites for use in aircraft (Njuguna and Pielichowski 2003, 2004a, b; Loidl et al. 2005; Lozano and Barrera 2001; Lebel et al. 2009; Inam et al. 2010; Grimmer and Dharan 2010; Kim et al. 2007; Zhou et al. 2008; Godara et al. 2009; Thostenson and Chou 2002–2006).

Since Tennent and Iijima discovered multi-walled nanotubes (MWNTs) in 1987 (Tennent 1987) and 1991 (Iijima 1991), respectively, and later (1993) Bethune et al. discovered single-walled nanotubes (SWNTs) (Bethune et al. 1993) it has been shown in many studies that CNTs are materials with extraordinary electrical, thermal and mechanical (high flexibility, strength and stiffness) properties. In addition, CNTs are characterised by high aspect ratios and magnetic properties. However, mechanical properties of carbon nanotubes are highly dependent upon the atomic structure of nanotubes and the number of walls. In fact, CNTs are much stronger than steel. In addition, their electrical conductivity is better than copper's electrical conductivity, and their thermal conductivity is higher than that of diamond (Li et al. 2000; Treacy et al. 1996; Yu et al. 2000; Salvetat et al. 1999; Ruoff and Lorents 1995; Guo and Guo 2003; Komarov and Mironov 2004; Chen and Huang 2006; Kis and Zettl 2008; Coleman et al. 2006a; Prylutskyy et al. 2000; Popov 2004; Dresselhaus et al. 2001; Ajayan 1999; Thostenson et al. 2001). Recent investigations have been focused on a wide number of applications for CNTs. Some of them include nanoelectronics, biomedical, field emission devices, composites, chemical sensors, biosensors, supporting substrates for heterogeneous catalysis, etc. Some of the applications that take advantage of the electrical properties of CNTs/polymer nanocomposites are super capacitors which are considered for applications such as intelligent structures for aerospace, electric vehicles, fuel cells, uninterruptible power supplies, shielding of electromagnetic interferences, photovoltaic devices such as more efficient solar cells, sensors, field-effect transistors and diodes that improve mechanical stability and conductivity of the devices (Dresselhaus et al. 2001; Ajayan 1999; Thostenson et al. 2001; Crawley 1994; Tahhan et al. 2003; Spinks and Wallace 2002; Park et al. 2008a; Yuan et al. 2008; Courty et al. 2003; Derycke et al. 2002; Thang et al. 2009; Ramasubramaniam et al. 2003; Hueso et al. 2007; Kuemmeth et al. 2008; Yu et al. 2009a; Toth et al. 2009; Jeong et al. 2010; Shui and Chung 1996; Baxendale 2006; Gou 2006; Tjong 2010; Lin et al. 2004; Yang et al. 2007; Lim et al. 2009; Wohlstadter et al. 2003; Kim et al. 2008; Foldvari and Bagonluri 2008a, b; Kumar and Ramaprabhu 2006; Zheng et al. 2007; Girishkumar et al. 2006; Toebes et al. 2004; Mu et al. 2005; Kim et al. 2009; Oliveira and Zarbin 2008; Niu et al. 2010; Baibarac and Gomez-Romero 2006; Kongkanand et al. 2006; Endo et al. 2004, 2008; Baughman et al. 2002; Ajayan and Zhou 2001; Capek 2009; Iyuke and Mahalik 2006). CNTs have been largely considered as prospective filler material for future polymer nanocomposites to enhance thermal stability and important material properties such as strength, conductivity, stiffness, tribological performances, and electromagnetic interference shielding of the polymer matrix (Thomassin et al. 2007; Moniruzzaman and Winey 2006; Breuer and Sundararaj 2004; Vail et al. 2009; Thostenson and Chou 2003, 2006; Lee et al. 2009; Chen et al. 2008; Tai et al. 2004; Gojny et al. 2004, 2005b;

Ogasawara et al. 2004; Safadi et al. 2002; Guo et al. 2005; Liu et al. 2004; Zhang et al. 2004a, 2006; Allaoui et al. 2002; Sumfleth et al. 2010; Kempel and Schlarb 2008; Martin et al. 2004a; Battisti et al. 2010; Thostenson et al. 2009; Krause et al. 2010; Nogales et al. 2004; Jin et al. 2001; Velasco-Santos et al. 2003; Nan et al. 2003; Biercuk et al. 2002; De Rosa et al. 2010; Hudziak et al. 2010; Valentini et al. 2004; Bokobza and Kolodziej 2006; Gauthier et al. 2005; Kalgaonkar and Jog 2008; Kim et al. 2010a). All these properties make CNTs an ideal reinforcement for nanocomposite matrix materials for aircraft components and other high demand, high performance applications. Note that for modern and efficient structural design of aerospace composites the materials used must be exploited to their maximum potential.

4.2 Problems Caused by Modifying Matrix Materials with CNTs

In order to reach the maximum benefits of CNTs in production of aerospace composite matrix systems it is very important to know which factors can influence their final material properties. The main background for nanocomposite production methods is a knowledge of nature, geometry, production processes and main characteristics of CNTs. In general, CNTs can be viewed as hollow cylinders formed by rolling graphite sheets, and their properties are defined by their atomic arrangement, diameter, length, and morphology. The atomic arrangement of a CNT can be classified as armchair, zig-zag or chiral shape depending on how the graphite walls of the CNT are rolled. In addition, depending on their atomic structure, CNTs can be metallic, semiconducting or semimetallic. CNTs can be classified into single-walled nanotubes (SWNTs), multi-walled nanotubes (MWNTs) and carbon nanofibres (CNFs). They usually have average diameter less than 200 nm. SWNTs are 1–2 nm in diameter and their length is in the micrometer scale. MWNTs basically consist of a group of coaxial SWNTs where each individual tube can have different chirality. The MWNT inner diameter can be of 2–10 nm while the exterior diameter can be of 20–70 nm. A typical length of a MWNT is about 5–50 μm. CNFs have a larger diameter ranging from 50 to 200 nm. The length of CNFs can be anywhere from several micrometers to several tens of centimetres. Compared to SWNTs and MWNTs, CNFs can be produced in higher volumes and at a lower cost. However, they usually contain much more defects than MWNTs. In addition, the strength of CNTs can be much greater than that of CNFs (Komarov and Mironov 2004; Chen and Huang 2006; Kis and Zettl 2008; Coleman et al. 2006a; Prylutskyy et al. 2000; Popov 2004; Dresselhaus et al. 2001; Ajayan 1999; Thostenson et al. 2001; Ajayan and Iijima 1992; Mordkovich 2003; Szleifer and Yerushalmi-Rozen 2005; Wu et al. 2008a; Hassanien et al. 1998).

In general, CNTs can be synthesized by several techniques. The main techniques are arc discharge, laser ablation, and chemical vapour deposition (CVD), etc. (Ando et al. 2002; Journet et al. 1997; Karthikeyan et al. 2009; Journet and Bernier 1998;

Zhang et al. 1998, Zeng et al. 2002; Varadan and Xie 2002; Kong et al. 1998; Eklund et al. 2002). Each technique has its specific merits and inevitable weaknesses. For the first two methods carbon vapour is produced by vaporisation of an electrode or target doped with a small amount of metallic catalyst particles. Both SWNTs and MWNTs could be produced by laser ablation method. The arc discharge technique results generally in MWNTs production. However, in some cases SWNTs can be found as well. The production yield of CNTs from both arc discharge and laser ablation is limited. However, the laser ablation can provide better control of the evaporation process and results in a higher purity of CNTs compared to arc discharge (Hornbostel et al. 2006). Here it should be noted that solar energy can be used for vaporisation of graphite in production of CNTs (Alvarez et al. 2000, 2001; Laplaze et al. 1998). However, the yield is usually low. The catalytic growth of CNTs by CVD is an effective route to produce larger numbers of nanotubes. CVD is a well established industrial process and the CNT production is easy to scale up. The carbon vapour source is derived from the chemical vapor decomposition of various hydrocarbon gases on transition metal catalyst. Use of alcohols as carbon source in CVD process has been also reported (Singjai et al. 2007; Nasibulin et al. 2006). Depending on the activation sources for the chemical reactions, the deposition process can be categorized into thermally activated, laser-assisted and plasma-assisted CVD. It should be noted that CNTs produced via thermally activated CVD have random and tangled structures of uncontrolled length and diameter. Formation of SWNTs or MWNTs via CVD route is governed by the size of catalyst particle, growth temperature and hydrocarbon source. SWNTs are preferably formed at 900–1,200 °C from chemically stable in this temperature range hydrocarbons (e.g. CO, CH_4) if the catalyst particle size is a few nanometres, whereas catalyst particles a few tens of nanometres and temperature range of 600–900 °C as well as use of unstable at higher temperatures hydrocarbons (e.g. acetylene, benzene, etc.) favour the formation of MWNTs (Tjong 2006; Alvarez et al. 2001; Harutyunyan et al. 2002; Ago et al. 2005). In addition, low temperature CVD (<400 °C) of high efficiency has been developed using oxidative dehydrogenation reaction of acetylene with CO_2 (Magrez et al. 2010). Produced by CVD technique CNTs are usually of high purity and can be relatively long. The lengths up to millimetre-scale are reported (Li et al. 2008; Pan et al. 1998).

In addition to above described techniques the production of CNTs via ball milling, diffusion flame synthesis, electrolysis, low temperature pyrolysis, homogeneous sonochemistry and catalyst arrays were developed (Maldonado and Stevenson 2004; Pierard et al. 2001; Chen et al. 1999; Vander Wal et al. 2000; Unrau et al. 2007, 2010; Hsu et al. 1996; Li et al. 1997; Vohs et al. 2004; Dai 2005; Katoh et al. 1999; Park et al. 2009). Ball milling is a simple method for producing CNTs. Graphite powder placed in a stainless steel container containing four hardened steel balls is milled for a long time at room temperature in argon atmosphere. Milled in such a manner, the powder is than annealed under inert gas flow at elevated temperatures (about 1,400 °C). The ball milling process forms nanotube nuclei, and annealing under purging initiates nanotube growth (Pierard et al. 2001;

Chen et al. 1999). However, the mechanism of this process is unknown. MWNTs are usually formed by this technique. In diffusion flame synthesis combustion of hydrocarbon gas is the source of both carbon and energy. Transition metal oxides are generally used as catalysts for growing CNTs in high-temperature diffusion flame furnaces (Vander Wal et al. 2000; Unrau et al. 2007, 2010). Both MWNTs and SWNTs can be produced using this method. This technique is capable of scale-up for high-volume industrial production of CNTs. Erosion of graphite rod cathodes under high current by electrolysis technique results in dispersion of nanoparticles in molten lithium chloride anode. Spiral and curled CNTs are usually extracted into a toluene phase from the anode (Hsu et al. 1996). Nano-sized silicon carbonitride has been used as the carbon source for production of capped MWNTs under pyrolysis at ~1,700 °C in a nitrogen-filled furnace. The tubes are formed on the nanopowder surface, therefore, a high amount of silicone carbonitride is found to be present in the CNT-hollows (Li et al. 1997). In order to reduce the growth temperature, the rational low temperature pyrolysis method has been proposed for CNT production using carbon halides instead of hydrocarbons as carbon source (Vohs et al. 2004; Dai 2005). The production of CNTs via homogeneous sonochemistry process takes place on the liquid-solid interface (e.g. liquid benzene derivatives/metal or salt particles) at hot spots created by sonication (temperatures more than 5,000 K could be reached). High crystalline nanotubes could be produced through such a technique (Katoh et al. 1999; Park et al. 2009). However, organic liquids can decompose and/or polymerise during this process. The production of MWNTS with uniform thickness, even morphology and good crystallinity, using as catalyst an array of porous anodic aluminium oxide template, and in situ production of SWNTs bundles and MWNTs by reducing a composite metal oxide powder using catalyst array technique are reported (Jeong et al. 2002; Orikasa et al. 2006). The production of MWNTs via carbonization of polymers under heat treatment (~900 °C) in inert atmosphere is possible as well (Wu et al. 2008a). Micro-emulsion core-shell poly (methyl methacrylate)-co-polyacrylonitrile (PMMA-PAN) copolymer particles blended and stretched with PMMA matrix into fibres are usually used for this purpose. The PMMA completely consumes and PAN carbonizes to MWNTs during this procedure. It is clear that initial size of stretched microparticles influences the final length and diameter of CNTs. Another process based on promotion of polymerization by chemically treated polymers at ~400 °C in an air-filled furnace for several hours allows production of MWNTs (Cho et al. 1996). Some CNTs with other morphologies than regular CNTs (e.g. bamboo-shape, Y-junctions, sea urchins, flowers, coils, etc.) have been prepared as well (Jang et al. 2002; Bredeau et al. 2008; Guojun et al. 2007). Catalyst supported metal oxides, zeolites and molecular-sieves can be also used for production of CNTs (Guojun et al. 2007; Takenaka et al. 2004; Ziebro et al. 2010; Fonseca et al. 1998; Rana et al. 2001). Development of "hetero-atomic" nanotubes such as boron-nitrogen (B-N), boron-carbon-nitrogen (B-C-N), and boron-carbon (B-C) nanotubes also has great interest in the scientific community (Sen et al. 1998; Kongsted et al. 2001; Stephan et al. 1998; Liu et al. 2008a). Note that incorporation of boron in CNTs usually increases their length and quality. So, it is obvious how different structure, morphology, geometry, etc. of CNTs can be produced by different techniques. The production methods for CNTs often result in products that have different diameters and lengths, as well as different levels of entanglements. For example, SWNTs usually associate in bundles and MWNTs are generally entangled in the form of curved agglomerates. Therefore, the dispersion parameters for CNTs can strongly depend on the production process used. In addition, CNTs usually contain different amounts of undesirable impurities of amorphous carbon, graphite particles, metal catalysts, etc. Therefore, purification of CNTs is needed prior to blending with polymers. Several purification methods such as filtration, chromatography, centrifugation, oxidation, chemical functionalisation and magnetic separation are reported as being efficient for this purpose (Ebbesen et al. 1994; Bonard et al. 1997; Yudasaka et al. 2000; Lian et al. 2004; Banerjee and Wong 2002; Duesberg et al. 1998; Bandow et al. 1998; Wiltshire et al. 2005; Liu et al. 2008b; Chen et al. 2010). Several other methods such as electrostatic plasma treatment and electric field manipulations have also been used for separation of individual CNTs from the larger agglomerates and impurities (Chen et al. 2010). Note that separation and assembly methods are often used in order to produce homogeneous individual dispersions/solutions of CNTs. However, stronger aggregation, local curvature of CNTs and formation of a higher quantity of defects or open-ended CNT-bundles usually support the purification processes.

Note that all the factors we have mentioned (nature, geometry, production and purification history, etc. of CNTs) can strongly influence the dispersability, quality of CNT dispersion and thus final material properties of nanocomposites. The background knowledge and basic investigation of all factors affecting processing and quality of CNT reinforced composite matrix systems allows us to understand better how they process. Therefore, many scientists are estimating the effective dispersion energy/force needed for optimal dispersion of different CNTs in various media (Barber et al. 2004; Nyden and Stoliarov 2008). This information is a very useful understanding of possible dispersion mechanism, as well as for selection or development of optimal dispersion methods and dispersion conditions producing CNT containing nanocomposites.

The production methods for nanocomposites attempt to overcome the challenge of transferring the exceptional properties of CNTs to a polymer material with an appropriate processing procedure. In order to utilise CNTs in different applications it is essential to disperse them both in the aggregated state and the nanoscale (in individual state). Many of the expected CNTs properties can be accomplished only if nanotubes are well dispersed or, even better, aligned in the polymer. The most used methods for orientation of CNTs in polymer matrix are shear flows, elongation flows, electric and magnetic fields (Paradise and Goswani 2006; Xiao and Zhang 2005; Xie et al. 2005; Zhu et al. 2006; Boccaccini et al. 2006). Improved dispersion and alignment of CNTs in some liquid crystalline polymers have been reported as well (Ji et al. 2010). It should be noted here that several important applications of CNT nanocomposites need formation of three-dimensional networks of nanotubes to improve the transport properties (electrical and thermal conductivities) of dispersion media, which can not be achieved by alignment of CNTs in one direction. However, when developing a manufacturing process for production of nanocomposites, dispersion and alignment of nanotubes in polymers are the two primary obstacles that are encountered.

The dispersion of CNTs is not a simple process since CNTs tend to agglomerate to each other due to the Van der Waal force attractions that exist between the tubes as a result of their significant surface areas and high aspect ratios (Cadek et al. 2004; Zhao et al. 1999; Thess et al. 1996). Their very stable chemical characteristics and lack of functional sites on the surface also complicate the dispersion issue. Moreover, the length of CNTs can range up to several millimetres, which is undesirable for efficient dispersion: the longer the nanotubes, the stronger are the interactions and entanglements between them. However, longer nanotubes could lead to better mechanical properties of composites even if dispersion is not good as desired. Therefore, many scientists are looking for a compromise ratio between quality of dispersion and length of nanotubes for improvements of optimal properties (Wang et al. 2007a; Usrey et al. 2009; Sinnott et al. 2003; Prolongo et al. 2008; Song and Youn 2005; Mohlala and Ray 2008; Koerner et al. 2005; Ma et al. 2009). The final length of dispersed CNTs can strongly influence the stability of dispersion as well. Another important factor, which influences the final material properties of nanocomposites by integrating high-strength nanotubes into polymers, is adhesion between CNTs and the polymer matrix. Knowledge of how nanotubes adhere to each other and to the dispersion media could lead to a better understanding of how to disperse them uniformly. Similar to the conventional fibre-reinforced composites, a load transfer across the CNT/matrix interface is required in order to increase the mechanical properties of reinforced nanocomposites (Ma et al. 2009; Du et al. 2007; Bal and Samal 2007; Khare and Bose 2005; Suhr and Koratkar 2008; Fraczek and Blazewicz 2009). Therefore, scientists pay great attention to the investigation of CNT aggregates and CNT/polymer interface (Nanda et al. 2008; Shaffer and Kinloch 2004; Meguid and Sun 2004; Xu et al. 2002; Andrews and Weisenberger 2004; Coleman et al. 2006b; Wagner 2002; Ma et al. 2010; Huxtable et al. 2003; Collison et al. 2010; Schadler et al. 1998; Qian et al. 2000; Zhou et al. 2004; Jiang et al. 2007). The interfacial load transfer can be governed by three mechanisms: Van der Waals forces, mechanical interlocking and covalent bonding. The Van der Waals forces between CNTs and matrix is the most common mechanism for interfacial load transfer. However, such forces are usually weak and CNTs do not bond well to the matrix, which results in relatively low load transfer efficiency (Jiang et al. 2007; Lau and Shi 2002; Li and Chou 2003; Odegard et al. 2003; Wagner et al. 1998; Desai and Haque 2005). Mechanical interlocking usually results from defects around the interface, therefore, it hardly ever occurs in CNTs because of their near to defect-free structure (Jiang et al. 2007). The covalent bonding mechanism requires functionalisation of the CNT/matrix interface, which can make the processing more difficult and, moreover, introduce defects to the CNTs, which results in worse mechanical properties of nanotubes (Jiang et al. 2007). Due to covalent functionalisation a reduction in thermal and electric properties of CNTs is often observed. Hence, noncovalent treatment of CNTs offers their functionalisation without affecting the electronic network of the tubes. The action mechanism of noncovalent treatments is usually based on the re-distribution of Van der Waals forces or on the π-stocking interactions (Sahoo et al. 2010; Hu et al. 2009).

Strongly entangled and associated CNTs need very high energy to be separated and uniformly dispersed in liquids, resins and melts. Usually, bad dispersion of CNTs in nanocomposites leads to only modest or poor mechanical properties improvement. In addition, the sedimentation of CNT-agglomerates or out-filtration of badly dispersed CNTs can occur by production of fibre reinforced polymer composites. Therefore, research has considered different methods to achieve good dispersion of CNTs in polymer matrices. However, there is one additional limitation to the use of CNTs as reinforcement for structural composite matrix materials: the viscosity of dispersion media strongly increases due to the significant area and high aspect ratio of CNTs. A better quality of dispersion results usually in higher viscosity, which can limit the processing procedure of structural composites (Pötschke et al. 2002; Chapartegui et al. 2010; Huang et al. 2006; Acevedo-Rullan et al. 2009; Kalyon et al. 2006; Abbasi et al. 2009; Du et al. 2004; Rahatekar et al. 2006). Note that very high requirements of safety, quality and reproducibility of structural nanocomposites for aerospace applications are necessary. Therefore, high quality homogeneous and reproducible dispersions with low concentrations of CNTs are preferred for production of aerospace composite matrix systems.

4.3 Mechanical Dispersion Methods and Dispersion Mechanisms

4.3.1 Pre-dispersion, Additives Assisted Dispersions, Doping with Nanoparticles

There are various production techniques that are being investigated for the production of nanocomposites. The ideal case is to obtain a stable dispersion of independent, separated nanotubes that further can be manipulated in order to have the preferred orientations of CNTs (one-, two- or three-dimensional) for production of fibres, flat sheets or bulk objects. There are two main approaches to nanotube dispersion: mechanical (physical) and chemical methods. In this chapter, chemical methods are mainly techniques affecting the chemical structure of CNTs (e.g. functionalisation, covalent bonding, and incorporation of other atoms in the carbon lattice) and will not be discussed here.

However, it should be noted that chemical methods are usually assisted by mechanical dispersion techniques. A physical dispersion route generally includes ultrasonication, high-shear and high-impact mixing, etc. (Hilding et al. 2003). On the other hand, physical dispersion methods are often supported by open-end or side-wall functionalisation of CNTs due to their breakage. Both physical and chemical approaches have been adapted to reduce the length of CNTs to certain extents that are suitable for blending (Wang et al. 2003).

The main task of this chapter is to present actual state of the art of mechanical (physical) dispersion methods for carbon nanotubes in aerospace composite matrix systems with and without other additives, such as solvents, emulsions, solutions of surfactants and salts, other nanoparticles, etc., which make the dispersion easier or protect CNTs from strong damages.

It should be noted that main mechanical dispersion methods are usually adapted to the type of polymer matrix used (thermoplastic, elastomeric, thermosetting) and thus to the respective processing methods.

Usual production techniques for thermoplastic materials are melt mixing, melt compounding, and melt spinning, etc. These techniques may be used along with conventional manufacturing processes such as extrusion, injection molding, and internal mixing. It was shown that use of the melt compounding process for CNTs reinforced thermoplasts has different success resulting in nanocomposites with poor dispersion and mechanical properties (e.g. Bhattacharyya et al. 2003) as well as in increase of mechanical properties compare to the neat polymer (e.g. Anoop et al. 2007). The melt processing can be used for thermosetting resins as well. However, such technique is worse in the case of thermoset nanomaterials production because of higher instability of their melt stability in the required processing windows. Therefore, blending processes are generally favoured for processing thermoplastic and elastomeric materials. The production of CNT modified nanocomposites via latex technique is also widely used for thermoplastic as well as elastomeric materials (Dalmas et al. 2005; Regev et al. 2004; Woo et al. 2009). The elastomeric nanomaterials with improved properties can be also obtained by direct mixing in blender or roll mills (Yue et al. 2006; Valentini et al. 2003; Lopez-Manchado et al. 2004; Xu et al. 2008). The use of high shear mixer for production of elastomeric polyurethane nanocomposites from reactive mixtures is also known (Chen et al. 2007). As it was mentioned in the "Introduction", epoxy resins are the favourite matrix materials for aerospace applications. Therefore, many scientists are working on the development of CNT/epoxy nanocomposites in order to find the optimum processing for improvement of properties as well as for the final applications. The most used mechanical dispersion techniques for such systems are shearing, milling, and ultrasonication. Many positive results have been reported concerning both optimised processing and improved material properties (e.g. Lau et al. 2005; Sandler et al. 1999; Leer et al. 2006).

Many useful additives were found to have an assisting effect on the dispersion of CNTs by producing nanocomposites. One of them is solution casting that consists in using a solvent to disperse CNTs and to blend this dispersion with a resins, polymers or their solutions, then evaporate the solvent (e.g. Ramamurthy et al. 2007). The advantages of this method are that it is simple, and the form of the produced nanocomposite depends only on the mould used. One important drawback of this technique is the economical aspect due to additional costs needed for the solvents and their evaporation. The studies of the interaction between SWNTs and different solvents are critical to understanding of CNT-solubilisation mechanism and allow optimising the processing procedures. One of the most important groups that interact directly via π-π-stocking forces and CH-π interactions with CNTs are aromatics. Note that π-π-stocking configurations are stronger than CH-π analogues. It was found that large numbers of aromatic compounds are capable to interact with CNTs, and thus improve their disentanglement. The promotion of CNT dispersion using polyaromatic compounds have been reported as well (Kar et al. 2008; Cheng et al. 2008, 2010; Chang et al. 2010; Inam et al. 2008; Moreno-Castilla 2004; Martin et al. 2004b; Saito et al. 2007; Nakashima et al. 2005; Lee et al. 2008; Jensen et al. 2000; Maeda et al. 2004; Sun et al. 2001; Chen et al. 2001; Backes et al. 2010). The improved degree of dispersion of SWNTs and better dispersion stability, compared to conventional dispersion, have been obtained using such modifications. In addition, non-chemical modification of CNT-surface with silane groups through physical interaction of the tubes with thiol groups of thiolated organosilanes have been reported as an efficient method for the promotion of SWNTs dispersion on the nanobundles level for the further sol-gel applications (Bottini et al. 2006). From several systematic studies of the dispersion of CNTs in different solvents it was found that polar forces and hydrogen bonding are dominant in the solubilisation of CNTs (Cheng et al. 2008). Additionally, it was concluded that a phenyl ring in chlorinated aromatic solvents is not a dominant factor for production of stable CNT dispersions (Cheng et al. 2010). The critical dispersion limits for several solvents were investigated as well. It was demonstrated that this parameter depends strongly on the conditions of the dispersion process (Cheng et al. 2010). This technique is efficient in production of all kinds of nanocomposites: thermosetting, thermoplastic and elastomeric materials as well. Moreover, if a monomer or co-monomer mixture is used as dispersion medium (gas or liquid), an evaporation stage results. The stabilisation of CNT dispersion is performed through the following in situ polymerisation, resulting in polymer/CNTs nanocomposite. This improves the processability and material properties of related nanocomposites (Chen et al. 2009; Datsyuk et al. 2004; Xia et al. 2003; Park et al. 2002; Li et al. 2006; Hasell et al. 2007; Vega et al. 2009). However, covalent bonding of a polymer is often observed by an in-situ polymerisation procedure, especially, if functionalised CNTs are used (Luo et al. 2010; Hu et al. 2007; Du et al. 2009). The stabilisation of CNT dispersions in polymer matrices, obtained by solvent casting procedure, using a coagulation method (precipitation of polymer/CNT solution in non-soluting liquids) is reported as well (Du et al. 2003; Giordano et al. 2007). Several studies have shown the efficiency of such technique toward increasing mechanical properties such as tensile strength and the Young modulus (e.g. Safadi et al. 2002). However, nanotubes also must be dispersed uniformly in a solution before being mixed with the polymer to make composite materials. Therefore, the use of mechanical dispersion methods is necessary in this case.

Another efficient supporting method which improved the separation of CNTs during dispersion using mechanical dispersion techniques, is treatment with alkali metal vapours (Bower et al. 1998), solutions of salts (Sabba and Thomas 2004; Chun et al. 2006; Mordkovich 2000), surfactants (Gong et al. 2000; Rastogi et al. 2008; Vaisman et al. 2006; Strano et al. 2003; Islam et al. 2003; Li et al. 2007; Matarredona et al. 2003; Tang et al. 2010; Blanch et al. 2010; Priya and Byrne 2008; Yu et al. 2009b; Paredes and Burghard 2004; Bystrzejewski et al. 2010; Chen et al. 2005), polymers (Manivannan et al. 2009; O'Connell et al. 2001; El-Hami and Matsushige

2004; Itzhak et al. 2010; Zhang et al. 2008a; Lee et al. 2007; Stürzl et al. 2009; Mottaghitalab et al. 2005; Schaefer et al. 2003; Strano 2006; Zheng et al. 2003; Yan et al. 2008; Wang et al. 2007b; Zou et al. 2008; Hirano et al. 2009), and emulsions (Dalmas et al. 2005; Woo et al. 2009; Xia et al. 2003). The main effect of such additives on the separation of CNTs from each other is in reducing the physical interactions between them and increasing compatibility and adhesion to the matrix materials. The action mechanism of each kind of efficient additives is briefly discussed below.

The use of salt solutions is generally efficient for water-borne systems. However, efficient promotion of CNT dispersion and improving of adhesion between CNTs and a polymer matrix using salts as additives in a melt processing technique is also possible. The action mechanism in this case is based on the salt cation-π-electron clouds of CNTs interactions, whereas organic acid anion interacts with the polymer matrix. Dispersion of CNTs by π-stacking interaction does not induce the degradation and destruction of the CNTs, thus does not influence their intrinsic properties. Disentanglement and dispersion of MWNTs in the individual state without fragmentation was achieved by doping of potassium cation into the MWNTs from the potassium-phenantrene-1,2-dimethoxyethane complex under moderate stirring (400 rotations per minute) at room temperature (Chun et al. 2006). It was found that MWNTs with a diameter of 14 nm (40-layer walls) do not react with cation containing intercalates, while smaller MWNTs (diameter of 5 nm; 4-layer walls) react strongly with intercalate, resulting in nanoflakes, due to the breaking up of the tubes (Mordkovich 2000).

Different surfactants (both neutral and ionic) have similar effects. Although the non-ionic surfactants are more efficient for dispersion of CNTs in water, ionic surfactants are better for dispersion of CNTs in aqueous polymer solutions. Usually surfactants are adsorbed or bonded through cation interaction with a CNT surface and stabilise the nanotubes against the strong Van der Waals interactions between them, hence prevent agglomeration. In addition, stabilisation of dispersion is usually observed (Rastogi et al. 2008; Vaisman et al. 2006). The micelle-like structuration of CNTs could be carried out in this case (Matarredona et al. 2003). However, an actual problem with surfactant induced dispersions is to find an effective method to remove the surfactant from the final product.

A very promising technique, improving dispersability of CNTs, is also their treatment with different polymers. In this way the immobilisation of polymer on the CNT surface is carried out via π-π interactions or adhesive forces and structure of nanotubes is not disturbed. In this case thermodynamically stable dispersions (even individual nanotubes) can be achieved (Lee et al. 2007). Formation of stable suspension of micelle-encapsulated CNTs can be reached in this case (Kang and Taton 2003). However, different polymers have different efficiency towards promotion of CNT separation, dispersion and dispersion stability. For example, Gum Arabic is reported as an excellent stabiliser for aqueous CNT-dispersions (Bandyopadhyaya et al. 2002), and, in contrast, poly(vinyl alcohol) is not efficient (Hilding et al. 2003). The supporting of CNT dispersion and improvement of dispersion stability with biopolymers is reported as well (Zheng et al. 2003; Yan et al. 2008). Use of self-assembling polymers as treatment for CNTs is also one efficient way to disperse them more easily (Dhullipudi et al. 2007). It was found that, by optimisation of the dispersing parameters, further centrifugation procedure and concentration ratios of the supporting polymer and SWNTs, it is possible to obtain near-to-monochiral nanotubes on a single ultrasonication and centrifugation step without any additional treatment (Stürzl et al. 2009). This indicated the fact that polymer wrapping is specific to a certain chiral angle, as to a nanotube diameter. The combination of benefits of both surfactant and polymer treatments of CNTs towards CNT dispersions was attempted using polymer emulsions as dispersive media (e.g. Woo et al. 2009). However, in some cases (depending on a mechanical method of dispersion) the polymers or surfactants can be adsorbed on the aggregated nanotubes instead of being adsorbed on individual nanotubes, and thus prohibiting efficient dispersion on a nanolevel.

A very interesting synergistic effect on the dispersion quality and final material properties was found by combined use of different nanofillers with CNTs (Sumfleth et al. 2009a; Fritzsche et al. 2009; Lorenz et al. 2009; Zhang et al. 2004b; Ning et al. 2003; Sumfleth et al. 2009b; Bhattacharya and Bhowmick 2010a, b; Wang et al. 2009; Peeterbroeck et al. 2004; Lau et al. 2006; He and Tian 2009). For example, combination of CNTs prepared by CVD technique on mesoporous molecular sieve impregnated with $Fe(NO_3)_3$ solution and fumed silica particles significantly enhanced their dispersion and interaction with the polymeric matrix resulting in great improvement of mechanical properties (Zhang et al. 2004b). Combined use of organoclays and MWNTs, processed by direct blending with polymer matrix, resulted in homogeneous dispersions and enhanced thermal, flame retardant and mechanical properties of ternary nanocomposites due to synergistic effect of simultaneous added clays and CNTs (Bhattacharya and Bhowmick 2010b; Wang et al. 2009; Peeterbroeck et al. 2004; Lau et al. 2006). A very interesting and promising technique is dispersion of CNTs with powders using a standard mechanical dispersion technique such as milling or ultrasonication in liquid media (Ning et al. 2003; Sumfleth et al. 2009b; Bhattacharya and Bhowmick 2010a). It was found that combined use of CNTs with other solids generally results in improved dispersions. Moreover, if soft particles such as aluminium powder are used as supporting source, the homogeneous distribution of CNTs and protective effect of Al against damage of CNTs are observed (Esawi and Morsi 2007).

The stabilisation of SWNT dispersion by use of highly charged nanoparticles such as ZrO_2 has been reported as well (Zhu et al. 2004). However, the mechanism of stabilisation is not investigated yet. Improved dispersability could be also reached by use of in-situ formation of inorganic nanoparticles on the CNT-surface using, for example, sol-gel or red-ox technique (Zhang et al. 2004b). A novel simple solubilisation process reducing the breakage of CNTs dramatically has been recently developed using π-stacking interaction of CNTs with graphene oxide (GO) (Zhang et al. 2010). Graphene is one-atom-thick two-dimensional novel carbon nanomaterial with excellent electronic, thermal and mechanical properties (Kim et al. 2010b). Its oxide form, consisting of two dimensional sheets, containing multiple aromatic regions and hydrophilic oxygen groups, is able to be exfoliated in water into individual graphene oxide sheets resulting in very stable suspensions (Park et al. 2008b; Wassei et al. 2010; Nguyen et al. 2009). Mixed together with graphene oxide,

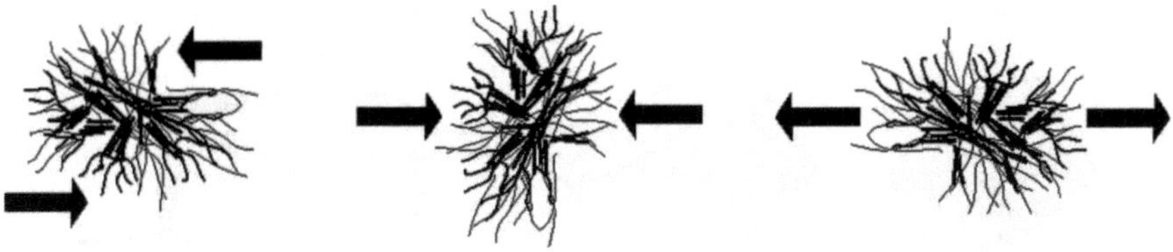

Fig. 4.1 Main stress forces acting on the CNT agglomerate

MWNTs under low-power ultrasonication form a CNT-GO complex and result in very stable dispersion if the diameter of the CNTs is large. When the diameter of the CNTs is less than a critical value, the π-stacking interactions between GO and CNTs become weaker than interactions between CNTs and thus agglomerate. Therefore, this technique is found to be promising for the fractionation of CNTs as well (Zhang et al. 2010).

4.3.2 High Shear Mixing

In polymer processing distributive (also called simple or extensive mixing) and dispersive (also called intensive mixing) mixing are usually distinguished. Distributive mixing aims to improve the simple spatial distribution of the components. In dispersive mixing cohesive resistances have to be overcome to achieve finer levels of dispersion. The CNTs consisting agglomerates need as cohesive component a certain minimum stress level to rupture the agglomerates. Dispersive mixing is usually more difficult to achieve than distributive mixing.

Three primary stressing mechanisms take place during dispersion procedure: shear, extension and impact (Fig. 4.1).

A polymer undergoes shear when one area of fluid flows with a different velocity than another one. High shear flow is not very efficient in achieving dispersive mixing because particles in the fluid are not only sheared they are also rotated. In an elongation flow particles undergo a stretching deformation without rotation. This is the reason why high shear mixers, where shearing forces are dominant, are relatively ineffective for dispersion of CNTs and are mostly used for distributive mixing.

Usually relatively high flow rates are required to generate high-shear forces in the fluids processed. This is also in accordance with requirements for well distribution. For this purpose a very broad variety of shear mixer devices is available. The main ones are impellers, rotors, rotor-stator combinations, pump mixers, mills or special dispersers. Impellers are usually different shaped blades fixed on a rotating shaft. Compared to impellers, rotors usually consist of round or cylindrical shapes with teeth. The series of impellers or high speed rotors could be used as well in order to

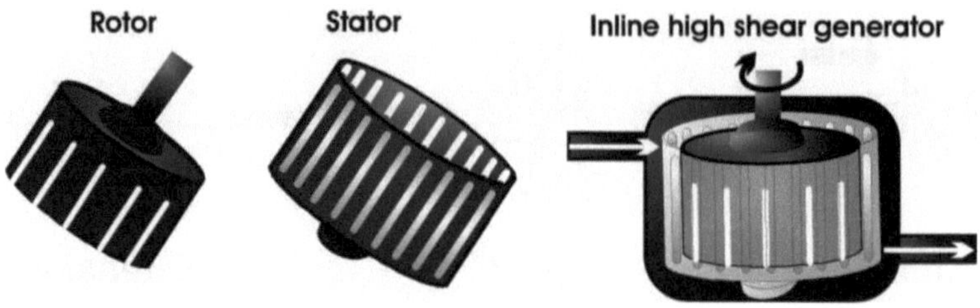

Fig. 4.2 An example of rotor-stator combinations

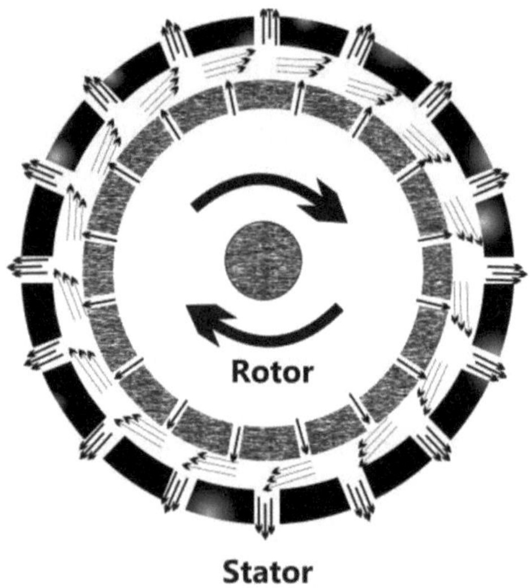

Fig. 4.3 Shear forces in a generator

increase the shearing. The main mechanism for creating the shear forces is that velocity of the fluid at the outside diameter of an impeller or rotor is higher than at the centre. In order to create extremely high shear zones the different combinations of the rotor with stator (Fig. 4.2) are used (common name is high-shear mixers). They can be designed in different forms such as axial- or radial-discharged mixers, toothed devices, colloidal mills etc. (Baldyga et al. 2008).

Usually, the gap between the rotor and the stator is very narrow. This causes much higher shearing of dispersion system compared to rotor alone (Fig. 4.3) (Pacek et al. 2007; Booker et al. 2010). Combined together rotor and stator are usually called a generator or a mixing head. Key design factors for such kind of high shear mixer are diameter, amount and rotation speed of the rotors, as well as the distance between rotor and stator. In addition, the number of rows of teeth, their angle, and the width of the openings between them are used as variables by mixer design. Therefore, this should be kept in mind by adopting the CNT nanocomposite processing.

Increased shearing could be reached by inline high shear rotor-stator mixers: the rotor-stator array is placed in housing with an inlet at one end and an outlet at the other. In this case mixing components are flowing through the generator in a continuous stream, which acts as a centrifugal pump. Such mixers offer a more controlled mixing environment, need less space and can be used in both batch and continuous processes. Equilibrium mixing can be achieved by passing the product through the inline shear mixer more than once. The equilibrium mixing means that characteristics of the mixture do not change significantly with the prolongation of processing time. The dispersions average particle size is usually used as parameter for equilibrium mixing. However, the viscosity of dispersing system is restricted by the centrifugal pumping action. In order to improve the shearing rates and increase the number of shearing events, as well as to reduce re-mixing cycles, the stator can be modified with holes or slots. Such mixers are usually referred to as ultra high shear inline mixers. Relative narrow particle size distribution can be obtained using ultra high shear inline mixers. However, they are more efficient for micro and sub-micrometre size of particles (Prolongo et al. 2008; Baldyga et al. 2008). Therefore, full deagglomeration of CNTs using this technique is difficult to achieve. High flow rates creating high-shear forces cause well distribution of agglomerated CNTs. However, their breakage is not avoided and presence of CNT agglomerates is usually observed in this case. Furthermore, optimal ratios between flow rates (shear forces) and distribution, deaglomeration, and breakage levels of CNTs should be adopted for each CNT/polymer combination. This is because blends of polymers of different nature with CNTs of different morphology as well as their concentrations have different viscosity, adhesion and interaction degrees. Therefore, the optimal processing procedures should be determined. Attention should be paid in order to achieve the optimum dispersion of the CNTs while minimising any potential breakage of the filler or destruction of polymer matrix. Note that it is very possible by applying high shear rates (Hilding et al. 2003). The popularity of high shear mixing is growing in many industries. High shear mixers are applied in industry to produce standard mixtures using the technique of equilibrium mixing. Note, that very often nanocomposite should be re-dispersed many times in order to achieve equilibrium mix. This is an additional drawback of this technique towards CNT-polymer nanocomposites processing. Note that high-shear mixing technique can be used for production of both compounds with low loadings of CNTs and masterbatches containing high concentrations of CNTs in a polymer matrix. Masterbatches can be diluted with a neat polymer and re-dispersed again in order to obtain better quality of nanocomposite. One additional drawback exists in using masterbatch processing: regular cleaning procedures. This problem is partially solved by high-shear mixer's design. Moreover, some high shear mixers are designed to run dry solving partially both the problem of high viscosity and the cleaning problem (Yang et al. 2005; Mu et al. 2008). However, it should be noted here that basic research is mainly concentrated on the "slow-shear" mixing technologies (e.g. axial and radial flow turbines) and thus high attention should be paid to scale-up in this technique to industrial use in order to avoid costly mistakes. As it was mentioned above, the selection of suitable mechanical methods for dispersion of CNTs in nanocomposite production depends strongly not only on the characteristics of CNTs, but also on the viscosity of materials. It is obvious that standard laboratory mixing techniques such as lab-mixers or high-speed dissolvers are not efficient enough to be used for high-viscosity materials. In addition, CNT agglomerates cannot be well dispersed by high-speed stirrers or dissolver discs nor for medium-viscosity materials (such as polyols, epoxy resins) either low-viscosity materials (such as water or organic solvents) even using long dispersion time. Therefore, they are generally used for the pre-dispersion of CNTs without their strong shortening or in applications which do not need the high quality dispersion of individual nanotubes. However, combined use of simple mixing technique with other high-power dispersion methods is often used (e.g. Mu et al. 2008; Zou et al. 2004). A more detailed discussion of such applications will be presented in this chapter later. It should be noted here that simple mixing technique (including magnet stirring, shaking or spin-mixing as well) has a potential to be used for the physically treated CNTs (e.g. doped with cation, surfactant or polymer assisted CNTs) if the breakage of CNTs should be maximal avoided.

High shear mixing is generally used for incorporation of CNTs in high-viscosity materials, such as thermoplasts and elastomers (Zou et al. 2004; Li and Shimizu 2007; Kasaliwal et al. 2010; Andrews et al. 2002; Oh et al. 2010; Tang et al. 2003; Kotsilkova et al. 2010; McClory et al. 2010; Thiebaud and Gelin 2010), and is less applied as dispersive technique for medium-viscosity and low-viscosity materials, however, is widely used as efficient pre-mixing and well-distribution technology. There are three main possibilities of incorporation of CNTs into high-viscosity melts: melt impregnation (direct mixing), solvent impregnation (using pre-dispersed in solvent CNTs) and in-situ polymerisation. All of these methods have been described briefly above. Melt-mixing of CNTs into thermoplastic polymers using conventional processing techniques, such as extrusion (especially twin screw extrusion) and moulding (especially injection moulding), are particularly desirable, because of the speed, simplicity and availability of these processes in the plastics industry. These methods are also beneficial because they are usually free of solvents and contaminants, which are present in solution processing methods and in-situ polymerization. Thermoplastic CNT-nanomaterials have a unique advantage, because in contrast to larger, microscale carbon fibres, less fibre breakage occurs, and a high ratio between length and diameter is maintained for CNTs. Use of high-shear mixing and longer processing times even may enhance dispersion, especially when coupled with elongation flow. Note that the most effective mechanism for dispersing is extensional stressing. Moreover, elongation flow during processing of the nanocomposite is usually yielding in alignment of nanotubes. Other methods that are being used for the production of aligned CNTs/polymer fibres are spinning, stretching, and melt mixing, etc. (Ranjan et al. 2010; Min et al. 2009; Haggenmueller et al. 2000; Chou et al. 2010; Jin et al. 1998). All of these methods were able to align CNTs and increase the young modulus of the pure polymer fibre. A more complex method such as a modified CVD, which integrates and growths CNTs aligned into a polymer substrate, is also known (Ng et al. 2002).

Narrow dies and nozzles are widely used for creating high shear flow without strong rotation of nanofillers (Sauter and Schuchmann 2007; Baldyga et al. 2009). In this case high tension stresses are initiated, which results in more efficient dispersion of CNTs. In order to promote deagglomeration of CNTs in nozzles surfactants are widely used as additives (Hilding et al. 2003). It should be noted that care must be taken in this case concerning the possible degradation of mixture components. The breakage of CNTs is increased by combination of shear and tension induced dispersion. This is the main idea realised in jet mixing and different homogenisation techniques, such as high pressure/shear and low pressure homogenisation methods.

The jet impinging mixer is a type of disperser which uses high pressure for creating high velocity fluid streams through the nozzle. Although jet imprinting dispersion technique is generally used for the deagglomeration of powders up to micro- and sub-micro-sizes, this method can be used in addition to other mixing procedures in order to create high shear, elongation, turbulence, cavitation, and impact forces (Pampuch 2004; Fauchais et al. 2010; Lind et al. 2010). This allows reaching higher deagglomeration level of nanoparticles. The simplest way to create jet dispersion supported with high-impact action mechanism is to pass the powder or suspension through the nozzle with high velocity and break up the jet by gas stream or by solid surface (outer counter-plates or rotating drums) (Sokolov and Yablokova 1996; Fonda et al. 1999; Nicolas 2002; Fauchais and Montavon 2010; Bricard and Friedel 1998; Schneider and Jensen 2008). The collision of CNT suspension with hard surfaces has much greater efficiency for CNT deagglomeration compared to gas flow. Usually high velocities and very small nozzles are needed for this purpose (Ng et al. 2002; Sauter and Schuchmann 2007). Jet imprinting technique is widely used for imprinting CNT-dispersions on different substrates (e.g. thin film transistors) (Takenobu et al. 2009; Fan et al. 2005). This technique can be also assisted by vacuum, electric field, plasma, etc. (Fauchais et al. 2010; Fauchais and Montavon 2010; Takenobu et al. 2009; Poppe et al. 1997).

Compared to high shear mixing, high shear homogenizing opens new pathways for fine dispersions with narrow particle size distributions. Its flexibility allows the introduction of exactly desired and validated shear forces. In practice high-shear homogenization with rotor-stator based systems means powder wetting and dispersion to achieve finely dispersed suspensions with smaller particle sizes. In standard rotor-stator based mixers the phenomena of pumping and shear energy creating are coupled. Therefore, with increasing rotation speed of the rotor, the shear rate and pumping capacity both increase accordingly. New high-shear homogenizers are designed to separate the pumping and shearing in two separate stages, by installing the pumping device in front of the rotor-stator machine (Fig. 4.4) (Fischer et al. 2009). This has several advantages such as separate control of the pumping and shearing and decreasing the temperature increment during mixing. The flexibility of controlling these parameters allows producing dispersions of better quality by lower power input compared to standard rotor-stator mixers. By controlling independently the flow rate and shear rate it is possible to work in every operating condition. The induction of powders in high shear homogenizers occurs between the pumping and the rotor-stator stages supported by a vessel vacuum device for immediate filler dispersion. The pump provides the rotor-stator with a constant fluid flow, while the vacuum at the induction valve pulls in the powder. The controlled

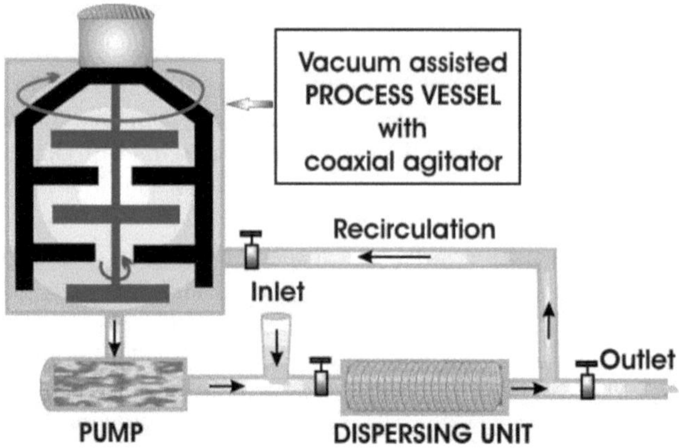

Fig. 4.4 High shear homogenizer working principle

flow conditions prevent a breakthrough of the powder in the opposite direction through the pumping impeller. Therefore filler is well dispersed and wetted with the liquid phase.

The high shear homogenizers can be divided into two types: high pressure homogenizers and low pressure homogenizers. In high pressure homogenization systems shear and elongation forces, turbulence and mechanical cavitation realise the dispersion of the particles through a sudden pressure drop of several hundred bars. A typical set up of a high-pressure homogenizer consists of a premix container, a high-pressure pump, a pressure measurement device and a dispersing unit. Typically high pressure dispersing units are based on valve systems, sometimes also on nozzle devices. Both technologies have their advantages and drawbacks: while the valve systems are very adjustable, they are not as efficient as nozzle systems. Nozzle-based dispersion devices have a superior dispersion efficiency compared to conventional valve systems. Due to the fixed geometry standard nozzle systems are difficult to adapt to new process conditions or products. The Low Pressure Homogenizer (LPH) is a novel dispersion system (product of firma Serendip AG, Switzerland) which combines the advantages of both technologies (valve system and a nozzle system) but avoids their disadvantages (e.g. adjusting, heat development) (Fischer et al. 2009). LPH is an outstanding device for the dispersion of emulsions and suspensions down to the nanometre range and under very gentle process conditions. Using this technique the particle agglomerates can be disintegrated and stabilised. The LPH dispersion device consists of a nozzle with continuously adjustable geometry of the dispersion zone. This allows an easy adaptation to different process conditions and recipes. High quality material of the nozzle material and high efficiency allows identical dispersion results at lower operating pressures compared to conventional systems.

Furthermore, another innovative concept has been realized (firma Serendip AG) in the Low Pressure Nanogenizer (LPN) (Fischer et al. 2008; Scheid and Fischer 2009; Fischer and Herzog 2010). The LPN is the device for the dispersion of even

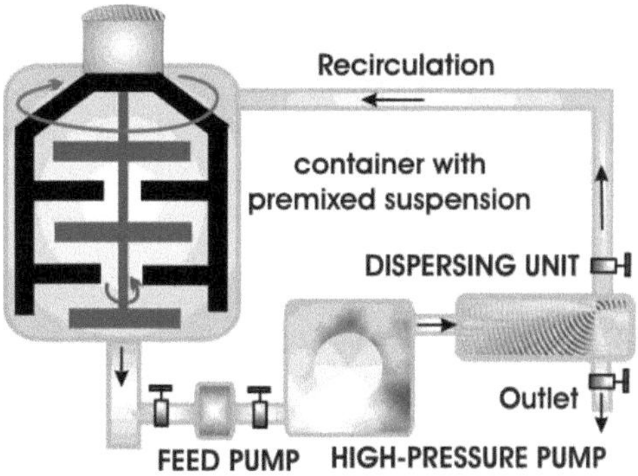

Fig. 4.5 Schematic presentation of working principal of LPN

abrasive and high viscose (up to 150 Pa s) suspensions down to the nanometre range and under very soft process conditions. LPN has very good dispersing performance at high viscosities, such as coatings, pigment slurries, resin or wax dispersions. On Fig. 4.5 the working principle of LPN is presented.

Usually, a product that has been pre-dispersed (e.g. in a dissolver) and brought to a certain level of fineness is used for further homogenization in LPN. Note that LPN can be connected to any preliminary stage. The pre-treated product is transferred to the LPN via the feed tank using a feed pump. Then a high-pressure pump of LPN is used for transporting the product into the dispersion device with constant volume flow rate. During the dispersion process the product is subjected to all primary mechanical forces for particle size reduction. The dispersion performance can be optimized by controlling the flow rate and pressure. High flow rates and turbulences are the critical parameters for high quality of the dispersion. After the treatment the product is discharged from the machine through the product outlet. If necessary, mixture can be re-circulated or re-dispersed in several passes.

Due to the intelligent and robust (extremely resistant to abrasion) design of LPN only a small part of the dispersion participates in abrasion. This guarantees long tool life times, particularly for abrasive pigment systems and slurries. Due to high efficiency of this technique, the heat development is much lower than with other dispersion technologies. The high efficiency allows much better dispersion results compared to conventional bead mills or other grinding systems. Therefore, LPN has a potential to replace such technologies in the future, especially for controllable, optimized and highly efficient dispersing CNTs in polymers under mild conditions. The labour, pilot and industrial LPN devices are already available on the market.

Another blending technology using Integral Pump Mixers (IPMs) provides a new approach to high-stress dispersive mixing (Maelstrom Advanced Process Technologies Ltd. 2007). They combine three primary stressing mechanisms to achieve both high dispersion (particle size reduction) and high distribution performances. The IPM mixing head comprises three elements: outer, central and inner (Fig. 4.6). They are fitted together inside one another. The outer and inner elements are locked together, and are stationary. The central element placed in between rotates on an axis which is offset from the axis of the inner and outer

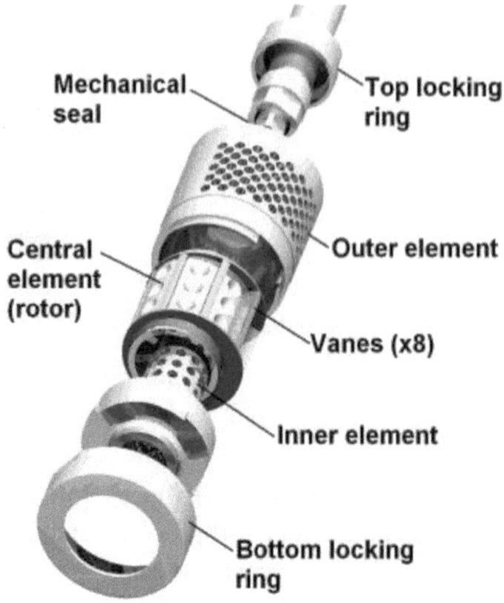

Fig. 4.6 IPM vane-type head assembly (Reprinted with permission from Maelstrom Advanced Process Technologies Ltd.: http://www.maelstrom-apt.com/ipm_tech.htm)

elements: the central element is mounted off-centre with respect to the inner and outer elements. The outer element has large inlet holes around part of its periphery, while the inner element has large outlet holes around its entire circumference. Vanes are fitted into slots on the central element. The vanes are trapped by the inner and outer elements and slide in their slots as the central element is rotated. This construction provides a set of cameras varying in volume as the central element is moved. The central element has nozzles between each pair of slots which are directed inwards towards the inner element. The main varying parameters for dispersive and distributive blending in IPM are nozzle geometry and diameter, and rotation speed of central element. IPM mixers are available in both batch and continuous forms although the internal construction of the mixing head is very similar in both types.

Integral Pump Mixers use internally generated positive displacement vane pumping action in order to generate internal pressure and to force fluid through small nozzles. This creates very high extending and shearing stresses. The chamber between the vanes and central and outer elements start to expand as they approach the inlet holes in the outer element causing decreased pressure. Therefore, fluid is drawn from the mixing vessel into the inlet holes and undergoes shearing by the vanes. When the last of the inlet holes is reached, sealing of mixture from the fluid outside the mixing head occurs. Then fluid inside undergoes increased pressure due to reduction in the chamber volume during the rotation of the central element around the high pressure side of the mixer. Therefore, the mixture is forced inwards through small nozzles in the central element to create very high extensional stressing. Then collision of the flow at high velocity through the nozzles with fluid on the wall of the internal element takes place, which provide a high degree of impact stressing. Then, fluid is pumped under low pressure into the chamber inside the inner element which is sealed at the top, and therefore passes out axially through the bottom of the mixing head. During its retention in the low pressure side of the mixer, the fluid experiences turbulent mixing and post-stress conditioning. Dynamic cutting and folding actions combined with vigorous turbulence using IPM technique provide good distributive mixing as well. The resulting dispersion has a narrower distribution of particle sizes when compared with a traditional rotor-stator mixer because all of the material is subjected to the same stress levels in the IPM mixing head. High stress mixing is very useful in applications where a reduction in particle size (dispersion) is required. The IPM concept offers improved mixing performance compare to high shear (rotor-stator) technology in most applications. IPM applying very high specific energy to the mixture allows achieving very good dispersions when compared with traditional high shear mixers. In addition, because of positive displacement pumping action IPM does not undergo the limitations of centrifugal pumping typical for rotor-stator mixers. Therefore, a much wider range of viscosities can be processed. Improved mixing effectiveness can be used either to process new or improved dispersion products or to perform the same operations faster and more efficiently.

4.3.3 Milling

Mills are available in various forms for both batch and continuous use. They are particularly suited to particle size reduction of solids which are suspended in fluids. However, throughput rates are generally low. Compressive and/or shear stresses are the main action mechanisms to create dispersions. Therefore, mills can be divided into two mixing strategies: high shear and high impact technologies.

4.3.3.1 High Shear Milling

The common high shear milling belongs usually to the two-roll and three-roll mills comprising rotating cylinders (two and three, respectively) to disperse materials between them. The rolls are usually supplied with heating and cooling systems. Such mills are often called roll calenders. Two-roll mills belong to the earliest group of machines used for processing natural rubber (since 1830) (Dumoulin 2003). The solvent assisted dispersion of CNTs within natural rubber by two-roll milling with moderate properties improvement has been reported as well (Sui et al. 2008). Currently this technique is also widely used for shaping high melt viscosity thermoplastic sheets and is particularly suitable for polymers with low thermal stability or which contain high amount of solid particles (Xu et al. 2008). This technique has one important benefit compared to other techniques such as extrusion: the calender is capable of conveying large rates of melt with a small input of mechanical energy. Co- and counter-rotation design of two-roll mills can be used. However, realisation of counter-rotation concept for two-roll calendar is low efficient because, during mixing procedure, stopping of the calenders and reversible

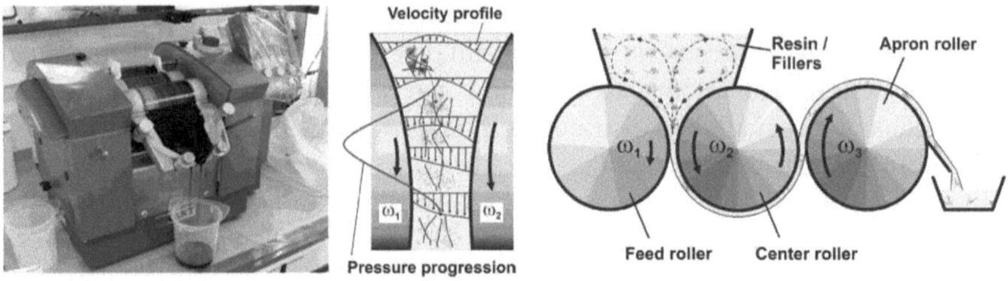

Fig. 4.7 Three-roll calender (*left*) and its working principle (*right*)

rotation in the opposite direction is needed. This problem can be solved in three-roll mills (Fig. 4.7) (Yasmin et al. 2003; Li et al. 1999). Four and more rolls can be used in the mill constructions as well (Osswald and Hernandez-Ortiz 2006). Note, that counter-rotating calenders are more effective in creating higher shear forces. The main process parameters that can be controlled by tree-roll milling are: distance between cylinders, their rotation speed, temperature and the pressure in the gap. The distance between rolls can be adjusted up to several micrometers. The pressure in the gap between rolls strongly depends on the diameter of rolls and their rotation speed as well as on the viscosity of the mixture. The maximal pressure is occurred slightly before the narrower distance between the cylinders and then is decreased, which promotes the better wetting of the fillers in the dispersion medium. Note that three-roll milling differs from other mills in that mostly pure shear stresses are created during processing.

Figure 4.7 shows the schematic of working principle of a three-roll mill. With different speed rotating cylinders pass the mixture between gaps with defined distance between them. The first and last cylinders are usually called feed and apron rolls, respectively. The mixture to be dispersed is placed between the feed and centre rolls. The rotation speed of each adjacent roll must be progressively higher to create high shear forces and efficient mixing. Due to differences in their rotation speeds, the cylinders introduce very high shear forces in the dispersion media between them. These forces cause dispersion of agglomerates and better distribution of particles in the mixture. The transfer of dispersion from the centre roll to the apron one is caused by adhesion of mixture and rolls due to surface tension of fluid under intensive shear forces. Additional particle dispersion and distribution is carried out in the second gap as well. The milled material is then separated from the apron cylinder by a knife pressed against it and removed into a container. Several mixing cycles may be repeated with the strategy of sequentially reducing the distance between rolls (up to minimal) in order to obtain optimal dispersion.

Three-roll milling is capable of dispersing CNTs homogeneously within thermosetting resins with low level of damages and ruptures on CNTs, compared to other wet techniques (Seyhan et al. 2009). However, the improvement of dispersion of MWNTs in a rubber system in accompaniment of their shortening by increasing the rotation speed and mixing time have been reported (Cho and Kim 2010). Therefore, care should be taken to the optimisation of three-roll mill dispersion processing for each polymer/CNT nanocomposite. An additional advantage of this technique is that more viscous mixtures can be processed, which is especially important when dispersing nanofillers with large surface areas (such as CNTs) are used, which increase viscosity significantly even with low loading. However, it should be noted that three-roll milling is more efficient for deagglomeration of the nanoclays and ceramic nanofillers, than for CNTs.

This technique is well adapted and used for the dispersion and better distribution of nanofillers (including CNTs) in epoxy and other thermosetting resins, which is beneficial for the application of this technique in the processing CNT nanomaterials for aerospace applications. Note that usually suspensions that are pre-dispersed by standard laboratory mixing or by high-shear mixing epoxy/CNT are used for further processing in calendar milling (Gojny et al. 2004; Kempel and Schlarb 2008; Sumfleth et al. 2009b). This technology achieved excellent dispersion results without strong reductions in lengths and the high aspect ratio of the nanotubes. This is important to enable a good load transfer from the polymer matrix and to help achieve a low percolation threshold in the conductivity of the resulting nanocomposites. One further advantage of the calendering method is the possibility of up-scaling the manufacturing process to meet technical demands.

4.3.3.2 High Impact Milling

Bead mills (other common names are pearl mills, ball mills, etc.) are another widely used type of mills. Usually they consist of a grinding chamber filled with hardened beads (e.g. zirconium dioxide, steel) and supported by a stirring mechanism (usually, rotor). Ball milling is a mechanical dispersion method which generates local high-impact areas between the balls resulting in a random crushing of the materials. Ball mills can be designed in horizontal or vertical construction. Much higher quantities of dispersed samples can be produced by ball milling compared to other dispersion techniques, which make this method very practical.

It is common method for the shortening and partial deagglomeration of CNTs (Ahn et al. 2007; Kukovecz et al. 2005; Smart et al. 2007; Shin et al. 2009; Konya et al. 2004). However, prolonged high-energy bead milling is able to transfer the CNTs into other forms of nanoparticles or even into amorphous graphite (Pierard et al. 2004; Li et al. 1999). Ball mills can be used for both dry grinding of CNTs (with or without presence of polymer) and wet dispersion of CNTs using fluids as dispersion medium (Ghose et al. 2006; Inkyo et al. 2008). The wet dispersion of CNTs is more widely used for nanocomposite production. This technique has one advantage because of easier stabilisation of dispersed nanoparticles by wetting them with the fluids. Additional use of additives for better stabilisation of obtained suspension is possible as well. Ball mills can be used in both continuous and discontinuous milling operations. High loading of nanofiller could be used in the ball milling process as well. This can be used for preparing the masterbatch suspensions.

In Fig. 4.8 the working principle and main action mechanism of ball milling are presented. Dispersion medium is moved by agitation causing the collision and

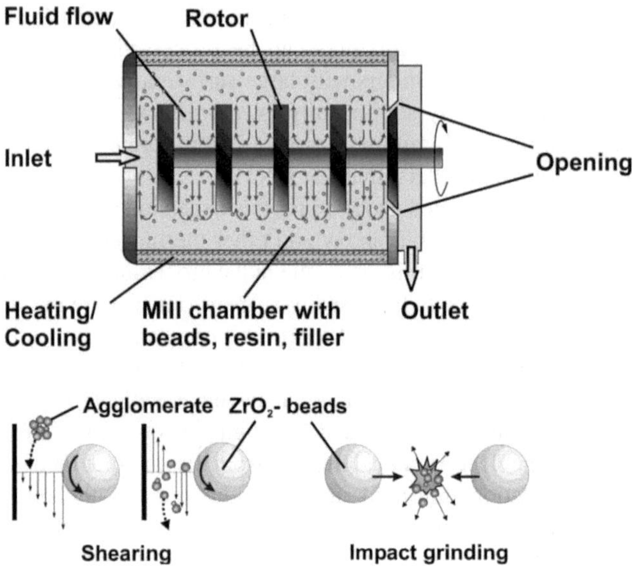

Fig. 4.8 Ball mill (*up*) and schematic of its main action mechanisms (*down*)

sliding of the milling beads on each other or on the rotor and vessel sides. Impact and shear forces created during the milling cause breakage and deagglomeration of nanoparticles.

The main controlling parameters for this technique are rotation speed, mixing time and diameter of milling beads. The higher rotation speed and bigger diameter of milling beads result in higher shear and impact forces. However, high-energy forces can cause undesirable damages of CNTs. Therefore, shorter milling times and smaller beads are most used under mild agitation for the shortening and dispersion of CNTs (Kukovecz et al. 2005; Inkyo et al. 2006; Inkyo and Tahara 2004).

The kind of beads used for milling as well as their hardness may influence the efficiency of ball mixing and quality of final suspension. The abrasion of beads also often occurs by the ball milling process. Therefore, small quantities of bead-material are usually presented in the final dispersion. However, such impurities can have a positive effect on the material properties of a nanocomposite as well. For example, positive influence of glass impurities on the mechanical and electrical properties of silicone/MWNT nanocomposite has been observed using glass beads for ball milling (Lim et al. 2010).

Because of lower impact energy smaller beads prevent strong damage of nanoparticles while being able to breaking up agglomerates. Note that efficiency of the ball milling depends also on the temperature, surface tension and viscosity of dispersion medium. Therefore, ball mills are often supplied with a cooling/heating system. The processing parameter (e.g. rotation speed and mixing time) should be optimised for each CNT/polymer system.

4.4 Rapid Expansion of Supercritical Suspension (RESS)

An innovative technique, which exploits the unique properties of supercritical fluids, based on a fast decrease of pressure in a gas stream of nanopowders, has been recently reported as efficient for synthesis and/or deagglomeration of nanoparticles, and their incorporation in polymer nanocomposites (Jung and Perrut 2001; Pourmortazavi and Hajimirsadeghi 2005; Wei et al. 2002; Horsch et al. 2006; Wu et al. 2008b; Chih and Cheng 2007; Bell et al. 2005; Thiering et al. 2001; To and Dave 2009; Hurst et al. 2009; Yang and Ozisik 2008; Bahrami and Ranjbarian 2007). This dispersion technique is known as Rapid Expansion of Supercritical Suspension (RESS). Supercritical CO_2 is generally used as supercritical fluid for this purpose. Decoration of CNTs by metallic nanoparticles using RESS technique has been also reported (Sun et al. 2007; Bayrakceken et al. 2007). Efficient deagglomeration of CNTs using RESS and their deposition on the dry polymer particle's surface in form of a coating for the further processing of nanocomposite has been observed as well (Narh et al. 2007). In addition to the dispersive efficiency of RESS there are other advantages such as low viscosity, high diffusivity and variable density of supercritical fluids. The principal of work of the RESS dispersion method is presented in Fig. 4.9.

At the first stage nanoparticle suspension in critical fluid should be prepared. This is done by charging the filler in a high-pressure vessel, followed by heating and pressuring with gas (e.g. CO_2) up to the supercritical point using a heating jacket and a supercritical gas pump. Then maintaining of this system for an appropriate time is needed for soaking the supercritical liquid into the solid phase. An additional stirring device can be also integrated into the high pressure vessel.

This allows more homogeneous distribution of agglomerates in a supercritical fluid. Obtained suspension is then passed through the nozzle with rapid depressure into the atmosphere pressure collector. The suspension undergoes pressure forces due

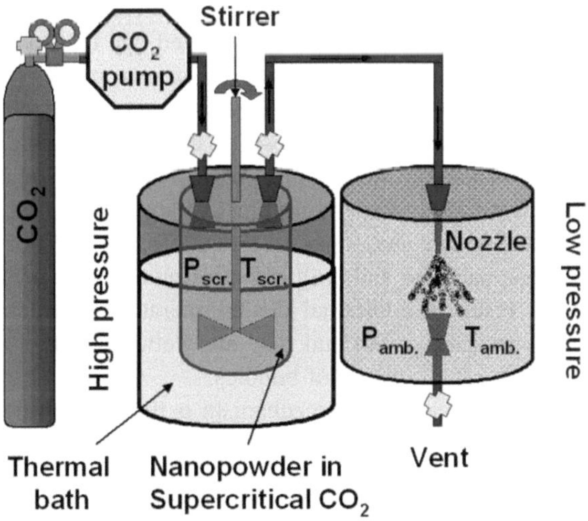

Fig. 4.9 Schematic of RESS procedure

to the rapid expansion of critical fluid resulting in deagglomeration of nanoparticles at this stage. The gas obtained from the supercritical fluid is vented through a filter from the collector in order to avoid pressure increase. The expanded nanofiller can be collected or directly mixed with resin or polymer matrix in wet or dry mixing conditions, which allows better stabilisation of separated particles and avoids re-agglomeration (especially in the case of CNTs).

4.5 Ultrasonication

One simple and most convenient method used for dispersion (deagglomeration) of CNTs in liquids, resins and polymers is the ultrasonication process. Usually, CNTs are first pre-mixed in dispersion media by a standard stirrer or high-shear mixer and then homogenised by ultrasound. There are three main physical phenomena of the dispersion procedure using ultrasonication: cavitation (formation and collapse of the bubbles), localised heating (up to temperatures higher than 5,000 K and pressure up to 500 atm) and formation of free radicals. Cavitation, overcoming the bounding forces between CNTs, is the action mechanism for the fracture and dispersion of solids, while two last phenomena reduce the efficiency of ultrasonication. The frequency of ultrasound is the key parameter determining the bubble size. Acoustic waves of the frequencies in the range from 10 kHz to 10 MHz are ultrasound waves (Suslick 1990). Low frequencies (~20 kHz) result in large bubbles and high energy forces occur at their collapse. The cavitation is reduced if the increase of frequency is due to formation of smaller bubbles. It is known that cavitation does not occur in many liquids if the frequency is higher than 2.5 MHz (Hilding et al. 2003). Ultrasonic dispersion is usually effective for the dispersion of nanotubes in liquids with viscosities up to 100 Pa · s. Cavitation is caused by regular changing in the increased-pressure and reduced-pressure phases during ultrasonication. During the increased-pressure phase an ultrasonicated fluid undergoes compressive forces. By changing the phase to the reduced pressure, cavitation bubbles are formed due to strong reduction of the local pressures under fluid vapour pressure. The next pressure phase change results in collapse of the bubbles, which causes high energy forces, that are capable of destroying the agglomerates. By collapse of cavitation bubbles, pressure-waves propagate in the dispersion media (Fig. 4.10). Nucleation of bubbles on the filler surface and their rapidly expanding and collapse cause local impact, tension and shear stresses, which can separate the CNTs. In addition, collisions between agglomerates, particles and walls of ultrasonic device are initiated by pressure-waves, which result in additional deagglomeration of fillers. If the volume fraction of nanoparticles is small and they are wetted by the fluid media, CNTs can remain separated after collapse of bubbles.

This common action mechanism is the basic principle of so-called "Hot-Spot" theory (Bittmann et al. 2009). However, another action mechanism model for ultrasonic dispersion is known as well. According to this model the real collapse

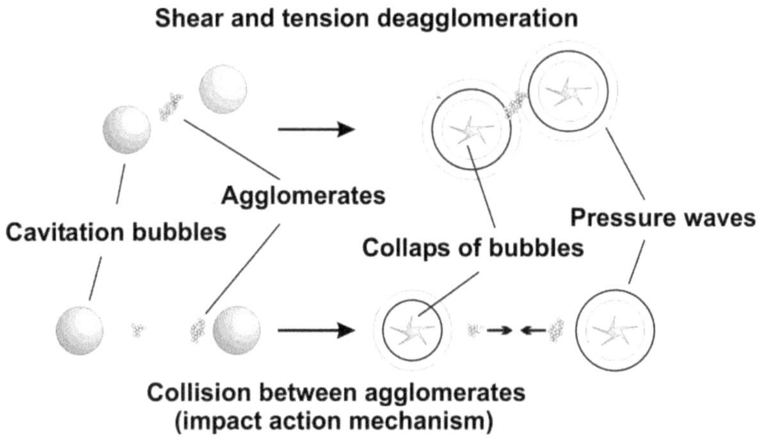

Fig. 4.10 Main action mechanisms of ultrasonication

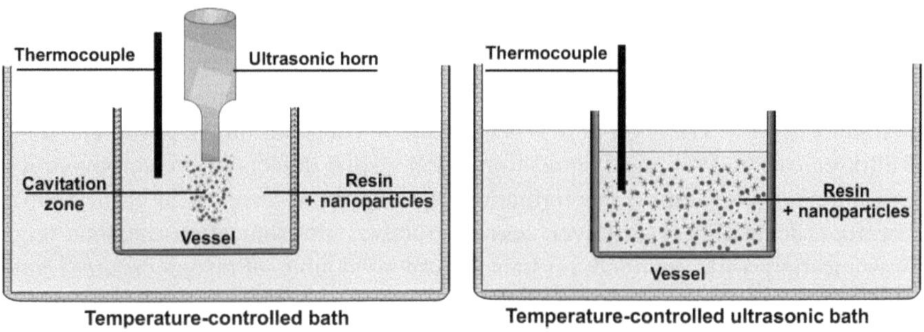

Fig. 4.11 Ultrasonic horn (*left*) and ultrasonic bath (*right*)

of cavitation bubbles does not take place, but splits in many smaller bubbles occurs, which is the action mechanism providing high stresses in the ultrasonicated media. This is because growing of cavitation bubbles is believed to be asymmetrical and easy splittable into smaller size. This model is based on the cyclic creating and dissipation of uncompensated electrical charges (through ions and dipols) on the cavitations, which results in stimulation and ionisation of adjacent molecules (Lepoint and Mullie 1994; Margulis 1994). The main difference between these two mechanisms is that in Hot-Spot theory, interaction between molecules, and in electrical theory between electrons and molecules, occurs.

Two major methods for CNT dispersions with ultrasonication procedure are commonly used: ultrasonic bath and ultrasonic horn or wand (Fig. 4.11). The ultrasonication baths are usually characterized by higher frequencies (40–50 kHz) than horns (25 kHz) (Hilding et al. 2003; Bittmann et al. 2009).

By use of an ultrasonic horn or wand for the dispersion process, the rapid oscillation of the horn or wand tip produces a conical cavitation zone of high energy in the dispersion media, which induces the flow that moves away from the tip and then recirculates through the conical zone. The size of this zone and speed of recirculation strongly depend on the boiling point, surface energy, and viscosity of dispersion media, as well as on the energy applied, geometry of vessel and placement of ultrasound source (Hilding et al. 2003). Opposite to the ultrasonic horn or wand, the ultrasonic bath goes not produce a local cavitation zone. Therefore, energy is more uniformly distributed through the dispersion media. Therefore, the ultrasonication dispersion procedure should be optimised for each dispersion system.

A proper ultrasonication procedure results very often in well dispersed nanotubes and better composite mechanical properties (e.g. Liao et al. 2004). However, these techniques have several important disadvantages. Ultrasonication can induce structural defects such as irreversible bending, buckling and fracture of graphene layers of CNTs. In addition, when the tube-walls are broken the formation of "worm-eaten" and "ragged" walls is very possible (Hilding et al. 2003; Lu et al. 1996). It was found that different mechanisms of CNT damaging are observed for SWNTs and MWNTs. The length decrease of SWNTs occurs only after the bundle size is reduced. However, shortened SWNTs rearrange into much bigger in diameter (~20 times) ropes (Hilding et al. 2003), MWNTs undergo, expansion and peeling. The fractionation of MWNT graphene layers is also occurring. The initiation of MWNT destruction with ultrasound starts on the external layers and transfers in the internal direction. So, MWNTs are not only getting shorter, but thinner as well. The level of MWNT destruction depends on the power and time of ultrasonication. Prolonged sonication increases the defects of the carbon structures ultimately leading to the formation of amorphous carbon (Lu et al. 1996). Therefore, development of novel, less destructive, ultrasonication methods (e.g. ultrasonication with diamond crystals or use of double ultrasonic source) and optimisation of ultrasonication procedure for dispersing CNTs in resins, solutions and polymers are of great interest in the scientific community (Hilding et al. 2003; Caneba et al. 2010). It was found that controlled mild ultrasonic treatment can result in minimised shortening of CNTs and is effective for dispersion of SWNTs even in water, which is usually difficult to reach due to inherent insolubility of SWNTs in common organic solvents and especially in water, caused by hydrophobic inert nature of SWNTs and their high capability of forming strong interacting bundles. Accurate control of the ultrasonication amplitude allows limiting damages of the SWNTs. Polymer assisted ultrasonication of SWNTs is capable of purifying them effectively. However, ultrasonication influences the yield of purified product. Sometimes, especially under non-controlled ultrasonication, up to 70% of starting material cannot be recovered. The quality of CNTs in this process also may be lowered (e.g. thermal stability) (Hilding et al. 2003). This technique is also often limited by small dispersion volumes. Despite this, the ultrasonication dispersion technique is one of the most convenient, cost-effective and widely used, even for industrial applications. Note that ultrasonication is very often used as the main dispersion process, even if a combination of different mechanical dispersive and distributive techniques is used in processing a CNT-nanocomposite.

4.6 Combined Dispersive Methods

Sequential use of different dispersion techniques is often applied in order to reach a higher level of deagglomeration and better dispersion quality of CNTs. The main reason to do this is to combine characteristics of each technique to dominate action mechanisms (e.g. shear, impact and tension). This has greater interest for scientific research than for industrial applications (due to the economics point of view). However, some industrial combined systems such as homogenisers, nanogenisers or ultrasound assisted high-shear mixers have been developed as well. Usually, combinations of high-energy and low-energy methods or combinations of different mild-condition techniques have scientific relevance towards influence of dispersion quality on the final material properties of CNT/polymer nanocomposites. For example, combined methods such as solvent casting, RESS or milling with following melt mixing (Mu et al. 2008; Ghose et al. 2006; Narh et al. 2007), high-shear dispersion with following three-roll milling (Kempel and Schlarb 2008; Sumfleth et al. 2009b), high speed stirring or high-shear mixing followed by ultrasonication (Yudasaka et al. 2000; Xie et al. 2005), etc. have been used for production of CNT-nanocomposites with improved material properties. Very interesting are also simultaneous combinations of different mixing techniques having different dominant action mechanisms, such as ultrasonic activated ball milling (Liang et al. 2009), high-shear assisted ball milling (Kempel and Schlarb 2008; Inkyo et al. 2006), etc. The most promising combinations of different mechanical dispersion methods (with use and without use of special additives) for dispersion of CNTs in epoxy resins are high-shear mixing/bead milling, high-shear mixing/ultrasonication, and bead milling/ultrasonication. However, it should be noted that even when combined mixing technique is used, processing parameters for each CNT/polymer should be adjusted. Moreover, the efficiency and quality of dispersion, as well as the final material properties of nanocomposites, will be different for each kind of both CNTs and polymer matrix.

4.7 Controlling Methods

The quality of CNT dispersion has a great importance because it can strongly influence the final properties of related nanocomposites. Three main phenomena should be taken into account analysing the CNT-suspensions: the level of deagglomeration, particle size distribution and storage stability.

The investigation of CNT dispersion can be done by visualisation of nanotubes and interface between them, as well as by determination of CNT-influence on the matrix. The characterisation of CNT dispersions needs instruments of high-resolution in order to image nanosized particles or detect the effects caused by nanofillers. Therefore, various microscopical, scattering and spectroscopical methods are widely used for this purpose (Belin and Epron 2005; Kao and Young 2004; Lucas and Young 2004). In order to determine in which dispersion state (e.g. isolated, bundles, aggregates, agglomerates) are CNTs in obtained suspension, several scattering techniques such as light scattering, neutron scattering or small-angle X-ray scattering have been used to investigate nanotube structures in suspension (Belin and Epron 2005; Fagan et al. 2006; Zhang et al. 2008b; Sun et al. 2008; Hartschuh et al. 2009). However, optical techniques are usually inefficient for the resolution of single CNTs. The average length of CNTs in a bulk sample can be measured using multiangle light scattering or dynamic light scattering (with following application of rheological models) techniques. However, it is difficult to estimate CNT length distribution by these methods. Photoluminescence is used for the characterisation of dispersion quality of semiconducting SWNTs (Belin and Epron 2005; Wang et al. 2004; Lefebvre et al. 2008; Maruyama et al. 2003). Another widely used instrument for characterisation of CNT suspensions is ultraviolet-visible-near infrared (UV-Vis-NIR) spectroscopy (Priya and Byrne 2009; Mathur et al. 2008; Grossiord et al. 2005). UV-Vis-NIR spectroscopy has been also used to investigate dynamics of exfoliation of CNTs in aqueous solutions (Grossiord et al. 2005; Yu et al. 2007). It was found that optical intensity of CNTs increase with their length (Fagan et al. 2007; Barone et al. 2005). Another widely used method for determination of CNT-diameter is Raman spectroscopy (resonant Raman scattering) allowing detection of vibrational modes (phonons), whose position on the Raman spectra strongly depends on the diameter of CNT (Thomsen and Reich 2007; Saito et al. 2008). For example, position of the breathing mode of SWNT bundles will be shifted in the Raman spectrum compared to that for individual SWNT. Therefore, this technique is often used for the mapping of nanotube diameter's distribution in bulk samples. Some positive results for the determination of CNT-dispersion level have been obtained using rheological characterisation as well (Zhang et al. 2008b).

Characterisation of CNT sizes (both diameter and length) and their distribution is best accomplished with imaging methods such as atomic force microscopy (AFM) and various electron microscopy techniques with or without accompaniment of energy dispersive X-ray spectroscopy (elemental analysis) (Bonifazi et al. 2006; Pötschke et al. 2004). The AFM is able to image the sample surface with a resolution of a few nanometres. It can be done in both topographical and phase contrast modes. Scanning Kelvin microscopy is able to measure the conductivity distribution in heterogeneous materials, therefore can be used for characterisation of CNT distribution in an insulating matrix (Prasse et al. 2001). Different scanning electron microscopy (SEM) techniques such as high-resolution SEM, field emission SEM, etc. are also widely used for imaging of CNT containing samples in order to investigate morphology and distribution of CNTs (Bonifazi et al. 2006; Pötschke et al. 2004; Prasse et al. 2001; Kovacs et al. 2007; Lillehei et al. 2009). Direct controlling of the CNT dispersion in volume can be done using transmission electron microscopy (TEM) (Belin and Epron 2005; Bonifazi et al. 2006; Pötschke et al. 2004). However, it should be noted that results, obtained from electron microscopy techniques can strongly depend on the sample preparation and on the contrast level between CNTs and matrixes. In addition, microscopical methods usually analyse only a small fraction of the total CNT sample and results can differ for different cross sections.

Differential mobility analysis has been reported to be a better alternative to other methods (e.g. light scattering, AFM) for faster determination of CNT length (<250 nm) distribution (Pease et al. 2009). This procedure is based on the separation and counting of CNT numbers on the condensation particle counter using a voltage sweep. However, this technique also needs previous purification and separation of agglomerates from CNT-suspension.

Most of the methods described above are capable of investigating stability of CNT suspensions in time, which influences their further storage, manipulating, processing and application conditions. However, there are only few works done in this field. In addition, systematic investigation of the influence of dispersion technique, mixing parameters, concentration of CNTs on the quality of obtained dispersions has not yet been performed. Therefore, investigation of CNT dispersions towards level of deagglomeration, size distribution and stability remains an actual and important research field.

4.8 Summary

The mechanical dispersion is the general tool for dispersion of nanofillers such as CNTs even if chemical functionalization of nanotubes is performed. Therefore, a great attention is paid for the improvement of existing mixing technologies or for development of new dispersion techniques. The most widely used physical dispersion techniques are ultrasonication, high-shear and high-impact mixing. Adaptation of these methods for dispersion of CNTs has been resulted in different variants of them. From the beginning high-powered dispersion conditions have been applied. However, it was found that strong damages or even destroying of CNTs occur during mixing. This often resulted in worse properties of dispersed CNTs, their insufficient for improvement length, high amount of impurities, etc. Therefore, uncial properties of CNTs could not be transferred on the maximal possible level into the composite materials. Later, the more effective dispersion of CNTs has been reached using the same techniques, which were modified and adapted for mild mixing conditions. The most promising nowadays are mild-condition ultrasonication (e.g. less destructive ultrasonication with diamond crystals or double-source ultrasonication), high-shear and low-pressure homogenisation, integral pump mixing (IPM), roll milling and jet mixing with rotating counter drums or combined use of them. Additionally, excellent dispersion method such as rapid expansion of supercritical suspensions (RESS) has been developed and adapted for composite production. It was also determinated that methods combined all possible action mechanisms (e.g. shearing, impact, extension, cavitation, etc.) are generally more sufficient for CNT-dispersion.

Other promising tools such as chemical functionalization and incorporation of other atoms in the lattice of CNTs have been found to be efficient for improved dispersability of CNTs. However, covalent changes in the CNT-structure results usually in creation of many undesirable defects and resulted functionalised nanotubes lost partially their extraordinary high characteristics. Compare to very popular chemical functionalization strategy, the physical treatment methods supporting better dispersability of CNTs have been found to be efficient. Various salts, surfactants, polyaromatic compounds, specific polymers (e.g. different block-copolymers or Gum Arabic) or even nanoparticles have been used as additives for improved deagglomeration of CNTs of various morphologies without damaging their chemical structure. This allows obtaining high-quality dispersions of CNTs with even higher lengths using mild-condition-powered mechanical dispersion methods. Due to better quality of dispersion of less damaged CNTs with higher length/diameter ratio the unique properties of nanotubes are served, which usually results in much better material properties (e.g. mechanical, transport properties) of related composites.

In such a way the maximal potential of CNTs could be achieved. Note that has a great importance, especially for aerospace applications. However, this problem is only partially solved. The systematically studies in the field of CNT-dispersions and their stability are very missing at the present day. Many industrial composite matrix systems are not investigated yet as potential media for CNT-based nanocomposites. Moreover, the relations between matrix and CNT types, their main characteristics and processing conditions differ for each system. Therefore, optimisation of CNT-dispersion procedures is needed for each nanocomposite. Additionally, the new production methods for CNTs are developing very fast, which results in higher and higher amount of potential systems to be studied. Note that application of CNTs in aerospace composite matrix systems is just at the starting stage and full potential of CNTs is not realised in structural composites yet. Therefore, further optimisation and development of dispersion methods is expected. Parallel to development of CNT-dispersion techniques the controlling and characterisation methods for the level of deagglomeration, quality of CNT-dispersions and related composites are under fast development now.

References

Abbasi, S., Carreau, P.J., Derdouri, A., Moan, M.: Rheological properties and percolation in suspensions of multiwalled carbon nanotubes in polycarbonate. Rheol. Acta. **48**, 943–959 (2009)

Abusafieh, A., Federico, D., Connell, S., Cohen, E.J., Willis, P.B.: Dimensional stability of CFRP composites for space based reflectors. Optomech. Design. Eng. **4444**, 9–16 (2001)

Acevedo-Rullan, A., Cameron-Soto, A.R., Aviles-Barreto, S.L.: Rheology and electrorheology of nanorod-loaded liquid crystalline polymers. In: Su, J., Wang, L.-P., Furuya, Y., Trolier-McKinstry, S., Leng, J. (eds.) Materials and Devices for Smart Systems III. Materials in Research Society Symposium Proceedings, vol. 1129, Warrendale, PA, 2009). doi: 10.1557/PROC-1129-V05-03 (2009)

Ago, H., Imamura, S., Okazaki, T., Saito, T., Yumura, M., Tsuji, M.: CVD growth of single-walled carbon nanotubes with narrow diameter distribution over Fe/MgO catalyst and their fluorescence spectroscopy. J. Phys. Chem. B **109**(20), 10035–10041 (2005)

Ahn, J.-H., Kim, Y.J., Hwang, S.J., Chung, H.: High energy ball milling of catalytically synthesized carbon nanotubes. Mater. Sci. Forum **534–536**, 193–196 (2007)

Ajayan, P.M.: Nanotubes from carbon. Chem. Rev. **99**, 1787–1799 (1999)

Ajayan, P.M., Iijima, S.: Smallest carbon nanotube. Nature **358**(6381), 23 (1992)

Ajayan, P.M., Zhou, O.Z.: Applications of carbon nanotubes. In: Dresselhaus, M.S., Dresselhaus, G., Avouris, Ph (eds.) Carbon Nanotubes. Topics in Applied Physics, vol. 80, pp. 391–425. Springer, Berlin (2001)

Allaoui, A., Bai, S., Cheng, H.M., Bai, J.B.: Mechanical and electrical properties of a MWNT/epoxy composite. Comput. Sci. Tech. **62**, 1993–1998 (2002)

Alvarez, L., Guillard, T., Sauvajol, J.L., Flamant, G., Laplaze, D.: Solar production of single-wall carbon nanotubes: growth mechanisms studied by electron microscopy and Raman spectroscopy. Appl. Phys. A **70**(2), 169–173 (2000)

Alvarez, L., Guillard, T., Sauvajol, J.L., Flamant, G., Laplaze, D.: Growth mechanisms and diameter evolution of single wall carbon nanotubes. Chem. Phys. Lett. **342**, 7–14 (2001)

Ando, Y., Zhao, X., Inoue, S., Iijima, S.: Mass production of multiwalled carbon nanotubes by hydrogen arc discharge. J. Cryst. Growth **237–239**, 1926–1930 (2002)

Andrews, R., Weisenberger, M.C.: Carbon nanotube polymer composites. Curr. Opin. Solid State Mater. Sci. **8**, 31–37 (2004)

Andrews, R., Jasques, D., Minot, M., Rantell, T.: Fabrication of carbon multiwall nanotube/polymer composites by shear mixing. Macromol. Mater. Eng. **287**, 395–403 (2002)

Anoop, A.K., Agarwal, U.S., Rani, J.: Carbon nanotubes-reinforced PET nanocomposite by melt-compounding. J. Appl. Polym. Sci. **104**, 3090–3095 (2007)

Backes, C., Mundloch, U., Ebel, A., Hauke, F., Hirsch, A.: Dispersion of HOPco® and CoMoCAT® single-walled nanotubes (SWNTs) by water soluble pyrene derivatives – depletion of small diameter SWNTs. Chem. Eur. J. **16**, 3314–3317 (2010)

Bahrami, M., Ranjbarian, S.: Production of micro- and nano-composite particles by supercritical carbon dioxide. J. Supercrit. Fluids. **40**, 263–283 (2007)

Baibarac, M., Gomez-Romero, P.: Nanocomposites based on conducting polymers and carbon nanotubes from fancy materials to functional applications. J. Nanosci. Nanotech. **6**(1), 1–14 (2006)

Bal, S., Samal, S.S.: Carbon nanotube reinforced polymer composites – a state of the art. Bull. Mater. Sci. **30**(4), 379–386 (2007)

Baldyga, J., Orciuch, W., Makowski, L., Malik, K., Özcan-Taskin, G., Eagles, W., Padron, G.: Dispersion of nanoparticle clusters in a rotor-stator mixer. Ind. Eng. Chem. Res. **47**, 3652–3663 (2008)

Baldyga, J., Makowski, L., Orciuch, W., Sauter, C., Schuchmann, H.P.: Agglomerate dispersion in cavitating flows. In: 13th European Conference on Mixing, London, 14–17 Apr 2009

Bandow, S., Asaka, S., Zhao, X., Ando, Y.: Purification and magnetic properties of carbon nanotubes. Appl. Phys. A: Mater. Sci. Proc. **67**, 23–27 (1998)

Bandyopadhyaya, R., Nativ-Roth, E., Regev, O., Yerushalmi-Rozen, R.: Stabilization of individual carbon nanotubes in aqueous solutions. Nano Lett. **2**(1), 25–28 (2002)

Banerjee, S., Wong, S.S.: Rational sidewall functionalization and purification of single-walled carbon nanotubes by solution-phase ozonolysis. J. Phys. Chem. B **106**(47), 12144–12151 (2002)

Barber, A.H., Cohen, S.R., Kenig, S., Wagner, H.D.: Interfacial fracture energy measurements for multi-walled carbon nanotubes pulled from a polymer matrix. Comput. Sci. Tech. **64**(15), 2283–2289 (2004)

Barone, P.W., Baik, S., Heller, D.A., Strano, M.S.: Near-infrared optical sensors based on single-walled carbon nanotubes. Nat. Mater. **4**, 86–92 (2005)

Battisti, A., Skordos, A.A., Partridge, I.K.: Percolation threshold of carbon nanotubes filled unsaturated polyesters. Comput. Sci. Tech. **70**(4), 633–637 (2010)

Bauer, M., Kahle, O., Landeck, S., Uhlig, C., Wurzel, R.: High performance composites using nanotechnology. Adv. Mater. Res. **32**, 149–152 (2008)

Baughman, R.H., Zakhidov, A.A., de Heer, W.A.: Carbon nanotubes – the route toward applications. Science **297**, 787–792 (2002)

Baxendale, M.: Carbon Nanotubes and Bucky Materials. In: Kasap, S., Capper, P. (eds.) Springer Handbook of Electronic and Photonic Materials, pp. 1147–1154. Springer, New York (2006)

Bayrakceken, A., Kitkamthorn, U., Aindow, M., Erkey, C.: Decoration of multi-wall carbon nanotubes with platinum nanoparticles using supercritical deposition with thermodynamic control of metal loading. Scr. Mater. **56**(2), 101–103 (2007)

Becker, O., Simon, G.P.: Epoxy layered silicate nanocomposites. Adv. Polym. Sci. **179**, 29–82 (2005)

Belin, T., Epron, F.: Characterization methods of carbon nanotubes: a review. Mater. Sci. Eng. B **119**, 105–118 (2005)

Bell, Ph.W., Anand, M., Fan, X., Enick, R.M., Roberts, C.B.: Stable dispersion of silver nanoparticles in carbon dioxide with fluorine-free ligands. Langmuir **21**, 11608–11613 (2005)

Bethune, D.S., Klang, C.-H., De Vries, M.S., Gorman, G., Savoy, R., Vazquez, J., Beyers, R.: Cobalt-catalysed growth of carbon nanotubes with single-atomic-layer walls. Nature **363**, 605–607 (1993)

Bhattacharya, M., Bhowmick, A.K.: Synergy in carbon black-filled natural rubber nanocomposites. Part I: Mechanical, dynamic mechanical properties, and morphology. J. Mater. Sci. (2010a). doi:10.1007/s10853-010-4699-6

Bhattacharya, M., Bhowmick, A.K.: Synergy in carbon black-filled natural rubber nanocomposites. Part II: Abrasion and viscoelasticity in tire like applications. J. Mater. Sci. (2010b). doi:10.1007/s10853-010-4700-4

Bhattacharyya, A.R., Sreekumar, T.V., Liu, T., Kumar, S., Ericson, L.M., Hauge, H., Smalley, R. E.: Crystallization and orientation studies in polypropylene/single wall carbon nanotube composites. Polymer **44**, 2373–2377 (2003)

Biercuk, M.J., Llaguno, M.C., Radosavljevic, M., Hyun, J.K., Johnson, A.T., Fischer, J.E.: Carbon nanotube composites for thermal management. Appl. Phys. Lett. **80**(15), 2767–2769 (2002)

Bittmann, B., Haupert, F., Schlarb, A.K.: Ultrasonic dispersion of inorganic nanoparticles in epoxy resin. Ultrason. Sonochem. **16**, 622–628 (2009)

Blanch, A.J., Lenehan, C.E., Quinton, J.S.: Optimizing surfactant concentrations for dispersion of single-walled carbon nanotubes in aqueous solution. J. Phys. Chem. B **114**(30), 9805–9811 (2010)

Boccaccini, A.R., Cho, J., Roether, J.A., Thomas, B.J.C., Minay, E.J., Shaffer, M.S.P.: Electrophoretic deposition of carbon nanotubes. Carbon **44**, 3149–3160 (2006)

Bokobza, L., Kolodziej, M.: On the use of carbon nanotubes as reinforcing fillers for elastomeric materials. Polym. Int. **55**, 1090–1098 (2006)

Bonard, J.-M., Stora, T., Salvetat, J.-P., Maier, F., Stöckli, T., Duschl, C., Forro, L., de Heer, W.A., Chatelain, A.: Purification and size-selection of carbon nanotubes. Adv. Mater. **9**(10), 827–831 (1997)

Bonifazi, D., Nacci, C., Marega, R., Campidelli, S., Ceballos, G., Modesti, S., Meneghetti, M., Prato, M.: Microscopic and spectroscopic characterization of paintbrush-like single-walled carbon nanotubes. Nano Lett. **6**(7), 1408–1414 (2006)

Booker, R.D., Green, M.J., Fan, H., Parra-Vasquez, A.N.G., Behabtu, N., Young, C.C., Hauge, R.H., Schmidt, H.K., Smalley, R.E., Hwang, W.-F., Pasquali, M.: High-shear treatment of single-walled carbon nanotube-superacid solutions as a pre-processing technique for the assembly of fibres and films. Proc. I Mech E. J. Nanoeng. Nanosyst.: Part N **222**, 101–109 (2010)

Bottini, M., Magrini, A., Rosato, N., Bergamaschi, A., Mustelin, T.: Dispersion of pristine single-walled carbon nanotubes in water by a thiolated organosilane: application in supramolecular nanoassemblies. J. Phys. Chem. B **110**(28), 13685–13688 (2006)

Bourbigot, S., Le Bras, M.: Zeolites: new synergistic agents for intumescent fire retardant thermoplastic formulations – criteria for the choice of the zeolite. Fire Mater. **20**, 145–154 (1996)

Bower, C., Suzuki, S., Tanigaki, K., Zhou, O.: Synthesis and structure of pristine and alkali-metal-intercalated single-walled carbon nanotubes. Appl. Phys. A. Mater. Sci. Proc. **67**, 47–52 (1998)

Bredeau, S., Peeterbroeck, S., Bonduel, D., Alexandre, M., Dubois, P.: From carbon nanotube coating to high-performance polymer nanocomposites. Polym. Int. **57**, 547–553 (2008)

Breuer, O., Sundararaj, U.: Big returns from small fibers: a review of polymer/carbon nanotube composites. Polym. Compos. **26**(6), 630–645 (2004)

Bricard, P., Friedel, L.: Two-phase jet dispersion. J. Hazard. Mater. **59**, 287–310 (1998)

Budnitskii, G.A., Matveev, V.S., Kazakov, M.E.: Properties and application of manmade fibres: carbon fibres and materials based on viscose fibres (review). Fibre Chem. **25**(5), 360–364 (1993)

Bystrzejewski, M., Huczko, A., Lange, H., Gemming, T., Büchner, B., Rümmeli, M.H.: Dispersion and diameter separation of multi-wall carbon nanotubes in aqueous solutions. J. Coll. Interf. Sci. **345**(2), 138–142 (2010)

Cadek, M., Coleman, J.N., Ryan, K.P., Nicolosi, V., Bister, G., Fonseca, A., Nagy, J.B., Szostak, K., Beguin, F., Blau, W.J.: Reinforcement of polymers with carbon nanotubes: the role of nanotube surface area. Nano Lett. **4**(2), 353–356 (2004)

Caneba, G.T., Dutta, C., Agrawal, V., Rao, M.: Novel ultrasonic dispersion of carbon nanotubes. J. Miner. Mater. Charact. Eng. **9**(3), 165–181 (2010)

Capek, I.: Dispersions, novel nanomaterial sensors and nanoconjugates based on carbon nanotubes. Adv. Coll. Interface Sci. **150**, 63–89 (2009)

Chang, C.-K., Hwang, J.-Y., Lai, W.-J., Chen, C.-W., Huang, C.-I., Chen, K.-H., Chen, L.-C.: Influence of solvent on the dispersion of single-walled carbon nanotubes in polymer matrix and the photovoltaic performance. J. Phys. Chem. C **114**(24), 10932–10936 (2010)

Chapartegui, M., Markaide, N., Florez, S., Elizetxea, C., Fernandez, M., Santamaria, A.: Specific rheological and electrical features of carbon nanotube dispersions in an epoxy matrix. Comput. Sci. Tech. **70**(5), 879–884 (2010)

Chen, Q., Huang, Z.: Carbon based nanostructures. In: Mahalik, N.P. (ed.) Micromanufacturing and Nanotechnology, pp. 247–274. Springer, Berlin (2006)

Chen, C., Tolle, T.B.: Aerospace applications for epoxy layered-silicate nanocomposite. In: Schwarz, J.A., Contescu, C.I., Putyera, K. (eds.) Dekker Encyclopedia of Nanoscience and Nanotechnology. Marcel Dekker, New York (2004)

Chen, Y., Gerald, J.F., Chadderton, L.T., Chaffron, L.: Nanoporous carbon produced by ball milling. Appl. Phys. Lett. **74**(19), 2782–2784 (1999)

Chen, R.J., Zhang, Y., Wang, D., Dai, H.: Noncovalent sidewall functionalization of single-walled carbon nanotubes for protein immobilization. J. Amer. Chem. Soc. **123**(16), 3838–3839 (2001)

Chen, C.S., Chen, X.H., Xu, L.S., Yang, Z., Li, W.H.: Modification of multi-walled carbon nanotubes with fatty acid and their tribological properties as lubricant additive. Carbon **43**, 1660–1666 (2005)

Chen, G.-X., Li, Y., Shimizu, H.: Ultrahigh-shear processing for the preparation of polymer/carbon nanotube composites. Carbon **45**, 2334–2340 (2007)

Chen, W., Lu, H., Nutt, S.R.: The influence of functionalized MWCNT reinforcement on the thermomechanical properties and morphology of epoxy nanocomposites. Comput. Sci. Tech. **68**, 2535–2542 (2008)

Chen, W.-C., Wang, R.K., Ziegler, K.J.: Coating individual single-walled carbon nanotubes with nylon 6,10 through emulsion polymerization. Appl. Mater. Eng. **1**(8), 1821–1826 (2009)

Chen, C., Ogino, A., Wang, X., Nagatsu, M.: Plasma treatment of multiwall carbon nanotubes for dispersion improvement in water. Appl. Phys. Lett. **96**(1–3), 131504 (2010)

Cheng, Q., Debnath, S., Gregan, E., Byrne, H.J.: Effect of solvent solubility parameters on the dispersion of single-walled carbon nanotubes. J. Phys. Chem. C **112**(51), 20154–20158 (2008)

Cheng, Q., Debnath, S., O'Neill, L., Hedderman, T.G., Gregan, E., Byrne, H.J.: Systematic study of the dispersion of SWNTs in organic solvents. J. Phys. Chem. C **114**(11), 4857–4863 (2010)

Chih, Y.-W., Cheng, W.-T.: Supercritical carbon dioxide-assisted synthesis of silver nano-particle in polyol process. Mater. Sci. Eng. B **145**, 67–75 (2007)

Cho, B.-M., Kim, G.-H.: Effect of the processing parameters on the surface resistivity of acrylonitrile-butadiene rubber/multiwalled carbon nanotube nanocomposites. J. Appl. Polym. Sci. **116**, 555–561 (2010)

Cho, W.-S., Hamada, E., Kondo, Y., Takayanagi, K.: Synthesis of carbon nanotubes from bulk polymer. Appl. Phys. Lett. **69**(2), 278–279 (1996)

Chou, T.-W., Gao, L., Thostenson, E.T., Zhang, Z., Byun, J.-H.: An assessment of the science and technology of carbon nanotube-based fibers and composites. Comput. Sci. Tech. **70**, 1–19 (2010)

Chun, K.-Y., Choi, S.K., Kang, H.J., Park, C.Y., Lee, C.J.: Highly dispersed multi-walled carbon nanotubes in ethanol using potassium doping. Carbon **44**, 1491–1495 (2006)

Coleman, J.N., Khan, U., Blau, W.J., Gun'ko, Y.K.: Small but strong: a review of the mechanical properties of carbon nanotube-polymer composites. Carbon **44**(9), 1624–1652 (2006a)

Coleman, J.N., Cadek, M., Ryan, K.P., Fonseca, A., Nagy, J.B., Blau, W.J., Ferreira, M.S.: Reinforcement of polymers with carbon nanotubes. The role of an ordered polymer interfacial region. Experiment and modelling. Polymer **47**, 8556–8561 (2006b)

Collison, C.J., Spencer, S., Preske, A., Palumbo, C., Helenic, A., Baily, R., Pellizzeri, S.: A new model for quantifying the extent of interaction between soluble polyphenylene-vinylenes and single-walled carbon nanotubes in solvent dispersions. J. Phys. Chem. B **114**(34), 11002–11009 (2010)

Courty, S., Mine, J., Tajbakhsh, A.R., Terentjev, E.M.: Nematic elastomers with aligned carbon nanotubes: new electromechanical actuators. Europhys. Lett. **64**(5), 654–660 (2003)

Crawley, E.F.: Intelligent structures for aerospace: technology overview and assessment. AIAA J. **32**(8), 1689–1699 (1994)

Dai, L.: Low-temperature, controlled synthesis of carbon nanotubes. Small **1**(3), 274–276 (2005)

Dalmas, F., Chazeau, L., Gauthier, C., Masenelli-Varlot, K., Dendievel, R., Cavaille, J.Y., Forro, L.: Multiwalled carbon nanotube/polymer nanocomposites: processing and properties. J. Polym. Sci.: Part B: Polymer Physics **43**, 1186–1197 (2005)

Datsyuk, V., Guerret-Piecourt, C., Dagreou, S., Billon, L., Dupin, J.-C., Flahaut, E., Peigney, A., Laurent, C.: Double walled carbon nanotube/polymer composites via in-situ nitroxide mediated polymerisation of amphiphilic block copolymers. Carbon **43**(4), 855–894 (2004)

De Rosa, I.M., Dinescu, A., Sarasini, F., Sarto, M.S., Tamburrano, A.: Effect of short carbon fibers and MWCNTs on microwave absorbing properties of polyester composites containing nickel-coated carbon fibers. Comput. Sci. Tech. **70**, 102–109 (2010)

Derycke, V., Martel, R., Appenzeller, J., Avouris, Ph: Controlling doping and carrier injection in carbon nanotube transistors. Appl. Phys. Lett. **80**(15), 2773–2775 (2002)

Desai, A.V., Haque, M.A.: Mechanics of the interface for carbon nanotube-polymer composites. Thin-Walled Structures **43**(11), 1787–1803 (2005)

Dhullipudi, R.B., Lvov, Y.M., Dobbins, T.A.: Dispersion of single walled carbon nanotubes by self assembly of polymers. Mater. Res. Soc. Symp. Proc. **947**, 0947-A03-58 (2007)

Dresselhaus, M.S., Dresselhaus, G., Avouris, Ph (eds.): Carbon Nanotubes: Synthesis, Structure, Properties, and Applications. Topics in Applied Physics, vol. 80, p. 425. Springer, Berlin (2001)

Du, F., Fischer, J.E., Winey, K.I.: Coagulation method for preparing single walled carbon nano tube/poly(methyl methacrylate) composites and their modulus, electrical conductivity, and thermal stability. J. Polym. Sci: Part B: Polym. Phys. **41**, 3333–3338 (2003)

Du, F., Scogna, R.C., Zhou, W., Brand, S., Fischer, J.E., Winey, K.I.: Nanotube networks in polymer nanocomposites: rheology and electrical conductivity. Macromolecules **37**, 9048–9055 (2004)

Du, J.-H., Bai, J., Cheng, H.-M.: The present status and key problems of carbon nanotube based on polymer composites. eXPRESS Polym. Lett. **1**(5), 253–273 (2007)

Du, Y., Hu, N., Zhou, H., Li, P., Zhang, P., Zhao, X., Dang, G., Chen, C.: Dispersion of multi-walled carbon nanotubes in poly(aryl ether ketone) obtained by in situ polymerization. Polym. Int. **58**, 832–837 (2009)

Duesberg, G.S., Muster, J., Krstic, V., Burghard, M., Roth, S.: Chromatographic size separation of single-wall carbon nanotubes. Appl. Phys. A: Mater. Sci. Proc. **67**, 117–119 (1998)

Dumoulin, M.M.: Polymer blends forming. Chapter 10. In: Utracki, L.A. (ed.) Polymer Blends Handbook, vol. 1, pp. 653–757. Kluwer Academic, Dordrecht (2003). 1500 p

Ebbesen, T.W., Ajayan, P.M., Hiura, H., Tanigaki, K.: Purification of nanotubes. Nature **367**(6463), 519 (1994)

Eklund, P.C., Pradhan, B.K., Kim, U.J., Xiong, Q., Fischer, J.E., Friedman, A.D., Holloway, B.C., Jordan, K., Smith, M.W.: Large-scale production of single-walled carbon nanotubes using ultrafast pulses from a free electron laser. Nano Lett. **2**(6), 561–566 (2002)

El-Hami, K., Matsushige, K.: Reduction and separation of carbon nanotube bundles using the P(VDF-TrFE) copolymer: nanobundles. J. Comp. Mater. **38**(16), 1443–1447 (2004)

Endo, M., Hayashi, T., Kim, Y.A., Terrones, M., Dresselhaus, M.S.: Applications of carbon nanotubes in the twenty-first century. Philos. Trans. R. Soc. Lond. A **362**, 2223–2238 (2004)

Endo, M., Strano, M.S., Ajayan, P.M.: Potential applications of carbon nanotubes. In: Jorio, A., Dresselhaus, G., Dresselhaus, M.S. (eds.) Carbon Nanotubes. Topics in Applied Physics, vol. 111, pp. 13–62. Springer, Berlin Heidelberg (2008)

Esawi, A., Morsi, K.: Dispersion of carbon nanotubes (CNTs) in aluminum powder. Compos.: Part A: Appl. Sci. Manuf. **38**(2), 646–650 (2007)

Fagan, J.A., Landi, B.J., Mandelbaum, I., Simpson, J.R., Bajpai, V., Bauer, B.J., Migler, K., Hight Walker, A.R., Raffaelle, R., Hobbie, E.K.: Comparative measures of single-wall carbon nanotube dispersion. J. Phys. Chem. B **110**(47), 23801–23805 (2006)

Fagan, J.A., Simpson, J.R., Bauer, B.J., De Paoli Lacerda, S.H., Becker, M.L., Chun, J., Migler, K. B., Hight Walker, A.R., Hobbie, E.K.: Length-dependent optical effects in single-wall carbon nanotubes. J. Am. Chem. Soc. **129**(34), 10607–10612 (2007)

Fan, Z., Wei, T., Luo, G., Wei, F.: Fabrication and characterization of multi-walled carbon nanotubes-based ink. J. Mater. Sci. **40**(18), 5075–5077 (2005)

Fauchais, P., Montavon, G.: Latest developments in suspension and liquid precursor thermal spraying. J. Therm. Spray Tech. **19**(1–2), 226–239 (2010)

Fauchais, P., Montavon, G., Bertrand, G.: From powders to thermally sprayed coatings. J. Therm. Spray Tech. **19**(1–2), 56–80 (2010)

Fiedler, B., Gojny, F.H., Wichmann, M.H.G., Nolte, M.C.M., Schulte, K.: Fundamental aspects of nano-reinforced composites. Comput. Sci. Tech. **66**, 3115–3125 (2006)

Firouzmanesh, M.R., Azar, A.A.: Study of thermal stability and ablation behavior of carbon/ epoxy-novolac composites. J. Appl. Polym. Sci. **88**, 2455–2461 (2003)

Fischer, L.: Advanced emulsification technologies. Narrow drop size distributions in the nano-meter region. Pharma Focus Asia: Manuf. **11** (2009) (http://www.pharmafocusasia.com/ manufacturing/emulsification_technologies.htm; http://www.pharmafocusasia.com/magazine/ archives/issue11.htm; http://www.pharmafocusasia.com/magazine/archives)

Fischer, L., Herzog, R.: A new dispersion technique as alternative to bead mills. Eur. Coat. J. **1**, 22–24 (2010)

Fischer, L., Scheid, S., Thomas, C.: Towards Nanoscaled Systems: New Dispersing Technologies. The Nanotechnology Conference and Trade Show, NSTI Nanotech, 1–5 June 2008

Foldvari, M., Bagonluri, M.: Carbon nanotubes as functional excipients for nanomedicines: I. Pharmaceutical properties. Nanomed.: NBM **4**, 173–182 (2008a)

Foldvari, M., Bagonluri, M.: Carbon nanotubes as functional excipients for nanomedicines: II. Drug delivery and biocompatibility issues. Nanomed.: NBM **4**, 183–200 (2008b)

Fonda, M., Petach, M., Rogers, C.F., Huntington, J., Stratton, D., Nishioka, K., Tipo, M.: Resuspension of particles by aerodynamic deagglomeration. Aerosol. Sci. Tech. **30**, 509–529 (1999)

Fonseca, A., Hernadi, K., Piedigrosso, P., Colomer, J.-F., Mukhopadhyay, K., Doome, R., Lazarescu, S., Biro, L.P., Lambin, Ph, Thiry, P.A., Bernaerts, D., Nagy, J.B.: Synthesis of single- and multi-wall carbon nanotubes over supported catalysts. Appl. Phys. A: Mater. Sci. Proc. **67**, 11–22 (1998)

Fraczek, A., Blazewicz, S.: Effect of dispersion of carbon nanotubes in polyacrylonitrile matrix on mechanical and thermal behavior of nanocomposites. 2nd National Conference on Nanotechnology 'NANO 2008'. J. Phys.: Conf. Ser. **146**, 012005 (6 pp) (2009)

Fritzsche, J., Lorenz, H., Klüppel, M.: CNT based elastomer-hybrid-nanocomposites with promising mechanical and electrical properties. Macromol. Mater. Eng. **294**, 551–560 (2009)
Gacitua, E.W., Ballerini, A.A., Zhang, J.: Polymer nanocomposites: synthetic and natural fillers. A review. Maderas: Ciencia y tecnologia **7**(3), 159–178 (2005)
Gao, Y., He, P., Lian, J., Schulz, M.J., Zhao, J., Wang, W., Wang, X., Zhang, J., Zhou, X., Shi, D.: Effects of surface modification, carbon nanofiber concentration, and dispersion time on the mechanical properties of carbon-nanofiber-polycarbonate composites. J. Appl. Polym. Sci. **103**, 3792–3797 (2007)
Garg, A.C., Mai, Y.-W.: Failure mechanism in toughened epoxy resins – a review. Comput. Sci. Tech. **31**, 179–223 (1988)
Gauthier, C., Chazeau, L., Prasse, T., Cavaille, J.Y.: Reinforcement effects of vapour grown carbon nanofibres as fillers in rubbery matrices. Comput. Sci. Tech. **65**, 335–343 (2005)
Ghose, S., Watson, K.A., Sun, K.J., Criss, J.M., Siochi, E.J., Connell, J.W.: High temperature resin/carbon nanotube composite fabrication. Comput. Sci. Tech. **66**, 1995–2002 (2006)
Gillham, J.K.: In: Seferis, J.C., Nicolais, L. (eds.) The Role of Polymer Matrix in Processing and Structural Properties of Composites, vol. 127. Plenum Press, New York (1983)
Gilman, J.W., Kashiwagi, T., Nyden, M., Brown, J.E.T., Jackson, C.L., Lomakin, S., Giannelis, E. P., Manias, E.: Flammability studies of polymer layered silicate nanocomposites: polyolefin, epoxy, and vinyl ester resins. Chapter 14. In: Ak-Malaika, S., Golovoy, A., Wilkie, C.A. (eds.) Chemistry and Technology of Polymer Additives, pp. 249–245. Blackwell Science Inc, Malden, MA (1999)
Giordano, A.N., Chaturvedi, H., Poler, J.C.: Critical coagulation concentrations for carbon nanotubes in nonaqueous solvent. J. Phys. Chem. C **111**(31), 11583–11589 (2007)
Girishkumar, G., Hall, T.D., Vinodgopal, K., Kamat, P.V.: Single wall carbon nanotube supports for portable direct methanol fuel cells. J. Phys. Chem. B. **110**(1), 107–114 (2006)
Godara, A., Mezzo, L., Luizi, F., Warrier, A., Lomov, S.V., van Vuure, A.W., Gorbatikh, L., Moldenaers, P., Verpoest, I.: Influence of carbon nanotube reinforced on the processing and the mechanical behaviour of carbon fiber/epoxy composites. Carbon **47**, 2914–2923 (2009)
Gojny, F.H., Wichmann, M.H.G., Köpke, U., Fiedler, B., Schulte, K.: Carbon nanotube-reinforced epoxy-composites: enhanced stiffness and fracture toughness at low nanotube content. Comput. Sci. Tech. **64**, 2363–2371 (2004)
Gojny, F.H., Wichmann, M.H.G., Fiedler, B., Bauhofer, W., Schulte, K.: Influence of nano-modification on the mechanical and electrical properties of conventional fibre-reinforced composites. Composites: Part A **36**, 1525–1535 (2005a)
Gojny, F.H., Wichmann, M.H.G., Fiedler, B., Schulte, K.: Influence of different carbon nanotubes on the mechanical properties of epoxy matrix composites – a comparative study. Comput. Sci. Tech. **65**, 2300–2313 (2005b)
Gong, X., Liu, J., Baskaran, S., Voise, R.D., Young, J.S.: Surfactant-assisted processing of carbon nanotube/polymer composites. Chem. Mater. **12**, 1049–1052 (2000)
Gou, J.: Single-walled nanotube bucky paper and composite. Polym. Int. **55**, 1283–1288 (2006)
Grimmer, C.S., Dharan, C.K.H.: Enhancement of delamination fatigue resistance in carbon nanotube reinforced glass fiber/polymer composites. Comput. Sci. Tech. **70**, 901–908 (2010)
Grossiord, N., Regev, O., Loos, J., Meuldijk, J., Koning, C.E.: Time-dependent study of the exfoliation process of carbon nanotubes in aqueous dispersions by using UV-visible spectroscopy. Anal. Chem. **77**, 5135–5139 (2005)
Guo, W., Guo, Y.: Giant axial electrostrictive deformation in carbon nanotubes. Phys. Rev. Lett. **91**(11), 115501–115504 (2003)
Guo, H., Sreekumar, T.V., Liu, T., Minus, M., Kumar, S.: Structure and properties of polyacrylonitrile/single wall carbon nanotube composite film. Polymer **46**, 3001–3005 (2005)
Guojun, Y., Sen, W., Jinlong, G., Dezhang, Z., Suixia, H.: Large-scale synthesis of bamboo-like carbon nanotubes over zeolites by catalytic decomposition of acetonitrile. Solid State Phenom. **121–123**, 167–170 (2007)

Gupta, N., Lin, T.C., Shapiro, M.: Clay/epoxy nanocomposites: processing and mechanical properties. JOM J. Miner. Met. Mater. Soc. **59**(3), 61–65 (2007)

Haggenmueller, R., Gommans, H.H., Rinzler, A.G., Fischer, J.E., Winey, K.I.: Aligned single-wall carbon nanotubes in composites by melt processing methods. Chem. Phys. Lett. **330**, 219–225 (2000)

Hanemann, T., Szabo, D.V.: Polymer-nanoparticle composites: from synthesis to modern applications. Materials **3**, 3468–3517 (2010)

Hartschuh, A., Qian, H., Georgi, C., Böhmler, M., Novotny, L.: Tip-enhanced near-field optical microscopy of carbon nanotubes. Anal. Bioanal. Chem. **394**(7), 1787–1795 (2009)

Harutyunyan, A.R., Pradhan, B.K., Kim, U.J., Chen, G., Eklund, P.C.: CVD synthesis of single wall carbon nanotubes under "soft" conditions. Nano Lett. **2**(5), 525–530 (2002)

Hasell, T., Yang, J., Wang, W., Li, J., Brown, P.D., Poliakoff, M., Lester, E., Howdle, S.M.: Preparation of polymer-nanoparticle composite beads by a nanoparticle-stabilised suspension polymerisation. J. Mater. Chem. **17**, 4382–4386 (2007)

Hassanien, A., Tokumoto, M., Kumazawa, Y., Kataura, H., Maniwa, Y., Suzuki, S., Achiba, Y.: Atomic structure and electronic properties of single-wall carbon nanotubes probed by scanning tunnelling microscope at room temperature. Appl. Phys. Lett. **73**(26), 3839–3841 (1998)

He, C.N., Tian, F.: A carbon nanotube- alumina network structure for fabricating epoxy composites. Scr. Mater. **61**, 285–288 (2009)

Hebert, M., Rousseau, C.-E., Shukla, A.: Shock loading and drop weight impact response of glass reinforced polymer composites. Comput. Struct. **84**(3), 199–208 (2008)

Hilding, J., Grulke, E.A., Zhang, Z.G., Lockwood, F.: Dispersion of carbon nanotubes in liquids. J. Disper. Sci. Technol. **24**(1), 1–41 (2003)

Hirano, A., Maeda, Y., Akasaka, T., Shiraki, K.: Synergistically enhanced dispersion of native protein-carbon nanotube conjugates by fluorualcohols in aqueous solution. Chem. Euro. J. **15**, 9905–9910 (2009)

Hornbostel, B., Haluska, M., Cech, J., Dettlaff, U., Roth, S.: ARC Discharge and Laser Ablation Synthesis of Singlewalled Carbon Nanotubes. In: Popov, V.N., Lambin, Ph (eds.) Carbon Nanotubes. NATO Sci. Ser. II: Math. Phys. Chem. **222**(1), 1–18. Springer, Dordrecht (2006)

Horsch, S., Serhatkulu, G., Gulari, E., Kannan, R.M.: Supercritical CO_2 dispersion of nano-clays and clay/polymer nanocomposites. Polymer **47**, 7485–7496 (2006)

Hshich, F.-Y., Beeson, H.D.: Flammability testing of flame-retarded epoxy composites and phenolics composites. Fire Mater. **21**, 41–49 (1997)

Hsu, W.K., Terrones, M., Hare, J.P., Terrones, H., Kroto, H.W., Walton, D.R.M.: Electrolytic formation of carbon nanostructures. Chem. Phys. Lett. **262**(1–2), 161–166 (1996)

Hu, N., Zhou, H., Dang, G., Rao, X., Chen, C., Zhang, W.: Efficient dispersion of multi-walled carbon nanotubes by in situ polymerization. Polym. Int. **56**, 655–659 (2007)

Hu, C.-Y., Xu, Y.-J., Duo, S.-W., Zhang, R.-F., Li, M.-S.: Non-covalent functionalization of carbon nanotubes with surfactants and polymers. J. Chin. Chem. Soc. **56**, 234–239 (2009)

Huang, X.: Fabrication and properties of carbon fibers. Materials **2**, 2369–2403 (2009)

Huang, Y.Y., Ahir, S.V., Terentyev, E.M.: Dispersion rheology of carbon nanotubes in a polymer matrix. Physical Review: B 73, 125422-1–125422-9 (2006)

Hudziak, S., Darfeuille, A., Zhang, R., Peijs, T., Mountjoy, G., Bertoni, G., Baxendale, M.: Magnetoresistive phenomena in an Fe-filled carbon nanotube/elastomer composite. Nanotechnology **21**, 8 pp (2010)

Hueso, L.E., Pruneda, J.M., Ferrari, V., Burnell, G., Valdes-Herrera, J.P., Simons, B.D., Littlewood, P.B., Artacho, E., Fert, A., Mathur, N.D.: Transformation of spin information into large electrical signals using carbon nanotubes. Nature **445**(25), 410–413 (2007)

Hurst, K.M., Roberts, C.B., Ashurst, W.R.: A gas-expanded liquid nanoparticle deposition technique for reducing the adhesion of silicone microstructures. Nanotechnology **20**, 185303 (9 pp) (2009)

Hussain, F., Hojjati, M., Okamoto, M., Gorga, R.E.: Review article: polymer-matrix nanocomposites, processing, manufacturing, and application: an overview. J. Comput. Math. **40**(17), 1511–1575 (2006)

Huxtable, S.T., Cahill, D.G., Shenogin, S., Xue, L., Ozisik, R., Barone, P., Usrey, M., Strano, M.S., Siddons, G., Shim, M., Keblinski, P.: Interfacial heat flow in carbon nanotube suspensions. Nat. Mater. **2**, 731–734 (2003)

Hwang, J.-F., Manson, J.A., Hertzberg, R.W., Miller, G.A., Sperling, L.H.: Fatigue crack propagation of rubber-toughened epoxies. Polym. Eng. Sci. **29**, 1477–1487 (1989)

Iijima, S.: Helical microtubules of graphitic carbon. Nature **354**, 56–58 (1991)

Inam, F., Yan, H., Reece, M.J., Peijs, T.: Dimethylformamide: an effective dispersant for making ceramic-carbon nanotube composites. Nanotechnology **19**, 195710 (5 pp) (2008)

Inam, F., Wong, D.W.Y., Kuwata, M., Peijs, T.Multiscale: Multiscale hybrid micro-nanocomposites based on carbon nanotubes and carbon fibers. J. Nanomater. (2010). doi:10.1155/2010/453420. Article ID: 453420, 12 p

Inkyo, M., Tahara, T.: Dispersion of agglomerated nanoparticles by micromedia mill, ultra apex mill. J. Soc. Powder Tech. Jap. **41**(8), 1–11 (2004)

Iyuke, S.E., Mahalik, N.P.: Carbon Nanotube Production and Applications: Basis of Nanotechnology. In: Mahalik, N.P. (ed.) Micromanufacturing and Nanotechnology, pp. 219–245. Springer, Berlin (2006)

Inkyo, M., Tokunaga, Y., Tahara, T., Iwaki, T., Iskandar, F., Hogan Jr., C.J., Okuyama, K.: Beads mill-assisted synthesis of Poly Methyl Methacrylate (PMMA)-TiO_2 nanoparticle composites. Ind. Eng. Chem. **47**, 2597–2604 (2008)

Islam, M.F., Rojas, E., Bergey, D.M., Johnson, A.T., Yodh, A.G.: High weight fraction surfactant solubilization of single-wall carbon nanotubes in water. Nano Lett. **3**(2), 269–273 (2003)

Itzhak, R., Raichman, D., Shahar, Z., Frey, G.L., Frey, J., Yerushalmi-Rozen, R.: Tailoring triblock copolymers for dispersion of individual, pristine, single-walled carbon nanotubes in organic solvents. J. Phys. Chem. C **114**(9), 3748–3753 (2010)

Iyuke, S.E., Mahalik, N.P.: Carbon Nanotube Production and Applications: Basis of Nanotechnology. In: Mahalik, N.P. (ed) Micromanufacturing and Nanotechnology, pp. 219–245. Springer, Berlin (2006)

Jang, J.W., Lee, D.K., Lee, C.E., Lee, T.J., Lee, C.J., Noh, S.J.: Growth-temperature induced metal-insulator transition in bamboo-shaped multiwalled carbon nanotubes. Solid State Commun. **124**, 147–150 (2002)

Jensen, L., Schmidt, O.H., Mikkelsen, K.V., Astrand, P.-O.: Static and frequency-dependent polarizability tensor for carbon nanotubes. J. Phys. Chem. B **104**(45), 10462–10466 (2000)

Jeong, S.-H., Lee, O.-J., Lee, K.-H.: Preparation of aligned carbon nanotubes with prescribed dimensions: template synthesis and sonication cutting approach. Chem. Mater. **14**(4), 1859–1862 (2002)

Jeong, K.-U., Lim, J.Y., Lee, J.-Y., Kang, S.L., Nah, C.: Polymer nanocomposites reinforced with multi-walled carbon nanotubes for semiconducting layers of high-voltage power cables. Polym. Int. **59**, 100–106 (2010)

Ji, Y., Huang, Y.Y., Rungsawang, R., Terenjev, E.M.: Dispersion and alignment of carbon nanotubes in liquid crystalline polymers and elastomers. Adv. Mater. **22**, 3436–3440 (2010)

Jiang, L., Tan, H., Wu, J., Huang, Y., Hwang, K.-C.: Continuum modeling of interfaces in polymer matrix composites reinforced by carbon nanotubes. Nano **2**(3), 139–148 (2007)

Jin, L., Bower, C., Zhou, O.: Alignment of carbon nanotubes in a polymer matrix by mechanical stretching. Appl. Phys. Lett. **73**(9), 1197–1199 (1998)

Jin, Z., Pramoda, K.P., Xu, G., Goh, S.H.: Dynamic mechanical behavior of melt-processed multi-walled carbon nanotube/poly(methyl methacrylate) composites. Chem. Phys. Lett. **337**, 43–47 (2001)

Journet, C., Bernier, P.: Production of carbon nanotubes. Appl. Phys. A. Mater. Sci. Proc. **67**, 1–9 (1998)

Journet, C., Maser, W.K., Bernier, P., Loiseau, A., Lamy de la Chapelle, M., Lefrant, S., Deniard, P., Lee, R., Fischer, J.E.: Large-scale production of single-walled carbon nanotubes by the electric-arc technique. Nature **388**, 756–758 (1997)

Jung, J., Perrut, M.: Particle design using supercritical fluids: literature and patent survey. J. Supercrit. Fluids **20**, 179–219 (2001)

Kalgaonkar, R.A., Jog, J.P.: Copolyester nanocomposites based on carbon nanotubes: reinforcement effect of carbon nanotubes on viscoelastic and dielectric properties of nanocomposites. Polym. Int. **57**, 114–123 (2008)

Kalyon, D.M., Dalwadi, D., Erol, M., Birinci, E., Tsenoglu, C.: Rheological behavior of concentrated suspensions as affected by the dynamics of the mixing process. Rheol. Acta **45**, 641–658 (2006)

Kandola, B.K., Horrocks, A.R., Myler, P., Blair, D.: Mechanical performance of heat/fire damaged novel flame retardant glass-reinforced epoxy composites. Composites: Part A **34**, 863–873 (2003)

Kang, Y., Taton, T.A.: Micelle-encapsulated carbon nanotubes: a route to nanotube composites. J. Am. Chem. Soc. **125**, 5650–5651 (2003)

Kao, C., Young, R.J.: A Raman spectroscopic investigation of heating effects and the deformation behaviour of epoxy/SWNT composites. Comput. Sci. Tech. **64**, 2291–2295 (2004)

Kar, T., Bettinger, H.F., Scheiner, S., Roy, A.K.: Noncovalent π-π stacking and CH-- π interactions of aromatics on the surface of single-wall carbon nanotubes: an MP2 study. J. Phys. Chem. C **112**(50), 20070–20075 (2008)

Karthikeyan, S., Mahalingam, P., Karthik, M.: Large scale synthesis of carbon nanotubes. E-J. Chem. **6**(1), 1–12 (2009)

Kasaliwal, G.R., Pegel, S., Göldel, A., Pötschke, P., Heinrich, G.: Analysis of agglomerate dispersion mechanisms of multiwalled carbon nanotubes during melt mixing in polycarbonate. Polymer **51**(12), 2708–2720 (2010)

Katoh, R., Tasaka, Y., Sekreta, E., Yumura, M., Ikazaki, F., Kakudate, Y., Fujiwara, S.: Sonochemical production of a carbon nanotube. Ultrasonics Sonochem. **6**, 185–187 (1999)

Kempel, F., Schlarb, A.K.: Strain sensing with nanoscale carbon fiber – epoxy composites. Appl. Mech. Mater. **13–14**, 239–245 (2008)

Khare, R., Bose, S.: Carbon nanotube based composites- a review. J. Miner. Mater. Charact. Eng. **4**(1), 31–46 (2005)

Kim, Y.A., Kamio, S., Tajiri, T., Hayashi, T., Song, S.M., Endo, M., Terrones, M., Dresselhaus, M.S.: Enhanced thermal conductivity of carbon fiber/phenolic resin composites by the introduction of carbon nanotubes. Appl. Phys. Lett. **9**, 093125 (2007)

Kim, J.-B., Premkumar, T., Giani, O., Robin, J.-J., Schue, F., Geckeler, K.E.: Nanocomposites of poly(L-lysine) and single-walled carbon nanotubes. Polym. Int. **57**, 311–315 (2008)

Kim, Y.I., Soundararajan, D., Park, C.W., Kim, S.H., Park, J.H., Ko, J.M.: Electrocatalytic properties of carbon nanofiber web-supported nanocrystalline Pt catalyst as applied to direct methanol fuel cell. Int. J. Electrochem. Sci. **4**, 1548–1559 (2009)

Kim, E.H., Lee, J.H., Jung, Y.G., Paik, U.: Enhancement of electro-optical properties in holographic polymer-dispersed liquid crystal films by incorporation of multiwalled carbon nanotubes into a polyurethane acrylate matrix. Polym. Int. **59**, 1289–1295 (2010a)

Kim, H., Abdala, A.A., Macosko, C.W.: Graphene/polymer nanocomposites. Macromolecules **43**, 6515–6530 (2010b)

Kis, A., Zettl, A.: Nanomechanics of carbon nanotubes. Philos. Trans. R. Soc. A. **366**, 1591–1611 (2008)

Koerner, H., Liu, W., Alexander, M., Mirau, P., Dowty, H., Vaia, R.A.: Deformation-morphology correlation in electrically conductive carbon nanotube-thermoplastic polyurethane nanocomposites. Polymer **46**(12), 4405–4420 (2005)

Komarov, F.F., Mironov, A.M.: Carbon nanotubes: present and future. Phys. Chem. Solid State. **5**(3), 411–429 (2004)

Kong, J., Cassel, A.M., Dai, H.: Chemical vapor deposition of methane for single-walled carbon nanotubes. Chem. Phys. Lett. **292**, 567–574 (1998)

Kongkanand, A., Vinodgopal, K., Kuwabata, S., Kamat, P.V.: Highly dispersed Pt catalysts on single-walled carbon nanotubes and their role in methanol oxidation. J. Phys. Chem. B **110**(33), 16185–16188 (2006)

Kongsted, J., Osted, A., Jensen, L., Astrand, P.-O., Mikkelsen, K.V.: Frequency-dependent polarizability of boron nitride nanotubes: a theoretical study. J. Phys. Chem. B. **105**(42), 10243–10248 (2001)

Konya, Z., Zhu, J., Niesz, K., Mehn, D., Kiricsi, I.: End morphology of ball milled carbon nanotubes. Carbon **42**(10), 2001–2008 (2004)

Kotsilkova, R., Ivanov, E., Krusteva, E., Silvestre, C., Cimmino, S., Duraccio, D.: Isotactic polypropylene composites reinforced with multiwall carbon nanotubes, Part 2: Thermal and mechanical properties related to the structure. J. Appl. Polym. Sci. **115**, 3576–3585 (2010)

Kovacs, J.Z., Andresen, K., Pauls, J.R., Garcia, C.P., Schossig, M., Schulte, K., Bauhofer, W.: Analyzing the quality of carbon nanotube dispersions in polymers using scanning electron microscopy. Carbon **45**(6), 1279–1288 (2007)

Krause, B., Ritschel, M., Täschner, Ch, Oswald, S., Gruner, W., Leonhardt, A., Pötschke, P.: Comparison of nanotubes produced by fixed bed and aerosol-CVD methods and their electrical percolation behaviour in melt mixed polyamide 6.6 composites. Comput. Sci. Tech. **70**, 151–160 (2010)

Krishnadas Nair, C.G.: Technology development and applications of composites at HAL. Bull. Mater. Sci. **17**(6), 1181–1196 (1994)

Kuemmeth, F., Ilani, S., Ralph, D.C., McEuen, P.L.: Coupling of spin and orbital motion of electrons in carbon nanotubes. Nature **452**(27), 448–452 (2008)

Kukovecz, A., Kanyo, T., Konya, Z., Kiricsi, I.: Long-time low-impact ball milling of multi-wall carbon nanotubes. Carbon **43**, 994–1000 (2005)

Kumar, M.K., Ramaprabhu, S.: Nanostructured Pt functionalized multiwalled carbon nanotube based hydrogen sensor. J. Phys. Chem. B. **110**(23), 11291–11298 (2006)

Laplaze, D., Bernier, P., Maser, W.K., Flamant, G., Guillard, T., Loiseau, A.: Carbon nanotubes: the solar approach. Carbon **36**(5–6), 685–688 (1998)

Lau, K.-T., Shi, S.-Q.: Failure mechanisms of carbon nanotube/epoxy composites pretreated in different temperature environments. Carbon **40**(15), 2965–2968 (2002)

Lau, K.-T., Lu, M., Lam, C.-K., Cheung, H.-Y., Sheng, F.-L., Li, H.-L.: Thermal and mechanical properties of single-walled carbon nanotube bundle-reinforced epoxy nanocomposites: the role of solvent for nanotube dispersion. Comput. Sci. Tech. **65**(5), 719–725 (2005)

Lau, K.-T., Gu, C., Hui, D.: A critical review on nanotube/nanoclay related polymer composite materials. Composites: Part B **37**, 425–436 (2006)

Le Bras, M., Bourbigot, S.: Mineral fillers in intumescent fire retardant formulations – criteria for the choice of a natural clay filler for the ammonium polyphosphate/pentaerythritol/polypropylene system. Fire Mater. **20**, 39–49 (1996)

Lebel, L.L., Aïssa, B., Paez, O.A., El Khakani, M.A., Therriault, D.: Three-dimensional micro structured nanocomposite beams by microfluidic infiltration. J. Micromech. Microeng. **19**, 125009, 7 pp (2009)

Lee, J.-H., Paik, U., Choi, J.-Y., Kim, K.K., Yoon, S.-M., Lee, J., Kim, B.-K., Kim, J.M., Park, M. H., Yang, C.W., An, K.H., Lee, Y.H.: Dispersion stability of single-walled carbon nanotubes using nafion in bisolvent. J. Phys. Chem. C **111**(6), 2477–2483 (2007)

Lee, J.-H., Yoon, S.-M., Kim, K.K., Cha, I.-S., Park, Y.J., Choi, J.-Y., Lee, Y.H., Paik, U.: Exfoliation of single-walled carbon nanotubes induced by the structural effect of perylene derivatives and their optoelectronic properties. J. Phys. Chem. C **112**(39), 15267–15273 (2008)

Lee, S.H., Kim, J.H., Choi, S.H., Kim, S.Y., Kim, K.W., Youn, J.R.: Effects of filler geometry on internal structure and physical properties of polycarbonate composites prepared with various carbon fillers. Polym. Int. **58**, 354–361 (2009)

Leer, C., Carneiro, O.S., Covas, J.A., Maia, J.M., van Hattum, F.W.J., Bernardo, C.A., Biro, L.P., Horvath, Z.E., Kiricsi, I.: Dispersion of carbon nanotubes in polycarbonate and its effect on the composite properties. Materials Science Forum **514–516**, 1125–1130 (2006)

Lefebvre, J., Maruyama, S., Finni, P.: Photoluminescence: science and applications. In: Jorio, A., Dresselhaus, G., Dresselhaus, M.S. (eds.) Carbon Nanotubes. Topics in Applied Physics, vol. 111, pp. 287–319. Springer, Berlin Heidelberg (2008)
Lepoint, T., Mullie, F.: What exactly is cavitation chemistry? Ultrason. Sonochem. **1**(1), 13–22 (1994)
Li, C., Chou, T.-W.: Multiscale modeling of carbon nanotube reinforced polymer composites. J. Nanosci. Nanotech. **3**(5), 423–430 (2003)
Li, Y., Shimizu, H.: High-shear processing induced homogeneous dispersion of pristine multiwalled carbon nanotubes in a thermoplastic elastomer. Polymer **48**(8), 2203–2207 (2007)
Li, Y.L., Yu, Y.D., Liang, Y.: A novel method for synthesis of carbon nanotubes: low temperature solid pyrolysis. J. Mater. Res. **12**(7), 1678–1680 (1997)
Li, Y.B., Wei, B.Q., Liang, J., Yu, Q., Wu, D.H.: Transformation of carbon nanotubes to nanoparticles by ball milling process. Carbon **37**, 493–497 (1999)
Li, F., Cheng, H.M., Bai, S., Su, G., Dresselhaus, M.S.: Tensile strength of single-walled carbon nanotubes directly measured from their macroscopic ropes. Appl. Phys. Lett. **77**(20), 3161–3163 (2000)
Li, X., Huang, Y., Li, J.: Study on synthesis and dispersion characteristics of MWNTs/PBO composites prepared by in-situ polymerization. Iranian Polym. J. **15**(4), 317–322 (2006)
Li, W.H., Lynam, C., Chen, J., Liu, B., Zhang, X.Z., Wallace, G.G.: Magnetorheology of single-walled nanotube dispersions. Mater. Lett. **61**(14–15), 3116–3118 (2007)
Li, X., Zhang, X., Ci, L., Shah, R., Wolfe, C., Kar, S., Talapatra, S., Ajayan, P.M.: Air-assisted growth of ultra-long carbon nanotube bundles. Nanotechnology **19**, 455609, 7 pp (2008)
Lian, Y., Maeda, Y., Wakahara, T., Akasaka, T., Kazaoui, S., Minami, N., Shimizu, T., Choi, N., Tokumoto, H.: Nondestructive and high-recovery-yield purification of single-walled carbon nanotubes by chemical functionalization. J. Phys. Chem. B **108**(26), 8848–8854 (2004)
Liang, Y., Ozawa, M., Krueger, A.: A general procedure to functionalize agglomerating nanoparticles demonstrated on nanodiamond. ACSNANO **3**(8), 2288–2296 (2009)
Liao, Y.-H., Marietta-Tondin, O., Liang, Z., Zhang, C., Wang, B.: Investigation of the dispersion process of SWNTs/SC-15 epoxy resin nanocomposites. Mater. Sci. Eng., A **385**, 175–181 (2004)
Lillehei, P.T., Kim, J.-W., Gibbons, L.J., Park, C.: A quantitative assessment of carbon nanotube dispersion in polymer matrices. Nanotechnology **20**, 325708 (7 pp) (2009)
Lim, B.K., Lee, S.H., Park, J.S., Kim, S.O.: Water-repellent macroporous carbon nanotube/elastomer nanocomposites by self-organized aqueous droplets. Macromol. Res. **17**(9), 666–671 (2009)
Lim, S.W., Park, E.Y., Cha, K.S., Lee, J., Shim, S.E.: Glass beads-assisted fine dispersion of multiwalled carbon nanotube in silicone matrix. Macromol. Res. **18**(8), 766–771 (2010)
Lin, Y., Taylor, S., Li, H., Fernando, K.A.S., Qu, L., Wang, W., Gu, L., Zhou, B., Sun, Y.-P.: Advances toward bioapplications of carbon nanotubes. J. Mater. Chem. **14**, 527–541 (2004)
Lind, T., Ammar, Y., Dehbi, A., Güntay, S.: De-agglomeration mechanisms of TiO_2 aerosol agglomerates in PWR steam generator tube rupture conditions. Nucl. Eng. Design **240**, 2046–2053 (2010)
Liu, T., Phang, I.Y., Shen, L., Chow, S.Y., Zhang, W.-D.: Morphology and mechanical properties of multiwalled carbon nanotubes reinforced nylon-6 composites. Macromolecules **37**, 7214–7222 (2004)
Liu, X.M., Romero, H.E., Gutierrez, H.R., Adu, K., Eklund, P.C.: Transparent boron-doped carbon nanotube films. Nano. Lett. **8**(9), 2613–2619 (2008a)
Liu, T., Luo, S., Xiao, Z., Zhang, C., Wang, B.: Preparative ultracentrifuge method for characterization of carbon nanotube dispersions. J. Phys. Chem. C **112**, 19193–19202 (2008b)
Loidl, D., Peterlik, H., Paris, O., Müller, M., Burghammer, M., Riekel, C.: Structure and mechanical properties of carbon fibres: a review of recent microbeam diffraction studies with synchrotron radiation. J. Synchrotron Rad. **12**, 758–764 (2005)

Lopez-Manchado, M.A., Biagiotti, J., Valentini, L., Kenny, J.M.: Dynamic mechanical and Raman spectroscopy studies on interaction between single-walled carbon nanotubes and natural rubber. J. Appl. Polym. Sci. **92**, 3394–3400 (2004)

Lorenz, H., Fritzsche, J., Das, A., Stöckelhuber, K.W., Jurk, R., Heinrich, G., Klüppel, M.: Advanced elastomer nano-composites based on CNT-hybrid filler systems. Comput. Sci. Tech. **69**(13), 2135–2143 (2009)

Lozano, K., Barrera, E.V.: Nanofiber-reinforced thermoplastic composites. I. Thermoanalytical and mechanical analyses. J. Appl. Polym. Sci. **79**, 125–133 (2001)

Lu, S.-Y., Hamerton, I.: Recent developments in the chemistry of halogen-free flame retardant polymers. Prog. Polym. Sci. **27**, 1661–1712 (2002)

Lu, K.L., Lago, R.M., Chen, Y.K., Green, M.H.L., Harris, P.J.F., Tsang, S.C.: Mechanical damage of carbon nanotubes by ultrasound. Carbon **34**(6), 814–816 (1996)

Lucas, M., Young, R.J.: Raman spectroscopic study of the effect of strain on the radial breathing modes of carbon nanotubes in epoxy/SWNT composites. Comput. Sci. Tech. **64**, 2297–2302 (2004)

Luo, P., Jiang, K., Wang, K., Yang, J., Xu, T., Chen, L., Fu, Q.: Plasma treatment-induced fluorine-functionalized multi-walled carbon nanotubes to modify poly(ethylene terephthalate) obtained via in situ polymerization. Polym. Int. **59**, 198–203 (2010)

Ma, W., Liu, L., Zhang, Z., Yang, R., Liu, G., Zhang, T., An, X., Yi, X., Ren, Y., Niu, Z., Li, J., Dong, H., Zhou, W., Ajayan, P.M., Xie, S.: High-strength composite fibers: realizing true potential of carbon nanotubes on polymer matrix through continuous reticulate architecture and molecular level couplings. Nano Lett. **9**(8), 2855–2861 (2009)

Ma, P.-C., Mo, S.-Y., Tang, B.-Z., Kim, J.-K.: Dispersion, interfacial interaction and re-agglomeration of functionalized carbon nanotubes in epoxy composites. Carbon **48**, 1824–1834 (2010)

Maeda, Y., Kimura, S.-I., Hirashima, Y., Kanda, M., Lian, Y., Wakahara, T., Akasaka, T., Hasegawa, T., Tokumoto, H., Shimizu, T., Kataura, H., Miyauchi, Y., Maruyama, S., Kobayashi, K., Nagase, S.: Dispersion of single-walled carbon nanotube bundles in nonaqueous solution. J. Phys. Chem. B **108**(48), 18395–18397 (2004)

Maelstrom Advanced Process Technologies Ltd.: In NanoPerspective – CPI 389 Newsletter, Summer, p. 3 (2007)

Magrez, A., Seo, J.W., Smajda, R., Korbely, B., Andresen, J.C., Mionic, M., Casimirius, S., Forro, L.: Low-temperature, highly efficient growth of carbon nanotubes on functional materials by an oxidative dehydrogenation reaction. ACS Nano. **4**(7), 3702–3708 (2010)

Mahmoodian, M., Pourabbas, B., Arya, A.B..: Preparation and characterization of Bis-GMA/TEGDMA/clay nanocomposites at low filler content regimes. J. Comput. Mater. **44**(11), 1379–1395 (2010)

Maldonado, S., Stevenson, K.J.: Direct preparation of carbon nanofiber electrodes via pyrolysis of Iron(II) phthalocyanine: electrocatalytic aspects for oxygen reduction. J. Phys. Chem. B **108**(31), 11375–11383 (2004)

Mangalgiri, P.D.: Composite materials for aerospace applications. Bull. Mater. Sci. **22**(3), 657–664 (1999)

Manivannan, S., Jeong, I.O., Ryu, J.H., Lee, C.S., Kim, K.S., Jang, J., Park, K.C.: Dispersion of single-walled carbon nanotubes in aqueous and organic solvents through a polymer wrapping functionalization. J. Mater. Sci. Mater. Electron. **20**, 223–229 (2009)

Manjunatha, C.M., Taylor, A.C., Kinloch, A.J., Sprenger, S.: The tensile fatigue behaviour of a silica nanoparticle-modified glass fibre reinforced epoxy composite. Comput. Sci. Tech. **70**, 193–199 (2010)

Manocha, L.M., Valand, J., Patel, N., Warrier, A., Manocha, S.: Nanocomposites for structural applications. Ind. J. Pure Appl. Phys. **44**, 135–142 (2006)

Margulis, M.A.: Fundamental problems of sonochemistry and cavitation. Ultrason. Sonochem. **1**(2), 1130–1139 (1994)

Martin, C.A., Sandler, J.K.W., Shaffer, M.S.P., Schwarz, M.-K., Bauhofer, W., Schulte, K., Windle, A.H.: Formation of percolating networks in multi-wall carbon-nanotube-epoxy composites. Comput. Sci. Tech. **64**, 2309–2316 (2004a)
Martin, R.B., Qu, L., Lin, Y., Harruf, B.A., Bunker, C.E., Gord, J.R., Allard, L.F., Sun, Y.-P.: Functionalized carbon nanotubes with tethered pyrenes: synthesis and photophysical properties. J. Phys. Chem. B **108**(31), 11447–11453 (2004b)
Maruyama, S., Miyauchi, Y., Murakami, Y., Chiashi, S.: Optical characterization of single-walled carbon nanotubes synthesized by catalytic decomposition of alcohol. New J. Phys. **5**, 149.1–149.12 (2003)
Matarredona, O., Rhoads, H., Li, Z., Harwell, J.H., Balzano, L., Resasco, D.E.: Dispersion of single-walled carbon nanotubes in aqueous solutions of the anionic suffactant NaDDBS. J. Phys. Chem: B **107**, 13357–13367 (2003)
Mathur, P.C., Bhatnagar, P.K., Singh, I., Bharadwaj, L.M., Pandey, R.: Optical characterization of single walled carbon nanotubes dispersed in sodium cholate and sodium dodecyl sulfate. In: Proceeding of 2nd IEEE International Nanoelectronics Conference (INEC 2008), pp. 968–971 (2008)
McClory, C., McNally, T., Baxendale, M., Potschke, P., Blau, W., Ruether, M.: Electrical and rheological percolation of PMMA/MWCNT nanocomposites as a function of CNT geometry and functionality. Eur. Polym. J. **46**, 854–868 (2010)
Meguid, S.A., Sun, Y.: On the tensile and shear strength of nano-reinforced composite interfaces. Mater. Design **25**, 289–296 (2004)
Min, B.G., Chae, H.G., Minus, M.L., Kumar, S.: Polymer/carbon nanotube composite fibers – an overview. In: Lee, K.-P., Gopalan, A.I., Marquis, F.D.S. (eds.) Functional Composites of Carbon Nanotubes and Application, pp. 43–73. Transworld Research Network, Trivandrum (2009)
Mohlala, M.S., Ray, S.S.: Preparation and characterization of polymer/multi-walled carbon nanotube nanocomposites. Solid State Phenom. **140**, 97–102 (2008)
Moniruzzaman, M., Winey, K.I.: Polymer nanocomposites containing carbon nanotubes. Macromolecules **39**, 5194–5205 (2006)
Mordkovich, V.Z.: Three types of behaviour of multiwall carbon nanotubes in reactions with intercalating agents. Mol. Cryst. Liq. Cryst. **340**, 775–780 (2000)
Mordkovich, V.Z.: Carbon nanofibers: a new ultrahigh-strength material for chemical technology. Theor. Found. Chem. Eng. **37**(5), 429–438 (2003)
Moreno-Castilla, C.: Adsorption of organic molecules from aqueous solutions on carbon materials. Carbon **42**, 83–94 (2004)
Morioka, K., Tomita, Y., Takigawa, K.: High-temperature fracture properties of CFRP composite for aerospace applications. Mat. Sci. Eng. A **319–321**, 675–678 (2001)
Mottaghitalab, V., Spinks, G.M., Wallace, G.G.: The influence of carbon nanotubes on mechanical and electrical properties of polyaniline fibers. Synth. Met. **152**, 77–80 (2005)
Mu, Y., Liang, H., Hu, J., Jiang, L., Wan, L.: Controllable Pt nanoparticle deposition on carbon nanotubes as an anode catalyst for direct methanol fuel cells. J. Phys. Chem. B **109**(47), 22212–22216 (2005)
Mu, M., Walker, A.M., Torkelson, J.M., Winey, K.I.: Cellular structures of carbon nanotubes in a polymer matrix improve properties relative to composites with dispersed nanotubes. Polymer **49**, 1332–1337 (2008)
Nakashima, N., Tanaka, Y., Tomonari, Y., Murakami, H., Kataura, H., Sakaue, T., Yoshikawa, K.: Helical superstructures of fullerene peapods and empty single-walled carbon nanotubes formed in water. J. Phys. Chem. B **109**(27), 13076–13082 (2005)
Nan, C.-W., Shi, Z., Lin, Y.: A simple model for thermal conductivity of carbon nanotube-based composites. Chem. Phys. Lett. **375**, 666–669 (2003)
Nanda, J., Maranville, C., Bollin, S.C., Sawall, D., Ohtani, H., Remillard, J.T., Ginder, J.M.: Thermal conductivity of single-wall carbon nanotube dispersions: role of interfacial effects. J. Phys. Chem. C **112**(3), 654–658 (2008)

Narh, K.A., Jallo, J.L., Dave, R.N.: Improving the dispersion of the carbon nanotubes in polymer based nanocomposites by dry coating. ANTEC 2199–2203 (2007)
Nasibulin, A.G., Moisala, A., Jiang, H., Kauppinen, E.I.: Carnon nanotube synthesis from alcohols by a novel aerosol method. J. Nanopart. Res. **8**, 465–475 (2006)
Ng, H.T., Foo, W.L., Li, J., Xu, G., Jaenicke, S., Chan, L., Li, S.F.Y.: Soft-lithography-mediated chemical vapor deposition of architectured carbon nanotube networks on elastomeric polymer. Langmuir **18**(1), 1–5 (2002)
Nguyen, D.A., Lee, Y.R., Raghu, A.V., Jeong, H.M., Shin, C.M., Kim, B.K.: Morphological and physical properties of a thermoplastic polyurethane reinforced with functionalized graphene sheet. Polym. Int. **58**, 412–417 (2009)
Nicolas, M.: Experimental study of gravity-driven dense suspension jets. Phys. Fluids **14**(10), 3570–3576 (2002)
Ning, J., Zhang, J., Pan, Y., Guo, J.: Fabrication and mechanical properties of SiO_2 matrix composites reinforced by carbon nanotube. Mater. Sci. Eng., A **357**(1–2), 392–396 (2003)
Niu, A., Han, Y., Wu, J., Yu, N., Xu, Q.: Synthesis of one-dimensional carbon nanomaterials wrapped by silver nanoparticles and their antibacterial behavior. J. Phys. Chem. C **114**(29), 12728–12735 (2010)
Njuguna, J., Pielichowski, K.: Polymer nanocomposites for aerospace applications: properties. Adv. Eng. Mater. **5**(11), 769–778 (2003)
Njuguna, J., Pielichowski, K.: Polymer nanocomposites for aerospace applications: fabrication. Adv. Eng. Mater. **6**(4), 193–203 (2004a)
Njuguna, J., Pielichowski, K.: Polymer nanocomposites for aerospace applications: characterization. Adv. Eng. Mater. **6**(4), 204–210 (2004b)
Njuguna, J., Pielichowski, K., Desai, S.: Nanofiller-reinforced polymer nanocomposites. Polym. Adv. Tech. **19**, 947–959 (2008)
Nogales, A., Broza, G., Roslaniec, Z., Schulte, K., Sics, I., Hsiao, B.S., Sanz, A., Garcia-Gutierrez, M.C., Rueda, D.R., Domingo, C., Ezquerra, T.A.: Low percolation threshold in nanocomposites based on oxidized single wall carbon nanotubes and poly(butylene terephthalate). Macromolecules **37**, 7669–7672 (2004)
Nurhaniza, M., Ariffin, M.K.A., Ali, A., Mustapha, F., Noraini, A.W.: Finite element analysis of composites materials for aerospace applications. 9th National Symposium on Polymeric Materials (NSPM 2009). IOP Conf. Series: Mater. Sci. Eng. **11**, 012010, 7 pp (2010)
Nyden, M.R., Stoliarov, S.I.: Calculations of the energy of mixing carbon nanotubes with polymers. Polymer **49**(2), 635–641 (2008)
O'Connell, M.J., Boul, P., Ericson, L.M., Huffman, C., Wang, Y., Haroz, E., Kuper, C., Tour, J., Ausman, K.D., Smalley, R.E.: Reversible water-solubilization of single-walled carbon nanotubes by polymer wrapping. Chem. Phys. Lett. **342**, 265–271 (2001)
O'Toole, B., Trabia, M.B., Thota, J., Wilcox, T., Nakelswamy, K.K.: Structural response of blast loaded composite containment vessels. SAMPE J. **42**(4), 6–13 (2006)
Odegard, G.M., Gates, T.S., Wise, K.E., Park, C., Siochi, E.J.: Constitutive modeling of nanotube-reinforced polymer composites. Comput. Sci. Tech. **63**(11), 1671–1687 (2003)
Ogasawara, T., Ishida, Y., Ishikawa, T., Yokota, R.: Characterization of multi-walled carbon nanotube/phenylethynyl terminated polyimide composites. Compos.: Part A **35**, 67–74 (2004)
Oh, J.S., Ahn, K.H., Hong, J.S.: Dispersion of entangled nanotube by melt extrusion. Korea-Austr. Rheol. J. **22**(2), 89–94 (2010)
Oliveira, M.M., Zarbin, A.J.G.: Carbon nanotubes decorated with both gold nanoparticles and polythiophene. J. Phys. Chem. C **112**(48), 18783–18786 (2008)
Orikasa, H., Inokuma, N., Okuba, S., Kitakami, O., Kyotani, T.: Template synthesis of water-dispersible carbon nano "test tubes" without any post-treatment. Chem. Mater. **18**, 1036–1040 (2006)
Osswald, T.A., Hernandez-Ortiz, J.P.: Polymer Processing: Modelling and Simulation, p. 606. Hanser Gardner Pub, Munich (2006)

Pacek, A., Baker, M., Utomo, A.T.: Characterisation of flow pattern in a rotor stator high shear mixer. In: Proceeding of European Congress of Chemical Engineering (ECCE-6) Copenhagen, 16–20 Sept 2007
Pampuch, R.: New materials and technologies. Bull. Polish Acad. Sci. Tech. Sci. **52**(4), 275–282 (2004)
Pan, Z.W., Xie, S.S., Chang, B.H., Wang, C.Y., Lu, L., Liu, W., Zhou, W.Y., Li, W.Z., Qian, L.X.: Very long carbon nanotubes. Nature **394**, 631–632 (1998)
Paradise, M., Goswani, T.: Carbon nanotubes – production and industrial applications. Mater. Design **28**, 1477–1489 (2006)
Paredes, J.I., Burghard, M.: Dispersion of individual single-walled carbon nanotubes of high length. Langmuir **20**, 5149–5152 (2004)
Park, C., Ounaies, Z., Watson, K.A., Crooks, R.E., Smith Jr, J., Lowther, S.E., Connell, J.W., Siochi, E.J., Harrison, J.S., St. Clair, T.L.: Dispersion of single wall carbon nanotubes by in situ polymerization under sonication. Chem. Phys. Lett. **364**, 303–308 (2002)
Park, C., Kang, J.H., Harrison, J.S., Costen, R.C., Lowther, S.E.: Actuating single wall carbon nanotube-polymer composites: intrinsic unimorphs. Adv. Mater. **20**, 2074–2079 (2008a)
Park, S., An, J., Piner, R.D., Jung, I., Yang, D., Velamakanni, A., Nguyen, S.T., Ruoff, R.S.: Aqueous suspension and characterization of chemically modified graphene sheets. Chem. Mater. **20**, 6592–6594 (2008b)
Park, H.S., Choi, B.G., Yang, S.H., Shin, W.H., Kang, J.K., Jung, D., Hong, W.H.: Ionic-liquid-assisted sonochemical synthesis of carbon-nanotube-based nanohybrids: control in the structures and interfacial characteristics. Small **5**(15), 1754–1760 (2009)
Pease III, L.F., Tsai, D.-H., Fagan, J.A., Bauer, B.J., Zangmeister, R.A., Tarlov, M.J., Zachariah, M.R.: Length distribution of single-walled carbon nanotubes in aqueous suspension measured by electrospray differential mobility analysis. Small **5**(24), 2894–2901 (2009)
Peeterbroeck, S., Alexandre, M., Nagy, J.B., Pirlot, C., Fonseca, A., Moreau, N., Philippin, G., Delhalle, J., Mekhalif, Z., Sporken, R., Beyer, G., Dubois, Ph: Polymer-layered silicate-carbon nanotube nanocomposites: unique nanofiller synergistic effect. Comput. Sci. Tech. **64**, 2317–2323 (2004)
Perez, R.M., Sandler, J.K., Altstädt, V., Hoffmann, T., Pospiech, D., Ciesielsci, M., Döring, M., Braun, U., Knoll, U., Schartel, B.: Effective halogen-free flame retardants for carbon fibre-reinforced epoxy composites. J. Mater. Sci. **41**, 4981–4984 (2006)
Pierard, N., Fonseca, A., Konya, Z., Willems, I., Van Tendeloo, G., Nagy, J.B.: Production of short carbon nanotubes with open tips by ball milling. Chem. Phys. Lett. **335**, 1–8 (2001)
Pierard, N., Fonseca, A., Colomer, J.-F., Bossuot, C., Benoit, J.-M., Van Tendeloo, G., Pirard, J.-P., Nagy, J.B.: Ball milling effect on the structure of single-wall carbon nanotubes. Carbon **42**, 1691–1697 (2004)
Popov, V.N.: Carbon nanotubes: properties and application. Mater. Sci. Eng. R. **43**, 61–102 (2004)
Poppe, T., Blum, J., Henning, Th: Generating a jet of deagglomerated small particles in vacuum. Rev. Sci. Instrum. **68**(6), 2529–2533 (1997)
Pötschke, P., Fornes, T.D., Paul, D.R.: Rheological behaviour of multiwalled carbon nanotube/ polycarbonate composites. Polymer **43**, 3247–3255 (2002)
Pötschke, P., Bhattacharyya, A.R., Janke, A.: Melt mixing of polycarbonate with multiwalled carbon nanotubes: microscopic studies on the state of dispersion. Eur. Polym. J. **40**, 137–148 (2004)
Pourmortazavi, S.M., Hajimirsadeghi, S.S.: Application of supercritical carbon dioxide in energetic materials processes: a review. Ind. Eng. Chem. Res. **44**, 6523–6533 (2005)
Prasse, T., Ivankov, A., Sandler, J., Schulte, K., Bauhofer, W.: Imaging of conductive filler networks in heterogeneous materials by scanning Kelvin microscopy. J. Appl. Polym. Sci. **82**, 3381–3386 (2001)
Priya, B.R., Byrne, H.J.: Investigation of sodium dodecyl benzene sulfonate assisted dispersion and debundling of single-wall carbon nanotubes. J. Phys. Chem. C **112**(2), 332–337 (2008)

Priya, B.R., Byrne, H.J.: Quantitative analyses of microwave-treated HiPco carbon nanotubes using absorption and Raman spectroscopy. J. Phys. Chem. C **113**(17), 7134–7138 (2009)
Prolongo, S.G., Buron, M., Gude, M.R., Chaos-Moran, R., Campo, M., Urena, A.: Effects of dispersion techniques of carbon nanofibers on the thermo-physical properties of epoxy nanoconposites. Comput. Sci. Tech. **68**, 2722–2730 (2008)
Prylutskyy, Yu I., Durov, S.S., Ogloblya, O.V., Buzaneva, E.V., Scharff, P.: Molecular dynamics simulation of mechanical, vibrational and electronic properties of carbon nanotubes. Comput. Mater. Sci. **17**, 352–355 (2000)
Qian, D., Dickey, E.C., Andrews, R., Rantell, T.: Load transfer and deformation mechanisms in carbon nanotube-polystyrene composites. Appl. Phys. Lett. **76**(20), 2868–2870 (2000)
Quilter, A.: Composites in aerospace applications. IHS White Paper **444**, 1–3 (2001)
Rahatekar, S.S., Koziol, K.K.K., Butler, S.A., Elliot, J.A., Shaffer, M.S.P., Mackley, M.R., Windle, A.H.: Optical microstructure and viscosity enhancement for an epoxy resin matrix containing multiwall carbon nanotubes. J. Rheol. **50**(5), 599–610 (2006)
Ramamurthy, P.C., Harrell, W.R., Gregory, R.V., Rao, A.M.: Integration and distribution of carbon nanotubes in solution-processed polyaniline/carbon nanotube composites. J. Electrochem. Soc. **154**(6), H495–H499 (2007)
Ramasubramaniam, R., Chen, J., Liu, H.: Homogeneous carbon nanotube/polymer composites for electrical application. Appl. Phys. Lett. **83**(14), 2928–2930 (2003)
Rana, R.K., Koltypin, Y., Gedanken, A.: Synthesis of carbon nanotubes from in situ generated cobalt nanoparticles and carbon monoxide. Chem. Phys. Lett. **344**, 256–262 (2001)
Ranjan, S., Jue, Z.-F., Chen, F.L.: Multi-walled carbon nanotube/polymer composite: a nano-enabled continuous fiber. J. Comp. Mater. **44**(11), 1305–1316 (2010)
Rastogi, R., Kaushal, R., Tripathi, S.K., Sharma, A.L., Kaur, I., Bharadwaj, L.M.: Comparative study of carbon nanotube dispersion using surfactants. J. Coll. Interface Sci. **328**(2), 421–428 (2008)
Regev, O., El Kati, P.N.B., Loos, J., Koning, C.E.: Preparation of conductive nanotube-polymer composites using latex technology. Adv. Mater. **16**(3), 248–251 (2004)
Ruoff, R.S., Lorents, D.C.: Mechanical and thermal properties of carbon nanotubes. Carbon **33**(7), 925–930 (1995)
Sabba, Y., Thomas, E.L.: High-concentration dispersion of single-wall carbon nanotubes. Macromolecules **37**, 4815–4820 (2004)
Safadi, B., Andrews, R., Grulke, E.A.: Multiwalled carbon nanotube polymer composites: synthesis and characterization of thin films. J. Appl. Polym. Sci. **84**, 2660–2669 (2002)
Sahoo, N.G., Rana, S., Cho, J.W., Li, L., Chan, S.H.: Polymer nanocomposites based on functionalized carbon nanotubes. Progr. Polym. Sci. **35**, 837–867 (2010)
Saito, K., Troiani, V., Qiu, H., Solladie, N., Sakata, T., Mori, H., Ohama, M., Fukuzumi, S.: Nondestructive formation of supramolecular nanohybrids of single-walled carbon nanotubes with flexible porphyrinic polypeptides. J. Phys. Chem. C **111**(3), 1194–1199 (2007)
Saito, R., Fantini, C., Jiang, J.: Excitonic states and resonance Raman spectroscopy of single-wall carbon nanotubes. In: Jorio, A., Dresselhaus, G., Dresselhaus, M.S. (eds.) Carbon Nanotubes. Topics in Applied Physics, vol. 111, pp. 251–286. Springer, Berlin, Heidelberg (2008)
Salamone, J.C. (ed.): Polymeric Materials Encyclopaedia, vol. 3, pp. 2204–2210. CRC Press, Boca Raton (1996)
Salvetat, J.-P., Bonard, J.-M., Thomson, N.H., Kulik, A.J., Forro, L., Benoit, W., Zuppiroli, L.: Mechanical properties of carbon nanotubes. Appl. Phys. A. **69**, 255–260 (1999)
Sandler, J., Shaffer, M.S.P., Prasse, T., Bauhofer, W., Schulte, K., Windle, A.H.: Development of a dispersion process for carbon nanotubes in an epoxy matrix and the resulting electrical properties. Polymer **40**, 5967–5971 (1999)
Sandler, J.K.W., Kirk, J.E., Kinloch, I.A., Shaffer, M.S.P., Windle, A.H.: Ultra-low electrical percolation threshold in carbon-nanotube-epoxy composites. Polymer **44**, 5893–5899 (2003)

Sands, J.M., Fink, B.K., McKnight, S.H., Newton, C.H., Gillespie Jr., J.W., Palmese, G.R.: Environmental issues for polymer matrix composites and structural adhesives. Clean Prod. Proc. **2**, 228–235 (2001)
Sauter, C., Schuchmann, H.P.: High pressure for dispersing and deagglomerating nanoparticles in aqueous solutions. Chem. Eng. Tech. **30**(10), 1401–1405 (2007)
Schadler, L.S., Giannaris, S.C., Ajayan, P.M.: Load transfer in carbon nanotube epoxy composites. Appl. Phys. Lett. **73**(26), 3842–3844 (1998)
Schaefer, D.W., Zhao, J., Brown, J.M., Anderson, D.P., Tomlin, D.W.: Morphology of dispersed carbon single-walled nanotubes. Chem. Phys. Lett. **375**, 369–375 (2003)
Scheid, S., Fischer, L.: Neue Dispergiertechnik als Alternative zu Techniken wie Perl- und Rührwerkskugelmühlen. Farbe und Lacke **9**(115), 30 (2009)
Schneider, T., Jensen, K.A.: Combined single-drop and rotating drum dustiness test of fine to nanosize powders using a small drum. Ann. Occup. Hyg. **52**(2), 23–34 (2008)
Sen, R., Satishkumar, B.C., Govindaraj, A., Harikumar, K.R., Raina, G., Zhang, J.-P., Cheetham, A.K., Rao, C.N.R.: B-C-N, C-N and B-N nanotubes produced by the pyrolysis of precursor molecules over Co catalysts. Chem. Phys. Lett. **287**, 671–676 (1998)
Seyhan, A.T., Tanoglu, M., Schulte, K.: Tensile mechanical behaviour and fracture toughness of MWCNT and DWCNT modified vinyl-ester/polyester hybrid nanocomposites produced by 3-roll milling. Mater. Sci. Eng., A **523**, 85–92 (2009)
Shaffer, M., Kinloch, I.A.: Prospects for nanotube and nanofibre composites. Comput. Sci. Tech. **64**(15), 2281–2282 (2004)
Shin, J., Kim, C., Geckeler, K.E.: Single-walled carbon naotube-polystyrene nanocomposites: dispersing nanotubes in organic media. Polym. Int. **58**, 579–583 (2009)
Shui, X., Chung, D.D.L.: A piezoresistive carbon filament polymer-matrix composite strain sensor. Smart Mater. Struct. **5**, 243–246 (1996)
Si, M., Zaitsev, V., Goldman, M., Frenkel, A., Peiffer, D.G., Weil, E., Sokolov, J.C., Rafailovich, M.H.: Self-extinguishing polymer/organoclay nanocomposites. Polym. Degr. Stab. **92**, 86–93 (2007)
Singjai, P., Thongtem, T., Kumfu, S., Thongtem, S.: Synthesis of CNTs via ethanol decomposition over ball-milled Fe_2O_3 coated copper sheets. Inorg. Mater. **43**(2), 143–147 (2007)
Sinnott, M., Cleary, P.: 3D DEM simulations of a high shear mixer. In: Third International Conference on CFD in the Materials and Process Industries, CSIRO, 10–12 Dec 2003. Melbourne, Australia, 217–222
Smart, S.K., Ren, W.C., Cheng, H.M., Lu, G.Q., Martin, D.J.: Shortened double-walled carbon nanotubes by high-energy ball milling. Int. J. Nanotechnol. **4**(5), 618–633 (2007)
Sokolov, V.N., Yablokova, M.A.: Apparatuses with gas injection and dispersion by turbulent liquid jets. Chem. Pet. Eng. **32**(3), 238–242 (1996)
Song, Y.S., Youn, J.R.: Influence of dispersion states of carbon nanotubes on physical properties of epoxy nanocomposites. Carbon **43**(7), 1378–1385 (2005)
Sorathia, U., Long, G., Gracik, T., Blum, M., Ness, J.: Screening tests for fire safety of composites for marine applications. Fire Mater. **25**, 215–222 (2001)
Spinks, G.M., Wallace, G.G.: Pneumatic carbon nanotube actuators. Adv. Mater. **14**(23), 1728–1732 (2002)
Stephan, O., Bando, Y., Loiseau, A., Willaime, F., Shramchenko, N., Tamiya, T., Sato, T.: Formation of small single-layer and nested BN cages under electron irradiation of nanotubes and bulk material. Appl. Phys. A: Mater. Sci. Proc. **67**, 107–111 (1998)
Strano, M.S.: Polymer-wrapped nanotubes. Nat. Mater. **5**, 433–434 (2006)
Strano, M.S., Moore, V.C., Miller, M.K., Allen, M.J., Haroz, E.H., Kittrell, C., Hauge, R.H., Smalley, R.E.: The role of surfactant adsorption during ultrasonication in the dispersion of single-walled carbon nanotubes. J. Nanosci. Nanotech. **3**, 81–86 (2003)
Stürzl, N., Hennrich, F., Lebedkin, S., Kappes, M.M.: Near monochiral single-walled carbon nanotube dispersions in organic solvents. J. Phys. Chem. C **113**(33), 14628–14632 (2009)

Suhr, J., Koratkar, N.A.: Energy dissipation in carbon nanotube composites: a review. J. Mater. Sci. **43**, 4370–4382 (2008)
Sui, G., Zhong, W.H., Yang, X.P., Yu, Y.H., Zhao, S.H.: Preparation and properties of natural rubber composites reinforced with pretreated carbon nanotubes. Polym. Adv. Tech. **19**, 1543–1549 (2008)
Sumfleth, J., Prado, L.A.S., Richau, S., Sriyai, M., Schulte, K.: Enhanced dispersion of MWCNTs and synergistic properties in multiphase epoxy nanocomposites by incorporation of inorganic nanoparticles. Solid State Phenom. **151**, 176–180 (2009a)
Sumfleth, J., Adroher, X.C., Schulte, K.: Synergistic effects in network formation and electrical properties of hybrid epoxy nanocomposites containing multi-wall carbon nanotubes and carbon black. J. Mater. Sci. **44**, 3241–3247 (2009b)
Sumfleth, J., Prehn, K., Wichmann, M.H.G., Wedekind, S., Schulte, K.: A comparative study of the electrical and mechanical properties of epoxy nanocomposites reinforced by CVD- and arc-grown multi-wall carbon nanotubes. Comput. Sci. Tech. **70**, 173–180 (2010)
Sun, Y., Wilson, S.R., Schuster, D.I.: High dissolution and strong light emission of carbon nanotubes in aromatic amine solvents. J. Am. Chem. Soc. **123**, 5348–5349 (2001)
Sun, Z., Zhang, X., Han, B., Wu, Y., An, G., Liu, Z., Miao, S., Miao, Z.: Coating carbon nanotubes with metal oxides in a supercritical carbon dioxide-ethanol solution. Carbon **45**(13), 2589–2596 (2007)
Sun, Z., Nicolosi, V., Rickard, D., Bergin, S.D., Aherne, D., Coleman, J.N.: Quantitative evaluation of surfactant-stabilized single-walled carbon nanotubes: dispersion quality and its correlation with zeta potential. J. Phys. Chem. C **112**(29), 10692–10699 (2008)
Suslick, K.S.: Sonochemistry. Science **247**(4949), 1439–1445 (1990)
Szleifer, I., Yerushalmi-Rozen, R.: Polymers and carbon nanotubes – dimensionality, interactions and nanotechnology. Polymer **46**(19), 7803–7818 (2005)
Tahhan, M., Truong, V.-T., Spinks, G.M., Wallace, G.G.: Carbon nanotube and polyaniline composite actuators. Smart Mater. Struct. **12**, 626–632 (2003)
Tai, N.-H., Yeh, M.-K., Liu, J.-H.: Enhancement of the mechanical properties of carbon nanotube/phenolic composites using a carbon nanotube network as the reinforcement. Carbon **42**, 2774–2777 (2004)
Takenaka, S., Ishida, M., Serizawa, M., Tanabe, E., Otsuka, K.: Formation of carbon nanofibers and carbon nanotubes through methane decomposition over supported cobalt catalysts. J. Phys. Chem. B **108**(31), 11464–11472 (2004)
Takenobu, T., Miura, N., Lu, S.-Y., Okimoto, H., Asano, T., Shiraishi, M., Iwasa, Y.: Ink-jet printing of carbon nanotube thin-film transistors on flexible plastic substrates. Appl. Phys. Express **2**, 025005-1–025005-3 (2009)
Tang, W., Santare, M.H., Advani, S.G.: Melt processing and mechanical property characterization of multi-walled carbon nanotube/high density polyethylene (MWNT/HDPE) composite films. Carbon **41**, 2779–2785 (2003)
Tang, Q.-Y., Chan, Y.-C., Wong, N.-B., Cheung, R.: Surfactant-assisted processing of polyimide/multiwall carbon nanotube nanocomposites for microelectronics applications. Polym. Int. **59**, 1240–1245 (2010)
Tekalur, S.A., Shukla, A., Shivakumar, K.: Blast resistance of polyurea based layered composite materials. Comput. Struct. **84**(3), 271–281 (2008a)
Tekalur, S.A., Shivakumar, K., Shukla, A.: Mechanical behavior and damage evolution in E-glass vinyl ester and carbon composites subjected to static and blast loads. Composites: Part B **39**(1), 57–60 (2008b)
Tekalur, S.A., Bogdanovich, A.E., Shukla, A.: Shock loading response of sandwich panels with 3-D woven E-glass composite skins and stitched foam core. Comput. Sci. Tech. **84**(3), 199–209 (2008c)
Tennent, H.G, Barber, J.J., Hoch, R..: Carbon fibrils, method for producing same and adhesive compositions containing same. US Patent 4663230 05/05/1987 (1996)

Thang, B.H., Hong, P.N., Khoi, P.H., Minh, P.N.: Application of multiwall carbon nanotubes for thermal dissipation in a micro-processor. J. Phys.: Conf. Ser. **187**, 012051, 6 pp (2009)
Thess, A., Lee, R., Nikolaev, P., Dai, H., Petit, P., Robert, J., Xu, C., Lee, Y.H., Kim, S.G., Rinzler, A.G., Colbert, D.T., Scuseria, G.E., Tomanek, D., Fischer, J.E., Smalley, R.E.: Crystalline ropes of metallic carbon nanotubes. Science **273**, 483–487 (1996)
Thiebaud, F., Gelin, J.C.: Characterization of rheological behaviors of polypropylene/carbon nanotubes composites and modeling their flow in a twin-screw mixer. Comput. Sci. Tech. **70**, 647–656 (2010)
Thiering, R., Dehghani, F., Foster, N.R.: Current issues relating to anti-solvent micronisation techniques and their extension to industrial scales. J. Supercrit. Fluids. **21**, 159–177 (2001)
Thomassin, J.-M., Lou, X., Pagnoulle, C., Saib, A., Bednarz, L., Huynen, I., Jerome, R., Detrembleur, C.: Multiwalled carbon nanotube/poly(ε-caprolactone)nanocomposites with exceptional electromagnetic interference shielding properties. J. Phys. Chem. C **111**(30), 11186–11192 (2007)
Thomsen, C., Reich, S.: Raman scattering in carbon nanotubes. In: Cardona, M., Merlin, R. (eds.) Light Scattering in Solid IX. Topics in Applied Physics, vol. 108, pp. 115–234. Springer, Berlin Heidelberg (2007)
Thostenson, E.T., Chou, T.-W.: Carbon nanotube-based composites for future air force and aerospace systems. Final report 2002–2005 for AFOSR GRANT F49620-02-1-0328 (2006)
Thostenson, E.T., Chou, T.-W.: On the elastic properties of carbon nanotube-based composites: modelling and characterization. J. Phys. D: Appl. Phys. **36**, 573–582 (2003)
Thostenson, E.T., Chou, T.-W.: Processing-structure-multi-functional property relationship in carbon nanotube/epoxy composites. Carbon **44**(14), 3022–3029 (2006)
Thostenson, E.T., Ren, Z., Chou, T.-W.: Advances in the science and technology of carbon nanotubes and their composites: a review. Comput. Sci. Tech. **61**, 1899–1912 (2001)
Thostenson, E.T., Li, C., Chou, T.-W.: Review: nanocomposites in context. Comput. Sci. Tech. **65**, 491–516 (2005)
Thostenson, E.T., Ziaee, S., Chou, T.-W.: Processing and electrical properties of carbon nanotube/vinyl ester nanocomposites. Comput. Sci. Tech. **69**, 801–804 (2009)
Tjong, S.C.: Structural and mechanical properties of polymer nanocomposites. Mater. Sci. Eng. R **53**, 73–197 (2006)
Tjong, S.C.: Editorial corner – a personal view. Carbon nanotubes – attractive nanofillers for forming bio-, functional and structural polymer composites. eXPRESS Polym. Lett. **4**(9), 516 (2010)
To, D., Dave, R.: Deagglomeration of nanoparticle aggregates via rapid expansion of supercritical or high-pressure suspensions. AICHE J. **55**(11), 2807–2826 (2009)
Toebes, M.L., van der Lee, M.K., Tang, L.M., in 't Veld, M.H.H., Bitter, J.H., van Dillen, A.J., de Jong, K.P.: Preparation of carbon nanofiber supported platinum and ruthenium catalysts: comparison of ion adsorption and homogeneous deposition precipitation. J. Phys. Chem. B. **108**(31), 11611–11619 (2004)
Toth, G., Mäklin, J., Halonen, N., Palosaari, J., Juuti, J., Jantunen, H., Kordas, K., Sawyer, W.G., Vajtai, R., Ajayan, P.M.: Carbon-nanotube-based electrical brush contacts. Adv. Mater. **21**, 1–5 (2009)
Treacy, M.M.J., Ebbesen, T.W., Gibson, J.M.: Exceptionally high Young's modulus observed for individual carbon nanotubes. Nature **381**, 678–680 (1996)
U.S. Department of Defense: Composite materials handbook. Polymer matrix composites. Materials usage, design, and analysis, vol. 3. Department of Defence Handbook, Department of Defense, United States of America (2002)
Unrau, C.J., Axelbaum, R.L., Biswas, P., Fraundorf, P.: Synthesis of single-walled carbon nanotubes in oxy-fuel inverse diffusion flames with online diagnostics. Proc. Combust. Inst. **31**(2), 1865–1872 (2007)
Unrau, C.J., Axelbaum, R.L., Fraundorf, P.: Single-walled carbon nanotube formation on iron oxide catalysts in diffusion flames. J. Nanopart. Res. **12**, 2125–2133 (2010)

Usrey, M.L., Chafee, A., Jeng, E.S., Strano, M.S.: Application of polymer solubility theory to solution phase dispersion of single-walled carbon nanotubes. J. Phys. Chem. C **113**(22), 9532–9540 (2009)
Vail, J.R., Burris, D.L., Sawyer, W.G.: Multifunctionality of single-walled carbon nanotube-polytetrafluoroethylene. Wear **267**, 619–624 (2009)
Vaisman, L., Wagner, H.D., Marom, G.: The role of surfactants in dispersion of carbon nanotubes. Adv. Coll. Interf. Sci. **128–130**, 37–46 (2006)
Valentini, L., Biagiotti, J., Kenny, J.M., Lopez Manchado, M.A.: Physical and mechanical behavior of single-walled carbon nanotube/polypropylene/ethylene-propylene-diene rubber nanocomposites. J. Appl. Polym. Sci. **89**, 2657–2663 (2003)
Valentini, L., Puglia, D., Frulloni, E., Armentano, I., Kenny, J.M., Santucci, S.: Dielectric behaviour of epoxy matrix/single-walled carbon nanotube composites. Comput. Sci. Tech. **64**, 23–33 (2004)
Vander Wal, R.L., Ticich, T.M., Curtis, V.E.: Diffusion flame synthesis of single-walled carbon nanotubes. Chem. Phys. Lett. **323**, 217–223 (2000)
Varadan, V.K., Xie, J.: Large-scale synthesis of multi-walled carbon nanotubes by microwave CVD. Smart Mater. Struct. **11**, 610–616 (2002)
Vega, J.F., Martinez-Salazar, J., Trujillo, M., Arnal, M.L., Müller, A.J., Bredeau, S., Dubois, Ph: Rheology, processing, tensile properties, and crystallization of polyethylene/carbon nanotube nanocomposites. Macromolecules **42**(13), 4719–4727 (2009)
Velasco-Santos, C., Martinez-Hernandez, A.L., Fisher, F., Ruoff, R., Castano, V.M.: Dynamical-mechanical and thermal analysis of carbon nanotube-methyl-ethyl methacrylate nanocomposites. J. Phys. D: Appl. Phys. **36**, 1423–1428 (2003)
Vlasveld, D.P.N., Daud, W., Nersee, H.E.N., Picken, S.J.: Continuous fibre composites with nanocomposite matrix: improvement of flexural and compressive strength at elevated temperatures. Composites: Part A **38**, 730–738 (2007)
Vohs, J.K., Brege, J.J., Raymond, J.E., Brown, A.E., Williams, G.L., Fahlman, B.D.: Low-temperature growth of carbon nanotubes from the catalytic decomposition of carbon tetrachloride. J. Am. Chem. Soc. **126**(32), 9936–9937 (2004)
Wagner, H.D.: Nanotube-polymer adhesion: a mechanics approach. Chem. Phys. Lett. **361**, 57–61 (2002)
Wagner, H.D., Lourie, O., Feldman, Y., Tenne, R.: Stress-induced fragmentation of multiwall nanotubes in a polymer matrix. Appl. Phys. Lett. **72**(2), 188–190 (1998)
Wang, Y., Wu, J., Wei, F.: A treatment method to give separated multi-walled carbon nanotubes with high purity, high crystallization and a large aspect ratio. Carbon **41**, 2939–2948 (2003)
Wang, F., Dukovic, G., Brus, L.E., Heinz, T.F.: Time-resolved fluorescence of carbon nanotubes and its implication for radiative lifetimes. Phys. Rev. Lett. **92**(17), 177401-1–177401-4 (2004)
Wang, S., Liang, Z., Wang, B., Zhang, C., Rahman, Z.: Precise cutting of single-walled carbon nanotubes. Nanotechnology **18**, 055301 (6 pp) (2007a)
Wang, Z., Liu, Q., Zhu, H., Liu, H., Chen, Y., Yang, M.: Dispersing multi-walled carbon nanotubes with water-soluble block copolymers and their use as supports for metal nanoparticles. Carbon **45**, 285–292 (2007b)
Wang, Z., Meng, X., Li, J., Du, X., Li, S., Jiang, Z., Tang, T.: A simple method for preparing carbon nanotubes/clay hybrids in water. J. Phys. Chem. B **113**(19), 8058–8064 (2009)
Wassei, J.K., Tung, V.C., Jonas, S.J., Cha, K., Dunn, B.S., Yang, Y., Kaner, R.B.: Stenciling graphene, carbon nanotubes, and fullerenes using elastomeric lift-Off membranes. Adv. Mater. **22**, 897–901 (2010)
Wei, D., Dave, R., Pfeffer, R.: Mixing and characterization of nanosized powders: an assessment of different techniques. J. Nanopart. Res. **4**, 21–41 (2002)
Williams, G., Trask, R., Bond, I.: A self-healing carbon fibre reinforced polymer for aerospace applications. Compos.: Part A: Appl. Sci. Manuf. **38**(6), 1525–1532 (2007)

Wiltshire, J.G., Li, L.J., Khobustov, A.N., Padbury, C.J., Briggs, G.A.D., Nicholas, R.J.: Magnetic separation of Fe catalyst from single-walled carbon nanotubes in an aqueous surfactant solution. Carbon **43**, 1151–1155 (2005)
Wohlstadter, J.N., Wilbur, J.L., Sigal, G.B., Biebuyck, H.A., Billadeau, M.A., Dong, L., Fischer, A.B.., Gudibande, S.R., Jameison, S.H., Kenten, J.H., Leginus, J., Leland, J.K., Massey, R.J., Wohlstadter, S.J.: Carbon nanotube-based biosensor. Adv. Mater. **15**(14), 1184–1187 (2003)
Woo, D.K., Kim, B.C., Lee, S.J.: Preparation and rheological behaviour of polystyrene/multi-walled carbon nanotube composites by latex technology. Korea Australia Rheology Journal **21**(3), 185–191 (2009)
Wu, S., Zhang, F., Yu, Y., Li, P., Yang, X., Lu, J., Ryu, S.: Preparation of -based carbon nanofibers by hot-stretching. Compos. Interf. **15**(7–9), 671–677 (2008a)
Wu, C.-I., Huang, J.-W., Wen, Y.-L., Wen, S.-B., Shen, Y.-H., Yeh, M.-Y.: Preparation of TiO_2 nanoparticles by supercritical carbon dioxide. Mater. Lett. **62**, 1923–1926 (2008b)
Xia, H., Wang, Q., Qui, G.: Polymer-encapsulated carbon nanotubes prepared through ultrasonically initiated in situ emulsion polymerization. Chem. Mater. **15**, 3879–3886 (2003)
Xiao, K.Q., Zhang, L.C.: Effective separation and alignment of long entangled carbon nanotubes in epoxy. J. Mater. Sci. **40**, 6513–6516 (2005)
Xie, X.-L., Mai, Y.-W., Zhou, X.-P.: Dispersion and alignment of carbon nanotubes in polymer matrix: a review. Mater. Sci. Eng. R **49**, 89–112 (2005)
Xu, X., Thwe, M.M., Shearwood, C., Liao, K.: Mechanical properties and interfacial characteristics of carbon-nanotube-reinforced epoxy thin films. Appl. Phys. Lett. **80**(15), 2833–2835 (2002)
Xu, G., Du, L., Wang, H., Xia, R., Meng, X., Zhu, Q.: Nonisothermal crystallization kinetics and thermomechanical properties of multiwalled carbon nanotube-reinforced poly(ε-caprolactone) composites. Polym. Int. **57**, 1052–1066 (2008)
Yan, L.Y., Poon, Y.F., Chan-Park, M.B., Chen, Y., Zhang, Q.: Individually dispersing single-walled carbon nanotubes with novel neutral pH water-soluble chitosan derivatives. J. Phys. Chem. C **112**(20), 7579–7587 (2008)
Yang, K., Ozisik, R.: Novel route to nanoparticle dispersion using supercritical carbon dioxide. J. Supercrit. Fluids. **43**, 515–523 (2008)
Yang, J., Sliva, A., Banerrjee, A., Dave, R.N., Pfeffer, R.: Dry particle coating for improving the flowability of cohesive powders. Powder Technol. **158**, 21–33 (2005)
Yang, K.-K., Wang, X.-L., Wang, Y.-Z.: Progress in nanocomposite of biodegradable polymer. J. Ind. Eng. Chem. **13**(4), 485–500 (2007)
Yasmin, A., Abot, J.L., Daniel, I.M.: Processing of clay/epoxy nanocomposites with a three-roll mill machine. Mater. Res. Soc. Symp. Proc. **740**, 13.7.1–13.7.6 (2003)
Yasmin, A., Luo, J.-J., Daniel, I.M.: Processing of expanded graphite reinforced polymer nanocomposites. Comput. Sci. Tech. **66**, 1179–1186 (2006)
Young, R.J., Eichhorn, S.J.: Deformation mechanisms in polymer fibres and nanocomposites. Polymer **48**, 2–18 (2007)
Yu, M.F., Lourie, O., Dyer, M.J., Moloni, K., Kelly, T.F., Ruoff, R.S.: Strength and breaking mechanism of multiwalled carbon nanotubes under tensile load. Science **287**, 637–640 (2000)
Yu, J., Grossiord, N., Koning, C.E., Loos, J.: Controlling the dispersion of multi-wall carbon nanotubes in aqueous surfactant solution. Carbon **45**, 618–623 (2007)
Yu, C., Masarapu, C., Rong, J., Wei, B., Jiang, H.: Stretchable supercapacitors based on buckled single-walled carbon nanotube macrofilms. Adv. Mater. **21**, 4793–4797 (2009a)
Yu, J., Chen, Y., Cheng, B.M.: Dispersion of boron nitride nanotubes in aqueous solution with the help of ionic surfactants. Solid State Commun. **149**, 763–766 (2009b)
Yuan, W., Hu, L., Yu, Z., Lam, T., Biggs, J., Ha, S.M., Xi, D., Chen, B., Senesky, M.K., Grüner, G., Pei, Q.: Fault-tolerant dielectric elastomer actuators using single-walled carbon nanotube electrodes. Adv. Mater. **20**, 621–625 (2008)
Yudasaka, M., Zhang, M., Jabs, C., Iijima, S.: Effect of an organic polymer in purification and cutting of single-wall carbon nanotubes. Appl. Phys. A: Mater. Sci. Proc. **71**, 449–451 (2000)
Yue, D., Liu, Y., Shen, Z.: Study on preparation and properties of carbon nanotubes/rubber composites. J. Mater. Sci. **41**, 2541–2544 (2006)

Zeng, X., Sun, X., Cheng, G., Yan, X., Xu, X.: Production of multi-wall carbon nanotubes on a large scale. Physica B **323**, 330–332 (2002)

Zhang, Y., Gu, H., Iijima, S.: Single-wall carbon nanotubes synthesized by laser ablation in a nitrogen atmosphere. Appl. Phys. Lett. **73**(26), 3827–3829 (1998)

Zhang, W.D., Shen, L., Phang, I.Y., Liu, T.: Carbon nanotubes reinforced nylon-6 composite prepared by simple melt-compounding. Macromolecules **37**, 256–259 (2004a)

Zhang, W.-D., Phang, I.Y., Shen, L., Chow, S.Y., Liu, T.: Polymer nanocomposites using urchin-shaped carbon nanotube-silica hybrids as reinforcing fillers. Macromol. Rapid Commun. **25**, 1860–1864 (2004b)

Zhang, L.C., Zarudi, I., Xiao, K.Q.: Novel behaviour of friction and wear of epoxy composites reinforced by carbon nanotubes. Wear **261**, 806–811 (2006)

Zhang, J., Gao, L., Sun, J., Liu, Y., Wang, Y., Wang, J., Kajiura, H., Li, Y., Noda, K.: Dispersion of single-walled carbon nanotubes by nafion in water/ethanol for preparing transparent conducting films. J. Phys. Chem. C **112**(42), 16370–16376 (2008a)

Zhang, Q., Fang, F., Zhao, X., Li, Y., Zhu, M., Chen, D.: Use of dynamic rheological behavior to estimate the dispersion of carbon nanotubes in carbon nanotube/polymer composites. J. Phys. Chem. B **112**(40), 12606–12611 (2008b)

Zhang, C., Ren, L., Wang, X., Liu, T.: Graphene oxide-assisted dispersion of pristine multiwalled carbon nanotubes in aqueous media. J. Phys. Chem. C **114**(26), 11435–11440 (2010)

Zhao, X., Ohkohchi, M., Shimoyama, H., Ando, Y.: Morphology of carbon allotropes prepared by hydrogen arc discharge. J. Cryst. Growth **198–199**, 934–938 (1999)

Zheng, M., Jagota, A., Semke, E.D., Diner, B.A., McLean, R.S., Lustig, S.R., Richardson, R.E., Tassi, N.G.: DNA-assisted dispersion and separation of carbon nanotubes. Nat. Mater. **2**, 338–342 (2003)

Zheng, S.-F., Hu, J.-S., Zhong, L.-S., Wan, L.-J., Song, W.-G.: In Situ one-step method for preparing carbon nanotubes and Pt composite catalysts and their performance for methanol oxidation. J. Phys. Chem. C **111**(30), 11174–11179 (2007)

Zhou, X., Shin, E., Wang, K.W., Bakis, C.E.: Interfacial damping characteristics of carbon nanotube-based composites. Comput. Sci. Tech. **64**(15), 2425–2437 (2004)

Zhou, Y., Pervin, F., Lewis, L., Jeelani, S.: Fabrication and characterization of carbon/epoxy composites mixed with multi-walled carbon nanotubes. Mater. Sci. Eng., A **475**, 157–165 (2008)

Zhu, J., Yudasaka, M., Zhang, M., Iijima, S.: Dispersing carbon nanotubes in water: a noncovalent and nonorganic way. J. Phys. Chem. **108**, 11317–11320 (2004)

Zhu, Y.-F., Zhang, C., Wang, J.-D., Shi, L., Liang, J.: Influence of electric field on dispersion of carbon nanotubes in liquids. J. Disp. Sci. Tech. **27**(3), 371–375 (2006)

Ziebro, J., Lukasiewicz, I., Borowiak-Palen, E., Michalkiewicz, B.: Low temperature growth of carbon nanotubes from methane catalytic decomposition over nickel supported on a zeolite. Nanotechnology **21**, 145308, 6 pp (2010)

Zou, Y., Feng, Y., Wang, L., Liu, X.: Processing and properties of MWNT/HDPE composites. Carbon **42**(2), 271–277 (2004)

Zou, J., Liu, L., Chen, H., Khondaker, S.I., McCullough, R.D., Huo, Q., Zhai, L.: Dispersion of pristine carbon nanotubes using conjugated block copolymers. Adv. Mater. **20**, 2055–2060 (2008)

Chapter 5
Chemical Functionalization of Carbon Nanotubes for Dispersion in Epoxy Matrices

Dimitrios J. Giliopoulos, Kostas S. Triantafyllidis, and Dimitrios Gournis

Contents

Abstract The remarkable physical properties of carbon nanotubes and their versatile chemical reactivity leading to various types of surface organo-functionalization were the main reasons why CNTs have become one of the most important types of nano-additives for the development of novel polymer (including epoxy) nanocomposites with improved and sometimes unique properties. The present chapter deals with the organo-functionalization of carbon nanotubes and the preparation of the respective epoxy – CNT nanocomposites. The effect of functionalization on dispersion of CNTs and on the final properties of the nanocomposites is discussed, while emphasis is given on the reactivity of the functional groups and their participation in the curing process of epoxy resins.

D.J. Giliopoulos • K.S. Triantafyllidis
Department of Chemistry, Aristotle University of Thessaloniki,
University Campus, P.O. Box 116, 54124 Thessaloniki, Greece

D. Gournis (✉)
Department of Materials Science and Engineering, University of Ioannina,
Ioannina 45110, Greece
e-mail: dgourni@cc.uoi.g

A.S. Paipetis and V. Kostopoulos (eds.), *Carbon Nanotube Enhanced Aerospace Composite Materials*, Solid Mechanics and Its Applications 188,
DOI 10.1007/978-94-007-4246-8_5,

Keywords Epoxy polymers • Epoxy resins • Nanocomposites • Carbon nanotubes • Organic functionalization • Amine and carboxyl functionalized carbon nanotubes

5.1 Introduction

Reinforcement of engineering polymers with nanosized fillers has attracted increasing interest over the last 20 years, due to the unique properties of the resulting polymer nanocomposite materials compared to those of pristine polymers or conventional composites (Pinnavaia and Beall 2000; Rothon 2003; Ke and Stroeve 2005; Mittal 2010; Giannelis 1996; Moniruzzaman and Winey 2006). Homogeneous dispersion of the nano-additives and utilization of their high available surface area (per unit mass) for interaction with the polymer, are the key-objectives for the preparation of polymer nanocomposites with improved properties. Many different types of inorganic nanostructures have been studied as polymer nanofillers, including silica and carbon nanoparticles (Ke and Stroeve 2005; Rothon 2003; Merkel et al. 2002; Huang 2002; Sumita et al. 1991), layered materials (i.e., clays, LDHs) (Giannelis 1996, 1998; Leroux and Besse 2001; Giannelis et al. 1999; Triantafyllidis et al. 2002a), carbon nanotubes (Ajayan et al. 2000; Moniruzzaman and Winey 2006) and nanofibers (Choi et al. 2005b), and more recently mesostructured silicas (Park et al. 2006) and graphene (Ramanathan et al. 2008). The selection of the most appropriate inorganic nano-additive depends on the specific requirements of the targeted application, regarding mechanical strength, thermal stability and thermal expansion, electrical/thermal conductivity, gas permeability, etc. The single and multi-wall carbon nanotubes (CNTs) have been widely applied for the preparation of nanocomposites, due to their impressive properties and mainly for improving the mechanical properties and electrical conductivity of polymers. The main drawback, however, is the difficulty of dispersing the individual nanotubes homogeneously within the polymer matrix in order to form a continuous network, which is necessary mainly for inducing electrical conductivity to the polymers.

In order to enhance the chemical compatibilization of carbon nanotubes with polymers, functionalization of the nanotube surface is usually applied, aiming mainly at: (i) an increase of the organophilicity of carbon nanotube surfaces and the loosing of nanotubes' bundles (usually formed when pristine CNTs are mixed with the (pre)polymers), thus leading to homogeneous dispersion of nanotubes in the polymer matrix and, (ii) the formation of chemical bonds between the functional groups of the modified carbon nanotubes and those of the polymer (if any), aiming at better interfacial properties in the nanocomposites.

Epoxy polymers are being extensively used in various aerospace applications. The present chapter deals with the organo-functionalization of carbon nanotubes and the preparation of the respective epoxy – CNT nanocomposites. The effect of functionalization on dispersion of CNTs and on the final properties of the nanocomposites will be discussed, while emphasis will be given on the reactivity of the functional groups and their participation in the curing process of epoxy resins.

5.2 Carbon Nanotubes – An Overview: Structure, Properties, Synthetic Methods, Chemical Functionalization

Carbon nanotubes (CNTs) constitute an outstanding material for use among others in the aerospace, textile, electronics, biomedical and plastics industry, since this material possesses high chemical and thermal stability, mechanical strength, stiffness and elasticity, and electrical and thermal conductivity as well as low density and weight (Coleman et al. 2006a; Baughman et al. 2002; Gao et al. 2004). The combination of all these superior properties in a single material has increased the interest of utilizing CNTs as nano-additives for the reinforcement of polymers. The polymer-CNT nanocomposites exhibit unique properties that derive from the structure and morphology of CNTs, provided that homogeneous dispersion and strong interfacial interaction with the polymer matrix have been accomplished. (Li et al. 2004; Dalton et al. 2003)

CNTs can be synthesized as multi-wall (MWCNTs), double-wall (DWCNTs) and single-wall (SWCNTs). Three techniques are mainly used nowadays to synthesize CNTs: laser ablation (Guo et al. 1995; Thess et al. 1996), arc discharge (Iijima 1991; Ebbesen and Ajayan 1992) and catalytic chemical vapour deposition (CVD) (Gournis et al. 2002; Tsoufis et al. 2007; Chen et al. 2002; Wei et al. 2002). Each method has its own advantages and limitations while among them, only CVD allows large scale production of CNTs at relatively low cost (Hafner et al. 1998). In brief, the method comprises the catalytic decomposition of hydrocarbon gases (methane, ethane, acetylene), at rather high temperatures, over catalytically active metallic centers (commonly transition metal nanoparticles, such as Fe, Co, Ni) embedded in solid supports. Solid supports already employed comprise, among others, zeolites, mesoporous silica, clays, graphite, MgO etc. (Tsoufis et al. 2007; Maccallini et al. 2010; Tsoufis et al. 2008; Gournis et al. 2002; Jiang et al. 2010; Policicchio et al. 2007; Triantafyllidis et al. 2008). The use of bi-metallic systems of transition metal oxides results in higher yields of synthesized CNTs compared to monometallic, since the synergistic action of the two metals enhances the total catalytic activity. (Qian et al. 2003; Qingwen et al. 2002) Finally, acetylene is more reactive than other hydrocarbons at the same reaction temperature, leading to CNTs of good quality, while in addition, it suppresses the formation of carbon nanoshells which poison the catalytic sites (Soneda et al. 2002).

CNTs cannot be easily dispersed in common solvents and have the tendency to aggregate into dense bundles of nanotubes, due to the intrinsic van der Waals attraction of the nanotubes to each other, which is associated with their high aspect ratio (up to 1,000) (Zhu et al. 2003). In recent years, many research efforts have focused on the development of methodologies for the chemical modification of CNTs in order to facilitate the disaggregation of individual nanotubes in solutions, as well as for producing CNT derivatives with even more attractive functional features (Tasis et al. 2006). In accordance with their behavior in various solvents, upon mixing with (pre)polymers the CNTs form micro-sized bundles which create defect sites in the polymer network instead of providing the benefits that can

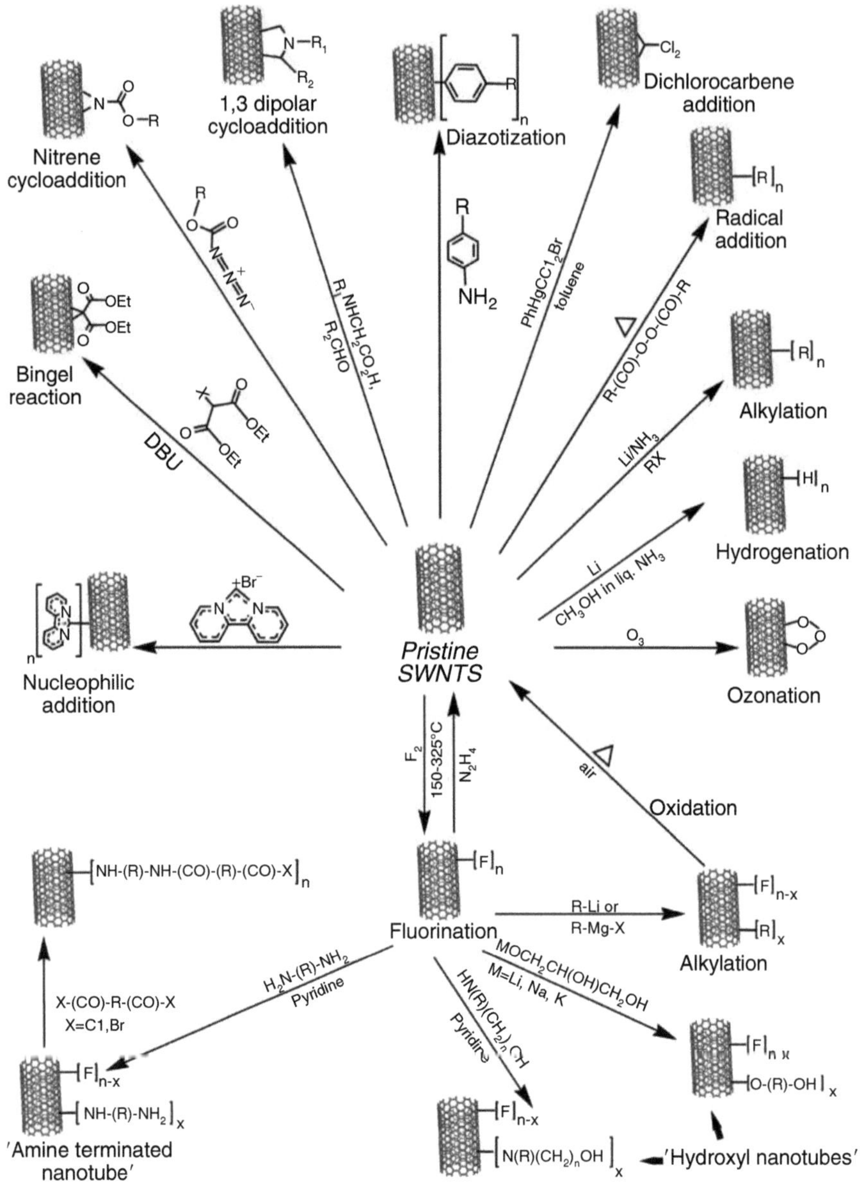

Fig. 5.1 Schematic describing various covalent sidewall functionalization reactions of SWNTs (Reproduced with permission from Gusev et al. 2000. Copyright Wiley-VCH)

be attained by the individual nanotubes (Qian et al. 2000). Several techniques have been applied to achieve homogeneous dispersion of CNTs in polymers, focusing mainly on the optimization of physical blending or chemical organo-functionalization of CNTs (Xie et al. 2005). Side-wall chemical functionalization is one of the most effective methods for homogeneous dispersion of CNTs in polymer

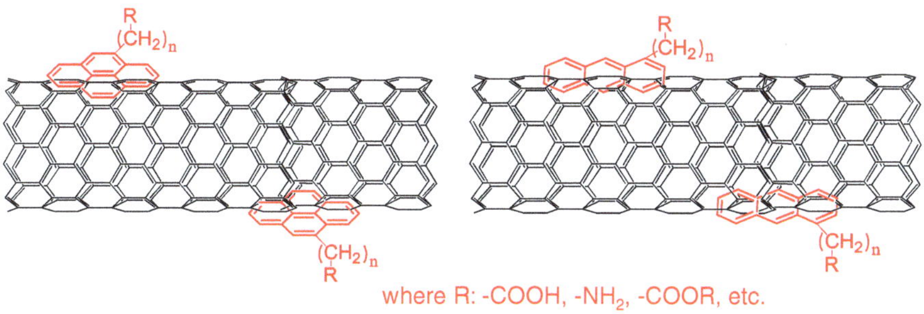

Fig. 5.2 Schematic representation of noncovalent sidewall functionalization of CNTs with pyrene (*left*) and anthracene (*right*) derivatives

matrices since strong interface adhesion is achieved between the functionalized carbon nanotubes and the surrounding polymer chains (Li et al. 2005b; Lin et al. 2003; Viswanathan et al. 2003; Qin et al. 2004; Zhang et al. 2004). In general, two main paths are usually followed for the functionalization of carbon nanotubes: (a) the covalent attachment of chemical groups, through reactions on the conjugated skeleton of CNTs, and (b) the noncovalent supramolecular adsorption or wrapping of various functional molecules on the surface of nanotubes.

An enormous number of reports can be found in the literature concerning chemical functionalization of CNTs while many review articles and books appeared the last decade which present and critically analyze all the chemical routes concerning this issue (see for example Karousis et al. 2010; Zhao and Stoddart 2009; Balasubramanian and Burghard 2005; Banerjee et al. 2005; Tasis et al. 2006; Mittal 2011). The covalent functionalization of CNTs leads to the attachment of functional groups on tube ends or sidewalls (Fig. 5.1). The covalent approach includes among others oxidation reactions, esterification-amidation reactions on oxidized CNTs, treatment with ionic liquids, complexation reactions on oxidized CNTs, halogenation, cycloaddition reactions, radical additions, nucleophilic additions, electrophilic additions, ozonolysis, electrochemical modifications, plasma-activation, mechanochemical functionalizations, and polymer grafting (Karousis et al. 2010). On the other hand, in the noncovalent functionalization, the CNT surface can be modified via van der Waals forces and π–π interactions, by adsorption or wrapping of polynuclear aromatic compounds (e.g. phenyl, naphthalene, phenanthrene, pyrene and anthracene derivatives, see Fig. 5.2) and other substances (like surfactants, macrocyclic host molecules, ionic liquids, dyes, alkoxysilanes, phosphines, etc.), polymers (epoxy, acrylic, aliphatic, conjugated, etc.) or biomolecules (e.g. proteins). Furthermore, the chemical modification of CNTs includes also the endohedral filling of CNTs with fullerenes and inorganic or organic substances (Karousis et al. 2010) as well as the decoration of CNT with metal or semiconductor nanoparticles (NPs). In the latter case, two main pathways have been developed including: (a) *in-situ* formation of metal NPs directly on CNT surfaces and (b) connection of preformed NPs to modified CNTs (for a review see Georgakilas et al. 2007).

5.3 Epoxy Resins/Polymers

Epoxy polymers (also called polyepoxides) are thermosetting polymers that derive from the reaction (crosslinking and polymerization) of an epoxy resin with a curing agent (also called hardener). Epoxy polymers exhibit very good mechanical and adhesives properties, high thermal and dimensional stability, resistance to many solvents, electrical insulation and relatively high barrier properties. Owing to their properties, epoxy polymers can find application as fiber reinforced pipe and composites, tooling and molding compounds, construction, electrical and aerospace adhesives, electrical castings and laminates, chemical resistant solids, tank linings, flooring, etc. (Irfan 1998; Harper 2000).

The most common epoxy resins are the glycidyl epoxy resins, such as diglycidyl ether of bisphenol A (DGEBA) and novolac epoxy resins. There are also aliphatic or cycloaliphatic epoxy resins (non-glycidyl resins). The DGEBA epoxy resin is produced from the reaction of epichlorydrin with bisphenol A in the presence of a basic catalyst (i.e. NaOH), as is schematically presented in Fig. 5.3:

The novolac epoxy resins are synthesized via reaction of phenolic novolac resin with epichlorohydrin in presence of NaOH, while the phenolic novolac resins are formed by the reaction of phenols with formaldehyde in the presence of an acidic catalyst. The structure of a cresol novolac epoxy resin is shown in Fig. 5.4:

Various types of curing agents can be used for the cross-linking/polymerization of the epoxy resin monomers. The most commonly used curing agents are polyamines, polyamides, anhydrides, isocyanates and polymercaptans. The cure kinetics, the glass transition temperature (T_g) of the cured system and in general the structure and properties of the produced epoxy polymer depend greatly on the nature of the curing agent, i.e. chemical structure (aromatic, cycloaliphatic, aliphatic), molecular weight, and degree of functionality.

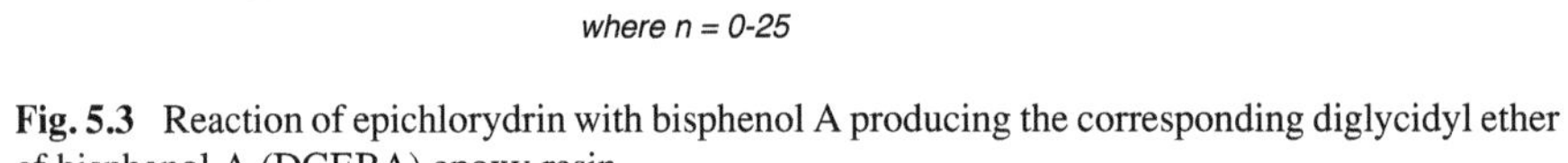

Fig. 5.3 Reaction of epichlorydrin with bisphenol A producing the corresponding diglycidyl ether of bisphenol A (DGEBA) epoxy resin

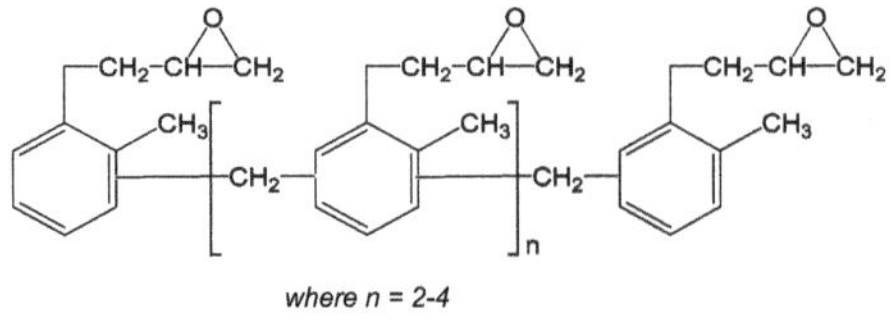

Fig. 5.4 Chemical structure of cresol novolac epoxy resin

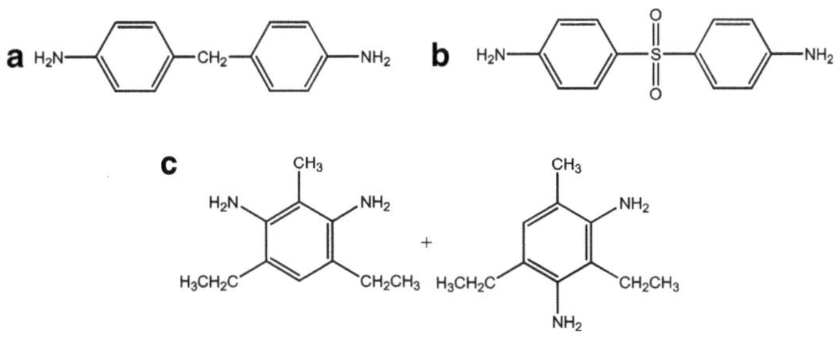

Fig. 5.5 Aromatic amine curing agents: (**a**) DDM, (**b**) DDS and (**c**) DDT

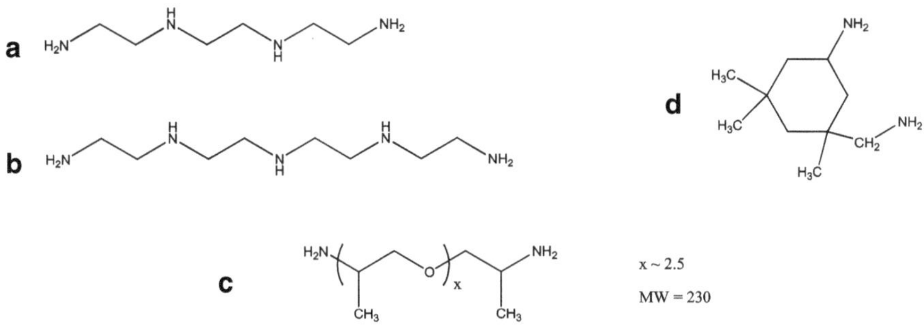

Fig. 5.6 Aliphatic amines used as epoxy resin curing agents: (**a**) TETA, (**b**) TEPA, (**c**) Jeffamine D-230 and cycloaliphatic amine: (**d**) IPD

Amine-based curing agents are usually preferred over the other types of curing agents because they are more reactive and can initiate the cross-linking reaction even at room temperature. *Aromatic diamines* can be used for the production of *high* T_g *epoxy polymers* which can be utilized in various aerospace applications. Such diamines are: (a) 4,4′-diaminediphenylmethane (DDM), (b) 4,4′-diaminodiphenylsulphone (DDS), and (c) 1,5-diamine-2,4-diethyltoluene (DDT) (see Fig. 5.5 for their structures). Epoxy polymers cured with DDT exhibit high T_g of about 180°C, while the use of DDS leads to polymer matrices with even higher T_g (over 200°C). DDM has similar chemical structure with DDS but gives products with lower, i.e. $T_g \approx 150°C$. However, DDM exhibits higher reactivity than DDS, which allows the curing to take place at lower temperature.

The use of *aliphatic or cycloaliphatic diamines* as curing agents provides epoxy polymers with *high or medium-to-high* T_g. Such diamines can be: (a) isophorone diamine (IPD), (b) triethylenetetramine (TETA) or tetraethylenepentamine (TEPA), and (c) short etheramines (e.g. Jeffamine D-230) (see Fig. 5.6 for their structure). The T_g of epoxy polymers cured with IPD is about 150°C, with TETA and TEPA about 125°C, and with α,ω-polypropylene oxide diamine (Jeffamine) D-230 about 80°C.

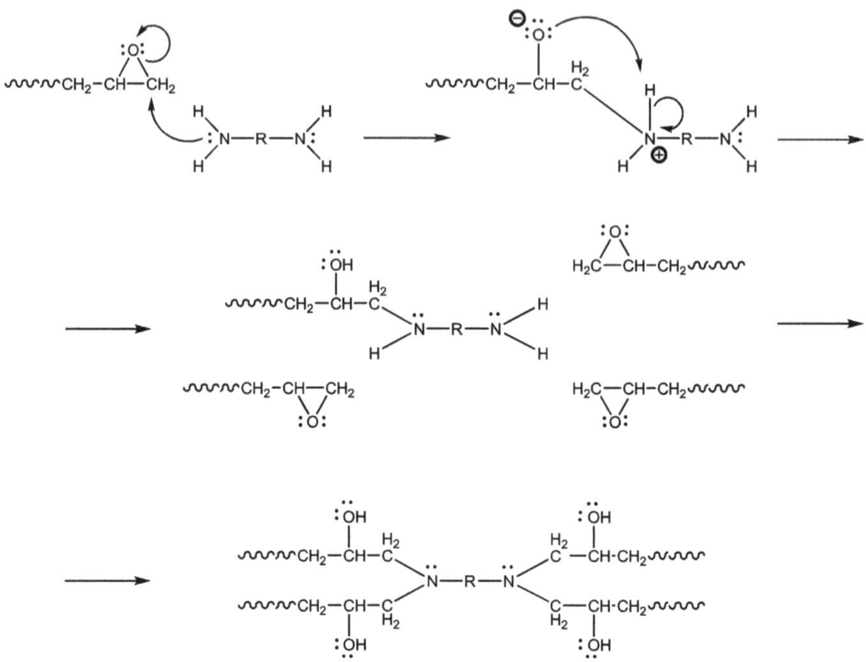

Fig. 5.7 Cross-linking of epoxy resin with a diamine curing agent

The cross-linking reaction of an epoxy resin with a polyamine (i.e. diamine) towards the formation of the corresponding epoxy polymer matrix is shown below in Fig. 5.7:

In addition to the type of curing agent, other parameters that affect the properties of the epoxy polymers are the molecular weight and the epoxy equivalent weight (EEW), i.e. the moles of epoxide groups per 100 g of epoxy resin, and the mixing ratio of resin and curing agent. In general, the selection of the type of epoxy resin and curing agent and of the curing parameters is based on the targeted application of the epoxy polymer and the related specific property requirements. For example, epoxy polymers based on DGEBA resins are commonly used to fabricate high strength pipes and composites reinforced with fibers (glass, graphite, aramid, carbon, etc.) utilizing their low viscosity for more appropriate mixing. They also exhibit excellent electrical insulation properties for use in electrical encapsulations, laminates and molding compounds. In addition, they offer high protection to metal surfaces against attack from acids, bases, solvents and fuel. Some of the already exceptional properties (i.e., temperature, chemical and solvent resistance) of the DGEBA resins are further improved in novolac epoxy resins due to the higher cross-link density which is attributed to the presence of multiple epoxide groups on their backbone chain.

Despite their impressive properties, the epoxy polymers can be further optimized for more demanding applications, by mixing with various types of inorganic additives or fillers, leading to the formation of either conventional *epoxy composites* or the more advanced *epoxy nanocomposites*. The latter are produced when at least one dimension of the additives is in the order of a few nanometers. Representative types of inorganic nano-additives used in epoxy nanocomposites are: silica, carbon and metal oxide nanoparticles (i.e., fumed silica, carbon black, metal oxide) (Schueler et al. 1997; Vassileva and Friedrich 2003, 2006; Singha and Thomas 2009; Tuncer et al. 2007; Preghenella et al. 2005), silica (Gonon et al. 2001; Ragosta et al. 2005), opal (Bogomolov et al. 2003), layered materials (clays, LDHs) (Lan and Pinnavaia 1994; Hsueh and Chen 2003; Xidas and Triantafyllidis 2010), carbon nanotubes (CNTs) (Njuguna and Pielichowski 2003; Njuguna and Pielichowski 2004a, b), glass and carbon (nano)fibers (Gusev et al. 2000; Iglesias et al. 2002; Kupke 1998; Choi et al. 2000; Tsantzalis et al. 2007), graphite (Li et al. 2005a) and graphene (Yang et al. 2009). The aim of using inorganic nano-additives is mainly to increase the toughness of epoxy polymers and to utilize the unique properties of the additives in order to induce better or even new characteristics to epoxy, by adding significantly lower amounts of additive in the polymer compared to the micro-sized fillers used in conventional composites.

Epoxy polymers and their nanocomposites are used in various aerospace applications as structural parts (i.e. in fuel tanks and pipes), adhesives and coatings (Bhowmik et al. 2009; De Fenzo et al. 2009; Prolongo et al. 2009; Njuguna and Pielichowski 2004a, b). The epoxy (nano)composites used in aerospace applications, should be able to exhibit their exceptional properties (i.e. mechanical strength, thermal and dimensional stability, tuned conductivity, low gas permeability, etc.) under real space environment conditions, such as intense thermal shocks due to the rapid change of temperature under high vacuum conditions. Carbon (nano)fibers have been widely tested as epoxy (nano)additives for aerospace applications, and more recently, after the discovery of carbon nanotubes, there has been a systematic effort to effectively disperse the CNTs in epoxies in order to prepare high-performance materials for various applications, including aerospace materials.

5.4 Dispersion of Functionalized Carbon Nanotubes in Epoxy Matrices

The formation of a "true" nanocomposite phase when mixing inorganic nano-additives with polymers depends greatly on the degree of miscibility of the inorganic phase with the polymer matrix. By improving the dispersion of the individual nano-objects, the benefits that result from the high surface to volume ratio of the nano-additives, compared to conventional micro-sized fillers, are maximized (Rothon 2003; Ramanathan et al. 2005).

Generally, the methodologies applied for the dispersion of carbon nanotubes in polymer matrices can be categorized in *physical* and *chemical*. The physical methods that have been used are the mechanical stirring/shear mixing, extrusion and ultrasound agitation (sonication), while the chemical methods are usually related with the chemical organo-functionalization of the CNTs. As has been discussed in the previous paragraphs, surface organo-functionalization of CNTs aims both at increasing the organophilicity of nanotubes as well as introducing surface functional groups (i.e. carboxyl or amino-groups) that can react with the functional groups on polymers' backbone, such as the oxirane rings of epoxy resins, creating strong interactions between the polymer and the surface of CNTs.

Carbon nanotubes have been used by many researchers, with or without functionalization, in order to further improve the properties of epoxy polymers or to induce additional properties, such as electrical and thermal conductivity. Representative examples of various epoxy-CNT nanocomposite materials are discussed in the next paragraphs, focusing on the type of physical and/or chemical methods that have been studied for enhancing the dispersion of CNTs within the epoxy matrix and on the resulting effects on the properties of the nanocomposites.

5.4.1 Epoxy Nanocomposites with Unmodified Carbon Nanotubes

Following the discovery of carbon nanotubes in the early 1990s (Iijima 1991; Ebbesen and Ajayan 1992), one of the first attempts to prepare a polymer-CNT nanocomposite was reported by Ajayan et al. in 1994–1995 (Ajayan 1995; Ajayan et al. 1994) who tried to align unmodified CNTs by dispersing them in an epoxy resin and cutting thin sections/films of the cured epoxy nanocomposite using a diamond knife. Early studies on the tensile properties of a UV-cured urethane/diacrylate thin polymer film containing CNTs showed that the multi-wall CNT-matrix stress transfer efficiency was at least one order of magnitude higher than in conventional carbon fiber-based composites (Wagner et al. 1998). It was also shown that the compression modulus is higher than the tensile modulus in epoxy nanocomposites with multi-wall CNTs, indicating that load transfer to the nanotubes in the composite is much higher in compression (Schadler et al. 1998). A percolation-dominated electrical conductivity was also identified by studying the properties of a conjugated-polymer-carbon-nanotube composite using poly (*p*-phenylenevinylene-co-2,5-dioctoxy-*m*-phenylenevinylene, PMPV) as the polymer matrix. It was shown that increasing the content of CNTs from 0 to 8% mass fraction, the conductivity was dramatically increased by up to ten orders of magnitude (Coleman et al. 1998; Curran et al. 1998). Following these initial reports on polymer-CNT nanocomposites, using unmodified CNTs, the number of published works from the year 2000 and onwards has started to increase rapidly. Epoxy-based nanocomposites were amongst the ones studied to a great extent (Xu et al. 2002; Lau et al. 2003; Sandler et al. 2003; Gojny et al. 2004; Song and Youn 2005; Martin et al. 2005; Lau et al. 2005; Grossiord et al. 2006; Moisala et al. 2006; Li et al. 2007; Liu and Grunlan 2007; Wang et al. 2008; Hernandez-Perez et al. 2008; Cebeci et al. 2009; Vavouliotis et al. 2010).

Although the properties of epoxy-based nanocomposites with unmodified CNTs were improved in general, it was shown that the tendency of carbon nanotubes to aggregate into bundles within the polymer matrix, limited the benefits that homogeneously dispersed well-separated carbon nanotubes could offer. As is discussed above, carbon nanotubes are held together with Van der Waals forces. Although these forces are considered very weak, due to the high aspect ratio of carbon nanotubes as well as their high polarizability, a large amount of energy is required to disaggregate carbon nanotube bundles within solvents or (pre)polymers (Grady 2010).

The as received/produced carbon nanotubes may contain a small amount of impurities such as amorphous carbon and traces of catalyst used for the production of nanotubes. These impurities can be removed by treating carbon nanotubes with acids and this procedure results in the formation of carboxylic groups on the surface of nanotubes (Rinzler et al. 1998; Hirsch 2002; Sun et al. 2002b). The carboxylic groups can be regarded as reactive/functional surface groups, especially in the case of epoxy resin polymerization, where the acidic protons of carboxyls can initiate the cross-linking reaction (Zhu et al. 2003). Thus, the so-called "unmodified" or "non-functionalized" CNTs, bear reactive surface carboxyls directly attached to the carbon nanotube walls. The discrimination between "unmodified" and "carboxylated-functionalized" CNTs depends on the intensity of the acid-treatment process and the resulting concentration of surface carboxyl groups (Zhu et al. 2009).

The use of a solvent, usually combined with mechanical stirring and/or sonication, has been widely applied as a method for improving the dispersion of nanotubes within polymers (Ma et al. 2010b). Organic solvents such as DMF, acetone, ethanol, toluene, chloroform and others, enhanced the dispersion of CNTs, which however were still in the form of bundles with few isolated nanotubes present in the epoxy polymer matrix (Lau et al. 2005; Song and Youn 2005; Thakre et al. 2010; Xu et al. 2002; Loos et al. 2008). A comparison study on the effect of the type of solvent (DMF, acetone, ethanol) showed that although the benefit induced in dispersion was similar for the three solvents compared to the epoxy nanocomposite prepared in their absence, a difference in the size of bundles was observed, with ethanol and DMF favoring the formation of smaller bundles (20–30 nm) compared to those formed by the use of acetone (40–50 nm) (Lau et al. 2005). Combination of solvents has also been found to be very effective, as in the case of trifluoroacetic acid (TFA) which was used as a co-solvent with *N,N*-dimethylformamide, dichloromethane, *n*-hexanol, toluene, tetrahydrofuran, and acetonitrile (Chen et al. 2007). As it can be seen in Fig. 5.8, addition of 10 vol.% of TFA in the above solvents resulted in significant improvement of dispersion of multi-wall CNTs in the respective solutions, especially in the case of dichloromethane, toluene, tetrahydrofuran, and acetonitrile.

The use of solvents in the preparation of epoxy-CNT nanocomposites usually plays a double role, i.e. facilitates the disaggregation of carbon nanotubes and lowers the viscosity of the epoxy resin (Loos et al. 2008). Both effects usually lead to better mixing characteristics and improved dispersion in the final nanocomposite. Despite however the benefits gained in dispersion by the use of solvents, the presence of solvent traces in the cured nanocomposites has been identified in almost all related

Fig. 5.8 Photographs of the as-received MWCNTs dispersed in (from *left* to *right*) *N*,*N*-dimethylformamide, dichloromethane, *n*-hexanol, toluene, tetrahydrofuran, and acetonitrile. (**a**) Without the addition of TFA; (**b**) with addition of 10 vol.% of TFA (Reproduced from Ramanathan et al. 2005 with permission from IOP Publishing Ltd.)

studies and in some cases it has been shown to affect the structure and performance properties of the materials (Lau et al. 2005; Allaoui and El Bounia 2009; Loos et al. 2008). This issue becomes even more important in real applications, such as in aerospace, where ultra-clean environments are usually required and desorption of traces of chemicals from structural parts, coatings, joints, etc. when exposed to relatively high temperatures should be stringently avoided.

The properties of epoxy-CNT nanocomposites that have been mostly studied are the mechanical strength, stiffness and modulus, electrical and thermal conductivity, thermal stability, as well as viscoelastic and rheological characteristics. Based on the up to date findings, it can be suggested that the effect of CNTs addition to various polymers, including epoxies, on mechanical properties is not as high as expected based on the intrinsic properties of carbon nanotubes, and it is highly dependent on dispersion and interfacial characteristics of the nanocomposites (Coleman et al. 2006b). In many studies on epoxy-CNT nanocomposites, the changes in mechanical properties are marginal and close to experimental error of the measurements. In addition, differences in the performance regarding tensile properties (Young's modulus, strength and elongation at break), flexural properties (storage and loss modulus, T_g) and impact properties (strength), have been often observed within the same nanocomposite systems (Song and Youn 2005; Loos et al. 2008). What seems to be of high importance is the interface between the carbon nanotubes and the polymer. Poor load transfer between nanotubes (in bundles) and between nanotubes and surrounding polymer chains may result in interfacial slippage and reduced performance (Schadler et al. 1998; Ajayan et al. 2000). Improvement of interfacial shear without sacrificing the mechanical strength and stiffness could result in very high mechanical damping, which is very important for many commercial applications (Suhr et al. 2005).

On the other hand, a significant improvement of electrical conductivity has been observed in the majority of related studies, irrespective of the effects on mechanical and thermal properties (Thakre et al. 2010; Song and Youn 2005). A better dispersion of CNTs within the epoxy matrix results usually in higher increase of electrical and thermal conductivity, associated with low values of percolation

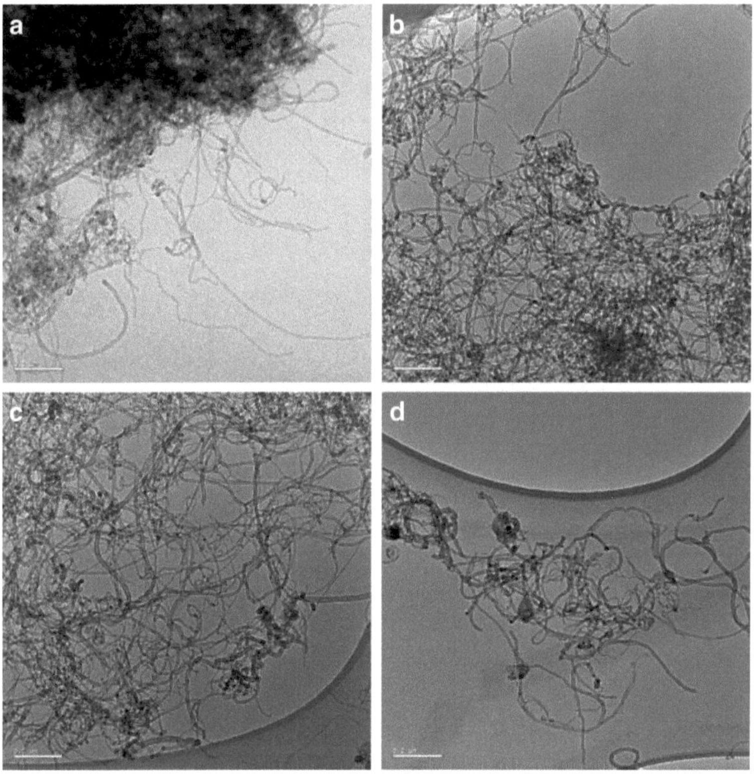

Fig. 5.9 TEM images of CNT agglomerates dispersed according to Conditions A, B, C, and D described in the text (scale bar = 0.2 μm) (Reproduced with permission from Njuguna and Pielichowski (2004b). Copyright Wiley-VCH)

threshold (usually below 0.5 wt.%) (Song and Youn 2005; Thakre et al. 2010). The combined effect of CNTs' aspect ratio and dispersion on percolation threshold and conductivity has been studied in relation with the method applied for preparation of the epoxy-CNT nanocomposites (Li et al. 2007). It was shown that the use of solvent (acetone) resulted in the formation of nanocomposites containing loosely entangled and uniformly dispersed CNT agglomerates (Fig. 5.9, condition B), which induced high electrical conductivity (Fig. 5.10), compared to the shear mixing in the absence of solvent (condition A). On the other hand, the use of UV/O_3 treatment (condition C) and ball-milling followed by UV/O_3 and silane treatment of nanotubes (condition D), induced better dispersion but at the same time damaged the structure of nanotubes. These latter nanocomposites exhibited a moderate increase of electrical conductivity. It was shown that if the CNT aspect ratio is too low, the formation of a conduction network requires a very high CNT content, regardless of the degree of CNT dispersion. In addition, it was suggested that the formed silane coating around the CNTs acted as a physical barrier to electrical conduction. Another methodology to improve electrical conductivity of epoxy

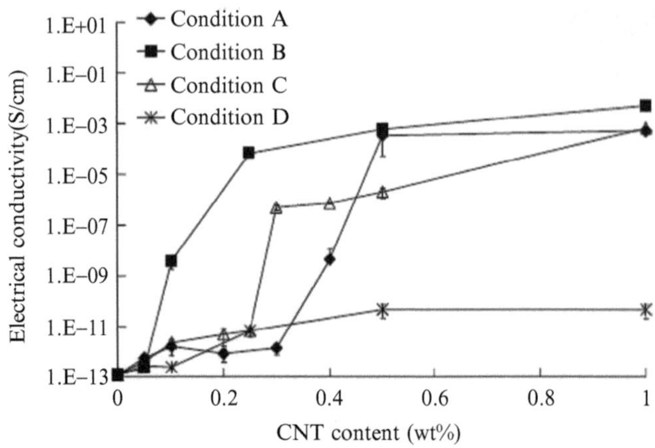

Fig. 5.10 DC electrical conductivities of nanocomposites as a function of CNT content (Reproduced with permission from Njuguna and Pielichowski (2004b). Copyright Wiley-VCH)

polymers was based on the simultaneous use of clays and CNTs as polymer additives (Liu and Grunlan 2007). It was shown that the addition of clay effectively improves the dispersion of single-wall CNTs in the epoxy matrix, leading to the formation of a continuous three-dimensional network of nanotubes. The epoxy-clay/CNT nanocomposite exhibited improved electrical conductivity and lower percolation threshold compared to the respective epoxy-CNT nanocomposite.

5.4.2 Epoxy Nanocomposites with Organically Modified Carbon Nanotubes

The chemical modification of carbon nanotubes was suggested from the early steps in polymer-CNT nanocomposite research, as an effective way for homogeneous dispersion of CNTs in the polymer matrix and for enhancing the interfacial interactions in the nanocomposites (Gong et al. 2000; Tiano et al. 2000; Jin et al. 2002; Czerw et al. 2001). The homogeneous dispersion is favored when the organophilicity of the carbon nanotube surface increases via covalent or noncovalent chemical modification with various organic molecules. An overview of the most common functionalization possibilities of CNTs with methods including defect and covalent sidewall functionalization, as well as noncovalent exo- and endohedral functionalization, is schematically presented in Fig. 5.11. Furthermore, the surface of CNTs can be enriched in functional groups which can be either bonded directly to the carbon walls (i.e. –OH and –COOH groups) or can be part of the covalently or noncovalently bonded organic moieties (see Sect. 5.2). These functional groups can react with functional groups of the polymer matrix, thus leading to enhanced interaction between the two phases. The case of epoxy resin polymerization (see

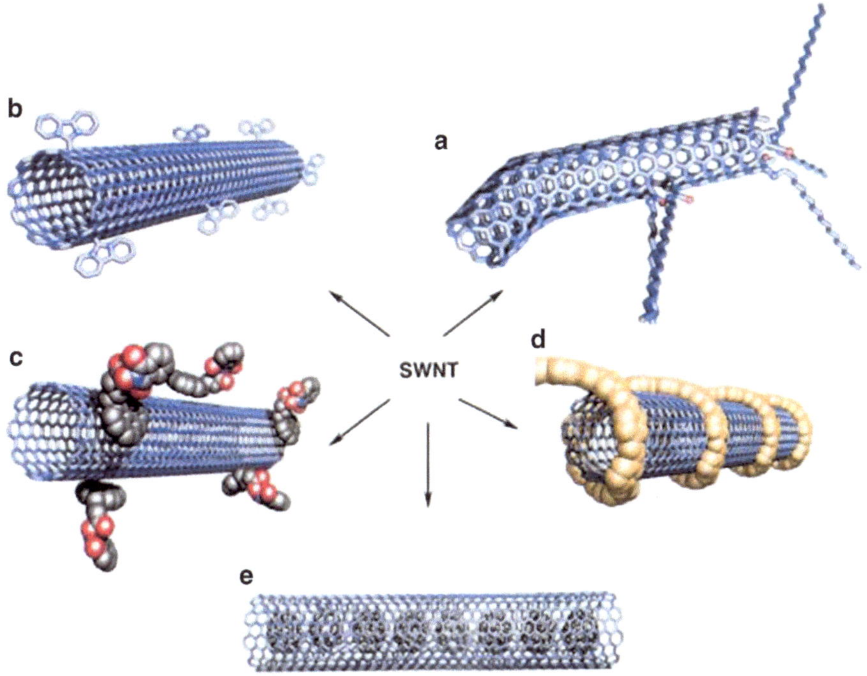

Fig. 5.11 Functionalization possibilities for SWNTs: (**a**) defect-group functionalization, (**b**) covalent sidewall functionalization, (**c**) noncovalent exohedral functionalization with surfactants, (**d**) noncovalent exohedral functionalization with polymers, and (**e**) endohedral functionalization with, for example, C_{60}. For methods (**b**)–(**e**), the tubes are drawn in idealized fashion, but defects are found in real situations (Reproduced with permission from Qian et al. 2000. Copyright Wiley-VCH)

Fig. 5.7) is a typical example where reactive/functional groups, such as $-NH_2$, –COOH, and –OH, can promote the cross-linking reaction on the surface of CNTs.

In the case of epoxy-CNT nanocomposites, it has been shown that the use of organophilic CNTs resulted in better dispersion of nanotubes accompanied with property improvement. The surface of nanotubes was rendered organophilic by different methods, i.e. by attaching organic moieties on the surface of carbon nanotubes (Tseng et al. 2007; Xu et al. 2010), wrapping long-chained organic molecules (usually polymers) around the nanotubes (González-Domínguez et al. 2010), or using surfactants that can act as coupling agents between carbon nanotubes and epoxy polymers (Gong et al. 2000), or as dispersants that prevent carbon nanotubes from re-aggregating upon mixing with the epoxy (pre)polymer (Cho and Daniel 2008). Functionalization of carbon nanotubes with organic molecules enhances their dispersion inside solvents and/or polymer matrices by introducing steric repulsive forces between carbon nanotubes that overcome the van der Waals coupling forces (Gong et al. 2000). Although it would be expected that the introduction of long-chained molecules inside polymer matrices could have

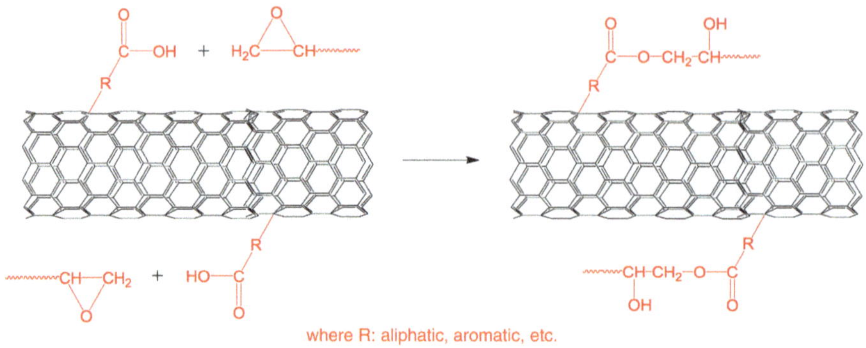

Fig. 5.12 Cross-linking of epoxy resin initiated by the –COOH groups attached on CNTs

a plasticizing effect on the final epoxy nanocomposite, however the concentration of the organic molecules attached on CNTs is usually very low compared to the bulk polymer mass and the effect is negligible (Gong et al. 2000). As a result, the glass transition temperature (T_g) of epoxy nanocomposites with organically functionalized CNTs increases with nanotube loading due to the confinement of polymeric chains mobility by the well dispersed nanotubes (Gong et al. 2000; Cho and Daniel 2008; Tseng et al. 2007; Xu et al. 2010). Nanoscale dispersion of carbon nanotubes also results in larger interaction area between the nanotubes surface and the polymer matrix, which in turn improves the mechanical properties of the epoxy polymer (Geng et al. 2008; Cho and Daniel 2008; Tseng et al. 2007; Xu et al. 2010).

5.4.3 *Epoxy Nanocomposites with Carboxyl Functionalized Carbon Nanotubes*

Carboxyl functionalized carbon nanotubes can be prepared via relatively intense, in comparison to the milder purification step, oxidation treatment with acids (i.e., H_2SO_4 and/or HNO_3 (Bahr and Tour 2002; Banerjee et al. 2005)), plasma induced oxidation (Bubert et al. 2003; Kim et al. 2006), UV/ozone oxidation (Li et al. 2007; Simmons et al. 2006) etc. The carboxyl groups formed by this procedure are attached directly on the nanotube walls. Another way to enrich the surface of CNTs with carboxyl groups is to modify the nanotubes with carboxylated derivatives of various organic molecules, as is discussed above. The existence of surface carboxyl groups enhances the dispersion of nanotubes in polar media, but on the other hand, they may inhibit their disaggregation (and consequently their dispersion in solvents or polymers) due to hydrogen bonds that can be formed among the carboxyl groups of nanotubes (Kukovecz et al. 2002; Banerjee and Wong 2002). Carboxyl groups act as proton donors in the epoxide ring opening reaction (see Fig. 5.12), leading to enhanced polymerization close to the surface of nanotubes and improved interfacial bonding. The reactivity of a carboxyl group

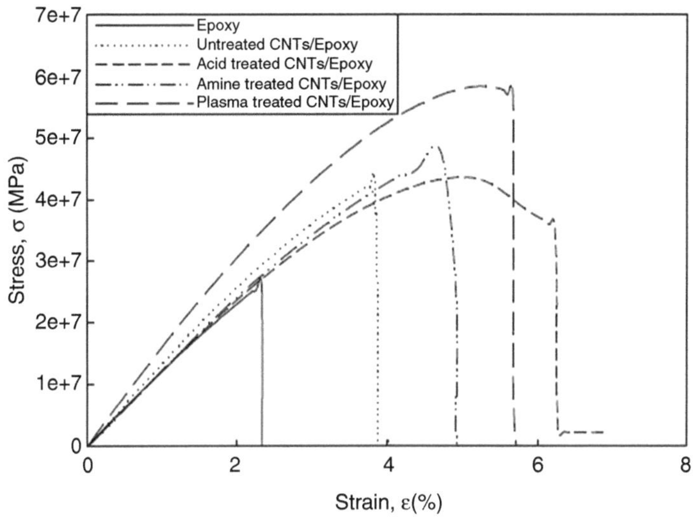

Fig. 5.13 Stress-strain curves of cured epoxy and composites containing 1 wt.% CNTs (Reprinted from Sun et al. 2002a with permission from Elsevier. Copyright 2006)

in epoxy polymerization is known from the well established curing mechanism of epoxy resins with carboxylic anhydrides (Weiss 1957; Ke et al. 2000; Miyagawa and Drzal 2004).

Carboxylated carbon nanotubes have been widely used for the preparation of epoxy nanocomposites by using solvents to facilitate their dispersion (Pizzutto et al. 2010; Suave et al. 2009; Bae et al. 2002; Larsen 2009; Zhang et al. 2008; Kim et al. 2005) or via direct mixing of the functionalized nanotubes with the epoxy resin (Fu et al. 2009; Montazeri et al. 2010a; Zhou et al. 2009; Ganguli et al. 2006; Guo et al. 2009; Larsen 2009; Zhang et al. 2008; Montazeri et al. 2010b, c). In most cases, the dispersion of the carboxylated nanotubes was better compared to the unmodified nanotubes. In addition, strong interfacial interactions were suggested based on the observed improvement of the nanocomposite properties. In a comparative study, carboxylated and unmodified carbon nanotubes were dispersed inside the epoxy matrix with and without the help of acetone and it was found that functionalized carbon nanotubes were better dispersed than unmodified ones (Pizzutto et al. 2010). The effect of the mixing procedure (i.e. mechanical stirring, sonication, shear mixing) was also studied, still however, the functionalized nanotubes were always better dispersed compared to the pristine ones (Fu et al. 2009; Ganguli et al. 2006).

The mechanical properties of the epoxy-carboxylated CNT nanocomposites are usually improved compared to the properties of pristine epoxy polymer or of nanocomposites with unmodified CNTs (Pizzutto et al. 2010; Montazeri et al. 2010a; Suave et al. 2009; Choi et al. 2005a; Larsen 2009; Zhang et al. 2008; Ganguli et al. 2006; Guo et al. 2009; Montazeri et al. 2010c). Addition of plasma treated/oxidized CNTs in epoxy polymer resulted in significant improvement of both stress and elongation at break compared to the pristine polymer (Fig. 5.13) (Kim et al. 2006).

The enhanced tensile properties of epoxy polymers by the addition of carboxylated carbon nanotubes is attributed to the increased stress-transfer from the matrix to the nanotubes through covalent bonds, whose formation is confirmed by the lack of pulled-out nanotubes at the fractured surfaces of the nanocomposite samples (Montazeri et al. 2010a). Epoxy nanocomposites with carboxylated carbon nanotubes exhibit also improved thermal and thermomechanical properties (Zhou et al. 2009; Suave et al. 2009; Choi et al. 2005a). This could be related to the reactivity of carboxyl groups acting as curing catalyst promoting vitrification (Zhou et al. 2009) and increasing cross-linking (Bae et al. 2002), leading to a more homogenous and dense network formation (Suave et al. 2009).

The good dispersion and interfacial properties provided by the carboxylated CNTs, induced also improvements in electrical conductivity of the epoxy polymer in analogy with unmodified nanotubes (Kim et al. 2005; Choi et al. 2005a). It was shown that electrical conductivity is dependent on the intensity of the oxidation process of nanotubes (Kim et al. 2005). Treatment of CNTs with a mixture of H_2O_2/NH_4OH resulted in epoxy nanocomposites with higher conductivity compared to those prepared by HNO_3 – treated nanotubes, due to the harsh conditions generated by the concentrated HNO_3 which partially damaged the structure of nanotube walls. Electrical conductivity of epoxy nanocomposites also depends on the loading of carboxylated carbon nanotubes, with higher loadings leading to increased conductivity (Choi et al. 2005a).

5.4.4 Epoxy Nanocomposites with Amine Functionalized Carbon Nanotubes

Amine functionalized carbon nanotubes are usually derived from carboxylated carbon nanotubes through different reactions (e.g. amidation, esterification, etc.) or by wrapping alkyl-amines with varying backbone size around the nanotubes (Zhu et al. 2004; Tasis et al. 2006; Sun et al. 2002a). The combination of surface organophilicity provided by organic moieties and of high reactivity of amine functional groups in epoxide ring opening, offers the most ideal system for enhanced dispersion and improved interfacial bonding in epoxy nanocomposites. A representative example of the above mechanism is described schematically in Fig. 5.14 where amino-functionalized single-wall CNTs react with the epoxide groups of the resin and become an integral structural component of the cured epoxy polymer network (Zhu et al. 2004). Pre-mixing of amine-modified nanotubes with the epoxy resin (before curing agent addition) provides sufficient time for the amine groups to interact with the epoxide rings and initiate polymerization close to the surface of nanotubes, thus leading to improved interfacial properties.

As in the case of unmodified and carboxylated carbon nanotubes, the dispersion of amine-functionalized CNTs in the epoxy matrix has been studied by applying different experimental procedures, including the use of solvents (Zhu et al. 2004;

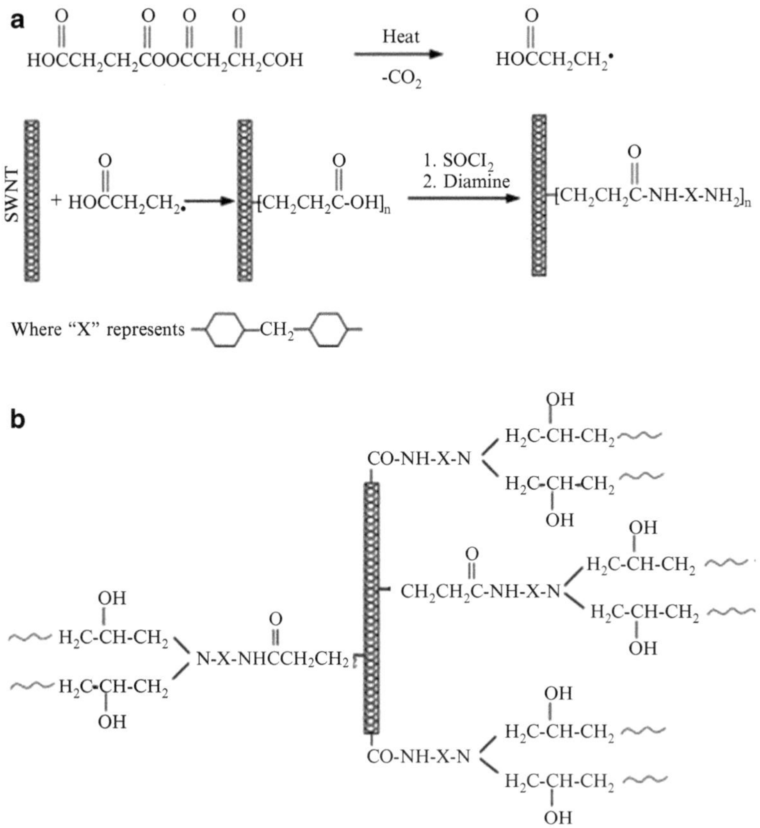

Fig. 5.14 (a) Reaction scheme for the functionalization of SWNTs (COOH groups at open ends are not shown here), and (b) integration of the functionalized SWNTs into epoxy (Reproduced with permission from Wang et al. 2008. Copyright Wiley-VCH)

Yang et al. 2008; Ma et al. 2010a; Prolongo et al. 2008; Kim et al. 2006; Ahn et al. 2008; Chen et al. 2008b) or via direct mixing (Wang et al. 2006a; Gojny et al. 2006; Zheng et al. 2006, 2010; Chen et al. 2008a; Shen et al. 2007a, b). The amine-functionalized carbon nanotubes seem to better disperse in polar solvents, such as chloroform, compared to unmodified carbon nanotubes (Prolongo et al. 2008). Good dispersion is preserved even after solvent removal resulting in a homogenous mixture with strong interfacial bonding between the nanotubes and the epoxy matrix (Zhu et al. 2004; Prolongo et al. 2008; Kim et al. 2006; Chen et al. 2008b). In the absence of solvents, the dispersion of nanotubes can be promoted by the presence of amine-bearing surface organic molecules (Wang et al. 2006a). Through a different approach, it was shown that Jeffamine T-403 (polyetheramine) can be used both as curing agent and polar solvent in order to improve the dispersion of unmodified and amino-functionalized CNTs within the epoxy matrix (Gojny and Schulte 2004). The dual use of polyetheramines as organic modifiers of inorganic nanoadditives and as curing agents has also been previously demonstrated for epoxy nanocomposites with organo-clays (Triantafyllidis et al. 2002a, b).

The strong interfacial bonding between the amino-functionalized carbon nanotubes and the epoxy polymer was confirmed by careful analysis of SEM and/or TEM images of fractured surfaces of the nanocomposite samples. The observation of broken nanotubes, instead of pulled-out nanotubes, was attributed to the strong adhesion of the nanotubes with the epoxy matrix (Chen et al. 2008a; Shen et al. 2007a). The homogeneous dispersion and the good interfacial properties in the epoxy nanocomposites prepared with amino-functionalized nanotubes, usually induce higher improvement of mechanical properties to epoxy polymers, compared to unmodified carbon nanotubes, mainly due to enhanced stress transfer from the polymer matrix to carbon nanotubes in the former nanocomposites (Zhu et al. 2004; Ma et al. 2010a; Kim et al. 2006; Ahn et al. 2008; Chen et al. 2008a, b; Gojny and Schulte 2004; Wang et al. 2006a, b; Shen et al. 2007a). Thermal and thermomechanical properties of epoxy polymers are also improved by the addition of amine functionalized carbon nanotubes (Chen et al. 2008b; Gojny and Schulte 2004; Wang et al. 2006a; Shen et al. 2007a; Yang et al. 2008). The glass transition temperature (T_g) of epoxy nanocomposites with amine-functionalized CNTs is higher compared to nanocomposites with non-functionalized nanotubes, while higher content of functionalized CNTs induces further increase of the T_g values (Chen et al. 2008b; Gojny and Schulte 2004). It was also shown that the T_g of epoxy nanocomposites depends on the type of the amine functional group, i.e. aromatic amines induce higher T_g compared to non-aromatic amines used as carbon nanotube modifiers (Shen et al. 2007a). The electrical conductivity of epoxy-CNT nanocomposites prepared by amine-functionalized CNTs is also improved (Valentini et al. 2004). However, the effect of functionalization on the structural integrity and on the aspect ratio of nanotubes can be decisive for the changes in the conductivity properties of the nanocomposites (Gojny et al. 2006), in accordance to the epoxy-carboxylated CNT nanocomposites.

5.5 Concluding Remarks

The remarkable physical properties of carbon nanotubes and their versatile chemical reactivity leading to various types of surface organo-functionalization were the main reasons why CNTs have become one of the most important types of nanoadditives for the development of novel polymer (including epoxy) nanocomposites with improved and sometimes unique properties. Although pristine/unmodified CNTs are capable of inducing noticeable improvements in the mechanical, conductivity and viscoelastic properties of epoxy polymers, it has been clearly demonstrated on the basis of up to date results that the chemical modification of nanotubes can maximize the benefits that carbon nanotubes can offer. This is accomplished via two routes: (a) improved dispersion of disaggregated nanotubes in the epoxy matrix, which is favored by the use of organophilic CNTs, i.e. nanotubes with attached organic moieties on their surface, and (b) increased interfacial bonding between nanotubes and epoxy matrix, which can be provided by reactive functional groups, such as –COOH and $-NH_2$ groups, attached on the surface of nanotubes (or on the organic moieties attached on nanotubes). However, further optimization of the functionalization methods of carbon nanotubes and mainly of the epoxy resin-CNT processing techniques is required in order to achieve more homogeneous dispersion of nanotubes throughout the whole epoxy polymer network. Finally, a crucial parameter with regard to the practical use of epoxy-CNT nanocomposites in various applications, including the aerospace industry, is the scaling-up of all the relevant synthetic-preparation procedures, from the large scale effective chemical organo-funationalization of CNTs to the manufacture of large polymeric structural parts using established industrial procedures, such as extrusion (vacuum assisted), resin transfer molding or filament winding.

References

Ahn, S.N., Lee, H.J., Kim, B.J., Tan, L.S., Baek, J.B.: Epoxy/amine-functionalized short-length vapor-grown carbon nanofiber composites. J. Polym. Sci. Pt A – Polym. Chem. **46**(22), 7473–7482 (2008). doi:10.1002/pola.23052

Ajayan, P.M.: Aligned carbon nanotubes in a thin polymer film. Adv. Mater. **7**(5), 489–491 (1995)

Ajayan, P.M., Stephan, O., Colliex, C., Trauth, D.: Aligned carbon nanotube arrays formed by cutting a polymer resin-nanotube composite. Science **265**(5176), 1212–1214 (1994)

Ajayan, P.M., Schadler, L.S., Giannaris, C., Rubio, A.: Single-walled carbon nanotube-polymer composites: strength and weakness. Adv. Mater. **12**(10), 750–753 (2000)

Allaoui, A., El Bounia, N.: How carbon nanotubes affect the cure kinetics and glass transition temperature of their epoxy composites? – a review. Expr. Polym. Lett. **3**(9), 588–594 (2009). doi:10.3144/expresspolymlett.2009.73

Bae, J., Jang, J., Yoon, S.H.: Cure behavior of the liquid-crystalline epoxy/carbon nanotube system and the effect of surface treatment of carbon fillers on cure reaction. Macromol. Chem. Phys. **203**(15), 2196–2204 (2002)

Bahr, J.L., Tour, J.M.: Covalent chemistry of single-wall carbon nanotubes. J. Mater. Chem. **12**(7), 1952–1958 (2002). doi:10.1039/b201013p

Balasubramanian, K., Burghard, M.: Chemically functionalized carbon nanotubes. Small **1**(2), 180–192 (2005)

Banerjee, S., Wong, S.S.: Rational sidewall functionalization and purification of single-walled carbon nanotubes by solution-phase ozonolysis. J. Phys. Chem. B **106**(47), 12144–12151 (2002)

Banerjee, S., Hemraj-Benny, T., Wong, S.S.: Covalent surface chemistry of single-walled carbon nanotubes. Adv. Mater. **17**(1), 17–29 (2005)

Baughman, R.H., Zakhidov, A.A., De Heer, W.A.: Carbon nanotubes – the route toward applications. Science **297**(5582), 787–792 (2002)

Bhowmik, S., Benedictus, R., Poulis, J.A., Bonin, H.W., Bui, V.T.: High-performance nano-adhesive bonding of titanium for aerospace and space applications. Int. J. Adhes. Adhes. **29**(3), 259–267 (2009)

Bogomolov, V.N., Kartenko, N.F., Kurdyukov, D.A., Parfen'eva, L.S., Smirnov, I.A., Sharenkova, N.V., Misiorek, H., Jezowski, A.: Thermal conductivity of the opal-epoxy resin nanocomposite. Phys. Solid State **45**(5), 957–960 (2003)

Bubert, H., Haiber, S., Brandl, W., Marginean, G., Heintze, M., Brüser, V.: Characterization of the uppermost layer of plasma-treated carbon nanotubes. Diam. Relat. Mater. **12**(3–7), 811–815 (2003)

Cebeci, H., de Villoria, R.G., Hart, A.J., Wardle, B.L.: Multifunctional properties of high volume fraction aligned carbon nanotube polymer composites with controlled morphology. Compos. Sci. Technol. **69**(15–16), 2649–2656 (2009). doi:10.1016/j.compscitech.2009.08.006

Chen, B., Parker, G., Han, J., Meyyappan, M., Cassell, A.M.: Heterogeneous single-walled carbon nanotube catalyst discovery and optimization. Chem. Mat. **14**(4), 1891–1896 (2002)

Chen, H., Muthuraman, H., Stokes, P., Zou, J.H., Liu, X., Wang, J.H., Huo, Q., Khondaker, S.I., Zhai, L.: Dispersion of carbon nanotubes and polymer nanocomposite fabrication using trifluoroacetic acid as a co-solvent. Nanotechnology **18**(41), 415606 (2007). doi:10.1088/0957-4484/18/41/415606

Chen, W., Lu, H.B., Nutt, S.R.: The influence of functionalized MWCNT reinforcement on the thermomechanical properties and morphology of epoxy nanocomposites. Compos. Sci. Technol. **68**(12), 2535–2542 (2008a). doi:10.1016/j.compscitech.2008.05.011

Chen, X.H., Wang, J.F., Lin, M., Zhong, W.B., Feng, T., Chen, J.H., Xue, F.: Mechanical and thermal properties of epoxy nanocomposites reinforced with amino-functionalized multi-walled carbon nanotubes. Mater. Sci. Eng. A **492**(1–2), 236–242 (2008b). doi:10.1016/j.msea.2008.04.044

Cho, J., Daniel, I.M.: Reinforcement of carbon/epoxy composites with multi-wall carbon nanotubes and dispersion enhancing block copolymers. Scr. Mater. **58**(7), 533–536 (2008). doi:10.1016/j.scriptamat.2007.11.011

Choi, M.H., Jeon, B.H., Chung, I.J.: The effect of coupling agent on electrical and mechanical properties of carbon fiber/phenolic resin composites. Polymer **41**(9), 3243–3252 (2000)

Choi, Y.K., Gotoh, Y., Sugimoto, K.I., Song, S.M., Yanagisawa, T., Endo, M.: Processing and characterization of epoxy nanocomposites reinforced by cup-stacked carbon nanotubes. Polymer **46**(25), 11489–11498 (2005a). doi:10.1016/j.polymer.2005.10.028

Choi, Y.K., Sugimoto, K.I., Song, S.M., Gotoh, Y., Ohkoshi, Y., Endo, M.: Mechanical and physical properties of epoxy composites reinforced by vapor grown carbon nanofibers. Carbon **43**(10), 2199–2208 (2005b)

Coleman, J.N., Curran, S., Dalton, A.B.., Davey, A.P., McCarthy, B., Blau, W., Barklie, R.C.: Percolation-dominated conductivity in a conjugated-polymer-carbon-nanotube composite. Phys. Rev. B – Condens. Matter Mater. Phys. **58**(12) (1998)

Coleman, J.N., Khan, U., Blau, W.J., Gunko, Y.K.: Small but strong: a review of the mechanical properties of carbon nanotube-polymer composites. Carbon **44**(9), 1624–1652 (2006a)

Coleman, J.N., Khan, U., Gun'ko, Y.K.: Mechanical reinforcement of polymers using carbon nanotubes. Adv. Mater. **18**(6), 689–706 (2006b). doi:10.1002/adma.200501851

Curran, S.A., Ajayan, P.M., Blau, W.J., Carroll, D.L., Coleman, J.N., Dalton, A.B.., Davey, A.P., Drury, A., McCarthy, B., Maier, S., Strevens, A.: A composite from poly (m-phenylene-vinylene-co-2,5-dioctoxy-p-phenylenevinylene) and carbon nanotubes: a novel material for molecular optoelectronics. Adv. Mater. **10**(14), 1091–1093 (1998)

Czerw, R., Guo, Z.X., Ajayan, P.M., Sun, Y.P., Carroll, D.L.: Organization of polymers onto carbon nanotubes: a route to nanoscale assembly. Nano Lett. **1**(8), 423–427 (2001). doi:10.1021/nl015548y

Dalton, A.B.., Collins, S., Munoz, E., Razal, J.M., Ebron, V.H., Ferraris, J.P., Coleman, J.N., Kim, B.G., Baughman, R.H.: Super-tough carbon-nanotube fibres – these extraordinary composite fibres can be woven into electronic textiles. Nature **423**(6941), 703 (2003)

De Fenzo, A., Formicola, C., Antonucci, V., Zarrelli, M., Giordano, M.: Effects of zinc-based flame retardants on the degradation behaviour of an aerospace epoxy matrix. Polym. Degrad. Stab. **94**(9), 1354–1363 (2009)

Ebbesen, T.W., Ajayan, P.M.: Large-scale synthesis of carbon nanotubes. Nature **358**(6383), 220–222 (1992)

Fu, S.Y., Chen, Z.K., Hong, S., Han, C.C.: The reduction of carbon nanotube (CNT) length during the manufacture of CNT/polymer composites and a method to simultaneously determine the resulting CNT and interfacial strengths. Carbon **47**(14), 3192–3200 (2009). doi:10.1016/j.carbon.2009.07.028

Ganguli, S., Aglan, H., Dennig, P., Irvin, G.: Effect of loading and surface modification of MWCNTs on the fracture behavior of epoxy nanocomposites. J. Reinf. Plast. Compos. **25**(2), 175–188 (2006). doi:10.1177/0731684405056425

Gao, X.P., Zhang, Y., Chen, X., Pan, G.L., Yan, J., Wu, F., Yuan, H.T., Song, D.Y.: Carbon nanotubes filled with metallic nanowires. Carbon **42**(1), 47–52 (2004)

Geng, Y., Liu, M.Y., Li, J., Shi, X.M., Kim, J.K.: Effects of surfactant treatment on mechanical and electrical properties of CNT/epoxy nanocomposites. Compos. Pt A – Appl. Sci. Manuf. **39**(12), 1876–1883 (2008). doi:10.1016/j.compositesa.2008.09.009

Georgakilas, V., Gournis, D., Tzitzios, V., Pasquato, L., Guldi, D.M., Prato, M.: Decorating carbon nanotubes with metal or semiconductor nanoparticles. J. Mater. Chem. **17**(26), 2679–2694 (2007)

Giannelis, E.P.: Polymer layered silicate nanocomposites. Adv. Mater. **8**(1), 29–35 (1996)

Giannelis, E.P.: Polymer-layered silicate nanocomposites: synthesis, properties and applications. Appl. Organomet. Chem. **12**(10–11), 675–680 (1998)

Giannelis, E.P., Krishnamoorti, R., Manias, E.: Polymer-silicate nanocomposites: model systems for confined polymers and polymer brushes. In: Granick, S. (ed.) Polymers in confined environments. Advances in polymer science, vol. 138, pp. 107–147. Springer, Berlin (1999)

Gojny, F.H., Schulte, K.: Functionalisation effect on the thermo-mechanical behaviour of multi-wall carbon nanotube/epoxy-compo sites. Compos. Sci. Technol. **64**(15), 2303–2308 (2004). doi:10.1016/j.compscitech.2004.01.024

Gojny, F.H., Wichmann, M.H.G., Kopke, U., Fiedler, B., Schulte, K.: Carbon nanotube-reinforced epoxy-compo sites: enhanced stiffness and fracture toughness at low nanotube content. Compos. Sci. Technol. **64**(15), 2363–2371 (2004). doi:10.1016/j.compscitech.2004.04.002

Gojny, F.H., Wichmann, M.H.G., Fiedler, B., Kinloch, I.A., Bauhofer, W., Windle, A.H., Schulte, K.: Evaluation and identification of electrical and thermal conduction mechanisms in carbon nanotube/epoxy composites. Polymer **47**(6), 2036–2045 (2006). doi:10.1016/j.polymer.2006.01.029

Gong, X., Liu, J., Baskaran, S., Voise, R.D., Young, J.S.: Surfactant-assisted processing of carbon nanotube/polymer composites. Chem. Mater. **12**(4), 1049–1052 (2000). doi:10.1021/cm9906396

Gonon, P., Sylvestre, A., Teysseyre, J., Prior, C.: Dielectric properties of epoxy/silica composites used for microlectronic packaging, and their dependence on post-curing. J. Mater. Sci. Mater. Electron. **12**(2), 81–86 (2001)

González-Domínguez, J.M., Ansón-Casaos, A., Castell, P., Díez-Pascual, A.M., Naffakh, M., Ellis, G., Gómez, M.A., Teresa Martínez, M.: Integration of block copolymer-wrapped single-wall carbon nanotubes into a trifunctional epoxy resin. Influence on thermal performance. Polym. Degrad. Stab. **95**(10), 2065–2075 (2010)

Gournis, D., Karakassides, M.A., Bakas, T., Boukos, N., Petridis, D.: Catalytic synthesis of carbon nanotubes on clay minerals. Carbon **40**(14), 2641–2646 (2002)

Grady, B.P.: Recent developments concerning the dispersion of carbon nanotubes in polymers. Macromol. Rapid Commun. **31**(3), 247–257 (2010)

Grossiord, N., Loos, J., Regev, O., Koning, C.E.: Toolbox for dispersing carbon nanotubes into polymers to get conductive nanocomposites. Chem. Mater. **18**(5), 1089–1099 (2006). doi:10.1021/cm051881h

Guo, T., Nikolaev, P., Thess, A., Colbert, D.T., Smalley, R.E.: Catalytic growth of single-walled nanotubes by laser vaporization. Chem. Phys. Lett. **243**(1–2), 49–54 (1995)

Guo, P., Song, H.H., Chen, X.H.: Interfacial properties and microstructure of multiwalled carbon nanotubes/epoxy composites. Mater. Sci. Eng. A **517**(1–2), 17–23 (2009). doi:10.1016/j.msea.2009.03.064

Gusev, A.A., Hine, P.J., Ward, I.M.: Fiber packing and elastic properties of a transversely random unidirectional glass/epoxy composite. Compos. Sci. Technol. **60**(4), 535–541 (2000)

Hafner, J.H., Bronikowski, M.J., Azamian, B.R., Nikolaev, P., Rinzler, A.G., Colbert, D.T., Smith, K.A., Smalley, R.E.: Catalytic growth of single-wall carbon nanotubes from metal particles. Chem. Phys. Lett. **296**(1–2), 195–202 (1998)

Harper, C.A. (ed.): Modern plastics handbook. McGraw-Hill, New York (2000)

Hernandez-Perez, A., Aviles, F., May-Pat, A., Valadez-Gonzalez, A., Herrera-Franco, P.J., Bartolo-Perez, P.: Effective properties of multiwalled carbon nanotube/epoxy composites using two different tubes. Compos. Sci. Technol. **68**(6), 1422–1431 (2008). doi:10.1016/j.compscitech.2007.11.001

Hirsch, A.: Functionalization of single-walled carbon nanotubes. Angew. Chem. Int. Ed. **41**(11), 1853–1859 (2002)

Hsueh, H.B., Chen, C.Y.: Preparation and properties of LDHs/epoxy nanocomposites. Polymer **44**(18), 5275–5283 (2003)

Huang, J.C.: Carbon black filled conducting polymers and polymer blends. Adv. Polym. Technol. **21**(4), 299–313 (2002)

Iglesias, J.G., González-Benito, J., Aznar, A.J., Bravo, J., Baselga, J.: Effect of glass fiber surface treatments on mechanical strength of epoxy based composite materials. J. Colloid Interface Sci. **250**(1), 251–260 (2002)

Iijima, S.: Helical microtubules of graphitic carbon. Nature **354**(6348), 56–58 (1991)

Irfan, M.H.: Chemistry and technology of thermosetting polymers in construction applications. Springer, Dordrecht (1998)

Jiang, S., Zhu, L., Ma, Y., Wang, X., Liu, J., Zhu, J., Fan, Y., Zou, Z., Hu, Z.: Direct immobilization of Pt-Ru alloy nanoparticles on nitrogen-doped carbon nanotubes with superior electrocatalytic performance. J. Power Sources **195**(22), 7578–7582 (2010)

Jin, Z.X., Pramoda, K.P., Goh, S.H., Xu, G.Q.: Poly (vinylidene fluoride)-assisted melt-blending of multi-walled carbon nanotube/poly (methyl methacrylate) composites. Mater. Res. Bull. **37**(2), 271–278 (2002)

Karousis, N., Tagmatarchis, N., Tasis, D.: Current progress on the chemical modification of carbon nanotubes. Chem. Rev. **110**(9), 5366–5397 (2010)

Ke, Y.C., Stroeve, P.: Polymer-layered silicate and silica nanocomposites. Elsevier, Amsterdam/Boston (2005)

Ke, Y., Lü, J., Yi, X., Zhao, J., Qi, Z.: Effects of promoter and curing process on exfoliation behavior of epoxy/clay nanocomposites. J. Appl. Polym. Sci. **78**(4), 808–815 (2000)

Kim, Y.J., Shin, T.S., Choi, H.D., Kwon, J.H., Chung, Y.C., Yoon, H.G.: Electrical conductivity of chemically modified multiwalled carbon nanotube/epoxy composites. Carbon **43**(1), 23–30 (2005). doi:10.1016/j.carbon.2004.08.015

Kim, J.A., Seong, D.G., Kang, T.J., Youn, J.R.: Effects of surface modification on rheological and mechanical properties of CNT/epoxy composites. Carbon **44**(10), 1898–1905 (2006). doi:10.1016/j.carbon.2006.02.026

Kukovecz, A., Kramberger, C., Holzinger, M., Kuzmany, H., Schalko, J., Mannsberger, M., Hirsch, A.: On the stacking behavior of functionalized single-wall carbon nanotubes. J. Phys. Chem. B **106**(25), 6374–6380 (2002)

Kupke, M.: Electrically conductive glass fibre reinforced epoxy resin. Mater. Res. Innov. **2**(3), 164–169 (1998)

Lan, T., Pinnavaia, T.J.: Clay-reinforced epoxy nanocomposites. Chem. Mater. **6**(12), 2216–2219 (1994)

Larsen, R.: Modification of single-walled carbon nanotubes and its influence on the interfacial strength in polymeric composites. J. Mater. Sci. **44**(3), 799–807 (2009). doi:10.1007/s10853-008-3155-3

Lau, K.T., Shi, S.Q., Zhou, L.M., Cheng, H.M.: Micro-hardness and flexural properties of randomly-oriented carbon nanotube composites. J. Compos. Mater. **37**(4), 365–376 (2003). doi:10.1106/002199803031043

Lau, K.T., Lu, M., Lam, C.K., Cheung, H.Y., Sheng, F.L., Li, H.L.: Thermal and mechanical properties of single-walled carbon nanotube bundle-reinforced epoxy nanocomposites: the role of solvent for nanotube dispersion. Compos. Sci. Technol. **65**(5), 719–725 (2005). doi:10.1016/j.compscitech.2004.10.005

Leroux, F., Besse, J.: Polymer interleaved layered double hydroxide: a new emerging class of nanocomposites. Chem. Mater. **13**(10), 3507–3515 (2001)

Li, Y.L., Kinloch, I.A., Windle, A.H.: Direct spinning of carbon nanotube fibers from chemical vapor deposition synthesis. Science **304**(5668), 276–278 (2004)

Li, J., Kim, J.K., Lung Sham, M.: Conductive graphite nanoplatelet/epoxy nanocomposites: effects of exfoliation and UV/ozone treatment of graphite. Scr. Mater. **53**(2), 235–240 (2005a)

Li, S.P., Qin, Y.J., Shi, J.H., Guo, Z.X., Yongfang, L., Zhu, D.B.: Electrical properties of soluble carbon nanotube/polymer composite films. Chem. Mater. **17**(1), 130–135 (2005b)

Li, J., Ma, P.C., Chow, W.S., To, C.K., Tang, B.Z., Kim, J.K.: Correlations between percolation threshold, dispersion state, and aspect ratio of carbon nanotubes. Adv. Funct. Mater. **17**(16), 3207–3215 (2007). doi:10.1002/adfm.200700065

Lin, Y., Zhou, B., Fernando, K.A.S., Liu, P., Allard, L.F., Sun, Y.P.: Polymeric carbon nanocomposites from carbon nanotubes functionalized with matrix polymer. Macromolecules **36**(19), 7199–7204 (2003)

Liu, L., Grunlan, J.C.: Clay assisted dispersion of carbon nanotubes in conductive epoxy nanocomposites. Adv. Funct. Mater. **17**, 2343–2348 (2007). doi:10.1002/adfm.200600785

Loos, M.R., Coelho, L.A.F., Pezzin, S.H., Amico, S.C.: Effect of carbon nanotubes addition on the mechanical and thermal properties of epoxy matrices. Mater. Res. Ibero-Am. J. Mater. **11**(3), 347–352 (2008)

Ma, P.C., Mo, S.Y., Tang, B.Z., Kim, J.K.: Dispersion, interfacial interaction and re-agglomeration of functionalized carbon nanotubes in epoxy composites. Carbon **48**(6), 1824–1834 (2010a)

Ma, P.C., Siddiqui, N.A., Marom, G., Kim, J.K.: Dispersion and functionalization of carbon nanotubes for polymer-based nanocomposites: a review. Compos. Pt A: Appl. Sci. Manuf. **41**(10), 1345–1367 (2010b)

Maccallini, E., Tsoufis, T., Policicchio, A., La Rosa, S., Caruso, T., Chiarello, G., Colavita, E., Formoso, V., Gournis, D., Agostino, R.G.: A spectro-microscopic investigation of Fe-Co bimetallic catalysts supported on MgO for the production of thin carbon nanotubes. Carbon **48**(12), 3434–3445 (2010)

Martin, C.A., Sandler, J.K.W., Windle, A.H., Schwarz, M.K., Bauhofer, W., Schulte, K., Shaffer, M.S.P.: Electric field-induced aligned multi-wall carbon nanotube networks in epoxy composites. Polymer **46**(3), 877–886 (2005). doi:10.1016/j.polymer.2004.11.081

Merkel, T.C., Freeman, B.D., Spontak, R.J., He, Z., Pinnau, I., Meakin, P., Hill, A.J.: Ultrapermeable, reverse-selective nanocomposite membranes. Science **296**(5567), 519–522 (2002)

Mittal, V. (ed.): Polymer nanotube nanocomposites: synthesis, properties, and applications. Scrivener Publishing LLC, Salem (2010)

Mittal, V. (ed.): Surface modification of nanotube fillers. Wiley-VCH, Weinheim (2011)

Miyagawa, H., Drzal, L.T.: Thermo-physical and impact properties of epoxy nanocomposites reinforced by single-wall carbon nanotubes. Polymer **45**(15), 5163–5170 (2004)

Moisala, A., Li, Q., Kinloch, I.A., Windle, A.H.: Thermal and electrical conductivity of single- and multi-walled carbon nanotube-epoxy composites. Compos. Sci. Technol. **66**, 1285–1288 (2006). doi:10.1016/j.compscitech.2005.10.016

Moniruzzaman, M., Winey, K.I.: Polymer nanocomposites containing carbon nanotubes. Macromolecules **39**(16), 5194–5205 (2006). doi:10.1021/ma060733p

Montazeri, A., Javadpour, J., Khavandi, A., Tcharkhtchi, A., Mohajeri, A.: Mechanical properties of multi-walled carbon nanotube/epoxy composites. Mater. Des. **31**(9), 4202–4208 (2010a)

Montazeri, A., Khavandi, A., Javadpour, J., Tcharkhtchi, A.: The effect of curing cycle on the mechanical properties of MWNT/Epoxy nanocomposite. Int. J. Polym. Anal. Charact. **15**(3), 182–190 (2010b)

Montazeri, A., Khavandi, A., Javadpour, J., Tcharkhtchi, A.: Viscoelastic properties of multi-walled carbon nanotube/epoxy composites using two different curing cycles. Mater. Des. **31**(7), 3383–3388 (2010c)

Njuguna, B., Pielichowski, K.: Polymer nanocomposites for aerospace applications: properties. Adv. Eng. Mater. **5**(11), 769–778 (2003)

Njuguna, J., Pielichowski, K.: Polymer nanocomposites for aerospace applications: characterization. Adv. Eng. Mater. **6**(4), 204–210 (2004a)

Njuguna, J., Pielichowski, K.: Polymer nanocomposites for aerospace applications: fabrication. Adv. Eng. Mater. **6**(4), 193–203 (2004b)

Park, I., Peng, H.G., Gidley, D.W., Xue, S.Q., Pinnavaia, T.J.: Epoxy-silica mesocomposites with enhanced tensile properties and oxygen permeability. Chem. Mater. **18**(3), 650–656 (2006). doi:10.1021/cm051768r

Pinnavaia, T.J., Beall, G.W. (eds.): Polymer-clay nanocomposites. Wiley, New York (2000)

Pizzutto, C.E., Suave, J., Bertholdi, J., Pezzin, S.H., Coelho, L.A.F., Amico, S.C.: Mechanical and dilatometric properties of carboxylated SWCNT/epoxy composites: effects of the dispersion in the resin and in the hardener. J. Reinf. Plast. Compos. **29**(4), 524–530 (2010)

Policicchio, A., Caruso, T., Chiarello, G., Colavita, E., Formoso, V., Agostino, R.G., Tsoufis, T., Gournis, D., La Rosa, S.: Electronic, chemical and structural characterization of CNTs grown by acetylene decomposition over MgO supported Fe-Co bimetallic catalysts. Surf. Sci. **601**(13), 2823–2827 (2007)

Preghenella, M., Pegoretti, A., Migliaresi, C.: Thermo-mechanical characterization of fumed silica-epoxy nanocomposites. Polymer **46**(26), 12065–12072 (2005). doi:10.1016/j.polymer.2005.10.098

Prolongo, S.G., Buron, M., Gude, M.R., Chaos-Moran, R., Campo, M., Urena, A.: Effects of dispersion techniques of carbon nanofibers on the thermo-physical properties of epoxy nanocomposites. Compos. Sci. Technol. **68**(13), 2722–2730 (2008). doi:10.1016/j.compscitech.2008.05.015

Prolongo, S.G., Gude, M.R., Sanchez, J., Ureña, A.: Nanoreinforced epoxy adhesives for aerospace industry. J. Adhes. **85**(4–5), 180–199 (2009)

Qian, D., Dickey, E.C., Andrews, R., Rantell, T.: Load transfer and deformation mechanisms in carbon nanotube- polystyrene composites. Appl. Phys. Lett. **76**(20), 2868–2870 (2000)

Qian, W., Liu, T., Wang, Z., Yu, H., Li, Z., Wei, F., Luo, G.: Effect of adding nickel to iron-alumina catalysts on the morphology of as-grown carbon nanotubes. Carbon **41**(13), 2487–2493 (2003)

Qin, S.H., Oin, D.Q., Ford, W.T., Resasco, D.E., Herrera, J.E.: Polymer brushes on single-walled carbon nanotubes by atom transfer radical polymerization of n-butyl methacrylate. J. Am. Chem. Soc. **126**(1), 170–176 (2004)

Qingwen, L., Hao, Y., Yan, C., Jin, Z., Zhongfan, L.: A scalable CVD synthesis of high-purity single-walled carbon nanotubes with porous MgO as support material. J. Mater. Chem. **12**(4), 1179–1183 (2002)

Ragosta, G., Abbate, M., Musto, P., Scarinzi, G., Mascia, L.: Epoxy-silica particulate nanocomposites: chemical interactions, reinforcement and fracture toughness. Polymer **46**(23), 10506–10516 (2005). doi:10.1016/j.polymer.2005.08.028

Ramanathan, T., Liu, H., Brinson, L.C.: Functionalized SWNT/polymer nanocomposites for dramatic property improvement. J. Polymer Sci., Pt B: Polymer Phys. **43**(17), 2269–2279 (2005). doi:10.1002/polb.20510

Ramanathan, T., Abdala, A.A., Stankovich, S., Dikin, D.A., Herrera-Alonso, M., Piner, R.D., Adamson, D.H., Schniepp, H.C., Chen, X., Ruoff, R.S., Nguyen, S.T., Aksay, I.A., Prud'Homme, R.K., Brinson, L.C.: Functionalized graphene sheets for polymer nanocomposites. Nat. Nanotechnol. **3**(6), 327–331 (2008)

Rinzler, A.G., Liu, J., Dai, H., Nikolaev, P., Huffman, C.B., Rodriguez-Macias, F.J., Boul, P.J., Lu, A.H., Heymann, D., Colbert, D.T., Lee, R.S., Fischer, J.E., Rao, A.M., Eklund, P.C., Smalley, R.E.: Large-scale purification of single-wall carbon nanotubes: process, product, and characterization. Appl. Phys. A – Mater. Sci. Process. **67**(1), 29–37 (1998)

Rothon, R.N. (ed.): Particulate-filled polymer composites, 2nd edn. Smithers Rapra Technology, Shrewsbury (2003)

Sandler, J.K.W., Kirk, J.E., Kinloch, I.A., Shaffer, M.S.P., Windle, A.H.: Ultra-low electrical percolation threshold in carbon-nanotube-epoxy composites. Polymer **44**(19), 5893–5899 (2003). doi:10.1016/s0032-3861(03)00539-1

Schadler, L.S., Giannaris, S.C., Ajayan, P.M.: Load transfer in carbon nanotube epoxy composites. Appl. Phys. Lett. **73**(26), 3842–3844 (1998)

Schueler, R., Petermann, J., Schulte, K., Wentzel, H.P.: Agglomeration and electrical percolation behavior of carbon black dispersed in epoxy resin. J. Appl. Polym. Sci. **63**(13), 1741–1746 (1997)

Shen, J.F., Huang, W.S., Wu, L.P., Hu, Y.Z., Ye, M.X.: The reinforcement role of different amino-functionalized multi-walled carbon nanotubes in epoxy nanocomposites. Compos. Sci. Technol. **67**, 3041–3050 (2007a). doi:10.1016/j.compscitech.2007.04.025

Shen, J.F., Huang, W.S., Wu, L.P., Hu, Y.Z., Ye, M.X.: Thermo-physical properties of epoxy nanocomposites reinforced with amino-functionalized multi-walled carbon nanotubes. Compos. Pt A – Appl. Sci. Manuf. **38**, 1331–1336 (2007b). doi:10.1016/j.compositesa.2006.10.012

Simmons, J.M., Nichols, B.M., Baker, S.E., Marcus, M.S., Castellini, O.M., Lee, C.S., Hamers, R.J., Eriksson, M.A.: Effect of ozone oxidation on single-walled carbon nanotubes. J. Phys. Chem. B **110**(14), 7113–7118 (2006)

Singha, S., Thomas, M.J.: Influence of filler loading on dielectric properties of epoxy-ZnO nanocomposites. IEEE Trans. Dielectr. Electr. Insul. **16**(2), 531–542 (2009)

Soneda, Y., Duclaux, L., Beguin, F.: Synthesis of high quality multi-walled carbon nanotubes from the decomposition of acetylene on iron-group metal catalysts supported on MgO. Carbon **40**(6), 965–969 (2002)

Song, Y.S., Youn, J.R.: Influence of dispersion states of carbon nanotubes on physical properties of epoxy nanocomposites. Carbon **43**(7), 1378–1385 (2005). doi:10.1016/j.carbon.2005.01.007

Suave, J., Coelho, L.A.F., Amico, S.C., Pezzin, S.H.: Effect of sonication on thermo-mechanical properties of epoxy nanocomposites with carboxylated-SWNT. Mater. Sci. Eng. A **509**(1–2), 57–62 (2009). doi:10.1016/j.msea.2009.01.036

Suhr, J., Koratkar, N., Keblinski, P., Ajayan, P.: Viscoelasticity in carbon nanotube composites. Nat. Mater. **4**(2), 134–137 (2005). doi:10.1038/nmat1293

Sumita, M., Sakata, K., Asai, S., Miyasaka, K., Nakagawa, H.: Dispersion of fillers and the electrical conductivity of polymer blends filled with carbon black. Polym. Bull. **25**(2), 265–271 (1991)

Sun, Y.P., Fu, K., Lin, Y., Huang, W.: Functionalized carbon nanotubes: properties and applications. Acc. Chem. Res. **35**(12), 1096–1104 (2002a)

Sun, Y.P., Fu, K.F., Lin, Y., Huang, W.J.: Functionalized carbon nanotubes: properties and applications. Acc. Chem. Res. **35**(12), 1096–1104 (2002b). doi:10.1021/ar010160v

Tasis, D., Tagmatarchis, N., Bianco, A., Prato, M.: Chemistry of carbon nanotubes. Chem. Rev. **106**(3), 1105–1136 (2006)

Thakre, P.R., Bisrat, Y., Lagoudas, D.C.: Electrical and mechanical properties of carbon nanotube-epoxy nanocomposites. J. Appl. Polym. Sci. **116**(1), 191–202 (2010)

Thess, A., Lee, R., Nikolaev, P., Dai, H.J., Petit, P., Robert, J., Xu, C.H., Lee, Y.H., Kim, S.G., Rinzler, A.G., Colbert, D.T., Scuseria, G.E., Tomanek, D., Fischer, J.E., Smalley, R.E.: Crystalline ropes of metallic carbon nanotubes. Science **273**(5274), 483–487 (1996)

Tiano, T., Roylance, M., Gassner, J.: Functionalization of single-wall nanotubes for improved structural composites. In: Adams, P.J., Elsworth, S.A., Petkauskos, M.J., Walton, T.C. (eds.) Revolutionary Materials: Technology and Economics. International SAMPE Technical Conference Series, vol. 32, pp. 192–199. Soc Advancement Material & Process Engineering, Azusa (2000)

Triantafyllidis, K.S., LeBaron, P.C., Pinnavaia, T.J.: Homostructured mixed inorganic-organic ion clays: a new approach to epoxy polymer-exfoliated clay nanocomposites with a reduced organic modifier content. Chem. Mater. **14**(10), 4088–4095 (2002a). doi:10.1021/cm0202862

Triantafyllidis, K.S., LeBaron, P.C., Pinnavaia, T.J.: Thermoset epoxy-clay nanocomposites: the dual role of alpha, omega-diamines as clay surface modifiers and polymer curing agents. J. Solid State Chem. **167**(2), 354–362 (2002b). doi:10.1006/jssc.2001.9541

Triantafyllidis, K.S., Karakoulia, S.A., Gournis, D., Delimitis, A., Nalbandian, L., Maccallini, E., Rudolf, P.: Formation of carbon nanotubes on iron/cobalt oxides supported on zeolite-Y: effect of zeolite textural properties and particle morphology. Microporous Mesoporous Mater. **110**(1), 128–140 (2008)

Tsantzalis, S., Tsotra, P., Karapappas, P., Vavouliotis, A., Fanis, N., Kostopoulos, V., Friedrich, K.: Influence of carbon nanofibers and piezoelectric particles on the thermomechanical behavior of epoxy mixtures. Solid State Phenom. **121–123**, 1419–1424 (2007)

Tseng, C.H., Wang, C.C., Chen, C.Y.: Functionalizing carbon nanotubes by plasma modification for the preparation of covalent-integrated epoxy composites. Chem. Mater. **19**(2), 308–315 (2007)

Tsoufis, T., Xidas, P., Jankovic, L., Gournis, D., Saranti, A., Bakas, T., Karakassides, M.A.: Catalytic production of carbon nanotubes over Fe-Ni bimetallic catalysts supported on MgO. Diam. Relat. Mater. **16**(1), 155–160 (2007)

Tsoufis, T., Jankovic, L., Gournis, D., Trikalitis, P.N., Bakas, T.: Evaluation of first-row transition metal oxides supported on clay minerals for catalytic growth of carbon nanostructures. Mater. Sci. Eng. B **152**(1–3), 44–49 (2008)

Tuncer, E., Sauers, I., James, D.R., Ellis, A.R., Paranthaman, M.P., Aytug, T., Sathyamurthy, S., More, K.L., Li, J., Goyal, A.: Electrical properties of epoxy resin based nano-composites. Nanotechnology **18**(2) (2007). doi:02570310.1088/0957-4484/18/2/025703

Valentini, L., Armentano, I., Puglia, D., Kenny, J.M.: Dynamics of amine functionalized nanotubes/epoxy composites by dielectric relaxation spectroscopy. Carbon **42**(2), 323–329 (2004). doi:10.1016/j.carbon.2003.10.039

Vassileva, E., Friedrich, K.: Epoxy/alumina nanoparticle composites. I. Dynamic mechanical behavior. J. Appl. Polym. Sci. **89**(14), 3774–3785 (2003). doi:10.1002/app. 12463

Vassileva, E., Friedrich, K.: Epoxy/alumina nanoparticle composites. II. Influence of silane coupling agent treatment on mechanical performance and wear resistance. J. Appl. Polym. Sci. **101**(6), 4410–4417 (2006). doi:10.1002/app. 23297

Vavouliotis, A., Fiamegou, E., Karapappas, P., Psarras, G.C., Kostopoulos, V.: DC and AC conductivity in epoxy resin/multiwall carbon nanotubes percolative system. Polym. Compos. **31**(11), 1874–1880 (2010). doi:10.1002/pc.20981

Viswanathan, G., Chakrapani, N., Yang, H.C., Wei, B.Q., Chung, H.S., Cho, K.W., Ryu, C.Y., Ajayan, P.M.: Single-step in situ synthesis of polymer-grafted single-wall nanotube composites. J. Am. Chem. Soc. **125**(31), 9258–9259 (2003)

Wagner, H.D., Lourie, O., Feldman, Y., Tenne, R.: Stress-induced fragmentation of multiwall carbon nanotubes in a polymer matrix. Appl. Phys. Lett. **72**(2), 188–190 (1998)

Wang, J.G., Fang, Z.P., Gu, A.J., Xu, L.H., Liu, F.: Effect of amino-functionalization of multi-walled carbon nanotubes on the dispersion with epoxy resin matrix. J. Appl. Polym. Sci. **100**(1), 97–104 (2006a). doi:10.1002/app. 22647

Wang, S.R., Liang, Z.Y., Liu, T., Wang, B., Zhang, C.: Effective amino-functionalization of carbon nanotubes for reinforcing epoxy polymer composites. Nanotechnology **17**(6), 1551–1557 (2006b). doi:10.1088/0957-4484/17/6/003

Wang, Q., Dai, J.F., Li, W.X., Wei, Z.Q., Jiang, J.L.: The effects of CNT alignment on electrical conductivity and mechanical properties of SWNT/epoxy nanocomposites. Compos. Sci. Technol. **68**(7–8), 1644–1648 (2008). doi:10.1016/j.compscitech.2008.02.024

Wei, B.Q., Vajtai, R., Jung, Y., Ward, J., Zhang, R., Ramanath, G., Ajayan, P.M.: Organized assembly of carbon nanotubes. Nature **416**(6880), 495–496 (2002)

Weiss, H.: Anhydride curing agents for epoxy resins. Ind. Eng. Chem. **49**(7), 1089–1090 (1957). doi:10.1021/ie50571a024

Xidas, P.I., Triantafyllidis, K.S.: Effect of the type of alkylammonium ion clay modifier on the structure and thermal/mechanical properties of glassy and rubbery epoxy-clay nanocomposites. Eur. Polym. J. **46**(3), 404–417 (2010). doi:10.1016/j.eurpolymj.2009.11.004

Xie, X.L., Mai, Y.W., Zhou, X.P.: Dispersion and alignment of carbon nanotubes in polymer matrix: a review. Mater. Sci. Eng. R-Rep. **49**(4), 89–112 (2005)

Xu, X.J., Thwe, M.M., Shearwood, C., Liao, K.: Mechanical properties and interfacial characteristics of carbon-nanotube-reinforced epoxy thin films. Appl. Phys. Lett. **81**(15), 2833–2835 (2002). doi:10.1063/1.1511532

Xu, L., Fang, Z., Song, P., Peng, M.: Functionalization of carbon nanotubes by corona-discharge induced graft polymerization for the reinforcement of epoxy nanocomposites. Plasma Process. Polym. **7**(9–10), 785–793 (2010)

Yang, K., Gu, M.Y., Jin, Y.P.: Cure behavior and thermal stability analysis of multiwalled carbon nanotube/epoxy resin nanocomposites. J. Appl. Polym. Sci. **110**(5), 2980–2988 (2008). doi:10.1002/app. 28898

Yang, H., Shan, C., Li, F., Zhang, Q., Han, D., Niu, L.: Convenient preparation of tunably loaded chemically converted graphene oxide/epoxy resin nanocomposites from graphene oxide sheets through two-phase extraction. J. Mater. Chem. **19**(46), 8856–8860 (2009)

Zhang, W.D., Shen, L., Phang, I.Y., Liu, T.: Carbon nanotubes reinforced Nylon-6 composite prepared by simple melt-compounding. Macromolecules **37**(2), 256–259 (2004)

Zhang, X.H., Zhang, Z.H., Xu, W.J., Chen, F.C., Deng, J.R., Deng, X.: Toughening of cycloaliphatic epoxy resin by multiwalled carbon nanotubes. J. Appl. Polym. Sci. **110**(3), 1351–1357 (2008). doi:10.1002/app. 28590

Zhao, Y.L., Stoddart, J.F.: Noncovalent functionalization of single-walled carbon nanotubes. Acc. Chem. Res. **42**(8), 1161–1171 (2009)

Zheng, Y.P., Zhang, A.B.., Chen, Q.H., Zhang, J.X., Ning, R.C.: Functionalized effect on carbon nanotube/epoxy nano-composites. Mater. Sci. Eng. A **435**, 145–149 (2006). doi:10.1016/j.msea.2006.07.106

Zheng, Y.P., Zhang, J.X., Yu, P.Y., Shi, W., Wang, B.: Investigation of the microstructure of epoxy resin/MWNTs nanocomposite by the positron annihilation technique. Polym. – Plast. Technol. Eng. **49**(10), 1016–1021 (2010)

Zhou, T.L., Wang, X., Wang, T.C.: Cure reaction of multi-walled carbon nanotubes/diglycidyl ether of bisphenol A/2-ethyl-4-methylimidazole (MWCNTs/DGEBA/EMI-2,4) nanocomposites: effect of carboxylic functionalization of MWCNTs. Polym. Int. **58**(4), 445–452 (2009). doi:10.1002/pi.2558

Zhu, J., Kim, J.D., Peng, H.Q., Margrave, J.L., Khabashesku, V.N., Barrera, E.V.: Improving the dispersion and integration of single-walled carbon nanotubes in epoxy composites through functionalization. Nano Lett. **3**(8), 1107–1113 (2003). doi:10.1021/nl0342489

Zhu, J., Peng, H.Q., Rodriguez-Macias, F., Margrave, J.L., Khabashesku, V.N., Imam, A.M., Lozano, K., Barrera, E.V.: Reinforcing epoxy polymer composites through covalent integration of functionalized nanotubes. Adv. Funct. Mater. **14**(7), 643–648 (2004). doi:10.1002/adfm.200305162

Zhu, Y.F., Ma, C., Zhang, W., Zhang, R.P., Koratkar, N., Liang, J.: Alignment of multiwalled carbon nanotubes in bulk epoxy composites via electric field. J. Appl. Phys. **105**(5) (2009). doi:10.1063/1.3080243

Chapter 6
Stress Induced Changes in the Raman Spectrum of Carbon Nanostructures and Their Composites

A.S. Paipetis

Contents

Abstract Raman spectroscopy of Carbon nanostructures is fundamental in characterising the morphology and the interaction of the nanostructure with the environment. This work provides an outline of the Raman Vibrational modes for graphitic structures starting from graphite fibres, to single-wall carbon nanotubes to multiwall carbon nanotubes and finally to Single- and Multi-layer Graphene. Following a brief outline of the dependence of the force constant on applied deformation, the stress induced changes in the Raman spectrum of graphitic structures are subsequently discussed with a view to elucidating the reinforcing ability of the CNTs in a matrix and assessing the stress transfer at the CNT matrix interface. The possibilities of employing CNTs as stress sensors in composite materials are also presented.

Keywords Carbon Nanotubes • Raman Spectroscopy • Stress monitoring • Nanocomposites

---

A.S. Paipetis (✉)
Department of Materials Science and Engineering, University of Ioannina, Ioannina, Greece
e-mail: paipetis@cc.uoi.gr

A.S. Paipetis and V. Kostopoulos (eds.), *Carbon Nanotube Enhanced Aerospace Composite Materials*, Solid Mechanics and Its Applications 188,
DOI 10.1007/978-94-007-4246-8_6,

6.1 Introduction

The scope of this work is to provide an overview of the stress-induced changes in the Raman spectrum of graphitic structures. As there has been limited work done in the field of hybrid composite systems, this chapter will focus on aspects related to the ability of employing the Raman technique for monitoring systems that are comprised of graphitic carbon nanostructures and their composites. This will be performed in order to provide an insight into the capabilities of the methodology as a means of sensing internal stresses and assessing the stress transfer in nano-reinforced composites.

Raman spectroscopy has always been an invaluable tool for evaluating the structure and properties of graphitic materials. Since the reference work by Tuinstra and Koenig on graphite fibres (Tuinstra and Koenig 1970), there is an immense volume of research effort that focuses on the interpretation of the Raman spectrum of sp^2 carbon allotropes. This research interest has been further boosted due to the study of fullerenes (Kuzmany et al. 2004), carbon nanotubes (CNTs) (Dresselhaus et al. 2005), and recently graphene (Malard et al. 2009b). These graphitic structures are extremely promising in that they offer exceptional mechanical, thermal and electronic properties.

It is noteworthy that the in-plane elastic modulus of graphene is regarded to be the highest of all known materials, on the order of 1 TPa (Sengupta et al. 2011). The Raman spectrum of graphite provides direct information on the C–C bond which exhibits this extraordinary stiffness; note that the E_{2g} in plane vibration is the only allowable Raman Vibrational mode in graphite. In this respect, the importance of Raman monitoring of sp^2 carbon is by definition justified. All deviations from the planar hexagonal array of the infinite graphitic sheet or graphene directly affect all the aforementioned properties but at the same time give rise to other modes in a Raman spectrum (Malard et al. 2009b). This further enhances the capability of Raman Spectroscopy to monitor the sp^2 morphology, as well as the effect that any external field is expected to have on it.

The tubular morphology of single wall CNTs (SWCNTs) is of particular interest since all these "carbon molecules" may possess unique spectral signatures depending particularly on their diameter and chirality (Dresselhaus et al. 2010). These spectral features distinguish them from the typical spectrum of carbon fibres and allow for their identification. Of particular interest are the resonance effects which are related to the dispersive nature of vibrational modes such as the Radial Breathing Mode (RBM), which are also significantly affected by the deformation that an external field may induce to the SWCNT (Dresselhaus et al. 2007). The lifting of degeneracy of the E_{2g} band in the axial and circumferential direction also leads to an alteration of the typical graphitic lines. The unique structure of double wall CNTs (DWCNTs) is mirrored in their Raman spectrum and is directly related to phenomena such as interlayer interaction and interlayer stress transfer. The Raman spectrum of multiwall CNTs (MWCNTs) is closer to the Raman Spectrum of graphite fibres, but still direct information about the stress transfer efficiency of MWCNT reinforced composites may be derived. Finally, the simplicity of graphene is unique in providing insight into all vibrational modes including the so called "disorder induced" Raman vibrational modes as well as elucidating the interlayer interaction I multi-layer graphene.

Due to the anharmonicity of the C–C bond (Wool 1980), the applied stress on all aforementioned structures is directly related to shifts in vibrational frequencies or changes in the intensities due to resonance phenomena. Splitting of bands like the graphitic (G) or the second-order graphitic (G′) line is also observed (Frank et al. 2011a). The calibration of stress-induced shifts with applied strain provides direct information about the stress transfer. The derived stress dependence of Raman bands in the ideal case of Graphene may also directly link the translation of far-field stress to the C–C bond stretching, which is in fact the essence of reinforcement for nano-reinforced materials (Frank et al. 2011b). In this respect, Raman Spectroscopy is unique in providing directly stress information. In addition, the capability of employing graphitic structures as stress sensors within structural composites is also significant (Zhao et al. 2002), since the Raman Spectrum of CNTs may provide information on interfacial stress transfer (Sureeyatanapas et al. 2010) even stress concentrations around notches (Zhao and Wagner 2003).

Summarizing, Raman Spectroscopy is an invaluable tool in characterizing sp^2 graphitic structures and their composites. On the other hand, the outburst of research activity in these graphitic structures has immensely increased the research in the field of Raman spectroscopy of such materials and, as a result, the capabilities of Raman spectroscopy for stress monitoring and sensing in sp^2 structures seem to ever increase as the volume of related research is rising.

6.2 The Raman Spectrum of Graphitic Structures

Common to all graphitic structures is the Graphitic or G band which corresponds to the in-plane lattice vibrations of the plane graphitic crystal (Vidano et al. 1981). The aforementioned band is one of the two Raman active E_{2g} vibrational modes for graphite together with the band observed at approximately 50 cm^{-1} which corresponds to a rigid layer shearing of the graphitic lattice (Nemanich et al. 1977). All deviations from planar geometry and symmetry result in alterations in the G band. As a result, the G band can be used to probe any divergence from the flat geometric structure of graphene. These divergences may comprise the strain induced by external forces, by layer interaction in a graphene with few layers or in multi-wall nanotubes, or even by the curvature of the side wall in tubular structures. In the latter case, more Raman active modes are present, characterizing the diameter, the chirality rendering thus the Raman signature of every tubular geometry almost unique (Dresselhaus et al. 2010). However, the overlapping tubular geometry of multi-wall CNTs makes these spectral features less distinguishable (Malard et al. 2009b). The strong feature at approximately 2,700 cm^{-1} is also characteristic of sp^2 carbon, and is characterized by its dispersive properties, or by its dependence on excitation frequency. The so-called G′-band is a second-order double resonance process and as such is very sensitive to the morphology of the structure providing information on the number of concentric tubes for multiwall nanotubes (Malard et al. 2009b). The origin of this second order feature has been greatly argued, but for reasons of consistency it will be referred to in this work as the "G′ band". Interestingly enough, with an increase in the number of walls of the CNTs or equivalently with an increase in the number of layers, more double resonance scattering processes occur, and the final spectral feature converges to that of graphite, where only two peaks are observed.

Apart from the graphitic lines of the Raman Spectrum, the presence of the "disorder induced" lines is evident in most sp^2 morphologies. These two lines are observed in the first order spectrum in the vicinity of 1,360 and 1,620 cm^{-1} and are denoted as the D and the D′ line respectively (Lespade et al. 1984). The disorder induced lines are strongly interrelated and have been directly associated with experimentally induced disorder in the graphene layers or the average defect distance (Lucchese et al. 2010) as well as the graphitization temperature which is directly related to stiffness (Huang and Young 1995) or crystallite size. In the case of high modulus carbon fibres, the disorder induced bands are reported to vary along the fibre length for an individual fibre (Paipetis and Galiotis 1996) or even along the fibre cross section (Katagiri et al. 1988). The work on carbon fibres strongly suggests that apart from the amount of crystal boundary that is inversely proportional to crystal size as suggested in early works (Tuinstra and Koenig 1970), lattice orientation is also responsible for the presence or not of the disorder lines. Moreover, the chirality of the nanotubes may enhance or diminish the disorder lines in the case of armchair nanotube morphology and zigzag nanotube morphology, respectively (Cançado et al. 2004).

SWCNTs are specially challenging in that their diameter is by definition smaller than that of the typical excitation wavelength. The typical Raman spectrum of SWCNTs is depicted in Fig. 6.1. Although the excitation volume is very small, resonance phenomena are responsible for intense and sharp Raman bands (Dresselhaus et al. 2007). In the case of SWCNTs, the specificity in Raman vibrational activity is summarised in the aforementioned splitting of the G line and the presence of the Radial Breathing Mode (RBM) which is both dispersive and characteristic of the carbon nanotube diameter. The RBM corresponds to the out-of-plane stretching or the radial breathing of the graphitic structure (Dresselhaus et al. 2005). The RBM frequency is inversely proportional to the SWCNT diameter, a property which stems directly from the moment of inertia or the carbon mass distribution around the nanotube axis. Fundamental to the characterisation of SWCNT using Raman spectroscopy is the dependence of the RBM on the excitation frequency, which gives rise to the Kataura plot (Kataura et al. 1999), see Fig. 6.2. This corresponds to different resonant properties of nanotubes of different diameters, which in their turn allow for the probing of the existence of "single molecule" structures within the bulk of multiple nanotube morphologies (Dresselhaus et al. 2005). In other words, the spectral information contained in a spectrum for a given excitation energy corresponds to the fraction of the nanotubes that are in resonance with the specific

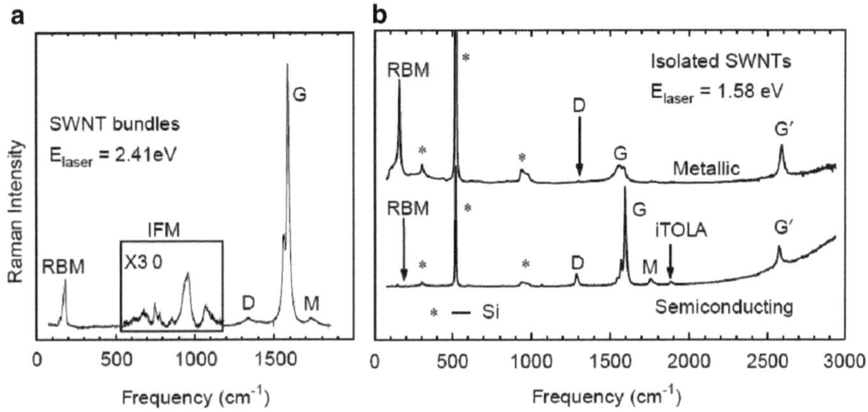

Fig. 6.1 (**a**) Raman spectra from SWNT bundles (**b**) Raman spectra from a metallic (*top*) and a semiconducting (*bottom*) SWNT at the single nanotube level (Reprinted from (Dresselhaus et al. 2005), with permission from Elsevier)

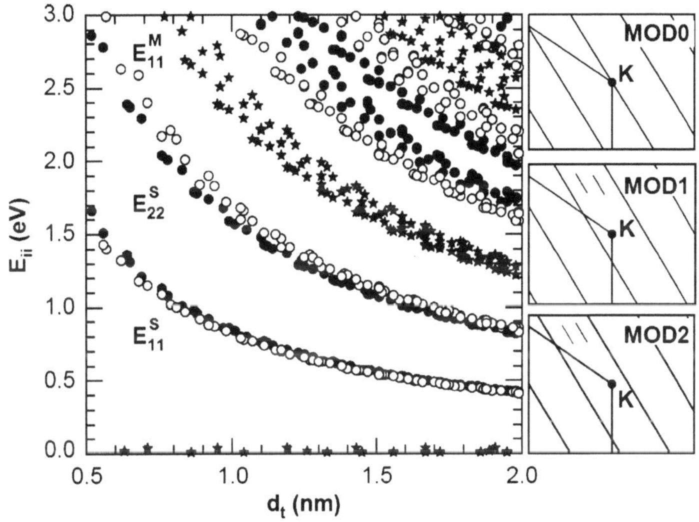

Fig. 6.2 The Kataura plot shows the transition energies vs. SWCNT diameter. The *right* panels show schematic figures defining the SWCNT classes (Reprinted from (Dresselhaus et al. 2005), with permission from Elsevier)

laser line (Milnera et al. 2000) which can even be assigned to a specific chirality (Jorio et al. 2001). The RBM feature is associated with small nanotube diameters, and thus is disappearing for Multi-Wall Nanotubes, although its presence has been reported under good resonance conditions (Benoit et al. 2002).

As aforementioned, SWCNTs present distinct features in the G line. This is attributed to the lifting of the degeneracy of the E_{2g} due to its tubular symmetry.

Typical of the Raman spectrum of SWNTs is the presence of the G^- and the G^+ lines which correspond to the axial and circumferential vibrations of the rolled graphene sheet. Whereas G^+ is sensitive to the presence of dopants (Dresselhaus and Dresselhaus 1981), G^- is sensitive to the nature of the tube i.e. metallic or semiconducting but not on the chirality (Pimenta et al. 1998), (Brown et al. 2001). Both G lines are reported to be formed from three peaks from different symmetries which are polarisation dependent (Jorio et al. 2000, 2003), raising the number of vibrational modes that form the G line to 6 (2A, 2E1 and 2E2).

Finally, the second-order Raman spectrum is dominated by the feature at approximately 2,700 cm^{-1}. The so-called G′ line is a double resonance process and is related to the D band at 1,350 cm^{-1}, or the disorder induced band. The D band is a single phonon process and the G′ prime band is a dual phonon double resonance process. In the case of graphite, the D band can be fitted with two Lorentzian distributions, whereas the G′ band can be fitted with one Lorentzian (Cançado et al. 2002), a fact that renders the specific line especially attractive for stress monitoring. As in the case of the D line, the G′ band is related to diameter and chirality. This is attributed to the fact that as the planar structure is converted to a tubular one in the case of CNTs, the bandwidth of these spectral lines are directly affected (Jorio et al. 2002). However, as has been reported (Souza Filho et al. 2002), whereas in the majority of graphitic structures, the G′ appears as a single distribution, in the case of individual SWCNTs, a two-peak vibrational activity has been reported. This vibrational activity is related to the electronic properties of the nanotube and is associated to distinct resonance processes sufficiently separated in resonance energy. Other cases where the G′ line presents a morphology that diverges from a single distribution are attributed to interlayer coupling (Malard et al. 2009a), tunnelling effects (Cui et al. 2009) etc. Other vibrational modes are also reported which include overtones and combination modes like the 1,750 cm^{-1} band, the iTOLA combination mode at the area of 1,800–2,000 cm^{-1}, or the intermediate frequency modes IFMs with frequencies that range between the RBM and the G mode (Fig. 6.1).

As previously stated, for multi-wall CNTs (MWCNTs), all the aforementioned features that distinguish SWCNTs from other graphitic structures tend to disappear and the spectrum converges to that of turbostratic graphite (Cançado et al. 2008). However, specific features have been reported which include the splitting of the G band which relates to the presence of very small diameter inner tubes (Benoit et al. 2002), the dual morphology of the G′ band which relates to the circumferential deformation of the tube particularly for double-wall CNTs (Bandow et al. 2004), or the presence of the RBM under specific resonance conditions (Pfeiffer et al. 2003).

Last but not least, graphene has recently been in the focal spot of the international research community. Being the simplest sp^2 Carbon structure or else the basic building unit of any graphitic structure, graphene is ideal for evaluation and further investigation using Raman Spectroscopy. Monolayer graphene is unique in that the G′ line is remarkably more intense than the G line and this can be understood in terms of a triple resonance process (Malard et al. 2009b). As more graphene layers are added to the structure, the G′ line transforms from a simple

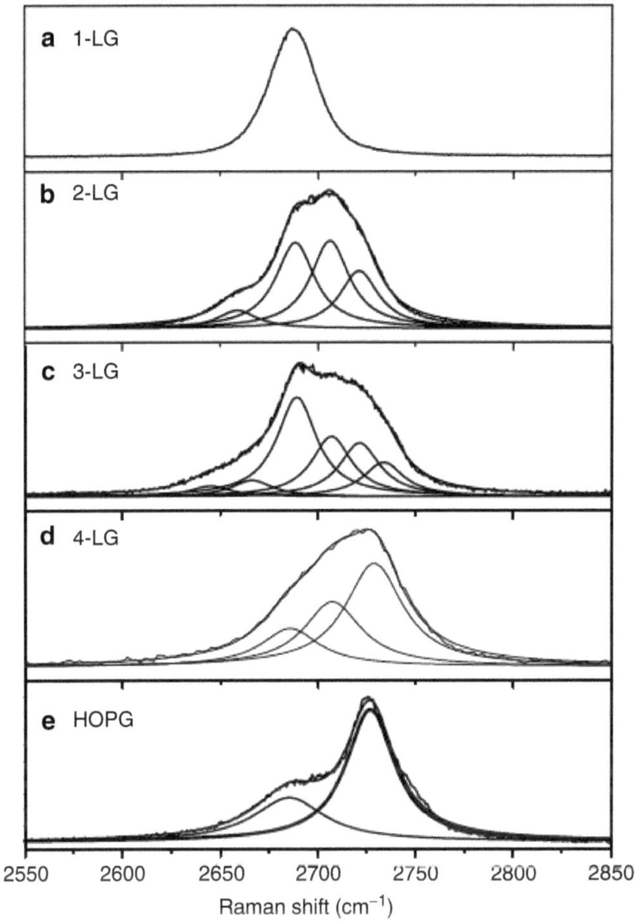

Fig. 6.3 The measured G′ Raman band with 2.41 eV laser energy for (**a**) 1-LG, (**b**) 2-LG, (**c**) 3-LG, (**d**) 4-LG, (**e**) HOPG (Reprinted from Malard et al. (2009b), with permission from Elsevier)

Lorentzian to a more complex peak, where more than one lines appear to coexist due to the interlayer interaction (Fig. 6.3). The complexity of the peak is attributed to the interlayer interaction and the random rotation along the c axis of the graphene layer (Malard et al. 2009a). Interestingly enough, turbostratic graphite where rotational effects are minimised, also exhibits a single Lorentzian morphology. However, this is upshifted in frequency and is wider and of lesser intensity than the corresponding G line (Cançado et al. 2008). Of particular interest is the emergence and morphology of the disorder lines, as different types of graphene edges are probed (Malard et al. 2009b).

6.3 Stress Dependence of the Raman Spectrum

6.3.1 The Principle

There are *3N-6* possible vibrations in the general case of a polyatomic molecule. Each one corresponds to an internal displacement co-ordinate. In the purely linear elastic case, the displacement may be regarded as directly proportional to the restoring force. The *3N-6* set of constants of proportionality are called the force constants. In this case, all vibrations are purely harmonic and the potential energy of the system is the sum of the quadratic terms whose coefficients are the force constants.

For one simple vibrational motion, according to the above,

$$F = k(x - x_0) \tag{6.1}$$

where F is the restoring force, k is the force constant, and $(x-x_0)$ is the distance from the equilibrium position x_0. Integrating Eq. (6.1) with respect to x provides the potential energy function U_p:

$$U_p = \frac{1}{2}k(x - x_0)^2 \tag{6.2}$$

which is the parabola shown in Fig. 6.4a. This oscillation is harmonic and its frequency v is independent on the distance from the position of equilibrium x_0. The quantum theory of the harmonic oscillator only allows one transition from one energy state to another $\Delta v = \pm 1$. These energy states are shown as dotted lines in Fig. 6.4a and are equidistant.

It is, however, well known that phenomena like overtones, combination bands, or difference bands (Colthup 1975) cannot be explained by the simplistic harmonic theory. In addition, concepts like bond breaking at high deformations demand a different approach to the potential energy function. Such an approximation is the *Morse function*, where the potential energy is a function of the dissociation energy D_e, or the energy required to break the bond:

$$U_p = D_e\left(1 - \mathrm{e}^{-b(x - x_0)}\right)^2 \tag{6.3}$$

where b is a constant.

In Fig. 6.4b, the potential energy of the anharmonic oscillator is depicted. The dotted lines represent the allowable energy levels. The quantum theory accounts for more than one transition between energy levels.

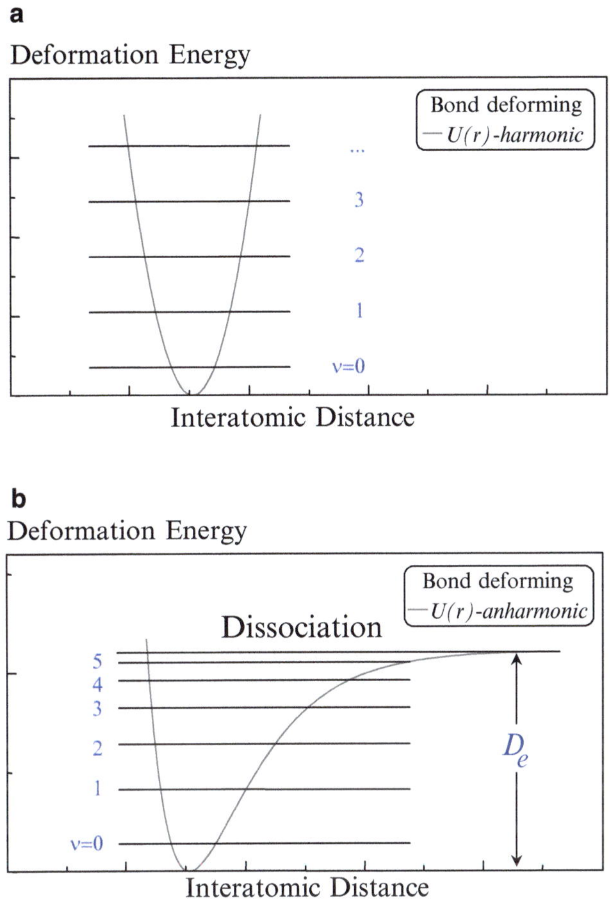

Fig. 6.4 (**a**) The potential energy function U_p for the Harmonic Oscillator. The *dotted lines* mark the allowable energy levels and are equidistant. (**b**) The potential energy function U_p for the Anharmonic Oscillator. The *dotted lines* mark the allowable energy levels and are no longer equidistant

6.3.2 Stress Dependence of the Vibrational Frequency

The second derivative of the Morse anharmonic potential energy function provides the equation for the force constant (Tashiro et al. 1990):

$$k = 2b^2 D_e (2e^{-2b(x-x_0)} - e^{-b(x-x_0)}). \tag{6.4}$$

As can be seen, the force constant is no longer a constant in the anharmonic case. Moreover, it is a function of the internuclear displacement and its dependence is depicted in Fig. 6.5a. For small positive internuclear displacements, $x = x - x_0 > 0$,

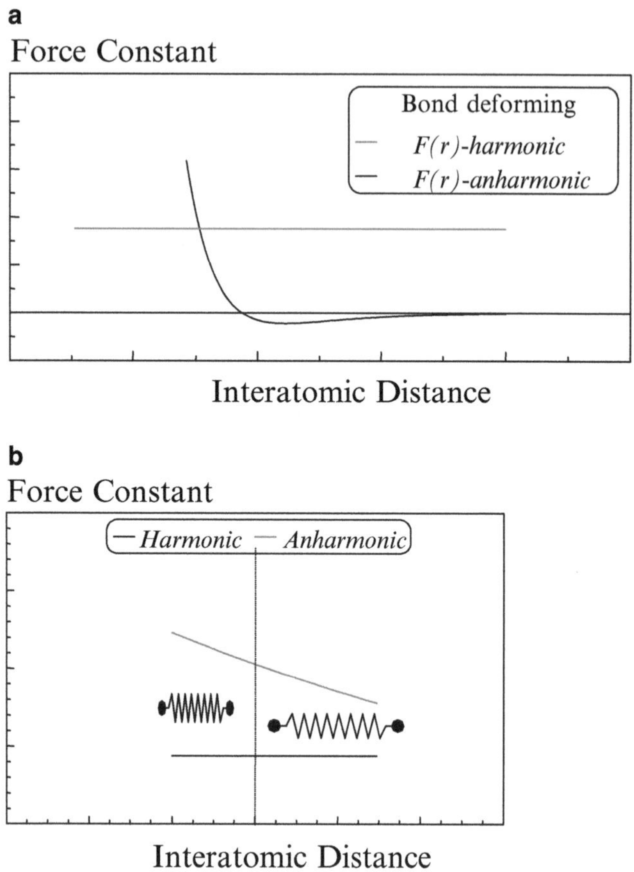

Fig. 6.5 (**a**) the variation of the force constant as a function of interatomic distance; (**b**) for small displacements, the stress dependence may be regarded to a good approximation as proportional to the molecular deformation

the force constant is monotonically decreasing. For positions near the equilibrium, the Morse function resembles the harmonic oscillator function, and the frequency ν can be regarded as proportional to $\sqrt{k}$ (Colthup 1975), (Tashiro et al. 1990). This results in a low frequency shift $\Delta\nu$ of the vibration. On the other hand, when the bond is compressed, that is when $\Delta x < 0$, the force constant increases causing a high frequency shift $\Delta\nu$.

The above principle provides the theoretical background for the frequency shift of distinct Raman bands when the molecule is subjected to external load. The theoretical calculation of the expected shift $\Delta\nu$ has been presented for simple molecules (Wool 1980; Tashiro et al. 1990). More complicated analyses include the lattice dynamical theory to predict stress induced shifts in polymer chains.

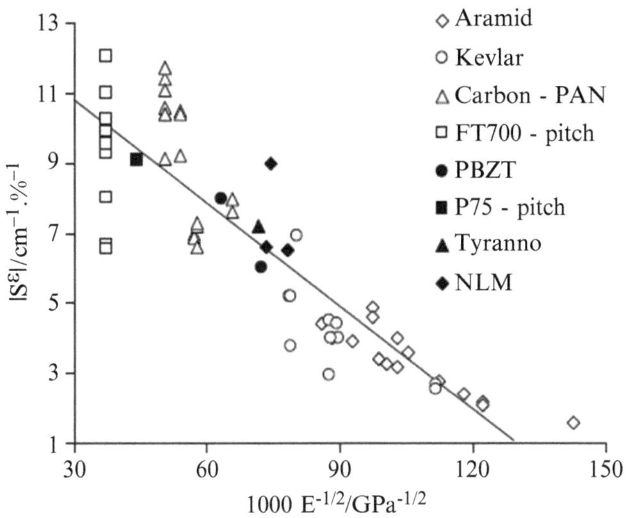

Fig. 6.6 Stress dependent shift of the G band vs. the inverse of Young's modulus square root (Reprinted from Gouadec and Colomban 2007, with permission from Elsevier)

For small displacements (Fig. 6.5b), the stress dependence may be regarded to a good approximation as proportional to the applied stress field σ. Bretzlaff and Wool (1983) propose the following:

$$\Delta v = a_\sigma \sigma Z \tag{6.5}$$

where a_σ is the proportionality constant.

The key feature that links the stress dependence of the molecule to any macroscopic deformation is whether this deformation affects the material at a molecular level. Whereas amorphous materials are not expected to show detectable stress sensitivity, highly crystalline materials, such as Kevlar® (Galiotis et al. 1985) or carbon (Robinson et al. 1987), are reported to exhibit measurable stress sensitivity. Provided that a suitable reference value is given for the unstressed material, experimental calibration curves may be employed to translate Raman frequency shifts to absolute strain. In most cases, a direct proportionality of the shift to the applied strain is adequate (Galiotis et al. 1983), although higher order dependence has been proposed in the literature to account for non-linear elastic behaviour (Melanitis et al. 1994).

What is of particular interest is that the shift of a strained bond is expected to be proportional to the bond deformation, or else that there is a direct relationship between the stress induced Raman shift and the bond stiffness or the Young's modulus of a macroscopic structure (Gouadec and Colomban 2007). These relationships allow for universal plots that correlate the strain induced Raman Shift with the moduli of known fibres, see Fig. 6.6.

6.4 Stress Induced Changes in the Raman Spectrum of Graphitic Structures

The application of stress in graphitic structures is limited by the size of the structure in that there must be an adequate means of transferring the stress. For this reason, although a lot of effort has been invested in the stress dependence of the Raman spectrum in structures of micron dimensional order such as carbon fibres, there is limited reported research on direct application of stress on CNTs. However, a lot of research effort has been associated with the behaviour of graphitic structures under pressure (hydrostatic or not). This section will focus on the direct stress application on graphitic structures.

Early works focus on the application of pressure on single crystal graphitic structures and reveal considerable frequency upshift of both E_{2g} Raman active modes (at ~50 and 1,580 cm^{-1}). Hanfland et al. report this upshift and correlate it to the structural deformation of the Graphitic crystal due to the strong anharmonicity of the C–C bond (Hanfland et al. 1989). In a recent work, Del Corro et al. employed a moissanite anvil cell coupled to a Raman microscope to monitor the evolution of distinct Raman bands of highly oriented pyrolytic graphite (HOPG) under non-hydrostatic pressure conditions. The employment of the moissanite cell instead of the typical diamond one allowed for monitoring of the shift of the D band (Del Corro et al. 2008). Figure 6.7 depicts the Raman spectra of a HOPG sample at three selected stresses and the Ratio of the D′/G band intensities as a function of the ratio of the D/G band intensities (Del Corro et al. 2011).

The stress induced changes in the Raman spectrum of Carbon fibres have been thoroughly studied, as Carbon Fibres are almost ideal for direct axial stress application. The first and second order Raman vibrational modes of high modulus fibres exhibit considerable shift with strain. As has been reported, the vibrational modes D, G and G′ exhibit strain dependence of approximately 7, 9 and 17 $cm^{-1}/\%$ respectively (Galiotis and Batchelder 1988). The considerable shift of the G′ band was as expected for a second-order feature and was correlated with the shift of the D band. The study of the spectroscopic behaviour of Carbon Fibres has been extended to compression via application of the Cantilever Beam Technique (Melanitis and Galiotis 1990). The deviation from non-linearity when the fibres are stressed in compression was attributed to the microscopic buckling of the graphene layers which was even reversible in the case of low modulus carbon fibres (Melanitis et al. 1994). For this reason, the technique may be employed for the determination of the compressive strength of individual fibres. As should be noted, the lower the graphitization of the Carbon Fibre, the harder it is to monitor the strain induced changes, as on one hand the disorder features increase in full width at Half Maximum (FWHM) and intensity relative to the G line and on the other hand the second order G′ line is not readily detectable (Melanitis et al. 1996). As postulated in the previous section, a direct relation between the modulus and the phonon frequency is expected, and therefore every Raman active bond should have a unique stress dependence. In this respect, Raman Frequency vs. stress calibration

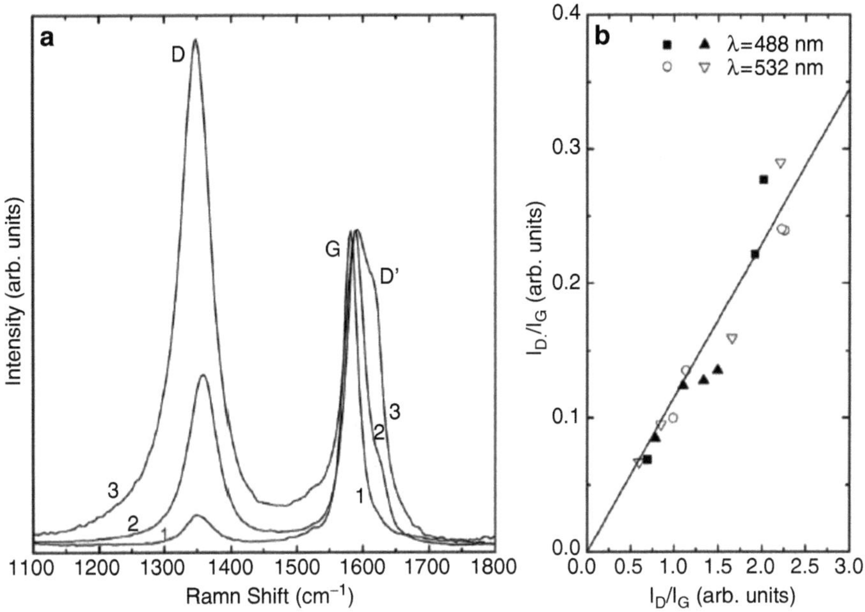

Fig. 6.7 (**a**) Raman spectra of a HOPG sample at three selected stresses (1, 2 and 3 denote local stresses of 1, 2 and 0 GPa, respectively). (**b**) Ratio of the D′/G band intensities as a function of the ratio of the D/G band intensities (Reprinted from Del Corro et al. (2011), with permission from Elsevier)

curves are more characteristic than Raman Frequency vs. strain calibrations (Paipetis and Galiotis 1996).

As aforementioned, the Raman study of sp^2 graphitic structures of nano scale such as CNTs, imposes restrictions in terms of the load application. Various researchers have focused on the induced Raman shifts with applied pressure. Sandler et al. (2003) report on the pressure dependence of the Raman modes of various carbon nano-structures such as different types of CNTs, graphite crystals and nano-fibres and compare their findings to the behaviour of high modulus carbon fibres. As expected, the Raman modes were found to shift reversibly to higher wave numbers with pressure. The authors used the polarization dependence of the strain induced Raman shift to predict the initial pressure dependence of all tested nanostructures and identified a reversible collapse condition for hollow nanostructures.

The Raman spectra of novel graphitic spheres identified as a side product of fullerene synthesis have been found to be similar to that of micro-crystalline graphite (Loa et al. 2001). The authors report that the silent low-frequency $B_{1g}(1)$ phonon of graphite becomes Raman active and that high pressure affects the G′ mode near 2,700 cm^{-1} which exhibits a peculiar dispersive behaviour.

The pressure evolution of the Raman spectrum of stacked-up carbon nanofibres which exhibit a unique morphology of stacked conical graphene cups along the fibre axis revealed a −3 cm^{-1}/GPa dependence for the D′ double resonance feature and −4.2 cm^{-1}/GPa for the graphitic G line (Papagelis et al. 2011). The authors report the merging of the G and D′ line which is indicative of the differential pressure dependence of the two Raman bands.

Carbon onions and nanocapsules exhibit similar behaviour to that of MWCNTs (Guo et al. 2009). The special characteristic of these structures is that they can sustain very high pressures prior to collapsing. However, differences in the pressure induced shifts between compression and decompression were attributed to structural damage of the nanostructures.

As aforementioned, the Raman Spectrum of more elaborate sp^2 structures possesses intrinsic characteristics. Apart from the inverse dependence of Raman bands to stress application, the Raman response of such structures may be more complicated. Relatively early studies report pressure induced changes in the Raman spectrum with an anomalous behavior in the pressure range of 10–16 GPa (Teredesai et al. 2000). The authors suggest that the intensity changes are attributed to pressure induced deviation from resonance conditions. The reported softening is attributed (i) to diameter dependent collapse of a fraction of the studied tubes and (ii) to the influence of the pressure medium which may penetrate the tube. The effect of the pressure medium is also identified in other studies (Dunstan and Ghandour 2009), where the nature of the pressure medium as well as the resonance effects with Raman excitation energy are reported to be of major importance.

The Raman spectroscopy of filled double-wall CNTs (DWCNTs) with trigonal Tellurium revealed red shift of the G′ band attributed to the softening of the C–C bonds upon capillary filling of the tubes. (Belandria et al. 2010). The authors employed the capillary filling of the tube to distinguish pressure induced changes between the inner and the outer wall and verified that in the presence of Te, the pressure coefficients of the G band of the internal and the external CNTs are larger than in the case of empty DWCNTs.

The Raman spectrum of metallic SWCNTs and DWCNTs under high pressure exhibits a variety of pressure induced changes. These include the deformation of SWCNTs and the DWCNT outer tubes, the quasi-isolation of the inner tubes as well as a narrowing of the characteristic Breit–Wigner–Fano Raman peak attributed to tube –tube interactions at high pressures (Christofilos et al. 2006). In Fig. 6.8, the distinct changes in the frequency area 1,350–1,700 cm^{-1} are depicted.

Venkateswaran et al. have studied the pressure dependence of RBM and G vibrational modes of purified and solubilised SWCNTs and reported that an abrupt drop in the intensity of these bands is seen near 2 GPa, which suggested a phase transition. The authors identified a 10 cm^{-1} upshift in the RBM of the purified SWCNTs compared to the as-received SWCNTs. Pressure induced changes were reversible and the pressure dependence of the RBM and G bands was significantly influenced by the changes in the electronic structure (Venkateswaran et al. 2001).

The second-order Raman G′ band of bundled DWCNTs and SWCNTs exhibited different pressure behaviour. The applied pressure induces a splitting of the G′ peak in the case of DWCNTs (Papagelis et al. 2007). In the latter case, the distinct components of the G′ vibrational mode are identified and associated with the inner and the outer tube diameter of the resonantly probed tubes and the strength of the inner-outer tube interaction. Moreover, the authors identify a dependence of the

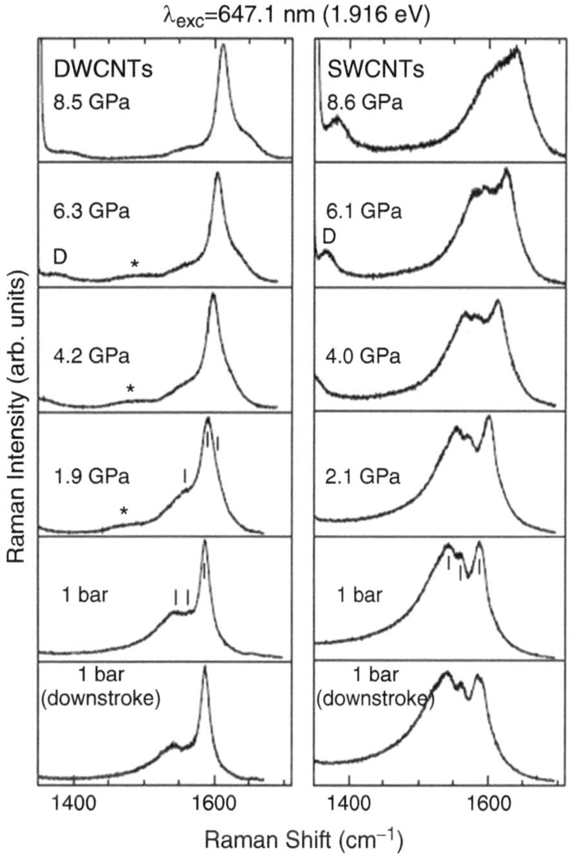

Fig. 6.8 Raman spectra of DWCNTs (*left panel*) and SWCNTs (*right panel*) in the G band frequency region at room temperature and for various pressures. *Vertical* lines denote the main G-band components, 'D' refers to the nanotube D-band, while asterisks mark a methanol–ethanol band (Reprinted from Christofilos et al. 2006, with permission from Elsevier)

pressure induced Raman Shift to the laser wavelength which they attribute to the sampling volume of the excitation wavelength. The effect of the inner–outer tube interaction has also been verified for increased laser powers (Puech et al. 2011).

The effect of high-pressure on the Raman Spectrum has also been studied in the case of monolayer, bilayer, and few-layer graphene samples supported on silicon (Proctor et al. 2009). The authors report that the pressure dependence tends to that of unsupported graphite with increasing graphene layers and attribute this finding to the fact that the compressive behaviour is dominated by the stiffness of the substrate.

Although there is comparatively extended research effort in the area of high pressure of sp^2 carbonaceous structures, a limited number of studies exist on the direct stress application on nano-scaled graphitic structures. Cronin et al. managed to strain individual SWCNTs by using an atomic force microscope tip and at the same time interrogating the strained tube with a Raman microprobe (Cronin et al. 2004). The SWCNT was deposited on SiO_2 and secured in place using electron beam lithography. The authors secured the uniformity in chirality and diameter by scanning along the length of the individual interrogated nanotube. The authors

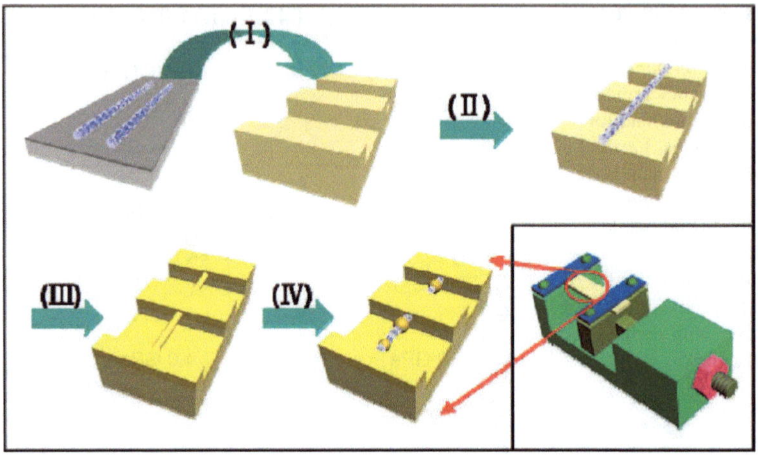

Fig. 6.9 Schematic drawing of the method for SWCNT tensile testing (Reprinted with permission from Gao et al. 2008. Copyright 2008 American Chemical Society)

report remarkable redshifts of the D, G and G′ bands with applied strain, i.e. 27, 14 and 40 cm^{-1} whereas no shift in the RBM is reported. The intensity of the RBM varies with strain due to relaxing of the resonance conditions. In a later work (Cronin et al. 2005), the authors report on the chirality dependence of the stress induced shift and report that semiconducting SWCNTs remain resonant with the window of applied strain whereas metallic SWCNTs move in and out of resonance with strain, indicating a strain induced shifting of the electronic subbands.

Elaborate studies of the chirality dependence on strain induced shift are performed in the work presented by Gao et al. (2008). In this work, individual SWCNTs are transferred on a polymethylsiloxane flexible scaffold and fixed in position by gold deposition (Fig. 6.9). Subsequent straining of the scaffold strains the individual SWCNTs. The authors report on the G line mode splitting due to applied strain (Fig. 6.10). Increasing Chiral angle is found to affect significantly the blue shift rate of the Raman G^- and G^+ line. The authors also report on the redshift of the IFM which is not consistent with the bond softening principle described in the previous section. Liu et al. (2009) provide an extensive overview of the effect of various stress fields on SWCNTs.

As mentioned above, graphene is the simplest of all sp^2 graphitic structures and at the same time their building unit. Recent experimental work on graphene under stress reveals the splitting of the G′ band (Fig. 6.11) which depends on the polarization of the excitation light, as well as the direction of stress (Frank et al. 2011a). The authors attribute the mode splitting to (i) the induced asymmetry of the Brillouin zone (ii) the additional contribution of the inner double resonance mechanism and (iii) the laser polarization with respect to the loading axis. The method to experimentally apply strain on the graphene layers was the cantilever beam method, where a clamped elastic beam is subjected to strain causing a strain gradient along its length. The graphene layer is attached to the surface of the beam. The support

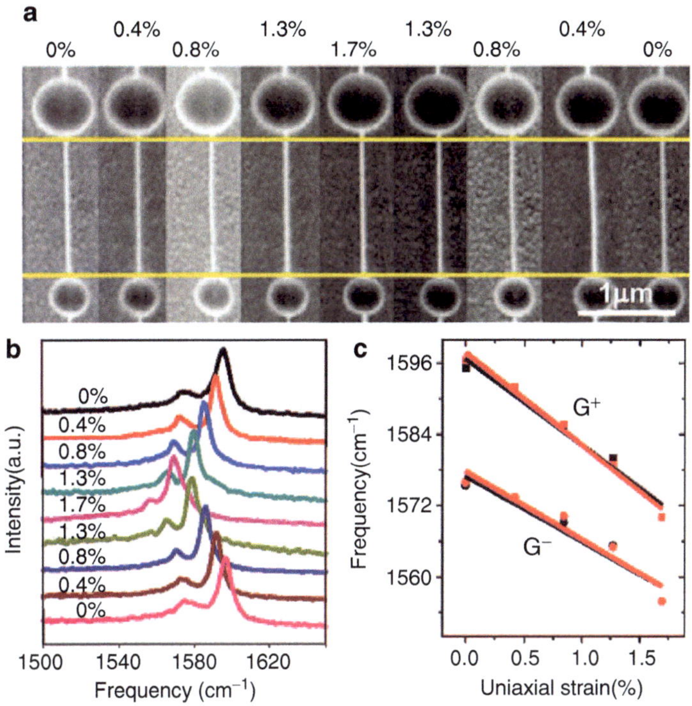

Fig. 6.10 (**a**) SEM image, (**b**) G-band spectra of (18, 5) SWNT when uniaxial strain first increases from 0 to 1.7% and then decreases to 0%. (**c**) G^+ and G^- frequencies variation as a function of uniaxial strain (Reproduced and reprinted with permission from Gao et al. 2008. Copyright 2008 American Chemical Society)

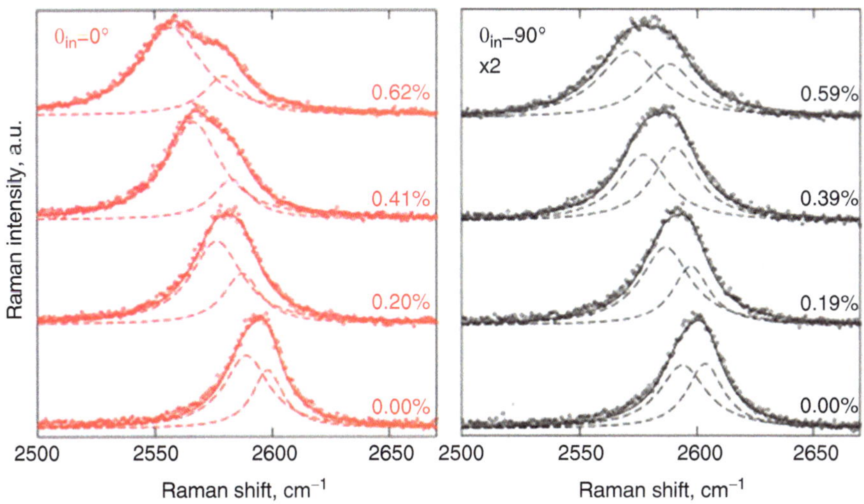

Fig. 6.11 Raman G′ band splitting under strain in graphene for parallel and vertical polarization with respect to the loading axis (Reprinted with permission from Frank et al. (2011a). Copyright 2008 American Chemical Society).

provided by this attachment allows for applying six times higher compressive strains than in the case of suspended graphene, prior to buckling (Frank et al. 2010). The study of graphene allows for the determination of a single stress factor or stress dependence of the Raman Frequency. This corresponds to the C–C bond stiffness and should be the limit of all sp^2 graphitic structures as stress is transferred from the far field to the C–C bond (Frank et al. 2011b). The universal value that is characteristic of the graphitic band was calculated to be approximately 5 cm^{-1}/MPa.

6.5 Stress Transfer Raman Studies in Composites Reinforced with sp^2 Graphitic Nanostructures

It is beyond any doubt that the stress transfer between the matrix and the reinforcing phase is of primary importance in the structural behaviour of reinforced composites. Raman microscopy has been employed for over two decades for monitoring the interface at microscopic level between graphitic materials and polymer matrices. The dependence of the Raman modes on applied stress allows for the local stress monitoring at a resolution which is practically only limited by the diffraction limit of the excitation wavelength λ, i.e. $\lambda/2$. The current section is focusing on composite materials reinforced with sp^2 morphologies, with regards to micromechanics of reinforcement. These materials may be categorized in terms of their dimensionality, or in other words the anisotropy of the nano-reinforcement which in its turn is controlled by the alignment of the nanophase in space. In this respect, macroscopic nano-graphitic structures may be 1D, like typical nano-fibres (Vigolo et al. 2000) which are by definition transversely isotropic, 2D, like bucky papers (Bahr et al. 2001) which may be employed to form orthotropic laminates, or 3D composite systems which are isotropic in the macro-scale (Coleman et al. 2006).

As aforementioned, the difficulty of handling and aligning the nano-dimensional phase is not an issue in the case of carbon fibres (diameter in the order of 10 μm and practically unlimited length). Carbon fibres are the first sp^2 carbon structures studied as reinforcing phase in structural components using Raman Microscopy. Aligned carbon fibres in model single fibre composites (Paipetis and Galiotis 2001), model multi-fibre composites (Galiotis et al. 1996) or even typical laminates (Chohan and Galiotis 1997) have been probed in order to study the efficiency of the stress transfer (Paipetis and Galiotis 1997), the effect of neighbouring fibres (Van Den Heuvel et al. 1997), and the stress redistribution after a fibre break (Marston et al. 1996). The acquired stress profiles have been associated with the integrity, the stiffness and the toughness of the interface (Yallee and Young 1998) and analytical models have been employed to model the axial and shear stress profiles at the interface (Paipetis et al. 1999).

The load transfer in CNT reinforced matrices is by far more complicated as it encompasses a variety of parameters which include dispersion, agglomeration, wetability, aspect ratio, alignment, and morphology. In other words there are a lot of prerequisites that have to be satisfied before the nanocomposite can be regarded as a macroscopically isotropic short fibre reinforced composite. The first published work on the Raman investigation of the load transfer in CNT reinforced composites (Schadler et al. 1998) reports a considerable shift in compression of the G′ band of multi-all nanotubes (approximately 7 cm^{-1}/%). This was not the same for the same nanocomposites in tension, where although the shift was slightly positive, the experimental scatter could not allow for direct conclusions. The authors attribute this observation to either poor bonding of the matrix and the nanotube surface or to weak bonding of the inner layers to the outer layer which leads to the sliding of the inner tubes with respect to the outer tube. The behaviour of the nano-composite in compression favours the second hypothesis, as geometrical constraints lead to the compressive deformation of all the tubular structure. Of course, the working hypothesis for this approach is that the stress induced Raman shift is averaged over the volume of the interrogated tubes. The reported stiffening of the epoxy matrix due to the MWCNT reinforcement is more prominent in compression than in tension but the difference is not as dramatic as in the case of the stress induced Raman Shift in tension and compression.

The first reported micromechanical test on individual CNTs is reported by Cooper et al., where CNTs bridging across holes in an epoxy matrix were pulled out from the epoxy matrix (Cooper et al. 2002). The authors employed the tip of a scanning probe microscope. A simultaneous recording of the applied forces allowed for a full force-displacement curve of the pull out process. The authors present a correlation between interfacial shear strength and the embedded length, to report that the interfacial shear strength falls with increasing embedded length, as is reported in single-fiber pullout tests. This is attributed to the fact that most of the shear stress transfer is occurring via the "ineffective length" (Pitkethly and Doble 1990). The authors claim that their findings support the hypothesis that the CNT polymer interface may be significantly stronger than the interface between fibre and matrix in typical systems such as glass or carbon fibre-reinforced composites. They attribute the enhanced adhesion to the existence of covalent bonds which arise from naturally occurring defect sites at the CNT wall.

Cooper et al. studied the deformation micromechanics of both SWCNTs and MWCNTs embedded in epoxy matrix using Raman Spectroscopy and confirmed the blueshift of the Raman G′ band with tension for all studied CNTs (Cooper et al. 2001). Interestingly enough, two different types of SWCNTs exhibit shifts that differ by an order of magnitude with macroscopically applied strain (Fig. 6.12). This difference is attributed either to lower stiffness or to poorer dispersion of the tubes that exhibit the low shift. The second postulation is indicative of the effect that the dispersion, or the initial CNT morphology may have in the final nanocomposite properties.

The authors assume that the nanotube-reinforced composites are short fibre reinforced composites with random reinforcement distribution and use well-known analytical formulations (Cox 1952; Evans and Gibson 1986) to derive equivalent nanotube moduli for 2D and 3D distributions. The maximum calculated modulus of the reinforcing phase is found to vary between approximately 80 and 800 for 2D

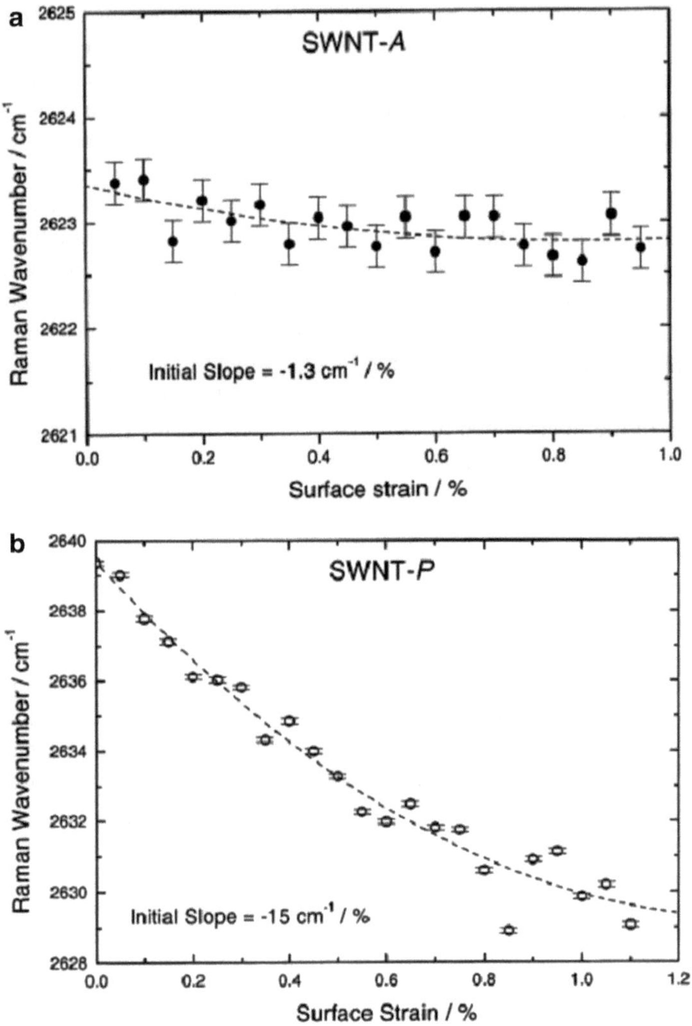

Fig. 6.12 The strain induced shift differs by one order of magnitude for two different types of SWCNTs (Reprinted from Cooper et al. 2002, with permission from Elsevier)

distribution of the reinforcement and between approximately 250 and 2,500 for 3D distribution of the reinforcement. The 2D calculated values are regarded as more reasonable and the authors conclude that the moduli of 300 GPa and 1 TPa for MWCNTs and SWCNTs are in line with experimental measurement. However the huge discrepancy between the two kinds of SWCNTs is not adequately addressed. According to later stress studies on the chirality dependence of the stress sensitivity of SWCNTs (Gao et al. 2008), this discrepancy could also be attributed to other reasons than agglomeration and dispersion. As aforementioned, these studies reveal major differences in stress dependence for different types of CNTs (the strain induced shifts

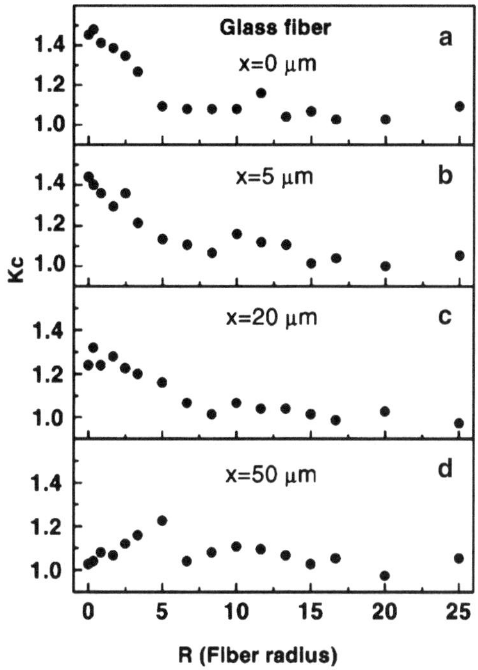

Fig. 6.13 Strain in the carbon fibre (*solid symbol*) and in the CNT modified matrix near the fibre edge (*open symbol*) measured simultaneously by microRaman spectroscopy at applied stress levels of (**a**) 3 MPa, (**b**) 7 MPa, and (**c**) 10 MPa (Reprinted from (Zhao and Wagner 2003), with permission from Elsevier)

range from -4 to -25 cm^{-1}/%, but they are calculated for the G^+ and G^- band and not for the G' as in the study by Cooper et al.).

Zhao et al. employed SWCNTs to monitor the stress concentrations around notches in nano-composites using polarized Raman Microscopy (Zhao et al. 2002) Using polarization studies with the polarization either vertical or perpendicular to the direction of the stress application they report that the polarization can be employed to interrogate nanotubes aligned in the polarization direction. In this case, the shifts of the G' band can be associated with the axial and transverse strain and calibration curves may be drawn. They observe a notable downshift with the polarization aligned to the stress application axis and an upshift in the transverse direction which they attribute to Poisson's contraction. The relative frequency shift when probing in the transverse direction away from the circular notch is associated with the stress concentration factor due to the notch. Similar stress concentration values were reported in a typical unidirectional aramid fibre composite laminate (Arjyal et al. 2000).

Furthermore, Zhao et al. employed the same technique to monitor the stress field around the break of a two-dimensional model composite (Zhao and Wagner 2003). In their study, they modify a polymer matrix using SWCNTs and make single fibre model composites both with E-glass fibre and high modulus carbon fibre. They are successful in creating a stress contour map around the stress discontinuity invoked by the fibre break. In the case of a high modulus carbon fibre, they perform simultaneous mapping of the fibre and the modified matrix to produce "mirror" strain profiles associated with the fibre failure (zero stress at the vicinity of the crack) and the surrounding matrix (stress concentration at the vicinity of the crack) (Fig. 6.13). The same strain sensing principle is also employed in the case of E-glass polypropylene interfaces (Barber et al. 2004) and the measured data are associated with interfacial shear strength values calculated from the classical fragmentation test employing the Kelly–Tyson "constant shear" model (Kelly and Tyson 1965).

Kao and Young studied the combined effect of laser heating and deformation in the Raman shift of the G′ band for SWCNTs in an epoxy matrix (Kao and Young 2004). As is reported in the case of carbon fibres (Everall et al. 1991), there is a downshift in the G′ line with increasing laser power. They employ a four-point bending device to apply uniform strain at distinct increments and monitor the induced shift seamlessly in tension and compression. Unlike in the case of MWCNTs (Schadler et al. 1998), there is continuity in the slope of the transition region between tension and compression. Additionally, they report a plateau in the induced shift both in tension and in compression around 0.5% which they attribute to buckling and interfacial failure for compression and tension respectively. They also report on the influence of thermal stresses induced by the differential contraction between the matrix and the SWCNTs. By comparing a cold and hot curing matrix system they indicate that the presence of residual stresses favours stress transfer.

In a more recent work, Kao and Young (2010) report a decrease in the stress transfer efficiency of CNT reinforced composites with cyclic loading. In particular they find that the stress sensitive G′ band is shifting with applied stress to lower wave numbers. The induced shift presents a hysteresis loop with cyclic loading and the authors correlate the hysteretic area to the dissipated energy (or to the induced damage at the interface), and normalize it to the total interfacial area between the nanotubes and the surrounding matrix. They employ reported experimental values from pull-out tests of individual nanotubes from a polymer matrix (Cooper et al. 2002) to evaluate interfacial damage of bundles and individual CNTs.

Cui et al. (2009) employed Raman Spectroscopy to study the effect of stress transfer in a double-walled carbon nanotube reinforced matrix. DWNTs are the simplest form of MWNTs. In this respect they can be employed to study both the polymer graphene interface as well as the wall to wall interface. In their study Cui et al. monitor the stress induced shift of the Raman bands of DWNTs during deformation and employ their findings to predict the behaviour of MWNTs. Cui et al. verify the splitting of the G′ band when a model composite system is subjected to tension and compression (Fig. 6.14). As the splitting is attributed to slippage of the inner wall of the DWCNT, the authors use the stress-induced shift of the outer and inner wall to predict effective reinforcement in MWCNT reinforced composites. They report poor inter-wall bonding or even that the inner nanotube wall is "virtually unstressed" to conclude that the effective Young's modulus in MWCNT reinforced systems is bound to be relatively low unless the inter-wall stress transfer is improved, potentially through the introduction of defects, or subsequent treatments such as radiation crosslinking (Peng et al. 2008).

Apart from the effect of applied strain on CNT nano-composites, the Raman shift has also been employed to study their residual strains in composites. Hadjiev et al. (2010) employed Raman Microscopy to measure residual strain in CNT reinforced epoxies. They took advantage of the difference in frequencies of the

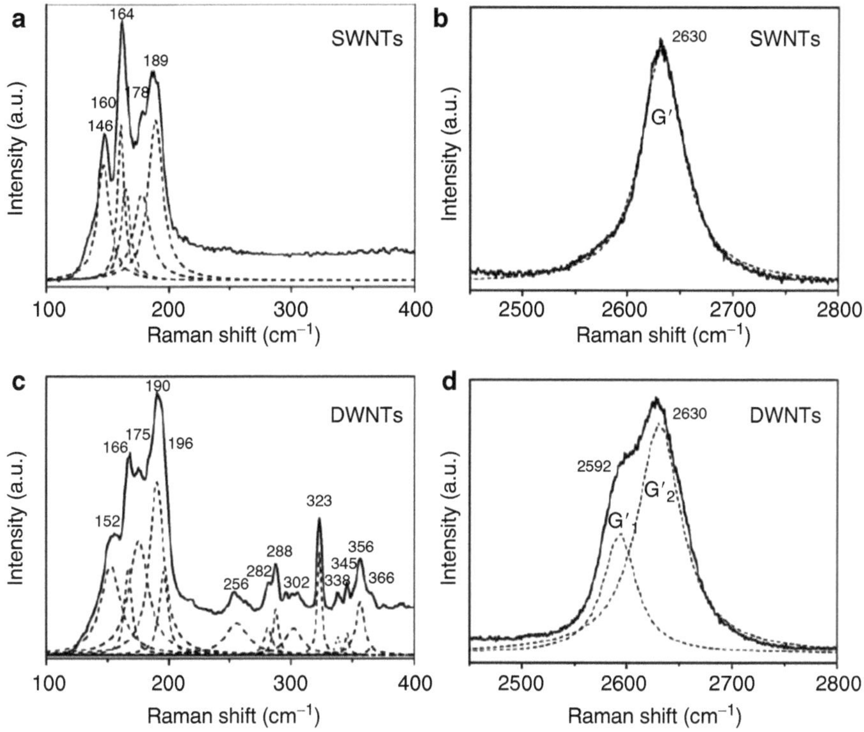

Fig. 6.14 Raman spectra of SWCNTs and DWCNTs obtained using a 633 nm HeNe laser. (**a**) Low-frequency region for the SWNTs, showing the RBMs. (**b**) The G′ region of the SWCNTs. (**c**) Low-frequency region for the DWNTs, showing the additional RBMs. (**d**) The G′ region of the DWNTs, showing splitting of the band (Reprinted from Cui et al. 2009, with permission from Elsevier)

CNT vibrational G^+ mode in the composite compared to that of relaxed CNTs to measure the local residual strains in the composites. They report considerable variation with both CNT functionalization and CNT concentration. More specifically, at room temperature and with the same local concentration of CNTs in the composite, the strains of oxidized and polyamidoamine-functionalized CNTs are found to be 2.5 times higher than that of the composite containing pristine CNTs. According to the authors, the higher residual strain of the composites loaded with functionalized CNTs is indicative of better stress transfer and integration in the epoxy matrix, which was verified by the improved tensile properties measured for the functionalized CNT composites. Interestingly enough, the residual strain is reported to depend on other parameters than the thermal coefficient mismatch between the CNTs and the epoxy, which in its turn is independent of post processing of the CNTs. However, the authors do not report the effect of post processing on the spectrum of CNTs prior to incorporation into the matrix. Lucas and Young (2007a) report on the spectral changes in the RBM and the G′ line induced by the thermal stresses in SWCNT reinforced composites. They employ epoxies cured at different temperatures to induce varied thermal stress field in the CNT reinforced polymer. They report that the relative intensities of the RBM vary with curing temperature, and correlate this variation to that induced by far-field strain application, when the composites are loaded in four-point flexure. They also report a red shift of the G′ band with increasing curing temperature and employ the measured shift to verify the reported thermal expansion coefficient of the epoxy matrix. Additionally, the effect of strain on RBM is studied using three different excitation wavelengths in order to study the well-known dispersive features of the RBM with applied strain (Lucas and Young 2007b). The authors report variations of between 10 and 200% of the RBM intensities over a range of strain between −0.6 and 0.7% depending on the nanotube diameter and its chirality accompanied by a shift of the G and G′ bands. They attribute these intensity changes to resonance effects and employ tight-binding calculations to predict intensity changes with uniaxial strain.

Comparative studies of the efficiency of CNTs as strain sensors have also been recently published (Sureeyatanapas et al. 2010). Both luminescence and Raman spectroscopy have been employed to monitor the stress build-up on the fibre of a single fibre fragmentation coupon. The authors combine single-walled nanotubes with a silane coating on the surface of samarium doped glass fibres. Thus, local strain could be simultaneously monitored using both techniques, despite the presence of this coating. Good agreement with shear-lag theory can be obtained using both techniques, during the fragmentation of the glass fibre. de la Vega et al. combine impedance spectroscopy with Raman spectroscopy to monitor the thermal stress built up during curing (de la Vega et al. 2009) and to simultaneously sense local and global strain in a carbon nanotube reinforced composite (de la Vega et al. 2011). In order to successfully monitor the stress built-up during curing, they correct the apparent Raman shifts with temperature. Although the Raman probing is sensitive enough to temperature and phase changes, there is no observable slope change in the conductivity of the cured samples reheated up to their ultimate processing temperature. Cured SWCNT epoxy composites above electrical percolation are simultaneously studied and a similar behavior is observed for both the Raman Shift and the electrical conductivity of the nano composite. Both techniques are found to undergo transitions beyond a critical strain level which according to the authors, coincides with the development of residual strain in the matrix, when the composites were subjected to cyclic loading.

Nano-reinforced composites with higher symmetry would include nano-reinforced fibres, which due to manufacturing processes favour alignment along the fibre axis. These nano-reinforced composites would be transversely anisotropic in terms of symmetry. Polarised Raman spectroscopy has been employed to characterize the alignment of the nano-reinforcement along the fibre axis, which would be a measure of the quality of the CNT-reinforced fibres. Chae et al. (2005) study polyacrylonitrile (PAN)/CNTs composite fibres, spun from solutions in dimethyl acetamide (DMAc), using SWCNTs, DWCNTs, MWCNTs, and vapour grown carbon nano-fibres (VGCNFs). The CNT content in all cases was 5 wt.%. In this case, Raman spectroscopy is employed for characterising the fibre morphology rather than the efficiency of the stress transfer or orientation. Chen and Tao (2006) report a manufacturing method of polymer nanocomposites with SWNTs by casting a suspension of SWNTs in a solution of thermoplastic polyurethane and tetrahydrofuran. In their case, the nano-reinforced composite is a thin film of well aligned SWCNTs. They achieve very good alignment and improved mechanical properties. Polarized Raman spectroscopy was employed to verify the achieved orientation. The authors attribute this orientation to the macroscopic alignment which results from solvent–polymer interaction induced orientation of soft segment chain during swelling and moisture curing. The study of stress transfer using Raman Spectroscopy in fibre geometries is reported by Lachman et al. (2009). They report on the strain sensitivity of the G′ Raman band of SWCNTs in polyvinyl alcohol-SWCNT composite fibres, with a view to employing such structures as strain or stress sensors when embedded in structural components. They observe higher shifts of the G′ Raman band when carboxylic functional groups are present at the nanotube surface and attribute this behaviour to improved interfacial adhesion. According to the authors, this enhancement of the interface increases the efficiency of such structures when used as stress sensors. However, they also report that improvements in interfacial adhesion do not lead to substantially better mechanical properties of the fibres. They explained this controversy by considering possible degradation of nanotubes during surface functionalisation. Their finding is important in that modification of CNTs with respect to achieving more efficient stress transfer may have adverse effect in other functionalities, such as reinforcement.

Deng et al. (2010) report on the manufacturing of Poly (p-phenylene terephthalamide)/single-walled carbon (PPTA/SWNT) composite fibres with different draw ratios using a dry-jet wet spinning process. The fibres were subsequently monitored using Raman spectroscopy. Raman scattering intensity mapping along the fibre is employed as a measure of the dispersion of the nano-reinforcement. The authors report that the nanotubes improve the polymer orientation in composite fibre with a draw ratio of 2 but degrade the orientation at higher draw ratios, and suggest that the reinforcement is more likely to be due to polymer chain orientation rather than nano-reinforcement. The interface of their studied system is reported to fail at far field strain higher than 0.35%. They also performed cyclic tests to assess the reversible deformation behaviour of the fibre as well as the gradual damage of the interface at high strains. They suggest that the hysteretic behaviour of the fibres in cyclic loading renders them useful in structural damping applications.

Blighe et al. (2011) measured the mechanical properties of coagulation-spun polymer–nanotube composite fibres with a volume fraction up to ~10%. They employed polarized Raman Spectroscopy to show (i) that orientation increases with drawing, indicating that significant nanotube alignment occurs and (ii) to demonstrate that the nanotube effective modulus also increases with drawing which suggests that the nanotube alignment in the fibres may be further improved. The authors introduce an empirical relationship between Krenchel's nanotube orientation efficiency factor (in other words the experimental deviation from the rule of mixtures) and calculate an orientation parameter via Raman Spectroscopy. They confirm that fibre modulus and fibre strength scales linearly with orientation and proceed to the calculation of the effective interfacial shear strength and critical length (40 MPa and 1,250 nm respectively).

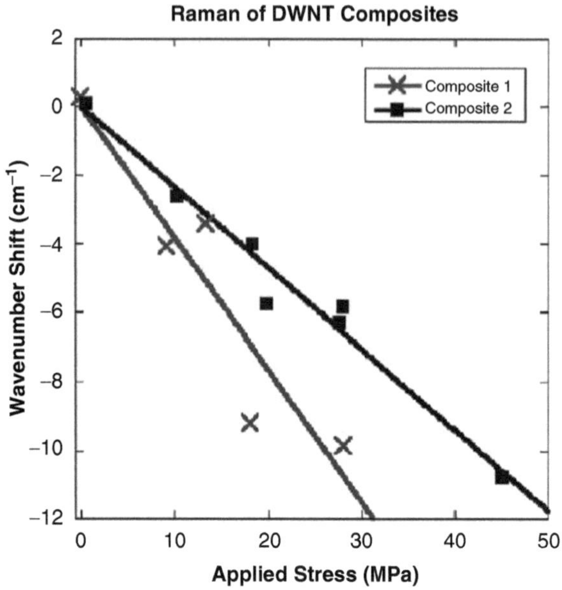

Fig. 6.15 Raman peak shift as a function of applied tensile stress for two nanotube composite samples (Reprinted from Brownlow et al. 2010, with permission from Elsevier)

An interesting study of the deformation of DWCNT/ epoxy composites employs a lamina configuration. Functionalized mats of DWCNTs are used to manufacture nano-reinforced composites where the distribution of the DWCNTs was practically 2D (Brownlow et al. 2010). The geometry of the structure can be regarded as transversely isotropic, but in contrast to the fibre geometry, the axis of symmetry is located vertical to the mat plane. The authors employ both FTIR and Raman. The FTIR technique is employed to estimate the average matrix stress, whereas the G′ peak shift is employed to monitor the stress build up in the composite. The authors report a large stress-induced shift in the G′ peak of 3.7 cm^{-1}/GPa which, compared to the "universal stress sensor" of approximately 5 cm^{-1}/GPa, is remarkably good and probably better than what a random 2D distribution of short reinforcing fibres should exhibit (Fig. 6.15). This experimental finding is indicative of the effect of the CNT distribution on the reinforcing ability of the nanophase.

Last but not least, very recent works focus on the interfacial shear stress transfer in model composites where graphene is embedded in a matrix and the Raman probe provides information on the stress built up on the sp^2 sheet. Srivastava et al. (2011) employed graphene as a filler and monitored the strain induced shifts of the G band shift of graphene platelets in polydimethyl-siloxane nanocomposites. In their study, they report large debonding strains of ~7% for graphene in the matrix, and a G band strain sensitivity of ~2.4 cm^{-1}/strain % which, compared to the measured shift of 0.1 cm^{-1}/% for single-walled carbon nanotube composites, suggests enhanced load-transfer. The surprising observation is that for strains higher than 2% the G line shifted to higher wave numbers reproducibly. The authors attribute this behaviour to the alignment of the polymer chains due to tension which results in lateral compression of the graphene platelets.

Gong et al. (2010) used Raman spectroscopy to monitor the stress transfer on a mechanically cleaved single graphene monolayer embedded in a thin polymer matrix layer. They monitored the G′ band shifts for their study. For strains up to 0.4%, the authors report a linear dependence of the G′ band. As the stress induced Raman shift exhibits a plateau at this strain level, the authors conclude that no further stress transfer can be sustained by the graphene/epoxy interface. The stress induced shift is measured to be as high as 60 cm^{-1}/% in the case of unloading of the strained sample. The authors employ the shear-lag model (Cox 1952) and the Kelly–Tyson formula (Kelly and Tyson 1965) to calculate interfacial shear strength of 2.3 and 0.3–0.8 MPa respectively which is a magnitude lower than the respective strength of carbon fibre epoxy interfaces. However, they suggest that the low values may be due to the fact that the interrogated graphene layer is shorter than the critical length required to build adequate axial stress.

Concluding, the incorporation of carbon nano-scaled structures in polymer matrices is very promising in providing stress transfer monitoring and stress sensing functionalities in the nano-reinforced composite using Raman Spectroscopy, as has already been performed in the case of carbon fibres. As should be noted the spectral signature and the stress induced shifts of any sp^2 structure is very dependent on the structure itself, the dispersion in the matrix, and the reinforcing symmetry. The uniqueness of the stress dependence of the Raman spectrum of any of these structures is complicating the task of calibrating the stress-induced changes in the Raman spectrum for the nano-reinforced composites. On the other hand, this uniqueness may be paramount in providing specific information about the stress transfer at the nanoscale. This is becoming more prominent as knowledge on the induced spectral changes is accumulating. Additionally, as the maximum value of the force constant should be the same for all sp^2 structures, this may provide a measure of the reinforcing ability of the nanophase as compared with the ultimate translation of the far field stress on the C–C bond.

6.6 Summary

The scope of this work is to provide an overview of the stress induced changes in the Raman spectrum of graphitic structures with a view to elucidating the reinforcing ability of the CNTs in a matrix and assess the stress transfer at the CNT matrix interface. At the same time the research effort towards employing CNTs as stress sensors in composite materials is presented.

To this end, an overview of the Raman Vibrational modes for all graphitic structures is presented starting from graphite fibres, to SWNTs to MWNTs and finally to Single and Multi-layer Graphene. The distinct differences in the Raman spectrum of these structures are highlighted.

Following this, the principle of the stress dependence of the Raman vibrational modes is presented in the general case. The basic principle of the anharmonic oscillator is presented in order to provide the reader an insight on the underlying principle of stress monitoring.

An overview of the induced changes in the Raman Spectrum of Graphite fibres, Nanotubes and Graphene is presented, either via pressure (hydrostatic or not) or direct stress application. An extensive literature survey is presented to cover all aspects of the changes in different Raman vibrational modes. Direct stress application on graphene has enabled the introduction of a unique stress sensor which characterizes all sp^2 graphitic structures and may characterize the reinforcing ability of the nanophase.

Finally, an extensive review of the Raman stress monitoring of nano-reinforced composites is presented. The review covers aspects relating to the reinforcing ability of the nanophase, the stress-sensing capability, as well as the stress transfer at the graphene/epoxy interface or even at the interface between the distinct layers in DWCNTs.

References

Arjyal, B.P., Katerelos, D.G., Filiou, C., Galiotis, C.: Measurement and modeling of stress concentration around a circular notch. Exp. Mech. **40**(3), 248–255 (2000)

Bahr, J.L., Yang, J., Kosynkin, D.V., Bronikowski, M.J., Smalley, R.E., Tour, J.M.: Functionalization of carbon nanotubes by electrochemical reduction of aryl diazonium salts: a bucky paper electrode. J. Am. Chem. Soc. **123**(27), 6536–6542 (2001)

Bandow, S., Hirahara, K., Hiraoka, T., Chen, G., Eklund, P.C., Iijima, S.: Turning peapods into double-walled carbon nanotubes. MRS Bull. **29**(4), 239–241, 260–264 (2004)

Barber, A.H., Zhao, Q., Wagner, H.D., Baillie, C.A.: Characterization of E-glass-polypropylene interfaces using carbon nanotubes as strain sensors. Compos. Sci. Technol. **64**(13–14), 1915–1919 (2004)

Belandria, E., Millot, M., Broto, J.M., Flahaut, E., Rodriguez, F., Valiente, R., Gonzalez, J.: Pressure dependence of Raman modes in double wall carbon nanotubes filled with 1D tellurium. Carbon **48**(9), 2566–2572 (2010). doi:10.1016/j.carbon.2010.03.036

Benoit, J.M., Buisson, J.P., Chauvet, O., Godon, C., Lefrant, S.: Low-frequency Raman studies of multiwalled carbon nanotubes: experiments and theory. Phys. Rev. B – Condens Matter Mater. Phys. **66**(7), 734171–734174 (2002)

Blighe, F.M., Young, K., Vilatela, J.J., Windle, A.H., Kinloch, I.A., Deng, L., Young, R.J., Coleman, J.N.: The effect of nanotube content and orientation on the mechanical properties of polymer-nanotube composite fibers: separating intrinsic reinforcement from orientational effects. Adv. Funct. Mater. **21**(2), 364–371 (2011)

Bretzlaff, R.S., Wool, R.P.: Frequency shifting and asymmetry in infrared bands of stressed polymers. Macromolecules **16**(12), 1907–1917 (1983)

Brown, S.D.M., Jorio, A., Corio, P., Dresselhaus, M.S., Dresselhaus, G., Saito, R., Kneipp, K.: Origin of the Breit-Wigner-Fano lineshape of the tangential G-band feature of metallic carbon nanotubes. Phys. Rev. B – Condens. Matter Mater. Phys. **63**(15), 1554141–1554148 (2001)

Brownlow, S.R., Moravsky, A.P., Kalugin, N.G., Majumdar, B.S.: Probing deformation of double-walled carbon nanotube (DWNT)/epoxy composites using FTIR and Raman techniques. Compos. Sci. Technol. **70**(10), 1460–1468 (2010)

Cançado, L.G., Pimenta, M.A., Saito, R., Jorio, A., Ladeira, L.O., Grueneis, A., Souza-Filho, A.G., Dresselhaus, G., Dresselhaus, M.S.: Stokes and anti-Stokes double resonance Raman scattering in two-dimensional graphite. Phys. Rev. B **66**(3), 035415 (2002)

Cançado, L.G., Pimenta, M.A., Neves, B.R.A., Medeiros-Ribeiro, G., Enoki, T., Kobayashi, Y., Takai, K., Fukui, K.I., Dresselhaus, M.S., Saito, R., Jorio, A.: Anisotropy of the Raman spectra of nanographite ribbons. Phys. Rev. Lett. **93**(4), 047403–047401 (2004)

Cançado, L.G., Takai, K., Enoki, T., Endo, M., Kim, Y.A., Mizusaki, H., Speziali, N.L., Jorio, A., Pimenta, M.A.: Measuring the degree of stacking order in graphite by Raman spectroscopy. Carbon **46**(2), 272–275 (2008)

Chae, H.G., Sreekumar, T.V., Uchida, T., Kumar, S.: A comparison of reinforcement efficiency of various types of carbon nanotubes in polyacrylonitrile fiber. Polymer **46**(24), 10925–10935 (2005). doi:10.1016/j.polymer.2005.08.092

Chen, W., Tao, X.: Production and characterization of polymer nanocomposite with aligned single wall carbon nanotubes. Appl. Surf. Sci. **252**(10), 3547–3552 (2006). doi:10.1016/j.apsusc.2005.05.028

Chohan, V., Galiotis, C.: Effects of interface, volume fraction and geometry on stress redistribution in polymer composites under tension. Compos. Sci. Technol. **57**(8), 1089–1101 (1997)

Christofilos, D., Arvanitidis, J., Tzampazis, C., Papagelis, K., Takenobu, T., Iwasa, Y., Kataura, H., Lioutas, C., Ves, S., Kourouklis, G.A.: Raman study of metallic carbon nanotubes at elevated pressure. Diam. Relat. Mater. **15**(4–8), 1075–1079 (2006). doi:10.1016/j.diamond.2005.11.027

Coleman, J.N., Khan, U., Blau, W.J., Gun'ko, Y.K.: Small but strong: a review of the mechanical properties of carbon nanotube-polymer composites. Carbon **44**(9), 1624–1652 (2006)

Colthup, N.B.: Introduction to infrared and Raman spectroscopy. Academic, London (1975)

Cooper, C.A., Young, R.J., Halsall, M.: Investigation into the deformation of carbon nanotubes and their composites through the use of Raman spectroscopy. Compos. Pt A: Appl. Sci. Manuf. **32**(3–4), 401–411 (2001)

Cooper, C.A., Cohen, S.R., Barber, A.H., Wagner, H.D.: Detachment of nanotubes from a polymer matrix. Appl. Phys. Lett. **81**(20), 3873–3875 (2002)

Cox, H.L.: The elasticity and strength of paper and other fibrous materials. Br. J. Appl. Phys. **3**(3), 72–79 (1952)

Cronin, S.B., Swan, A.K., Unlü, M.S., Goldberg, B.B., Dresselhaus, M.S., Tinkham, M.: Measuring the uniaxial strain of individual single-wall carbon nanotubes: resonance Raman spectra of atomic-force-microscope modified single-wall nanotubes. Phys. Rev. Lett. **93**(16), 167401 (2004)

Cronin, S.B., Swan, A.K., Unlü, M.S., Goldberg, B.B., Dresselhaus, M.S., Tinkham, M.: Resonant Raman spectroscopy of individual metallic and semiconducting single-wall carbon nanotubes under uniaxial strain. Phys. Rev. B **72**(3), 035425 (2005)

Cui, S., Kinloch, I.A., Young, R.J., Noé, L., Monthioux, M.: The effect of stress transfer within double-walled carbon nanotubes upon their ability to reinforce composites. Adv. Mater. **21**(35), 3591–3595 (2009)

de la Vega, A., Kovacs, J.Z., Bauhofer, W., Schulte, K.: Combined Raman and dielectric spectroscopy on the curing behaviour and stress build up of carbon nanotube-epoxy composites. Compos. Sci. Technol. **69**(10), 1540–1546 (2009)

de la Vega, A., Kinloch, I.A., Young, R.J., Bauhofer, W., Schulte, K.: Simultaneous global and local strain sensing in SWCNT-epoxy composites by Raman and impedance spectroscopy. Compos. Sci. Technol. **71**(2), 160–166 (2011)

Del Corro, E., Gonzalez, J., Taravillo, M., Escoffier, W., Baonza, V.G.: Graphite under non-hydrostatic conditions. High Press. Res. **28**(4), 583–586 (2008)

Del Corro, E., Taravillo, M., González, J., Baonza, V.G.: Raman characterization of carbon materials under non-hydrostatic conditions. Carbon **49**(3), 973–979 (2011). doi:10.1016/j.carbon.2010.09.064

Deng, L., Young, R.J., van der Zwaag, S., Picken, S.: Characterization of the adhesion of single-walled carbon nanotubes in poly(p-phenylene terephthalamide) composite fibres. Polymer **51**(9), 2033–2039 (2010)

Dresselhaus, M.S., Dresselhaus, G.: Intercalation compounds of graphite. Adv. Phys. **30**(2), 139–326 (1981)

Dresselhaus, M.S., Dresselhaus, G., Saito, R., Jorio, A.: Raman spectroscopy of carbon nanotubes. Phys. Rep. **409**(2), 47–99 (2005)

Dresselhaus, M.S., Dresselhaus, G., Jorio, A.: Raman spectroscopy of carbon nanotubes in 1997 and 2007. J. Phys. Chem. C **111**(48), 17887–17893 (2007)

Dresselhaus, M.S., Jorio, A., Hofmann, M., Dresselhaus, G., Saito, R.: Perspectives on carbon nanotubes and graphene Raman spectroscopy. Nano Lett. **10**(3), 751–758 (2010)

Dunstan, D.J., Ghandour, A.J.: High-pressure studies of carbon nanotubes. High Press. Res. **29**(4), 548–553 (2009)

Evans, K.E., Gibson, A.G.: Prediction of the maximum packing fraction achievable in randomly oriented short-fibre composites. Compos. Sci. Technol. **25**(2), 149–162 (1986)

Everall, N.J., Lumsdon, J., Christopher, D.J.: The effect of laser-induced heating upon the vibrational Raman spectra of graphites and carbon fibres. Carbon **29**(2), 133–137 (1991)

Frank, O., Tsoukleri, G., Parthenios, J., Papagelis, K., Riaz, I., Jalil, R., Novoselov, K.S., Galiotis, C.: Compression behavior of single-layer graphenes. ACS Nano **4**(6), 3131–3138 (2010)

Frank, O., Mohr, M., Maultzsch, J., Thomsen, C., Riaz, I., Jalil, R., Novoselov, K.S., Tsoukleri, G., Parthenios, J., Papagelis, K., Kavan, L., Galiotis, C.: Raman 2D-band splitting in graphene: theory and experiment. ACS Nano **5**(3), 2231–2239 (2011a)

Frank, O., Tsoukleri, G., Riaz, I., Papagelis, K., Parthenios, J., Ferrari, A.C., Geim, A.K., Novoselov, K.S., Galiotis, C.: Development of a universal stress sensor for graphene and carbon fibres. Nat. Commun. **2**(1) (2011b). doi:10.1038/ncomms1247

Galiotis, C., Batchelder, D.N.: Strain dependencies of the first- and second-order Raman spectra of carbon fibres. J. Mater. Sci. Lett. **7**, 545–547 (1988)

Galiotis, C., Young, R.J., Batchelder, D.N.: A resonance Raman spectroscopic study of the strength of the bonding between an epoxy resin and a polydiacetylene fibre. J. Mater. Sci. Lett. **2**(6), 263–266 (1983)

Galiotis, C., Robinson, I.M., Young, R.J., Smith, B.J.E., Batchelder, D.N.: Strain dependence of the Raman frequencies of a Kevlar 49 fibre. Polym. Commun. Guildford **26**(12), 354–355 (1985)

Galiotis, C., Chohan, V., Paipetis, A., Vlattas, C.: Interfacial measurements and fracture characteristics of single and multi-fiber composites by remote laser Raman microscopy. ASTM Spec. Tech. Publ. **1290**, 19–33 (1996)

Gao, B., Jiang, L., Ling, X., Zhang, J., Liu, Z.: Chirality-dependent Raman frequency variation of single-walled carbon nanotubes under uniaxial strain. J. Phys. Chem. C **112**(51), 20123–20125 (2008). doi:10.1021/jp809374j

Gong, L., Kinloch, I.A., Young, R.J., Riaz, I., Jalil, R., Novoselov, K.S.: Interfacial stress transfer in a graphene monolayer nanocomposite. Adv. Mater. **22**(24), 2694–2697 (2010)

Gouadec, G., Colomban, P.: Raman spectroscopy of nanomaterials: how spectra relate to disorder, particle size and mechanical properties. Prog. Cryst. Growth Charact. Mater. **53**(1), 1–56 (2007)

Guo, J.J., Liu, G.H., Wang, X.M., Fujita, T., Xu, B.S., Chen, M.W.: High-pressure Raman spectroscopy of carbon onions and nanocapsules. Appl. Phys. Lett. **95**(5), 51920 (2009)

Hadjiev, V.G., Warren, G.L., Sun, L., Davis, D.C., Lagoudas, D.C., Sue, H.J.: Raman microscopy of residual strains in carbon nanotube/epoxy composites. Carbon **48**(6), 1750–1756 (2010)

Hanfland, M., Beister, H., Syassen, K.: Graphite under pressure: equation of state and first-order Raman modes. Phys. Rev. B **39**(17), 12598 (1989)

Huang, Y., Young, R.J.: Effect of fibre microstructure upon the modulus of PAN- and pitch-based carbon fibres. Carbon **33**(2), 97–107 (1995)

Jorio, A., Dresselhaus, G., Dresselhaus, M.S., Souza, M., Dantas, M.S.S., Pimenta, M.A., Rao, A.M., Saito, R., Liu, C., Cheng, H.M.: Polarized Raman study of single-wall semiconducting carbon nanotubes. Phys. Rev. Lett. **85**(12), 2617–2620 (2000)

Jorio, A., Saito, R., Hafner, J.H., Lieber, C.M., Hunter, M., McClure, T., Dresselhaus, G., Dresselhaus, M.S.: Structural (n, m) determination of isolated single-wall carbon nanotubes by resonant Raman scattering. Phys. Rev. Lett. **86**(6), 1118–1121 (2001)

Jorio, A., Fantini, C., Dantas, M.S.S., Pimenta, M.A., Souza Filho, A.G., Samsonidze, G.G., Brar, V.W., Dresselhaus, G., Dresselhaus, M.S., Swan, A.K., Ünlü, M.S., Goldberg, B.B., Saito, R.: Linewidth of the Raman features of individual single-wall carbon nanotubes. Phys. Rev. B – Condens. Matter Mater. Phys. **66**(11), 1154111–1154118 (2002)

Jorio, A., Pimenta, M.A., Souza Filho, A.G., Samsonidze, G.G., Swan, A.K., Unlü, M.S., Goldberg, B.B., Saito, R., Dresselhaus, G., Dresselhaus, M.S.: Resonance Raman spectra of carbon nanotubes by cross-polarized light. Phys. Rev. Lett. **90**(10), 107403 (2003)

Kao, C.C., Young, R.J.: A Raman spectroscopic investigation of heating effects and the deformation behaviour of epoxy/SWNT composites. Compos. Sci. Technol. **64**(15), 2291–2295 (2004) (Special issue)

Kao, C.C., Young, R.J.: Assessment of interface damage during the deformation of carbon nanotube composites. J. Mater. Sci. **45**(6), 1425–1431 (2010)

Katagiri, G., Ishida, H., Ishitani, A.: Raman spectra of graphite edge planes. Carbon **26**(4), 565–571 (1988)

Kataura, H., Kumazawa, Y., Maniwa, Y., Umezu, I., Suzuki, S., Ohtsuka, Y., Achiba, Y.: Optical properties of single-wall carbon nanotubes. Synth. Met. **103**(1–3), 2555–2558 (1999)

Kelly, A., Tyson, W.R.: Tensile properties of fibre-reinforced metals: copper/tungsten and copper/molybdenum. J. Mech. Phys. Solid. **13**(6), 329–350 (1965)

Kuzmany, H., Pfeiffer, R., Hulman, M., Kramberger, C.: Raman spectroscopy of fullerenes and fullerene-nanotube composites. Philos. Trans. R. Soc. A: Math. Phys. Eng. Sci. **362**(1824), 2375–2406 (2004)

Lachman, N., Bartholome, C., Miaudet, P., Maugey, M., Poulin, P., Wagner, H.D.: Raman response of carbon nanotube/PVA fibers under strain. J. Phys. Chem. C **113**(12), 4751–4754 (2009). doi:10.1021/jp900355k

Lespade, P., Marchand, A., Couzi, M., Cruege, F.: Caracterisation de materiaux carbones par microspectrometrie Raman. Carbon **22**(4–5), 375–385 (1984). doi:10.1016/0008-6223(84)90009-5

Liu, Z., Zhang, J., Gao, B.: Raman spectroscopy of strained single-walled carbon nanotubes. Chem. Commun. **45**, 6902–6918 (2009). doi:10.1039/b914588e

Loa, I., Möschel, C., Reich, A., Assenmacher, W., Syassen, K., Jansen, M.: Novel graphitic spheres: Raman spectroscopy at high pressures. Phys. Status Solidi B Basic Res. **223**(1), 293–298 (2001)

Lucas, M., Young, R.J.: Effect of residual stresses upon the Raman radial breathing modes of nanotubes in epoxy composites. Compos. Sci. Technol. **67**(5), 840–843 (2007a)

Lucas, M., Young, R.J.: Unique identification of single-walled carbon nanotubes in composites. Compos. Sci. Technol. **67**(10), 2135–2149 (2007b)

Lucchese, M.M., Stavale, F., Ferreira, E.H.M., Vilani, C., Moutinho, M.V.O., Capaz, R.B., Achete, C.A., Jorio, A.: Quantifying ion-induced defects and Raman relaxation length in graphene. Carbon **48**(5), 1592–1597 (2010). doi:10.1016/j.carbon.2009.12.057

Malard, L.M., Guimarães, M.H.D., Mafra, D.L., Mazzoni, M.S.C., Jorio, A.: Group-theory analysis of electrons and phonons in N -layer graphene systems. Phys. Rev. B **79**(12), 125426 (2009a)

Malard, L.M., Pimenta, M.A., Dresselhaus, G., Dresselhaus, M.S.: Raman spectroscopy in graphene. Phys. Rep. **473**(5–6), 51–87 (2009b)

Marston, C., Gabbitas, B., Adams, J., Nutt, S., Marshall, P., Galiotis, C.: Failure characteristics in carbon/epoxy composite tows. Compos. Pt A: Appl. Sci. Manuf. **27**(12 Pt A), 1183–1194 (1996)

Melanitis, N., Galiotis, C.: Compressional behaviour of carbon fibres – Part 1 A Raman spectroscopic study. J. Mater. Sci. **25**(12), 5081–5090 (1990)

Melanitis, N., Tetlow, P.L., Galiotis, C., Smith, S.B.: Compressional behaviour of carbon fibres - Part II Modulus softening. J. Mater. Sci. **29**(3), 786–799 (1994)

Melanitis, N., Tetlow, P.L., Galiotis, C.: Characterization of PAN-based carbon fibres with Laser Raman Spectroscopy. Part I Effect of processing variables on Raman band profiles. J. Mater. Sci. **31**(4), 851–860 (1996)

Milnera, M., Kürti, J., Hulman, M., Kuzmany, H.: Periodic resonance excitation and intertube interaction from quasicontinuous distributed helicities in single-wall carbon nanotubes. Phys. Rev. Lett. **84**(6), 1324–1327 (2000)

Nemanich, R.J., Lucovsky, G., Solin, S.A.: Infrared active optical vibrations of graphite. Solid State Commun. **23**(2), 117–120 (1977)

Paipetis, A., Galiotis, C.: Effect of fibre sizing on the stress transfer efficiency in carbon/epoxy model composites. Compos. Pt A: Appl. Sci. Manuf. **27**(9), 755–767 (1996). doi:10.1016/1359-835x(96)00054-1

Paipetis, A., Galiotis, C.: A study of the stress-transfer characteristics in model composites as a function of material processing, fibre sizing and temperature of the environment. Compos. Sci. Technol. **57**(8), 827–838 (1997)

Paipetis, A., Galiotis, C.: Modelling the stress-transfer efficiency of carbon-epoxy interfaces. Proc. R. Soc. A: Math. Phys. Eng. Sci. **457**(2011), 1555–1577 (2001)

Paipetis, A., Galiotis, C., Liu, Y.C., Nairn, J.A.: Stress transfer from the matrix to the fibre in a fragmentation test: Raman experiments and analytical modeling. J. Compos. Mater. **33**(4), 377–399 (1999)

Papagelis, K., Andrikopoulos, K.S., Arvanitidis, J., Christofilos, D., Galiotis, C., Takenobu, T., Iwasa, Y., Kataura, H., Ves, S., Kourouklis, G.A.: High pressure Raman study of the second-order vibrational modes of single- and double-walled carbon nanotubes. Phys. Status Solidi B Basic Res. **244**(11), 4069–4073 (2007)

Papagelis, K., Arvanitidis, J., Christofilos, D., Souliou, S.M., Galiotis, C., Ves, S., Kourouklis, G. A.: High-pressure Raman study of stacked-cup carbon nanofibers. High Press. Res. **31**(1), 131–135 (2011)

Peng, B., Locascio, M., Zapol, P., Li, S.Y., Mielke, S.L., Schatz, G.C., Espinosa, H.D.: Measurements of near-ultimate strength for multiwalled carbon nanotubes and irradiation-induced crosslinking improvements. Nat. Nanotechnol. **3**(10), 626–631 (2008). doi:10.1038/nnano.2008.211

Pfeiffer, R., Kuzmany, H., Kramberger, C., Schaman, C., Pichler, T., Kataura, H., Achiba, Y., Kürti, J., Zólyomi, V., lyomi, V.: Unusual high degree of unperturbed environment in the interior of single-wall carbon nanotubes. Phys. Rev. Lett. **90**(22), 225501 (2003)

Pimenta, M.A., Marucci, A., Empedocles, S.A., Bawendi, M.G., Hanlon, E.B., Rao, A.M., Eklund, P.C., Smalley, R.E., Dresselhaus, G., Dresselhaus, M.S.: Raman modes of metallic carbon nanotubes. Phys. Rev. B – Condens. Matter Mater. Phys. **58**(24), R16016–R16019 (1998)

Pitkethly, M.J., Doble, J.B.: Characterizing the fibre/matrix interface of carbon fibre-reinforced composites using a single fibre pull-out test. Composites **21**(5), 389–395 (1990). doi:10.1016/0010-4361(90)90436-z

Proctor, J.E., Gregoryanz, E., Novoselov, K.S., Lotya, M., Coleman, J.N., Halsall, M.P.: High-pressure Raman spectroscopy of graphene. Phys. Rev. B – Condens. Matter Mater. Phys. **80**(7) (2009)

Puech, P., Nanot, S., Raquet, B., Broto, J.M., Millot, M., Anwar, A.W., Flahaut, E., Bacsa, W.: Comparative Raman spectroscopy of individual and bundled double wall carbon nanotubes. Phys. Status Solidi B Basic Res. **248**(4), 974–979 (2011)

Robinson, I.M., Zakikhani, M., Day, R.J., Young, R.J., Galiotis, C.: Strain dependence of the Raman frequencies for different types of carbon fibres. J. Mater. Sci. Lett. **6**(10), 1212–1214 (1987)

Sandler, J., Shaffer, M.S.P., Windle, A.H., Halsall, M.P., Montes-Morán, M.A., Cooper, C.A., Young, R.J.: Variations in the Raman peak shift as a function of hydrostatic pressure for various carbon nanostructures: A simple geometric effect. Physical Review B - Condensed Matter and Materials Physics. **67**(3), 354171–354178 (2003)

Schadler, L.S., Giannaris, S.C., Ajayan, P.M.: Load transfer in carbon nanotube epoxy composites. Appl. Phys. Lett. **73**(26), 3842–3844 (1998). doi:10.1063/1.122911

Sengupta, R., Bhattacharya, M., Bandyopadhyay, S., Bhowmick, A.K.: A review on the mechanical and electrical properties of graphite and modified graphite reinforced polymer composites. Prog. Polym. Sci. (Oxford) **36**(5), 638–670 (2011)

Souza Filho, A.G., Jorio, A., Samsonidze, G.G., Dresselhaus, G., Dresselhaus, M.S., Swan, A.K., Ünlü, M.S., Goldberg, B.B., Saito, R., Hafner, J.H., Lieber, C.M., Pimenta, M.A.: Probing the electronic trigonal warping effect in individual single-wall carbon nanotubes using phonon spectra. Chem. Phys. Lett. **354**(1–2), 62–68 (2002)

Srivastava, I., Mehta, R.J., Yu, Z.Z., Schadler, L., Koratkar, N.: Raman study of interfacial load transfer in graphene nanocomposites. Appl. Phys. Lett. **98**(6) (2011). doi:10.1063/1.3552685

Sureeyatanapas, P., Hejda, M., Eichhorn, S.J., Young, R.J.: Comparing single-walled carbon nanotubes and samarium oxide as strain sensors for model glass-fibre/epoxy composites. Compos. Sci. Technol. **70**(1), 88–93 (2010)

Tashiro, K., Wu, G., Kobayashi, M.: Quasiharmonic treatment of infrared and Raman vibrational frequency shifts induced by tensile deformation of polymer chains. J. Polymer Sci., Part B: Polymer Phys. **28**(13), 2527–2553 (1990)

Teredesai, P.V., Sood, A.K., Muthu, D.V.S., Sen, R., Govindaraj, A., Rao, C.N.R.: Pressure-induced reversible transformation in single-wall carbon nanotube bundles studied by Raman spectroscopy. Chem. Phys. Lett. **319**(3–4), 296–302 (2000)

Tuinstra, F., Koenig, J.L.: Characterization of graphite fiber surfaces with Raman spectroscopy. J. Compos. Mater. **4**, 494–499 (1970)

Van Den Heuvel, P.W.J., Peijs, T., Young, R.J.: Failure phenomena in two-dimensional multi-fibre microcomposites: 2. A Raman spectroscopic study of the influence of inter-fibre spacing on stress concentrations. Compos. Sci. Technol. **57**(8), 899–911 (1997)

Venkateswaran, U.D., Brandsen, E.A., Schlecht, U., Rao, A.M., Richter, E., Loa, I., Syassen, K., Eklund, P.C.: High pressure studies of the Raman-active phonons in carbon nanotubes. Phys. Status Solidi B **223**(1), 225–236 (2001). doi:10.1002/1521-3951(200101)223:1<225::aid-pssb225>3.0.co;2-6

Vidano, R.P., Fischbach, D.B., Willis, L.J., Loehr, T.M.: Observation of Raman band shifting with excitation wavelength for carbons and graphites. Solid State Commun. **39**(2), 341–344 (1981)

Vigolo, B., Penicaud, A., Coulon, C., Sauder, C., Pailler, R., Journet, C., Bernier, P., Poulin, P.: Macroscopic fibers and ribbons of oriented carbon nanotubes. Science **290**(5495), 1331–1334 (2000)

Wool, R.P.: Infrared studies of deformation in semicrystalline polymers. Polym. Eng. Sci. **20**(12), 805–815 (1980)

Yallee, R.B., Young, R.J.: Evaluation of interface fracture energy for single-fibre composites. Compos. Sci. Technol. **58**(12), 1907–1916 (1998)

Zhao, Q., Wagner, H.D.: Two-dimensional strain mapping in model fiber-polymer composites using nanotube Raman sensing. Compos. Pt A: Appl. Sci. Manuf. **34**(12), 1219–1225 (2003)

Zhao, Q., Frogley, M.D., Wagner, H.D.: Direction-sensitive strain-mapping with carbon nanotube sensors. Compos. Sci. Technol. **62**(1), 147–150 (2002)

Chapter 7
Mechanical and Electrical Response Models of Carbon Nanotubes

T.C. Theodosiou and D.A. Saravanos

Contents

Abstract Carbon nanotubes have remarkable mechanical and electrical properties. One promising feature is their electrical resistance that strongly depends on mechanical deformation. This, in combination with the fact that nanotubes can be dispersed into polymeric matrices, makes them ideal constituents for the development of novel multifunctional materials and devices. When dispersed into an insulating polymer, nanotubes are known to induce conductive behavior to the composite. This is attributed to the formation of conductive nanotube networks due to percolation. When a nanocomposite is mechanically deformed, load is

---

T.C. Theodosiou • D.A. Saravanos (✉)
Department of Mechanical Engineering & Aeronautics, University of Patras, Patras, Greece
e-mail: saravanos@mech.upatras.gr

A.S. Paipetis and V. Kostopoulos (eds.), *Carbon Nanotube Enhanced Aerospace Composite Materials*, Solid Mechanics and Its Applications 188,
DOI 10.1007/978-94-007-4246-8_7, © Springer Science+Business Media Dordrecht 2013

transferred to the nanotubes, as well. As they deform and rearrange, their electrical properties change and the percolation networks are distorted. This effect is studied in this chapter using three models: (i) an atomistic molecular mechanics approach for prediction of the mechanical response of carbon nanotubes, (ii) a subatomic tight-binding approach for prediction of the piezoresistive response of individual carbon nanotubes, and (iii) a homogenized microscale model for prediction of the piezoresistive response of carbon nanotube doped insulating polymers. Results seem to be in agreement with experimental results for small deformations.

Keywords Carbon nanotubes • Molecular mechanics • Tight binding model • Piezoresistive response • Homogenization

7.1 Mechanical Properties of Carbon Nanotubes

7.1.1 Introduction

The properties of carbon nanotubes come from interactions among atoms. Today there are various successful approaches at the atomistic level, including Molecular Mechanics (Burkert and Allinger 1982), Molecular Dynamics (Allen and Tildesley 1989; Rapaport 1995), Tight-Binding (Ashcroft and Mermin 1976b; Morse 1929), Ab-Initio (Levine 1991) etc. Each approach has advantages and disadvantages; usually, a computationally efficient method lacks in accuracy and vice versa. Molecular Mechanics/Dynamics works at the atomic level, while the others include electronic and subatomic models. The goal of this work is to predict the mechanical response of a nanotube at the atomic level, thus the use of subatomic modeling is not necessary. Molecular Dynamics can predict the time evolution of a molecular system and can be very useful for the study of liquids or melts. For the case of carbon nanotubes, however, Molecular Mechanics seems to be the most suitable, since the structure of CNTs is quite stable and does not change in time.

Molecular Mechanics methods, in fact, apply Newtonian Mechanics at atomic and molecular level. All methods have the same features:

- Every atom or group of atoms is represented as an individual particle;
- Each particle can interact with other particles within a finite radius;
- Chemical bonds can be represented as special springs; the equilibrium position is determined either theoretically or experimentally;
- Energy calculations are based on atomic positions, assuming the molecular system to be frozen, i.e. without any atomic vibrations.

It is clear that Molecular Mechanics is a rather simplified approach. The seeming lack of accuracy in the description of a molecular system makes this approach the most computationally effective for the following reasons:

- In contrast with subatomic models, the constitutive equations are familiar to a wider scientific circle;
- The required computational effort is minimal.

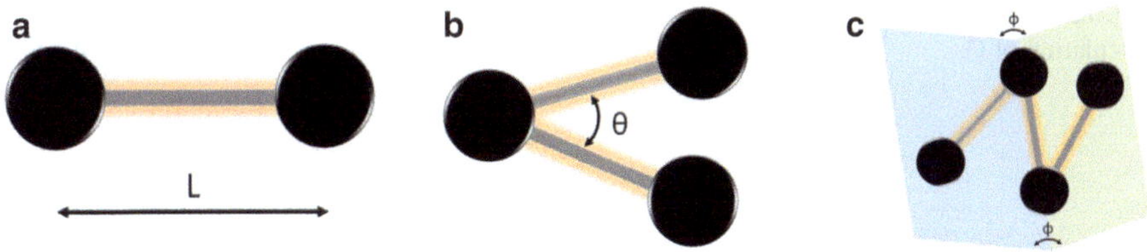

Fig. 7.1 Atomic interactions for molecular mechanics analyses: (**a**) bond stretching, (**b**) angle bending, (**c**) dihedral angles

7.1.2 The Brenner Model

Depending on the required accuracy, Molecular Mechanics can take into account a variety of atomic interactions, such as bond stretching, angle bending, dihedral angles etc. as shown in Fig. 7.1. The more interactions included, the more accurate the model. However, not all interactions are necessary; the Brenner model for Hydrocarbons suggests that a reasonably accurate description of the system can be obtained by including bond stretching and angle bending (Tersoff 1998; Brenner 1990). This approach is employed for the analysis of carbon nanotubes in this work. The main advantage of the Brenner model against other similar models is that it has been calibrated for use with carbon and organic molecules. This model has also been successfully employed by other researchers who used Molecular Dynamics for the analysis of carbon nanotubes (Luo et al. 1998).

In the context of the Brenner model, each bond is affected by its near-field environment. The interactions between each pair of atoms are expressed in terms of a repulsive and an attractive term, while an additional term is introduced in order to include the effect of the neighboring atoms. The mathematical formulation of the Brenner model for a system of N atoms is:

$$V = \sum_{i=1}^{N} \sum_{j>i}^{N} \left[V_R(r_{ij}) - \frac{1}{2} \cdot (B_{ij} + B_{ji}) \cdot V_A(r_{ij}) \right] \tag{7.1}$$

The terms V_R and V_A express the repulsive and attractive potential respectively between the atoms "i" and "j" and depend only on the interatomic distance r_{ij}. These terms can be further expanded to:

$$V_R(r_{ij}) = F_C(r_{ij}) \cdot \frac{D_e}{S-1} \cdot e^{-\sqrt{2S} \cdot (r_{ij} - r_e)} \tag{7.2}$$

$$V_A(r_{ij}) = F_C(r_{ij}) \cdot \frac{D_e \cdot S}{S-1} \cdot e^{-\sqrt{\frac{2}{S}} \cdot (r_{ij} - r_e)} \tag{7.3}$$

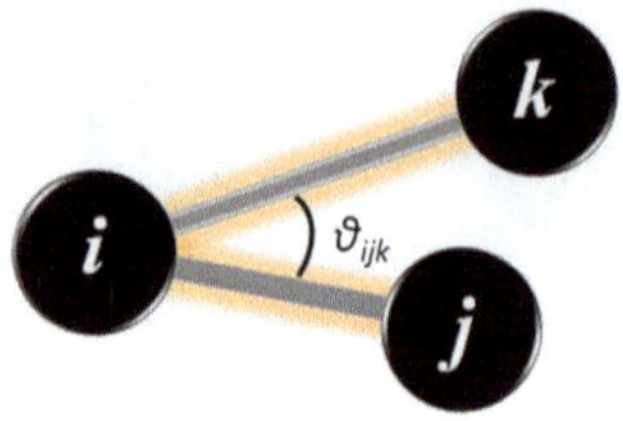

Fig. 7.2 Angle θ for the calculation of G

Table 7.1 Parameters of the Brenner potential

Parameter name	Value	Units
R_e	1.39	Å
D_e	6.00	eV
b	2.10	–
R_1	1.70	Å
R_2	2.00	Å
S	1.22	
δ	0.50	
a_0	0.00020813	
c_0	330	
d_0	3.50	

Term B expresses the effect of the neighboring atoms and its mathematical formulation is:

$$B_{ij} = \left[1 + \sum_{k \neq i,j}^{N} G(\theta_{ijk}) \cdot F_C(r_{ik})\right]^{-\delta} \tag{7.4}$$

where G is a function of the angle θ_{ijk} formed by atoms "i", "j" and "k" (Fig. 7.2) and is expressed as:

$$G(\theta) = \alpha_0 \cdot \left[1 + \frac{c_0^2}{d_0^2} - \frac{c_0^2}{d_0^2 + (1 + \cos\theta)^2}\right] \tag{7.5}$$

Finally, F_C is employed in order to determine the size of the near-field environment:

$$F_C(r) = \begin{cases} 1 & r \leqslant R_1 \\ \frac{1}{2} \cdot \left[1 + \cos\left(\pi \cdot \frac{r - R_1}{R_2 - R_1}\right)\right] & R_1 < r \leqslant R_2 \\ 0 & r > R_2 \end{cases} \tag{7.6}$$

The parameters required for the definition of Eqs. (7.1), (7.2), (7.3), (7.4), (7.5), and (7.6) are listed in Table 7.1.

7.1.3 Equations of Equilibrium

The equations of equilibrium for a carbon nanotube can be obtained in variational form from the minimization of the total potential energy:

$$\min(\Pi) = \min(V - \mathbf{F} \cdot \mathbf{u}) \tag{7.7}$$

where V is the total energy as calculated by (7.1), $\mathbf{F}$ is the vector of the external forces and $\mathbf{u}$ is the vector of atomic displacements, using extended vector notation.

Using a Taylor expansion series, the total energy can be recast as

$$\Pi = \Pi_0 + \frac{\partial \Pi}{\partial \mathbf{u}} \cdot d\mathbf{u} + \frac{1}{2} \cdot d\mathbf{u}^T \cdot \frac{\partial^2 \Pi}{\partial \mathbf{u}} \cdot d\mathbf{u} + \ldots \tag{7.8}$$

The following quantities can be then introduced:

$$\boldsymbol{\Psi} = \frac{\partial \Pi}{\partial \mathbf{u}} \tag{7.9}$$

$$[\mathbf{K}] = \frac{1}{2} \cdot \frac{\partial^2 \Pi}{\partial \mathbf{u}^2} \tag{7.10}$$

Therefore, Eq. (7.8) becomes:

$$\Pi = \Pi_0 + \boldsymbol{\Psi} \cdot d\mathbf{u} + d\mathbf{u}^T \cdot [\mathbf{K}] \cdot d\mathbf{u} \tag{7.11}$$

From Eqs. (7.7) and (7.9) it is clear that:

$$\boldsymbol{\Psi} = \frac{\partial \Pi}{\partial \mathbf{u}} = \frac{\partial}{\partial \mathbf{u}}(V - \mathbf{F} \cdot \mathbf{u}) = \frac{\partial V}{\partial \mathbf{u}} - \mathbf{F} \tag{7.12}$$

Vector $\boldsymbol{\Psi}$ expresses the equilibrium between internal and external forces and it is termed *Imbalance Vector*, while $[\mathbf{K}]$ in Eq. (7.10) is actually a linearized (tangential) stiffness matrix.

According to the Principle of Virtual Works, the work produced by a virtual displacement $d\mathbf{u}$ will be:

$$\delta\Pi = d\mathbf{u} \cdot \boldsymbol{\Psi} + d\mathbf{u} \cdot ([\mathbf{K}] \cdot d\mathbf{u}) \tag{7.13}$$

as derived from Eq. (7.11).

There are numerous methods to predict atomic positions in the equilibrium state. The simplest one is perhaps the Newton-Raphson method or one of its modified variants (Ypma 1995; Suli and Mayers 2003). First, the atomic positions are

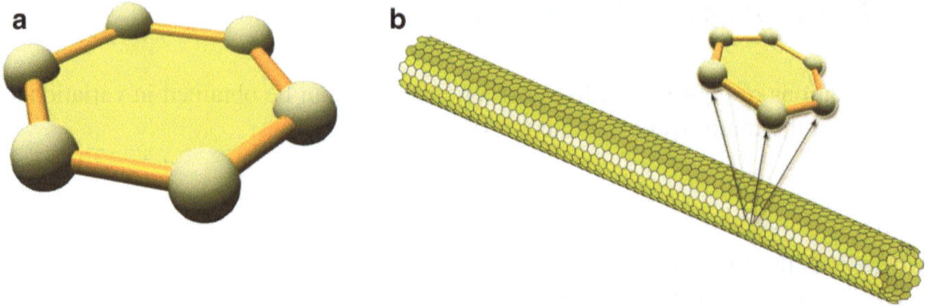

Fig. 7.3 Suggested configuration for the molecular finite element. (**a**) the molecular finite element and (**b**) a carbon nanotube represented as an assembly of molecular finite elements

roughly estimated. A better estimate is obtained if a corrective term $d\mathbf{u}$ is added to the vector of atomic positions. This term is calculated in (7.13) by setting $\delta\Pi = \Pi - \Pi_0 = 0$:

$$[\mathbf{K}] \cdot d\mathbf{u} = -\boldsymbol{\Psi} \tag{7.14}$$

Equation (7.14) can be used repeatedly until the optimal equilibrium state is obtained.

7.1.4 Finite Element Approach

If all possible atomic interactions are taken into account, the assembly and solution of Eq. (7.14) becomes very time-consuming, even with the introduction of the cut-off function – Eq. (7.6). In order to address this issue, it is proposed to assemble the total stiffness matrices from smaller ones, calculated for small portions of the carbon nanotube. Each portion of the nanotube can be treated as a special finite element; the internal energy of these finite elements can be calculated from Molecular Mechanics equations as described earlier, thus, they can be termed as Molecular Finite Elements.

The basic principle is that not all atomic interactions need to be considered. On the contrary, each atom interacts with other atoms within a finite area, as suggested by the Brenner model. The shape of this novel finite element should be defined by the maximum range of the Brenner potential (R_2) and the periodicity of the nanotube geometry. Although more than one configurations are possible, the 6-node hexagonal form seems to be the more efficient. It is obvious that each atom corresponds to a node of the molecular finite element (Fig. 7.3); each node has three degrees of freedom, that is movement along the three axes (x, y, z).

The effect of overlapping bonds of neighboring elements can be easily eliminated by introducing a scaling factor $a_c = 0.5$ into (7.2) and (7.3), that is:

$$V_R(r_{ij}) = a_c \cdot F_C(r_{ij}) \cdot \frac{D_e}{S-1} \cdot e^{-\sqrt{2S}\cdot(r_{ij}-r_e)} \tag{7.15}$$

$$V_A(r_{ij}) = a_c \cdot F_C(r_{ij}) \cdot \frac{D_e \cdot S}{S-1} \cdot e^{-\sqrt{\frac{2}{S}}\cdot(r_{ij}-r_e)} \tag{7.16}$$

Alternatively, the original formulation may be preserved but the overlap effect must be taken into account during post-processing, otherwise the nanotubes will have increased stiffness.

Following this approach Eqs. (7.9) and (7.10) may be applied on each individual finite element for the calculation of internal forces and the stiffness matrices:

$$\boldsymbol{\Psi}_e = \frac{\partial \Pi_e}{\partial \mathbf{u}} \tag{7.17}$$

$$[\bar{\mathbf{K}}_e] = \frac{1}{2} \cdot \frac{\partial^2 \Pi_e}{\partial \mathbf{u}^2} \tag{7.18}$$

where the subscript "*e*" implies calculations for an individual element. Since atoms and element nodes are matched one-to-one, the effect of curvature of the nanotube wall is automatically included into the calculations. After all imbalance vectors and stiffness matrices have been calculated, the global imbalance vector and the global stiffness matrix are assembled.

In contrast to other successful methods there is no need for assumptions like Periodic Boundary Conditions, Homogeneous Displacements etc. This makes the introduced approach applicable to any nanotube configuration. Moreover, the value of the scaling factor a_c in (7.15) and (7.16) can be appropriately modified in order to increase or decrease the strength of individual bonds, which implies the existence of lattice defects. The non-linear effects are taken into account through the use of Molecular Mechanics, while Finite Element Analysis allows for using various computational tools, such as optimized solvers for Eq. (7.14) and parallel processing systems that take advantage of modern technology capabilities.

7.1.5 *Effective Medium Response*

After completion of the numerical procedure, results can be post-processed and provide the effective mechanical properties of carbon nanotubes. Initially, the deformation gradient tensor has to be calculated (Arroyo and Belytschko 2002; Xiao and Belytschko 2004):

$$[\mathbf{F}] = \frac{\partial \mathbf{u}}{\partial \mathbf{U}} \tag{7.19}$$

where $\mathbf{u}$ is the vector of atomic positions at any deformed state and $\mathbf{U}$ is the respective vector at the initial equilibrium state. The strain tensor (Lagrange-Green) will be (Hutter and Johnk 2004):

$$[\varepsilon] = \frac{1}{2}\left[\mathbf{F}^T\mathbf{F} - \mathbf{I}\right] \tag{7.20}$$

where $[\mathbf{I}]$ is the identity tensor. The corresponding stress tensor will be:

$$\left[\sigma_{ij}\right] = \frac{1}{\pi \cdot d_t \cdot L_0 \cdot t} \cdot \frac{\partial \Pi}{\partial \varepsilon_{ij}} \tag{7.21}$$

where d_t is the nanotube diameter, L_0 the length and t the wall thickness, considered to be equal to the interlayer distance of graphite (Li and Chou 2003; Stankovich et al. 2007), $t = 3.4$ Å. Clearly, the quantity:

$$\Omega_0 = \pi \cdot d_t \cdot L_0 \cdot t \tag{7.22}$$

is the volume of the undeformed cylinder as if it were a continuous hollow cylinder.

Following this approach, every mechanical property can be easily calculated, e.g. the Young modulus and Poisson's ratio:

$$E_{11} = \frac{\sigma_{11}}{\varepsilon_{11}} \tag{7.23}$$

$$\nu = -\frac{\varepsilon_{12}}{\varepsilon_{11}} \tag{7.24}$$

where the subscripts "11" and "12" denote the axial and radial direction respectively.

7.1.6 Numerical Procedure

Implementation of the introduced methodology consists of three phases: (a) Pre-Processing, (b) Analysis and (c) Post-Processing.

Preprocessing actually concerns the definition of the problem; an initial configuration is roughly estimated and boundary conditions are applied. The exact atomic positions are not important at this phase because atoms will take their equilibrium positions during the second phase. Mechanical loading is applied incrementally in the form of atomic displacements or forces. The load increment is quite important because small increments increase the solution time, while larger increments lead to instabilities owing to the non-linear terms of the Molecular Mechanics models. In practice, load increments must lead to atomic displacements one order of magnitude lower than the atomic bond between two carbon atoms (α_{CC}).

Definition of the problem is followed by a numerical Newton-Raphson method. First the imbalance vector ($\boldsymbol{\Psi}_e$) and the tangential stiffness matrix $\left[\mathbf{K}_e\right]$ are calculated for every molecular finite element using (7.17), (7.18) respectively and the global imbalance vector ($\boldsymbol{\Psi}$) and stiffness matrix are assembled $\left[\mathbf{K}\right]$. If the magnitude of $\boldsymbol{\Psi}$ exceeds a critical value, then the current configuration is treated as unstable and the energy minimization procedure is applied. Eq. (7.14) is solved and the corrective vector $d\mathbf{u}$ is obtained. This vector is added to the vector of atomic positions $\mathbf{U}$, so that the new atomic locations represent a more stable configuration. The procedure is repeated until the magnitude of the imbalance vector converges to zero. Then, a load increment is applied and the procedure is repeated in order to obtain the equilibrium state under loading. The whole procedure is repeated for a predetermined range of load increments.

Finally, Eqs. (7.20), (7.21), (7.22), (7.23), and (7.24) are applied, as well as any other equation of Elasticity Theory, in order to obtain the effective mechanical properties.

7.1.6.1 Optimization

As the nanotube size grows, its degrees of freedom increase, as well, requiring more computational power. An extensive study of the solution process concluded that the most computationally demanding stage is the synthesis of the stiffness matrices. Consequently, any optimization effort should be focused on Eq. (7.18).

To elevate this issue, a modified Newton-Raphson method is employed. Clearly, both methods should lead to the very same result. Thus, it is suggested to use:

$$[\mathbf{K}_{ic}] \cdot d\mathbf{u} = -\psi \tag{7.25}$$

instead of (7.18). In this equation the tangential stiffness matrix is not calculated in every iteration, but only for the initial configuration – as implied by the subscript "ic" (Initial Configuration). In this way the most time-consuming part of the analysis appears only once during the whole procedure and this results in boosting the solution speed.

The basic concept of the Newton-Raphson method – modified or not – is that the stiffness matrix effects the direction along which an atom should move in order to be in a better equilibrium state. Keeping always the same stiffness matrix in (7.25), atoms are actually guided toward "wrong" equilibrium positions, as the stiffness matrix corresponds only to the initial configuration. Therefore, during each loading cycle additional iterations are required before the molecular system finds its equilibrium. Indeed, in some cases up to four times more iterations are required. However, these iterations involve only the calculation of first derivatives (imbalance vector) and the total time is dramatically lower than in the original approach.

Of course, it is always possible that the use of the "wrong" stiffness matrix could lead the nanotube to instability. This is easily identified by monitoring the magnitude of the imbalance vector; should it increase, the atomic positions are reset and a new stiffness matrix is calculated.

In any case, the required solution time is significantly less. Figure 7.4 depicts the time required to complete a numerical simulation of a mechanically loaded nanotube vs. the total degrees of freedom. It is quite clear that the optimized procedure is significantly faster.

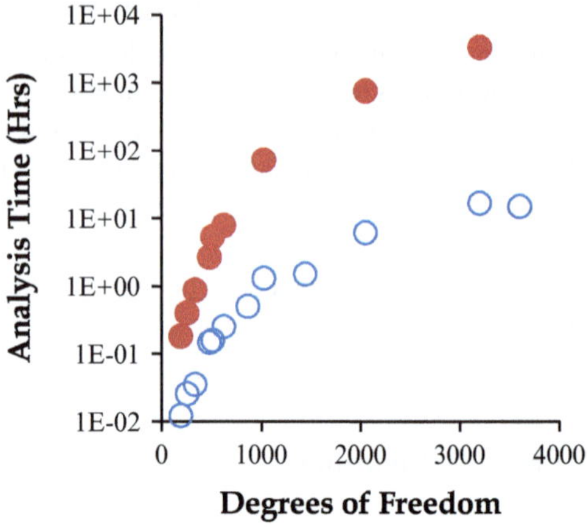

Fig. 7.4 Required time to obtain solution for the original (*filled markers*) and the optimized (*open markers*) procedure

7.1.7 Predictions and Validations

7.1.7.1 Prediction of Elastic Properties

The introduced methodology can be applied to practically any mechanical loading. Figure 7.5 depicts the variation of the total energy of a nanotube subjected to tension. It is clear that the energy variation can be well approximated by a second-order polynomial in the investigated strain range. Thus, owing to Eq. (7.21), nanotubes are expected to exhibit a linear stress-strain response. Indeed, as shown in Fig. 7.6, a linear response is observed, however, the aspect ratio (R) seems to play a significant role even if the nanotube helicity is the same.

$$R = \frac{\text{Length}}{\text{Diameter}} \tag{7.26}$$

The aspect ratio affects the slope of the stress-strain curve. Since a linear response is exhibited, this slope expresses in fact the Young modulus (Fig. 7.7). For small aspect ratios nanotubes exhibit increased stiffness, but as R increases, the Young modulus seems to converge. The interpolation curve seems to follow a Chapman equation:

$$\mathrm{E} = 0.6E_0 \cdot \left[1 - e^{-\frac{R_0}{10}R}\right]^{-\frac{1}{3}} \tag{7.27}$$

where (R_0, E_0) is the first pair of values in the diagram. This equation proved to approximate any nanotube configuration.

Equation (7.27) has the advantage of estimating the convergence value of a very large model – which would require extremely high analysis times – using input from small, computationally efficient, models.

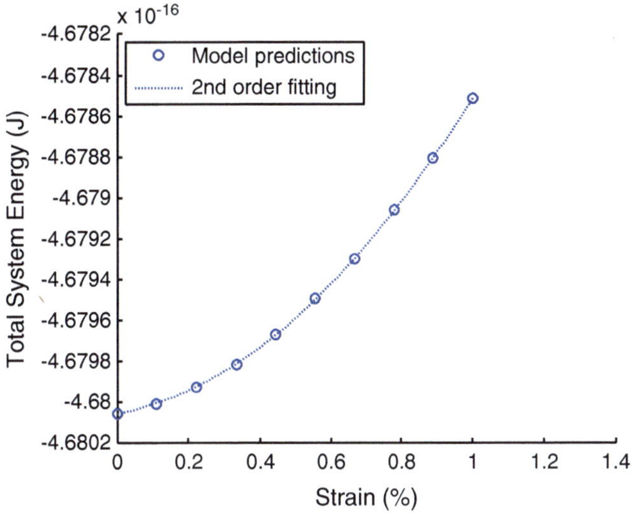

Fig. 7.5 Energy variation for a nanotube subjected to tension

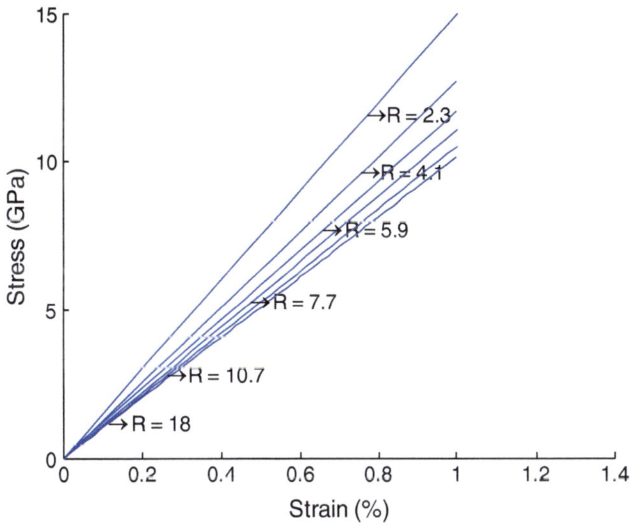

Fig. 7.6 Stress–strain curves for nanotubes of various aspect ratios

For the prediction of Poisson's ratio, Eq. (7.24) can be used. Results are depicted in Fig. 7.8.

Interestingly, Eq. (7.27) still approximates the convergence value, although a slight underestimation is observed.

The nanotube linear response is preserved in compression, as well (Fig. 7.9).

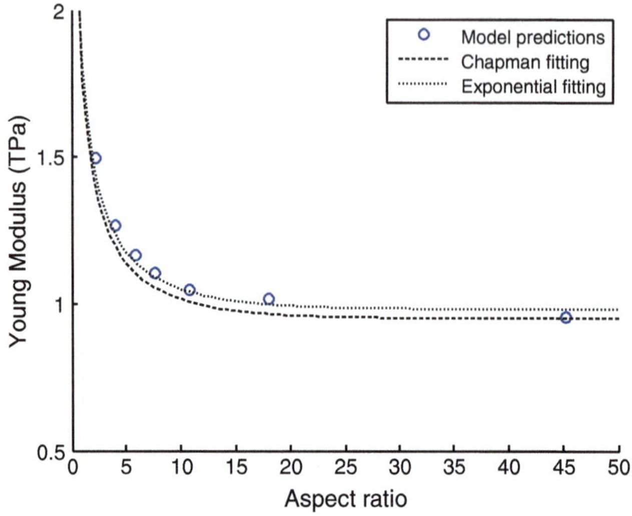

Fig. 7.7 Effect of the aspect ratio on the Young modulus predictions

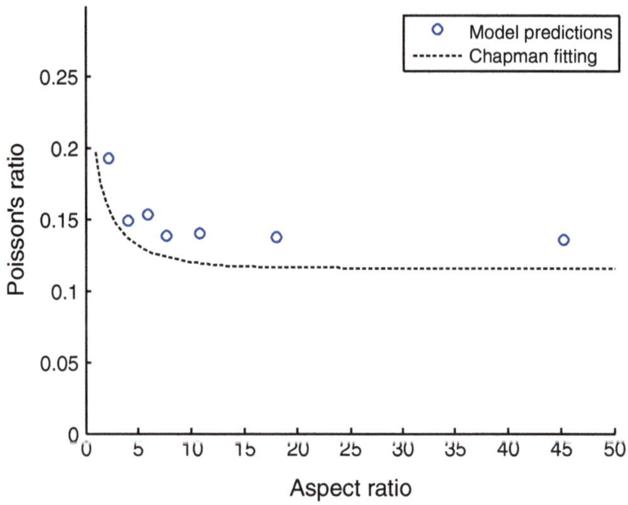

Fig. 7.8 Predictions for Poisson's ratio

7.1.7.2 Prediction of Failure

Failure can be predicted through the total energy diagram, at points where the energy curve is discontinuous (Fig. 7.10).

Interestingly, there is an indication of failure for strain near 25%, independent of the aspect ratio, leading to a stress value near 250 GPa. However, this is not total

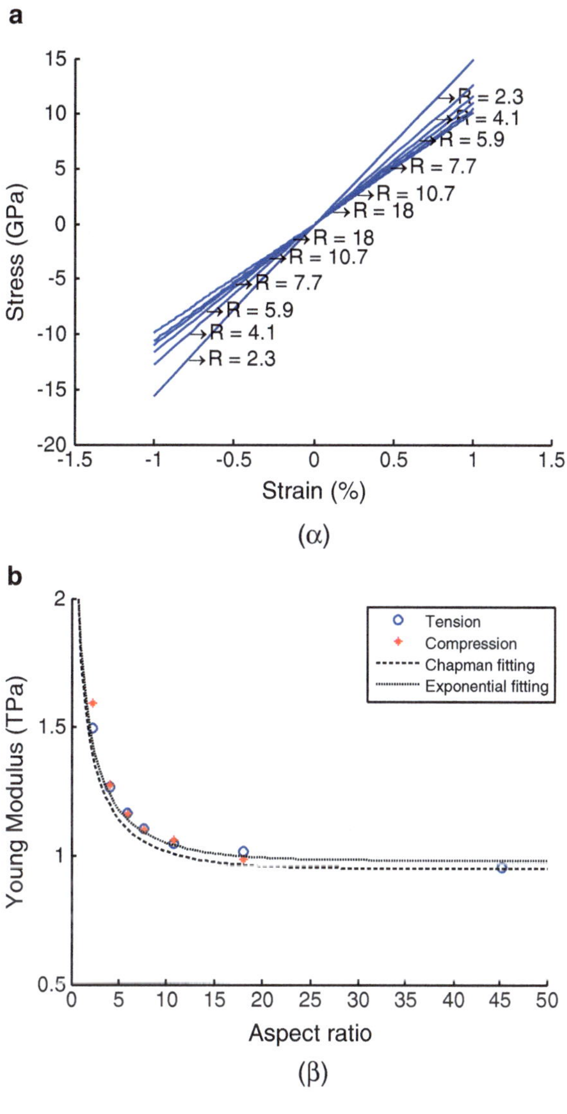

Fig. 7.9 Nanotube response in tension and compression. (**a**) Stress–strain curves and (**b**) Young modulus for nanotubes of various aspect ratios

failure, as the nanotube can still be loaded. Total failure seems to occur for strain around 30–35%, where the value of energy becomes constant.

For the case of compression, discontinuities appear sooner, in the range of −10% to −20% strain (Fig. 7.11). In this case failure occurs due to local buckling effects instead of bond breaks. Obviously, the mechanisms involved in tensile and compressive failure and quite different, thus, no direct correlation should be attempted.

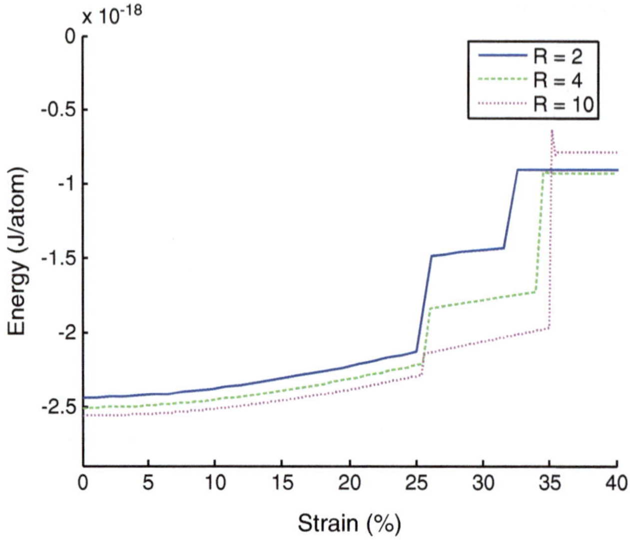

Fig. 7.10 Strain energy per atom and indication of failure for nanotubes of various aspect ratios

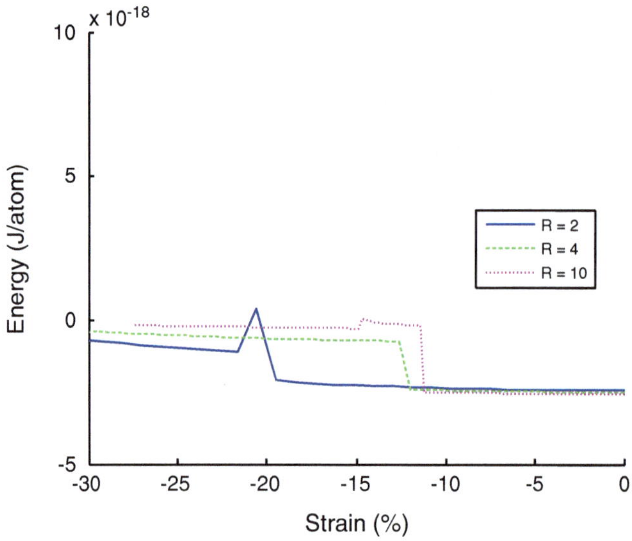

Fig. 7.11 Strain energy per atom for nanotubes of various aspect ratios (R)

7.1.7.3 Validations Summary

The present model validations are summarized in Table 7.2.

It is very clear that the predictions of the introduced model are reasonably close to reported values.

Table 7.2 Validations summary

		Present work	Theoretical predictions	Experimental measurements
Tensile modulus	E_T	0.965	0.91–1.23	0.95–1.25
Compressive modulus	E_C	0.936	0.91–1.23	–
Poisson's ratio	v	0.14	0.11–0.29	–
Shear modulus (GPa)	G	392	414–460	300–400
Failure strain	ε_f	25%	18.5–21%	–
Buckling strain	ε_b	−11%	−10 to −12%	–

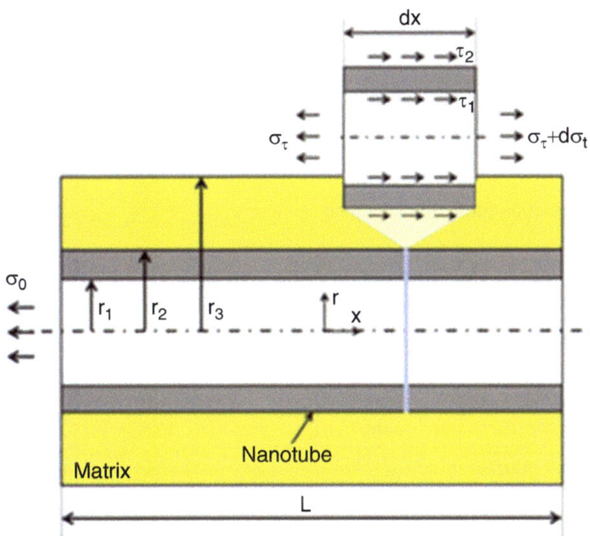

Fig. 7.12 Effective representation of matrix and nanotube in the context of a Shear-Lag model; the inset depicts an infinitesimal part of the nanotube

7.1.7.4 Load Transfer

For prediction of the response of a carbon nanotube doped polymer, the distribution of the load transferred from matrix to the nanotube is very important, especially for study of the electromechanical coupling effects discussed later. The load distribution can be approximated using a qualified Shear-Lag (Cox 1952; Nairn 1997) model. Of course, alternative and more accurate approaches have been proposed (Seidel and Lagoudas 2006; Seidel et al. 2008), but an extensive study of the load transfer mechanisms exceeds the scope of this work. In the context of Shear-Lag, the matrix and the nanotube are represented as two concentric hollow cylinders made of an effective continuous medium (Fig. 7.12). The effective Nanotube cylinder has the elastic properties calculated by the molecular finite element.

The composite consists of two phases (matrix – nanotube). The basic assumption is that no failure occurs at the matrix-nanotube interface; that is the displacements

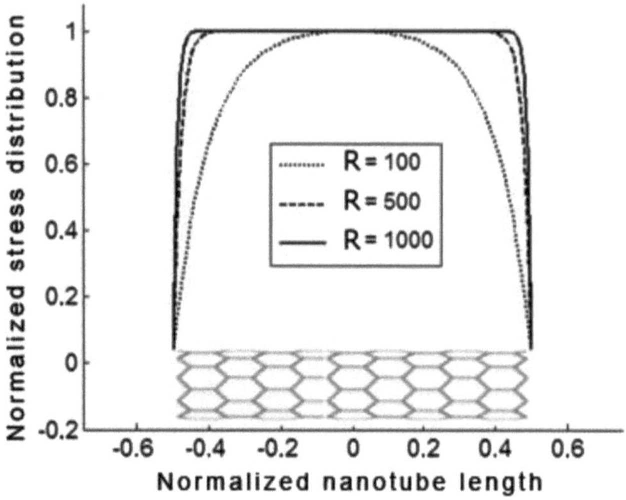

Fig. 7.13 Load distribution along the nanotube

should be continuous at the boundary of the two phases, otherwise there will be microcracks and local defects which exceed the scope of this work. Following this approach the stress transferred to the nanotubes is:

$$\sigma_t = \frac{(r_3-r_1)\cdot E_t\cdot\sigma_0}{(r_2-r_1)\cdot E_t+(r_3-r_2)E_m} + \frac{\cosh(C\cdot x)}{\cosh\left(C\cdot\frac{L}{2}\right)}\left[1-\frac{(r_3-r_2)\cdot E_t}{(r_2-r_1)\cdot E_t+(r_3-r_1)\cdot E_m}\right]\cdot\sigma_0$$
$$with \quad C=\sqrt{\frac{3\cdot[(r_2-r_1)\cdot E_t+(r_3-r_2)\cdot E_m]}{(r_3-r_2)^2\cdot(r_2-r_1)\cdot E_t\cdot(1+v_m)}} \tag{7.28}$$

where E_m, E_t are the elasticity modulii of the matrix (m: matrix) and the nanotube (t: tube) respectively, and σ_0 is the stress applied on the far field.

As depicted in Fig. 7.13 the aspect ratio (R) plays a significant role. For $R>500$ the load distribution is uniform along 95% of the length, while for $R>1000$ the distribution is uniform practically along the whole nanotube. This has been verified by other relevant studies (Xiao and Zhang 2004; Gao and Li 2005; Haque and Ramasetty 2005). The aspect ratio of real nanotubes is in the range of 10,000–50,000; thus, it can be assumed that the stress and strain are practically uniform along the nanotube.

7.2 Piezoresistive Properties of Carbon Nanotubes

7.2.1 Introduction

One of the most attractive features of carbon nanotubes in engineering applications is their conductive nature and their strain sensor and actuation potential which may be exploited towards the engineering of novel multifunctional materials and devices

(Avouris and Collins 1998; Avouris et al. 2005; Fraysse et al. 2002). Other studies have proved the coupling between mechanical and electrical properties, i.e. mechanical deformation can alter electronic properties, while electric fields can induce mechanical deformation (Guo and Guo 2003).

The macroscopic electrical properties of materials are determined by their electronic properties, which can be calculated from their electronic band structure (Bernholc et al. 2002), while the electronic band structure is strongly affected by mechanical deformation (Minot et al. 2004). The phenomenon of dependence of electric properties on the mechanical response resembles the behavior of piezoresistive sensors. Physical evidence of this electromechanical coupling effect has already been presented by various researchers (Liu et al. 2004; Gartstein et al. 2003).

Currently, there are numerous methods for the prediction of the piezoresistive response:

- **Boltzmann Transport Equation** (Huang 1987): Each system of particles tends to be in a state of thermodynamic equilibrium. The position of a particle in space and time can be predicted through a probability distribution function. The electron motion under the influence of electrostatic forces and fields can be described by considering the electron cloud as group of particles. The Boltzmann Transport Equation is an integrated approach that can take into account multiple phenomena. However, owing to this multiphysics approach, the probability distribution function is not always well defined and the solution of the equation is a rather time-consuming procedure. Nevertheless, it has already been employed successfully for the prediction of nanotube electrical properties (Pozdnyakov et al. 2006; Aksamija and Ravaioli 2008; Aksamija et al. 2009).
- Energy Methods: These methods are based on the study of the allowed and disallowed energy levels of the electrons. Various successful methods have been presented from time to time. Some of the most popular are:
  - **Ab-initio methods**: These methods are based on the analytical solution of Schrödinger's equation (Griffiths 2004) and the study of electrons in a field. They are assumed to be the most accurate, having, though, a considerable computational cost. These have been successfully employed (Ayuela et al. 2008; Reich et al. 2002).
  - **Plane Waves and Grids** (Ashcroft and Mermin 1976a): This is a series of methods suitable for periodic crystals, based on relatively simple numerical procedures. They use real-space calculations and can be applied on finite systems.
  - **Augmented Functions** (Slater 1937, 1953): They involve the study of electronic wavefunctions in areas near and far from the atomic nucleus. The existence of non-linearities is their major disadvantage.
  - **Tight-Binding** (Bloch 1928; Slater and Koster 1954): This approach practically simplifies the solution of Schrödinger's equation by using a set of approximate wavefunctions, based on the superposition of free electron wavefunctions. It is maybe the most popular approach for the following

two reasons: (a) it provides a simple representation of the electronic properties; and (b) it can lead to relatively accurate descriptions and energy calculations.

It has to be noted, of course, that all energy methods can lead to the same results, if applied properly.

In this work the Tight-Binding method will be employed for the reasons described. The piezoresistive response of carbon nanotubes is predicted employing a three-phase analysis: (a) the electronic band structure of a CNT is determined using a representative cell; (b) the electric conductance is calculated as a function of the electronic band structure using the Landauer formula (Landauer 1957; Bagwell and Orlando 1989) and the Wentzel-Kramers-Brillouin (WKB) and Miller-Good (MG) approximations; and (c) the total effect of mechanical deformation of the nanotube on its conductance/resistance is predicted by combining the results of the previous two phases.

7.2.2 *Electronic Band Structure*

7.2.2.1 General

When carbon atoms are far enough apart to have no interactions, the electrons occupy space depending on their energy and momentum (Orbitals); each electron lies on a specific energy level. As atoms get closer and form bonds; due to Pauli's Principle (Griffiths 2004; Massimi 2005), the Orbitals deform and the electrons become rearranged in a process called hybridization. This means that the energy level of each electron is slightly changed giving rise to numerous slightly different energy levels, the number of which depends on the number of the interacting electrons. The diagram depicting all allowed energy levels in space is the Electronic Band Structure diagram.

If the number of electrons is large enough, the allowed energy levels are so dense that they form continuous "allowed zones". Between the allowed zones are the so-called forbidden zones or band gaps. Depending on the size of a band gap a material can behave as a conductor, insulator or semiconductor. Currently, characterization of materials based on the band gap is rather a matter of conventions and assumptions, while various nomograms exist (Durrant 2000). In general, it can be said that at temperatures near absolute zero, a material behaves as an insulator when its band gap is higher than 0.3 eV while at room temperature, this threshold temperature is assumed near 3 eV.

The relative position atoms combined with the shape and orientation of the orbitals play a significant role for atomic interactions. Two general cases are identified (Muller 1994):

- Sigma-bonds (σ): It is the strongest type of covalent bond. Sigma-bonds are symmetric to rotation around the bond axis.

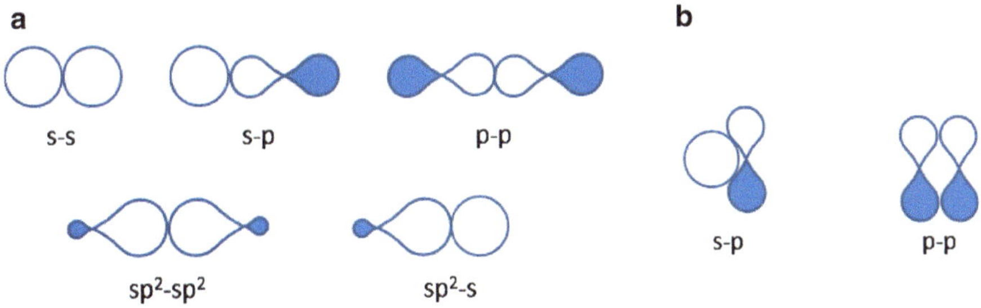

Fig. 7.14 Schematic representation of a various bond types. (**a**) sigma-bonds, (**b**) pi-bonds

– Pi-bonds (π): This bond is formed by two lobed orbitals, each contributing one electron. The two orbitals are anti-symmetric against the plane containing at least one of the two atoms. Pi-bonds are weaker than sigma-bonds.

Any other interaction can be obtained by superpositioning of the sigma and pi bonds taking into account their relative orientation (Kaxiras 2003) (Fig. 7.14). It should be noted that the $sp\pi$ bond is so weak that is assumed as non-existent; that is why only lobed orbitals can form pi-bonds according to the pi-bond definition.

For the case of graphite and nanotubes, the electrons involved are located in $2sp^2$ orbitals – that is sp^2 of the second orbit – while there is one more electron, which is moving independently inside a $2p$ orbital, perpendicular to the plane of the atomic bonds. Although all p orbitals are equivalent, it is assumed by convention that the $2sp^2$ orbital comes from superposition of the $2s$ and $2p_x$, $2p_y$.

7.2.2.2 Assumptions

Various approaches can be employed depending on the required accuracy. The first assumption found in literature is that the energy band structure of nanotubes is identical to that of graphite, taking into account that the nanotube is periodic only along its axial direction. A better approach is to include the wall curvature as well (Hamada et al. 1992; Saito et al. 1992). This means that the atomic distances and angles will eventually change and will not be equivalent anymore; in contrast to graphite, interaction will depend on the relative orientation of each atom's neighborhood. Including additional phenomena should lead to more realistic results, but there is no fully-established methodology for the moment. Nevertheless, as proved later, taking into account only these two effects leads to realistic results.

A very important assumption for the calculations is that only the $2p_z$ electrons contribute to conductivity. Conductivity implies that the electrons are able to move under the influence of an electric field. However, not all electrons are allowed to leave their atoms. For every hybrid sp^2 bond there are two possible energy states σ και σ^* with low and high energy respectively; σ electrons are

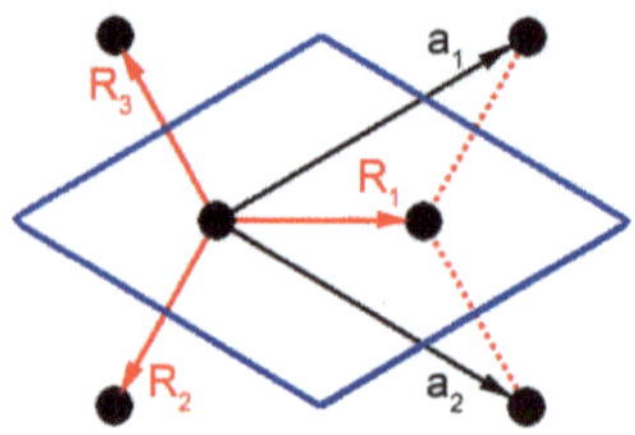

Fig. 7.15 Representative cell for the calculation of the secular equation matrices

bound to their atoms while σ^* electrons may travel through the lattice (Bruus 2004). The σ electrons are considered to be located in a *valence* or *bonding energy* zone, while the σ^* electrons are considered to be located in a *conductive* or *anti-bonding energy zone*. The forbidden zone between these two is so high that the electron cannot be so excited that it jumps from the valence zone to the conductive one. This assumption has been supported by the very first studies of graphite (Wallace 1947) and it is verified here as well.

Calculations are limited to nearest neighbor interactions. Following this assumption, the required computations are kept to a minimum. Involving long-range interactions would probably give better results, but would also require additional computational power. For the needs of the present work, this assumption proved to be valid.

Last but not least, it is assumed that mechanical load is uniformly applied and the nanotubes are uniformly deformed, preserving this way their symmetry and periodicity. This assumption has been verified by the Molecular Finite Element in combination with Shear-Lag, as discussed earlier.

7.2.2.3 Calculations

Thanks to the nanotube periodicity calculations need to be performed only on specific high symmetry sites taking advantage of the reciprocal space theory (Gibbs 1881; Gibbs and Wilson 1902). According to this, the Chiral ($\mathbf{C_h}$):

$$\mathbf{C_h} = n \cdot \mathbf{a_1} + m \cdot \mathbf{a_2} \equiv (n, m) \tag{7.29}$$

and Translational (**T**) vector:

$$\mathbf{T} = t_1 \cdot \mathbf{a_1} + t_2 \cdot \mathbf{a_2} \tag{7.30}$$

of the real space can be represented by vectors $\mathbf{K}_1, \mathbf{K}_2$ respectively, thus, any point $\mathbf{r}$ of the real lattice defined in terms of $\mathbf{C_h}$ and $\mathbf{T}$ can be reflected to the point $\mathbf{k}$ of the reciprocal lattice in terms of $\mathbf{K}_1, \mathbf{K}_2$. For a nanotube fragment containing N unit cells – like the one depicted in Fig. 7.15 – it has been proved that vectors differing by $N\mathbf{K_i}$ are equivalent (Saito et al. 1998). Thus, it is not necessary to perform calculations on every point of the nanotube lattice.

Table 7.3 Hamiltonian and Overlap matrix elements

Orbital energies (Saito et al. 1998)	$E_{2P} = 0$ eV	
Coupling parameters for the undeformed lattice (Saito et al. 1998)	$t_{ss} = -6.769$ eV	$s_{ss} = 0.212$ eV
	$t_{sp} = -5.580$ eV	$s_{sp} = 0.102$ eV
	$t_{pp\sigma} = -5.037$ eV	$s_{pp\sigma} = 0.146$ eV
	$t_{pp\pi} = -3.033$ eV	$s_{pp\pi} = 0.129$ eV
Coupling parameters for the undeformed lattice	$V_i(r) = t_i\left(\frac{r_0}{r}\right)^2$ with $i = ss, sp, pp\sigma, pp\pi$	
	$S_i(r) = s_i\left(\frac{r_0}{r}\right)^2$ with $i = ss, sp, pp\sigma, pp\pi$	
Hamiltonian and Overlap Matrix elements	$h_{zz} = \sum_{p \neq q} \left(\cos^2(\mathbf{R}_{pq},\hat{\mathbf{z}}) \cdot V_{pp\sigma}(R_{pq}) + \sin^2(\mathbf{R}_{pq},\hat{\mathbf{z}}) \cdot V_{pp\pi}(R_{pq})\right) \cdot e^{i\mathbf{k}\mathbf{R}_{pq}}$	
	$s_{zz} = \sum_{p \neq q} \left(\cos^2(\mathbf{R}_{pq},\hat{\mathbf{z}}) \cdot S_{pp\sigma}(R_{pq}) + \sin^2(\mathbf{R}_{pq},\hat{\mathbf{z}}) \cdot S_{pp\pi}(R_{pq})\right) \cdot e^{i\mathbf{k}\mathbf{R}_{pq}}$	

where r_o and r are the bond lengths of the undeformed and deformed lattice respectively (Fig. 7.15)

7.2.2.4 Energy Terms

The allowed energy levels can be calculated by solving the secular equation (Saito et al. 1998; Martin 2004):

$$|\mathbf{H} - E_n\mathbf{S}| = 0 \tag{7.31}$$

where **H** is the Hamiltonian and **S** is the Overlap matrix. Following the assumption that only the $2p_z$ interactions are needed for the calculations, Eq. (7.33) can be reduced to:

$$\mathbf{H} - \begin{bmatrix} E_{2p} & h_{zz} \\ h^*_{zz} & E_{2p} \end{bmatrix} \quad \mathbf{S} = \begin{bmatrix} 1 & s_{zz} \\ s^*_{zz} & 1 \end{bmatrix} \tag{7.32}$$

The terms required for the definition of the secular equation matrices are summarized in Table 7.3.

7.2.2.5 Effect of the Finite Perimeter

A carbon nanotube can be extremely long, thus calculations should be performed on a continuous path along $\mathbf{K}_2$. The length of this path is determined by the length of the translational vector in reciprocal space, that is $|\mathbf{T}|_{reciprocal} = |\mathbf{K}_2| = \frac{2\pi}{|\mathbf{T}|}$ according to the reciprocal lattice theory. On the other hand, as the nanotube perimeter is finite, calculations need to be performed only on points of symmetry, that is on points defined by $k_c \cdot \mathbf{K}_1$, with $k_c = 0, 1, \ldots, \mathrm{N} - 1$, so that all non-equivalent points are included.

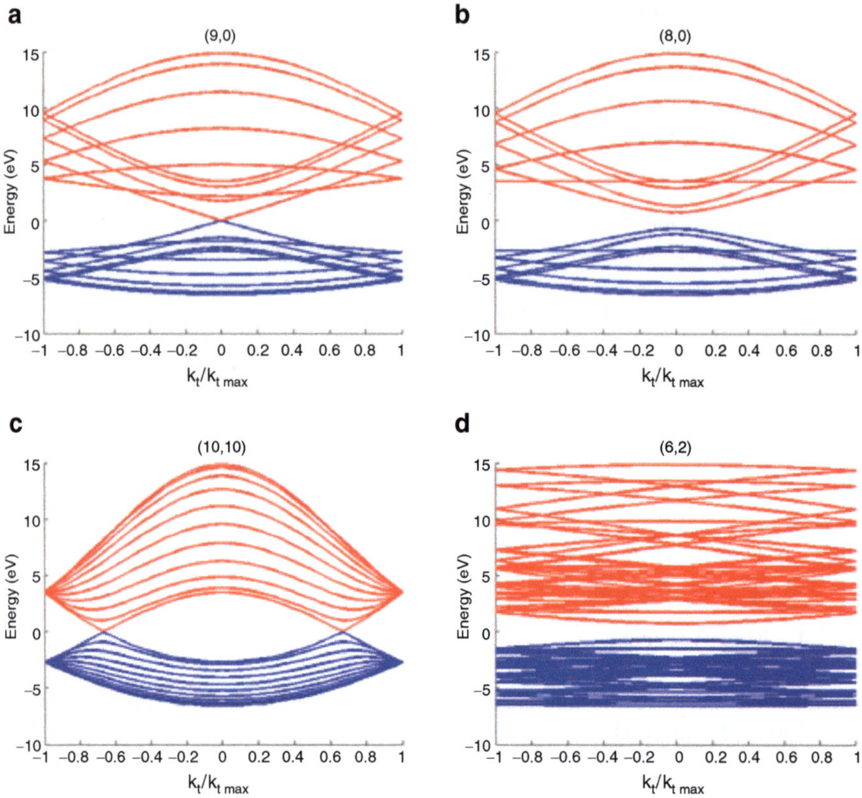

Fig. 7.16 Energy diagrams for nanotubes of various geometries

To sum up, calculation should be performed on the reciprocal lattice points defined by:

$$\mathbf{k} = k_c \cdot \mathbf{K_1} + k_t \cdot \frac{\mathbf{K_2}}{|\mathbf{K_2}|} \quad k_c = 0, 1, ..., \mathrm{N} - 1 \quad k_t \in \left[-\frac{\pi}{|\mathbf{T}|}, \frac{\pi}{|\mathbf{T}|}\right] \tag{7.33}$$

The range of k_t has been selected this way, in order to conform to conventions found in the literature; any other range of length $2\pi/|\mathbf{T}|$ is also acceptable and leads to the very same results.

The solution of Eq. (7.31) on every point **k** produces diagrams like the ones depicted in Fig. 7.16, each one having a lower part (valence band) and upper part (conductive band), possibly separated by a forbidden zone (band gap). Four different types of electronic band structure can be identified. Figure 7.16a depicts the energy diagram for a (9, 0) zig-zag nanotube. It is expected to exhibit conductive behavior as the band gap is zero. Figure 7.16b depicts the energy diagram for an (8, 0) nanotube. This is also a zig-zag nanotube but exhibits semiconductive behavior. The energy bands are separated by a forbidden zone, but the band gap is small enough so that under certain circumstances electrons can jump from the valence band to the conductivity zone. Figure 7.16c depicts the energy diagram of an armchair nanotube. It is noticeable that the two bands penetrate each other. Finally, Fig. 7.16d depicts the energy diagram for a nanotube with no special geometrical symmetry.

7.2.2.6 Effect of the Curvature

On a plane graphite sheet the vectors connecting each atom to its nearest neighbor (Fig. 7.15) can be calculated by:

$$\mathbf{R}_1 = \tfrac{1}{3}(\mathbf{a}_1 + \mathbf{a}_2), \quad \mathbf{R}_2 = -\tfrac{2}{3}\mathbf{a}_1 + \tfrac{1}{3}\mathbf{a}_2, \quad \mathbf{R}_3 = \tfrac{1}{3}\mathbf{a}_1 - \tfrac{2}{3}\mathbf{a}_2 \tag{7.34}$$

As the graphite sheet is rolled forming a nanotube, the interatomic distances and angles will eventually change due to curvature and the $2p_z$ orbitals will not be parallel anymore. This will affect the terms required for the assembly of Eq. (7.32). Assuming that the distances $\widehat{R}_i = |\boldsymbol{R}_i|$ as calculated by Eq. (7.34) are equal to the respective arc lengths of the curved graphitic lattice, the new interatomic distances $\bar{R}_i$ can be correlated to the chiral angle and the radius of the nanotube:

$$\bar{R}_i = \sqrt{2r^2(1 - \cos\varphi_i) + \widehat{R}_i^{\,2} \sin^2\left(\frac{\pi}{6} - \theta + (i-1)\frac{2\pi}{3}\right)} \tag{7.35}$$

where φ is the angle between the axis of the central atoms and the respective axes of all other atoms:

$$\varphi_i = \tfrac{\widehat{R}_i}{r}\left[\cos\left(\tfrac{\pi}{6} - \theta + (i-1)\tfrac{2\pi}{3}\right)\right], \quad i = 1, 2, 3 \tag{7.36}$$

r is the nanotube radius, θ is the chiral angle and i denotes each one of the three nearest-neighbor atoms.

Following the same approach, but using the modified geometrical quantities, the obtained energy diagrams provide a more accurate description of the electronic band structure. It is clear in Fig. 7.17 that the allowed energy levels have been shifted and a forbidden area has been induced.

7.2.3 Electrical Resistance

7.2.3.1 Calculations

Carbon nanotubes are molecular structures dominated by quantum phenomena, thus the classic electrical equations (Kirchhoff and Ohms's Laws etc.) are not applicable. Using, therefore, the Landauer equation (Landauer 1957; Bagwell and Orlando 1989) the electrical conductivity can be calculated in atomistic scale as:

$$G = \frac{2e^2}{\pi\hbar}T \tag{7.37}$$

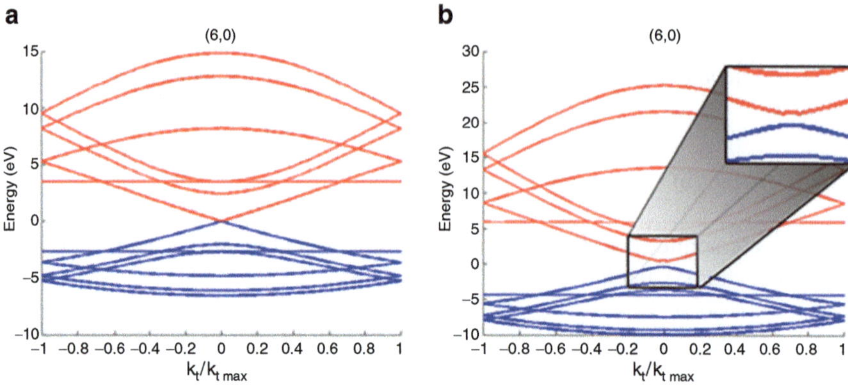

Fig. 7.17 Band structure for the same nanotube (**a**) without curvature effects and (**b**) with curvature effects. The inset shows the existence of a forbidden area between the two bands

where e is the electron charge and T is the *Transmission Probability*. The transmission probability practically correlates the macroscopic electrical conductivity with the atomistic properties of the nanotube, and it expresses the probability for an electron to "jump" from the bonding zone to the conductive one. The calculation of T requires the solution of:

$$T = \int_{-\infty}^{\infty} \frac{e^{-2\int_{x_1}^{x_2} \frac{\sqrt{2m}}{\hbar}\sqrt{-Q(x)}dx}}{1+\frac{1}{4}\left[e^{-2\int_{x_1}^{x_2} \frac{\sqrt{2m}}{\hbar}\sqrt{-Q(x)}dx}\right]^2} \cdot \left(\frac{1}{1+e^{\frac{E-\mu_s}{kT}}} + \frac{1}{1+e^{\frac{E-\mu_d}{kT}}}\right) dE \qquad (7.38)$$

which is hard to solve in a computationally efficient way. However, there is a number of approximate solutions that can be applied to the case of nanotube conductivity, such as the Wentzel-Kramers-Brillouin (WKB) method and the Miller-Good (MG) approximation (Razavy 2003), according to which:

$$T_{WKB} = e^{-A} \quad T_{MG} = \frac{1}{1+e^{A}} \quad A = 2\int_{x_1}^{x_2} \frac{\sqrt{2m}}{\hbar}\sqrt{-Q(x)}dx \qquad (7.39)$$

7.2.3.2 Predictions and Validations

Although Tight-Binding is well-established and widely used, its predictions are usually impossible to be correlated to experimental data for a number of reasons: (a) the most import limitation is the fact that the precise nanotube geometry is usually not known; (b) the theoretical analysis assumes that nanotubes are symmetrical

Table 7.4 Predictions for the electrical resistance

	n	m	d (Å)	R_{WKB} (kΩ)	R_{MG} (kΩ)
1	10	9	12.89	15.46	21.91
2	11	8	12.94	10.88	17.67
3	12	7	13.03	15.84	22.29
4	13	6	13.17	14.89	21.34
5	14	4	12.82	17.82	24.47
6	15	3	13.08	12.94	20.61
7	16	1	12.94	10.85	17.27
Average value:				14.10	20.84
Experimental:				20 kΩ	

periodic, having no structural defects, which practically happens almost never; (c) the interaction of the nanotube with its environment induces additional issues, etc.

From the seemingly present limited experimental data that can be directly correlated to the introduced approach, it is concluded that single-wall carbon nanotube structures have an electrical resistance in the range of 20–50 kΩ (Avouris et al. 2000). Experimental measurements on individual nanotubes showed a value for the electrical resistance around 20 kΩ (Zhou et al. 2000). Although the precise geometry was impossible to find, the diameter was estimated around 13 Å, which limits the candidate nanotubes to the ones listed in Table 7.4. The same table contains also the prediction of both WKB and MG; WKB seems to underestimate the value of resistance.

Javey et al. (2004) have measured the resistance for nanotubes of length (a) 10 nm, (b) 300 nm and (c) 3 μm. According to this report, longer nanotubes have higher resistance, although the theoretical prediction of the present study predicts always the same value. This divergence comes from the fact that as the length increases more defects are observed and additional quantum mechanical phenomena affect the nanotube response (Tian and Datta 1994; Datta 2004). For the 10 nm nanotube, which is assumed to be the most defect-free configuration, it is known that its diameter is around 1.5–2.5 nm and its resistance is measured around 40 kΩ. The resistance of the experimental setup is 15 kΩ, thus, the resistance of the nanotube should be 25 kΩ. The introduced approach leads to an estimation of 26 kΩ for the nanotube, which is very close to the measured value.

Therefore, at this point it can be stated that for nanotubes with prefect geometrical structure their electrical resistance can be calculated reasonably well based on their electronic band structure.

7.2.4 Strain Effects

The atoms of the nanotube are rearranged under the influence of mechanical deformation and get to a new state of equilibrium. In this section it is assumed that strain is uniformly applied along the nanotube, as predicted by the shear-lag model implemented earlier.

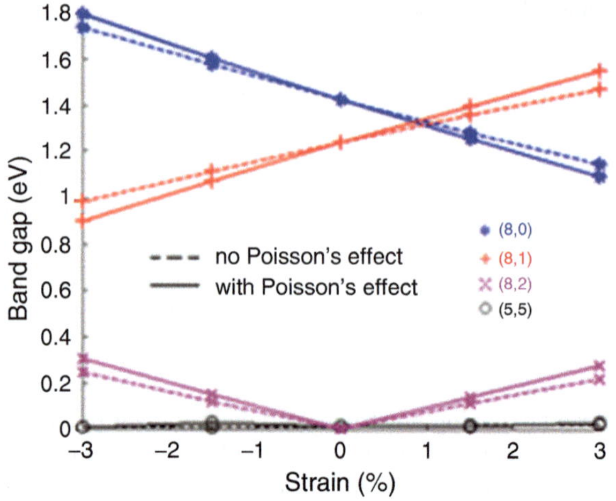

Fig. 7.18 Band gap variation under the influence of strain for various nanotube geometries. *Dashed lines* imply the absence of Poisson's effect

7.2.4.1 Electronic Band Structure

Earlier studies have described the effects of mechanical deformation using complicated relations of "deformed bond vectors" which describe the spatial rearrangement of atoms (Yang et al. 1999). In the present study, this procedure has been significantly simplified by applying strain to the lattice unit vectors and recalculating all geometrical quantities:

$$\mathbf{a}_{\mathbf{i\ def}} = (\mathbf{I} + \mathbf{E}) \cdot \mathbf{a}_{\mathbf{i}} \tag{7.40}$$

where **I** and **E** are the unit and strain tensor respectively:

$$\mathbf{E} = \begin{bmatrix} \varepsilon & \frac{1}{2}\gamma \\ \frac{1}{2}\gamma & -\nu \cdot \varepsilon \end{bmatrix} \tag{7.41}$$

where ε denotes the axial strain, γ is the torsional strain and ν is Poisson's ratio. The new electronic band structure can be calculated by implementing Eqs. (7.31), (7.32), (7.33), (7.34), (7.35), (7.36), (7.37), (7.38), and (7.39) using the "deformed unit vectors".

The band structure results are identical with the predictions of earlier studies, but their current formulation is significantly simpler. Another novelty is the implementation of Poisson's ratio (Fig. 7.18) which was absent in previous successful studies (Liu et al. 2004). The depicted nanotube geometries were specifically selected in order to be directly correlated with results found in the literature (Yang et al. 1999). More precise calculations take into account the wall curvature as well. As shown in Fig. 7.19 the non-parallel p_z interactions should not be neglected since they totally change the nanotube response.

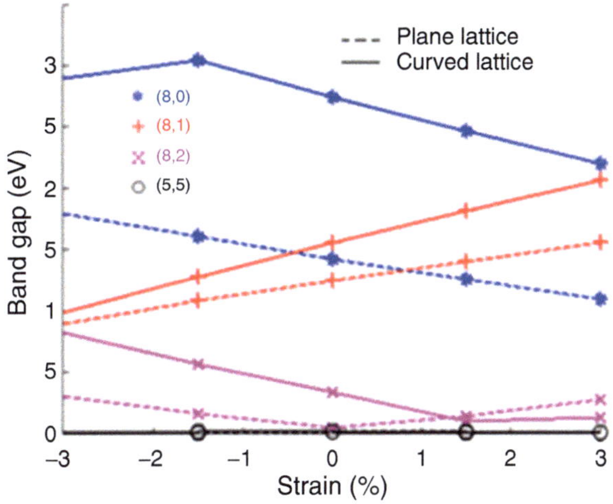

Fig. 7.19 The effect of wall curvature on the band gap variation

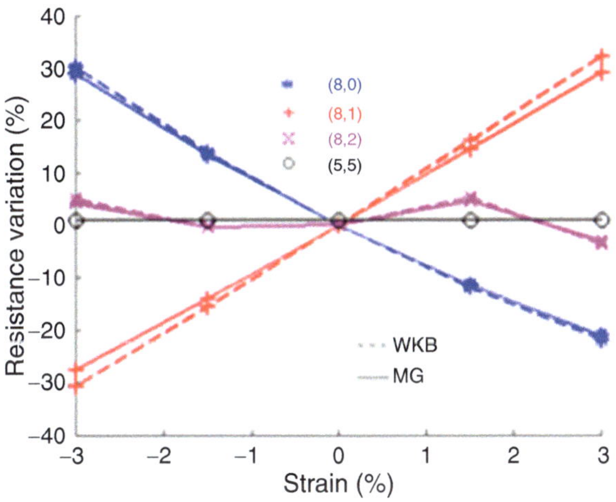

Fig. 7.20 Resistance variation due to mechanical deformation

7.2.4.2 Electrical Resistance

Both WKB and MG methods have been used in this study. Strain is applied on nanotubes of various geometries, taking into account both Poisson's effect and the wall curvature. Figure 7.20 depicts the variation of the electrical resistance due to mechanical deformation for the same nanotubes studied earlier in this work.

It is noticeable that although the resistance predictions of the two methods are not the same, the predictions for the resistance variations are almost identical. Additionally, it is clear that the predicted response follows in general the behavior of the band gap variation; this is actually expected since the band gap expresses the difficulty for the electrons to "flow" along the graphitic lattice.

7.3 Piezoresistive Properties of CNT-Doped Polymers

7.3.1 Introduction

Although carbon nanotubes exhibit these unique properties, they can't be individually exploited for the development of strain sensors due to technical difficulties. It is noticeable that much effort is spent for the development CNT manipulation systems (Liu et al. 2008; Deng et al. 2006). For this reason, carbon nanotubes are used mainly as reinforcement of inferior materials.

Polymeric and ceramic matrices usually exhibit non-conductive behavior since their conductivity does not exceed 10^{-10} S/m. Dispersion of a conductive material into an insulator can lead to the synthesis of a conductive composite material. The electrical properties of the composite depend obviously on the concentration of the conductive phase. At low content, the composite conductivity is similar to the one of pure matrix; however, after some critical threshold content, conductivity dramatically increases by orders of magnitude. This phenomenon has been attributed to the formation of conductive paths inside the insulating matrix; this phenomenon is well described by the *Percolation Network Theory* (Kesten 1982; Grimmett 1989). The formation of conductive networks highly depends on the geometry of the conductive dopants. The small size and high aspect ratio of the nanotubes contribute to keeping the threshold content to a minimum (Sandler et al. 2003; Moniruzzaman and Winey 2006).

When a CNT composite is subjected to mechanical loading, part of the load is transferred to the nanotubes. As proved earlier, mechanical deformation significantly changes their electrical properties; thus, the resistance of the nanocomposite is expected to change under mechanical deformation. This electromechanical coupling resembles the behavior of widely used piezoresistive sensors. Therefore, CNT composites are excellent candidates for the fabrication of a new generation of piezoresistive sensors (Sinha et al. 2006; Pham et al. 2008; Park et al. 2008).

The coupled piezoresistive response of nanocomposites occurs mainly due to two mechanisms:

- the effect of mechanical loading on the nanotubes properties and
- the effect of mechanical deformation on the conductive nanotube networks. Nanotube dislocations may lead to the formation of new conductive paths or distortion of existing ones.

The dominant mechanism is not currently known. On the contrary, numerous studies have supported one or the other (Dang et al. 2007; Alig et al. 2008; Dharap et al. 2004; Grow et al. 2005).

For the study of conductive networks there are various approaches that, however different, are based on the same principle:

- Initially, a representative volume (or unit cell) of the composite is considered having a finite number of dispersed nanotubes. Periodic Boundary Conditions

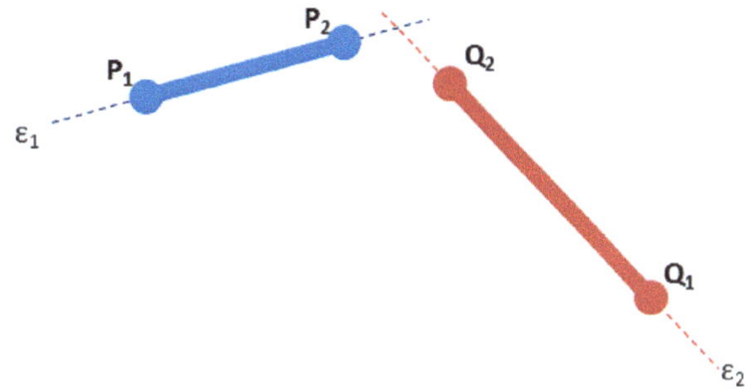

Fig. 7.21 Nanotubes represented as line segments

(PBCs) are applied to ensure that the properties of the composite are reflected in the properties of the representative volume.

- Then the unit cell is checked for nanotube *clusters*, that is groups of interconnected nanotubes. When a cluster extends through the whole unit cell, it is assumed that it extends through the whole composite due to the applied PBCs, which implies a conductive nature for the composite.

The variation of electrical properties of individual carbon nanotubes has been studied previously. However, at microscale (ply level) it is practically impossible to perform an analysis like this because a nanocomposite should contain a very high number of dispersed carbon nanotubes. Various approaches have been implemented to alleviate this issue; an effective material response is obtained based on experimental observations of the resistance variations due to mechanical loading (Loh et al. 2007; Kempel and Schlarb 2008).

7.3.2 *Conductive Networks*

The problem of contact between nanotubes in 3D space is not so simple; there are two equations – one for each line segment – but three unknown coordinates. Thus, a more advanced approach is required.

7.3.2.1 Geometrical Representation of Carbon Nanotubes

The main difference among the various approaches in literature is the geometrical representation of nanotubes. The simplest representation considers nanotubes as line segments arranged in 2D space (Pike and Seager 1974). According to this approach each nanotube is represented as a line segment of finite length and is placed in a random position with random orientation. The cluster identification is based on scanning for segment intersections (Fig. 7.21).

More advanced approaches consider (a) arcs of finite length and radius, (c) high order polynomials etc. All these work well for the study of nanocomposite films (Kumar et al. 2008; Ural et al. 2009), where due to small thickness, nanotubes are practically arranged in a 2D plane. A more realistic approach should require a 3D representation.

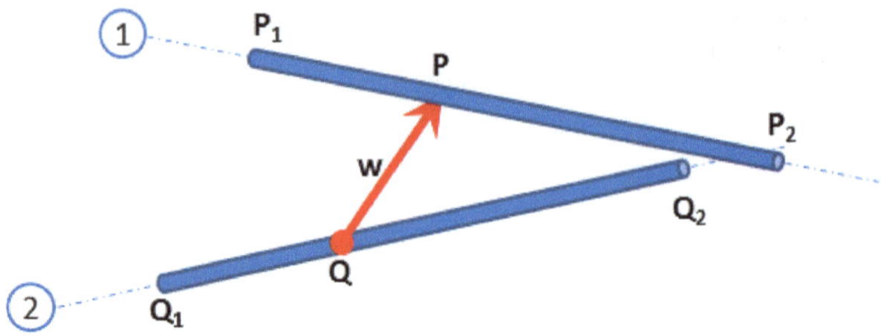

Fig. 7.22 Geometrical representation of carbon nanotubes in 3D space

In the present study, the analysis considers nanotubes as 3D line segments defined by the points $\mathbf{P}_1$, $\mathbf{P}_2$ and $\mathbf{Q}_1$, $\mathbf{Q}_2$ respectively (Fig. 7.22). Any random point of each nanotube is given by:

$$\mathbf{P}(t) = \mathbf{P}_1 + t \cdot (\mathbf{P}_2 - \mathbf{P}_1) \equiv \mathbf{P}_1 + t \cdot \mathbf{p}, \quad t \in [0, 1] \tag{7.42}$$

$$\mathbf{Q}(s) = \mathbf{Q}_1 + s \cdot (\mathbf{Q}_2 - \mathbf{Q}_1) \equiv \mathbf{Q}_1 + s \cdot \mathbf{q}, \quad s \in [0, 1] \tag{7.43}$$

where $\mathbf{p} = \mathbf{P}_2 - \mathbf{P}_1$ and $\mathbf{q} = \mathbf{Q}_2 - \mathbf{Q}_1$.

7.3.2.2 Nanotube Contact

The vector $\mathbf{w}$ that connects the two random points will obviously be:

$$\mathbf{w}(t, s) = \mathbf{P}(t) - \mathbf{Q}(s) \tag{7.44}$$

Given that the shortest distance of a point from a line is the perpendicular one, the distance of $\mathbf{P}$ from line "2" will be minimum when $\mathbf{w}$ is perpendicular to line "2". In the same way, the distance of $\mathbf{Q}$ from line "1" will be minimum when $\mathbf{w}$ is perpendicular to line "1". Thus, the points $\mathbf{P}_C = \mathbf{P}(t_C), \mathbf{Q}_C = \mathbf{Q}(s_C)$, for which the distance of the two nanotubes becomes minimum, will define the vector:

$$\begin{aligned} \mathbf{w}_C &= \mathbf{w}(t_C, s_C) = \mathbf{P}(t_C) - \mathbf{Q}_C(s) \\ &= (\mathbf{P}_1 - \mathbf{Q}_1) + t_C \cdot (\mathbf{P}_2 - \mathbf{P}_1) - s_C \cdot (\mathbf{Q}_2 - \mathbf{Q}_1) \end{aligned} \tag{7.45}$$

or simply:

$$\mathbf{w}_C = \mathbf{r} + t_C \cdot \mathbf{p} - s_C \cdot \mathbf{q}, \quad \begin{cases} \mathbf{r} = \mathbf{P}_1 - \mathbf{Q}_1 \\ \mathbf{p} = \mathbf{P}_2 - \mathbf{P}_1 \\ \mathbf{q} = \mathbf{Q}_2 - \mathbf{Q}_1 \end{cases} \tag{7.46}$$

The coefficients t_C, s_C are obtained by solving the system:

$$\left.\begin{aligned}\mathbf{w}_C\cdot\mathbf{p}=0\\ \mathbf{w}_C\cdot\mathbf{q}=0\end{aligned}\right\}\rightarrow\left.\begin{aligned}\mathbf{r}\cdot\mathbf{p}+t_C\cdot\mathbf{p}\cdot\mathbf{p}-s_C\cdot\mathbf{q}\cdot\mathbf{p}=0\\ \mathbf{r}\cdot\mathbf{q}+t_C\cdot\mathbf{p}\cdot\mathbf{q}-s\cdot\mathbf{q}\cdot\mathbf{q}=0\end{aligned}\right\}$$
$$\rightarrow\left.\begin{aligned}(\mathbf{p}\cdot\mathbf{p})t_C-(\mathbf{q}\cdot\mathbf{p})s_C=-\mathbf{r}\cdot\mathbf{p}\\ (\mathbf{p}\cdot\mathbf{q})t_C-(\mathbf{q}\cdot\mathbf{q})s=-\mathbf{r}\cdot\mathbf{q}\end{aligned}\right\} \tag{7.47}$$

The coefficients of the system can be expressed as:

$$a=\mathbf{p}\cdot\mathbf{p}\quad b=\mathbf{p}\cdot\mathbf{q}\quad c=\mathbf{q}\cdot\mathbf{q}\quad d=\mathbf{r}\cdot\mathbf{p}\quad e=\mathbf{r}\cdot\mathbf{q} \tag{7.48}$$

Then, the solution of the system will be:

$$t_C=\frac{b\cdot e-c\cdot d}{a\cdot c-b^2}\quad s_C=\frac{a\cdot e-b\cdot d}{a\cdot c-b^2} \tag{7.49}$$

with $t_C, s_C\in[0,1]$.

The denominator

$$a\cdot c-b^2=p^2\cdot q^2-[p\cdot q\cdot\cos(\mathbf{p},\mathbf{q})]^2=p^2.q^2\cdot\left[1-\cos^2(\mathbf{p},\mathbf{q})\right] \tag{7.50}$$

is always non-zero, unless the two nanotubes are perfectly parallel to each other.

Following this approach the points for minimum distance can be determined in computationally very efficient way. According to experimental observation, the minimum distance is not required to be zero, in order to consider the nanotubes in contact; on the contrary, if two nanotubes are separated by distance $d_{\min}\leqslant 1nm$ (Balberg 1987; Simoes et al. 2009), a conductive network can still be formed.

To sum up, this geometrical technique has two important advantages against classical plane analysis: (a) it can identify intersection points in 3D space using only the two line equations – i.e. it can solve a system of two equations with three unknowns – and (b) it can take into account tunneling phenomena among nanotubes, according to which direct contact among nanotubes is not required for the formation of conductive nanotube networks.

7.3.2.3 Numerical Procedure

The following numerical procedure can be employed for the study of conductive networks:

1. Initially, a representative volume is defined, termed as the *Simulation Box*, and periodic boundary conditions are applied along all three dimensions. The dimensions of the Simulation Box are in general in the order of the nanotube length, but this may change subject to the required accuracy.
2. A finite number of nanotubes are randomly placed into the Simulation Box. The number of nanotubes is correlated to a specific concentration.

3. The random system is checked for "infinite clusters", i.e. clusters that expand through the whole Simulation Box. Steps 2 and 3 are denoted as a *Throw*.
4. A high number of Throws takes place for each nanotube content and the probability for the existence of at least one infinite cluster is calculated as the ratio of the Throws with at least one infinite cluster over the total number of Throws. This probability is widely known as *Percolation Probability*:

$$p = \frac{N_{perc}}{N_{total}} \tag{7.51}$$

5. The number of nanotubes is increased – and correlated to a higher CNT content – and steps 2–5 are repeated for a predefined range of nanotube concentrations.
6. Finally, results are statistically studied; considering the needs of this work, the percolation probability and the average number of infinite clusters vs. the CNT content are pursued, $p = p(v_{CNT})$ and $C = C(v_{CNT})$.

It should be noted that the exact number of Throws is not strictly defined, but is rather a matter of trial and error. However, as the number of Throws increases, the percolation curves become smoother and after a threshold the number of Throws has practically no affect on them. In this work the convergence threshold is identified around 800 Throws.

7.3.2.4 Cluster Identification

The identification of clusters can be performed optically very easily. However, the high number of Throws is forbidding, thus, a computational procedure is required. This procedure is summarized as follows:

1. Each nanotube placed into the Simulation Box is identified by a unique number.
2. Nanotubes are examined in pairs for contact. The contact point must be inside the Simulation Box.
3. If two nanotubes are in contact, they are given a common identification number.
4. After all nanotubes have been checked and their identification numbers have been appropriately modified, the identification numbers of the nanotubes crossing each face of the Box are listed. If the same identification number is found in two opposite faces of the Simulation Box, then there is a cluster expanding through the whole Box, that is an infinite cluster.

Figure 7.23 depicts the identification of infinite clusters in plane and three-dimensional problems.

7.3.2.5 Predictions and Validations

The study of totally random and homogeneous dispersion leads to diagrams like Fig. 7.24.

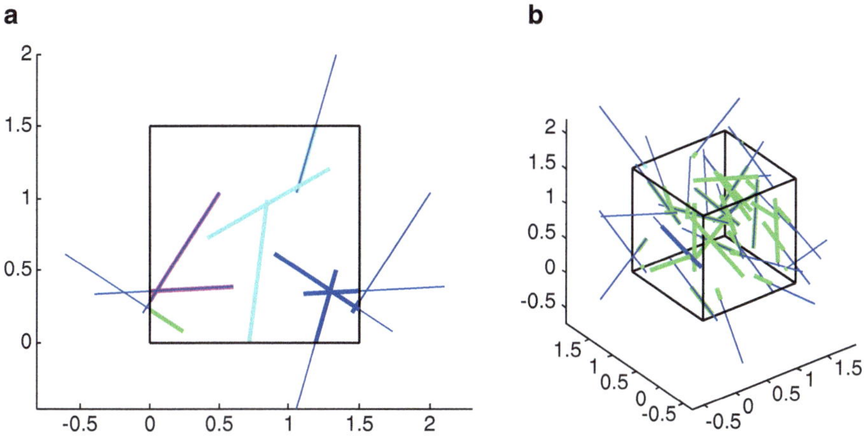

Fig. 7.23 Schematic representation of infinite clusters in (**a**) 2D and (**b**) 3D

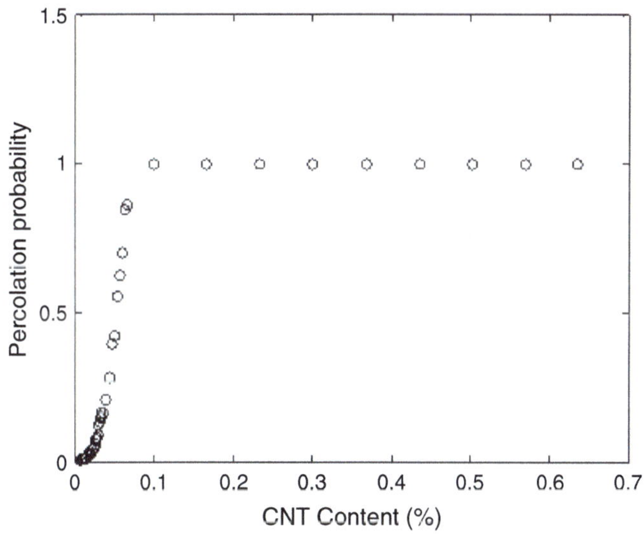

Fig. 7.24 Percolation probability vs. CNT content

It is clear that for CNT content around 0.06%v/v the percolation probability dramatically increases, while for values over 0.1% its value is always 1, i.e. the composite exhibits conductive behavior. This critical value is termed as the *Percolation Threshold*. If compared to other studies, the predictions of the introduced models seem to be realistic (Table 7.5).

Table 7.5 Prediction for percolation

Study	Percolation threshold (% v/v)
Present	0.06–0.10
Ramasubramaniam and Chen (2003)	0.01–0.04
Lu and Mai (2009)	0.11–0.24
Thostenson et al. (2009)	0.07
Chang et al. (2009)	0.19–0.37
Zhao et al. (2009)	0.17–0.43

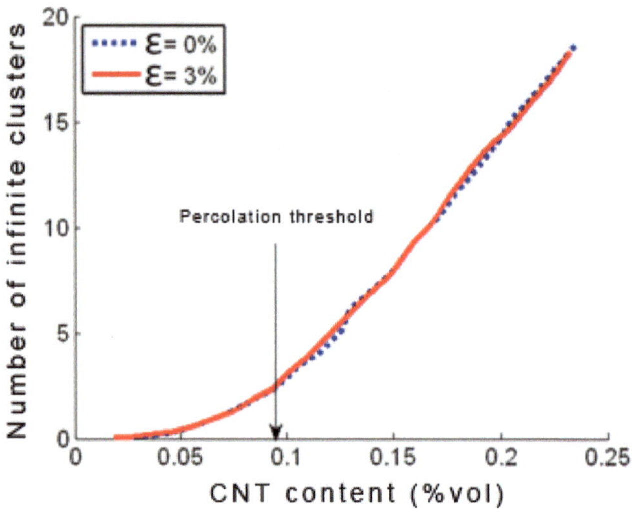

Fig. 7.25 Effect of strain on the average number of conductive paths

7.3.2.6 Effect of Strain

For the study of strain effects, the very same procedure is followed with the difference that, after each Throw, strain is incrementally applied and the identification of clusters takes place after the rearrangement of nanotubes. The variation of the average number of conductive paths inside the Simulation Box is depicted in Fig. 7.25.

Strain 3% is considered to be an extreme value, since failure occurs at much less strain in usual materials (Barber et al. 2003). Thus, even if linear elastic behavior was feasible till $\varepsilon = 3\%$, the nanotube dislocations are inadequate to induce any significant changes on the conductive network.

This conclusion is, however, not valid for all materials. A special case has been studied with nanotubes partially aligned along the loading axis assuming high strains. This configuration resembles various novel nanocomposites like *Carbon Nanotube Fiber (CNFs)* (Koziol et al. 2007), *Carbon Nanotube Carpets (CNCs)*

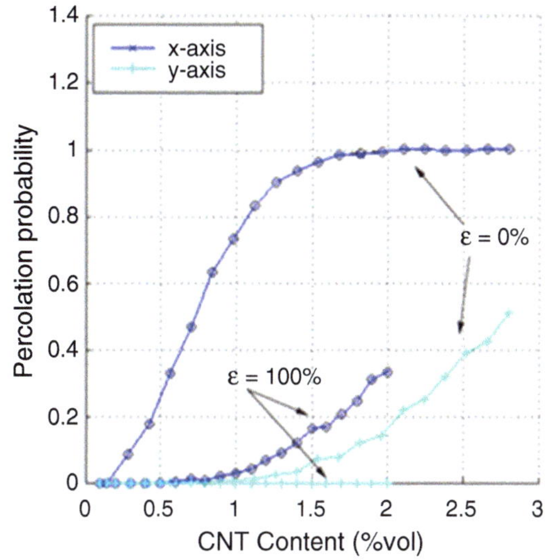

Fig. 7.26 Large strain effects on materials with partially aligned composites

(Kang et al. 2009; Nessim et al. 2008) etc. Figure 7.26 depicts the behavior of these materials. In this case, it is clear that large strains can cause significant distortions on the conductive network. The strain range that was inspected, although very high, is feasible for viscous materials; this behavior has been experimentally verified for CNT/PVA fibers (Alexopoulos et al. 2010). But since the structure and content of the tested specimens are not known, the correlation can currently be only qualitative.

In general, it can be said that for materials with good cohesion between the matrix and the nano-dopant, mechanical loading doesn't seem to affect the structure of conductive networks. This fact implies that the electromechanical coupling comes from the piezoresistive nature of nanotubes.

7.3.3 Effective Response

7.3.3.1 Effective Response of Nanotubes

It has already been demonstrated that the exact response of a nanotube depends on its electronic band structure and consequently its geometry. It is shown in Fig. 7.27 that electrical resistance can increase or decrease with strain. Based on this the effective response of carbon nanotubes can be obtained as an average response of an adequate number of nanotubes.

The study of every possible geometry is practically not possible, thus nanotubes are categorized based on their diameter, which is actually the only quantity that can be directly measured. Following this approach and assuming an average diameter around 2 nm for the chosen nanotubes, the effective response depicted in Fig. 7.28 has been identified.

Fig. 7.27 Resistance variation vs. strain

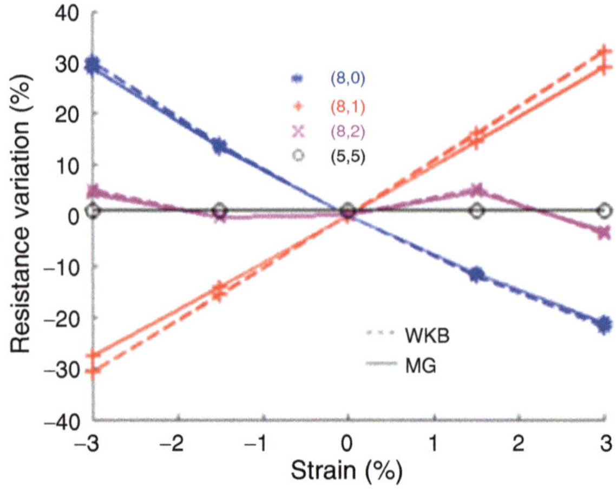

Table 7.6 Nanotubes of diameter 2 ± 0.01 nm

	n	m	d (nm)
1	19	10	1.998
2	22	6	1.999
3	24	3	2.007
4	25	1	1.998

This response has been identified by implementing the following procedure:

1. An average diameter is selected. In this example the diameter is 2 ± 0.01 nm. These values should be correlated to experimental measurements.
2. Nanotubes of diameter within the specified range are identified. For the mentioned values $d = 2 \pm 0.01nm$, the corresponding nanotubes are the ones enlisted in Table 7.6.
3. Each nanotube is incrementally loaded and for every strain value the band structure and resistance are calculated.
4. The average resistance value is calculated for every loading step.

It has to be noted that the diagrams in Fig. 7.28 correspond to the nanotubes in Table 7.6; selection of other nanotube configurations could lead to totally different results.

The strain-resistance variation diagram seems to be almost linear, i.e.

$$\frac{\Delta R}{R_0} = c_R \bar{\varepsilon} \tag{7.52}$$

where $\bar{\varepsilon}$ is the applied strain. ΔR is the resistance variation and R_0 the resistance of the undeformed nanotubes. The coefficient c_R is in fact the slope of the diagram in Fig. 7.28.

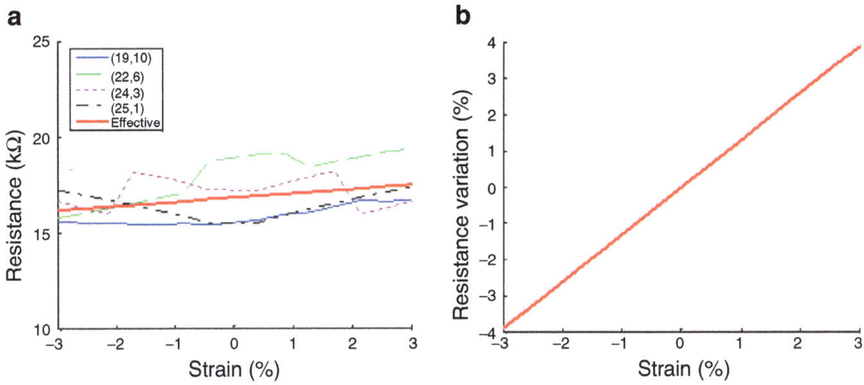

Fig. 7.28 Identification of an effective response: (**a**) electrical resistance of individual nanotubes; (**b**) variation of resistance vs. strain

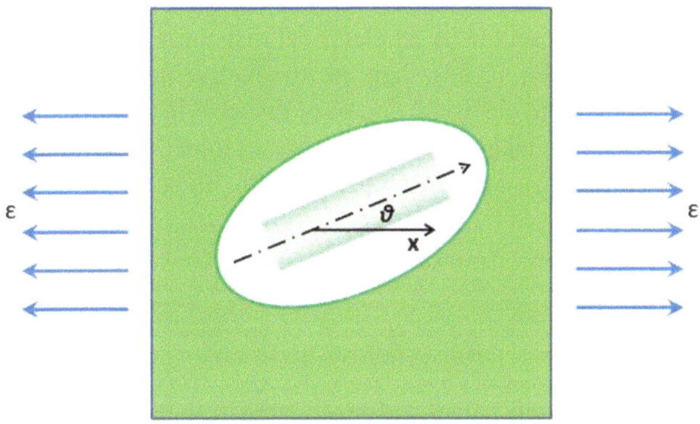

Fig. 7.29 Strain applied on randomly oriented nanotube

Considering the nanotube orientation, they are practically never fully aligned along the loading axis, thus, the load transferred to the nanotubes should not be equal to the load applied on the composite. Taking into account only the axial deformation of nanotubes:

$$\bar{\varepsilon} = \varepsilon \cdot \cos^2\varphi \tag{7.53}$$

where ε is the uniaxial strain and φ the angle between the axis of the nanotubes and the loading axis (Fig. 7.29).

Finally, the resistance variation of nanotubes will be:

$$\frac{\Delta R}{R_0} = c_R \cdot \varepsilon \cdot \cos^2\varphi \tag{7.54}$$

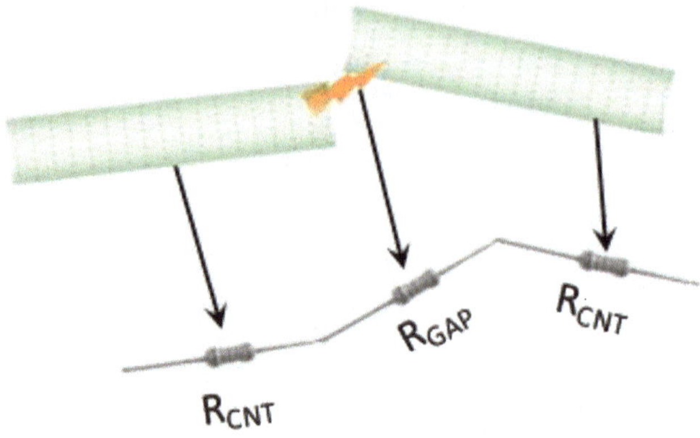

Fig. 7.30 Schematic representation of a nanotube system and its equivalent electrical circuit

7.3.3.2 Effective Response of Conductive Networks

If every nanotube is represented by an electrical resistor (R_{CNT}) whose value varies with strain according to Eq. (7.54), then the conductive network of nanotubes can be thought as an equivalent electrical circuit of resistors in parallel and in series. An additional term should be taken into account due to tunneling effects between non-contacting nanotubes (R_{GAP}). This representation is schematically depicted in Fig. 7.30.

Generally, the resistance variation should come from the contribution of both mechanisms:

$$\Delta R_{tot} = \Delta R_{CNT} + \Delta R_{GAP} \tag{7.55}$$

Since the total variation is calculated as superposition of all individual terms, the variation of each term can be studied independently.

7.3.3.3 Nanotube Resistance

Assuming the nanotubes as a circuit of resistors (R_i) in series, the total resistance will be:

$$R_{tot} = \sum_{i=1}^{N} R_i \tag{7.56}$$

The corresponding resistance variation of the conductive network will be:

$$\frac{\Delta R_{tot}}{R_{tot}^0}(\varepsilon) = \frac{\sum_{i=1}^{N} \left[R_i(\varepsilon) - R_i^0\right]}{\sum_{j=1}^{N} R_j^0} \tag{7.57}$$

where the exponent "0" denotes resistance at the unloaded state. Due to Eq. (7.54):

$$R_i(\varepsilon) - R_i^0 = \Delta R_i = R_i^0 \cdot c_R \cdot \varepsilon \cdot \cos^2\varphi_i \tag{7.58}$$

thus, Eq. (7.57) becomes:

$$\frac{\Delta R_{tot}}{R_{tot}^0} = c_R \cdot \varepsilon \cdot \frac{\sum_{i=1}^{N} R_i^0 \cdot \cos^2\varphi_i}{\sum_{j=1}^{N} R_j^0} \tag{7.59}$$

7.3.3.4 Gap Resistance

Considering the gap resistance (R_{GAP}) there is currently no well-established model. For the needs of this analysis, a cut-off function has been used, inspired from the commonly used cut-off function on molecular interactions, according to which the conductivity between two nanotubes smoothly decreases as a function of their intermediate distance (d) from an initial value G_0, when they are in contact, to 0:

$$f_C(d) = \frac{1}{2}\left[1 + \cos\left(\pi \frac{d}{d_{eff}}\right)\right] \tag{7.60}$$

where d_{eff} is the 1 nm limit introduced earlier and obtained from experimental observations. Following this approach, the electrical conductivity between two nanotubes will be:

$$G(d) = G_0 \cdot f_C(d) \tag{7.61}$$

and if strain is taken into account:

$$G(d, \varepsilon) = G \circ d(\varepsilon) = G_o \cdot f_C \circ d(\varepsilon)$$
$$\text{with } d(\varepsilon) = d_0 \cdot (1 + \varepsilon)$$

where d_0 is the initial distance between the nanotubes (Fig. 7.31). The respective resistance value can be obtained with inversion of Eq. (7.62):

$$R(d, \varepsilon) = \frac{1}{G \circ d(\varepsilon)} = \frac{R_0}{f_C \circ d(\varepsilon)} \tag{7.63}$$

where $R_0 = \frac{1}{G_0}$.

Since R_0, G_0 are generally not known, the resistance variation is used:

$$\frac{\Delta R}{R} = \frac{\frac{R_0}{f_C \circ d(\varepsilon)} - \frac{R_0}{f_C(d_0)}}{\frac{R_0}{f_C(d_0)}} = \frac{f_C(d_0)}{f_C \circ d(\varepsilon) - f_C(d_0)} \tag{7.64}$$

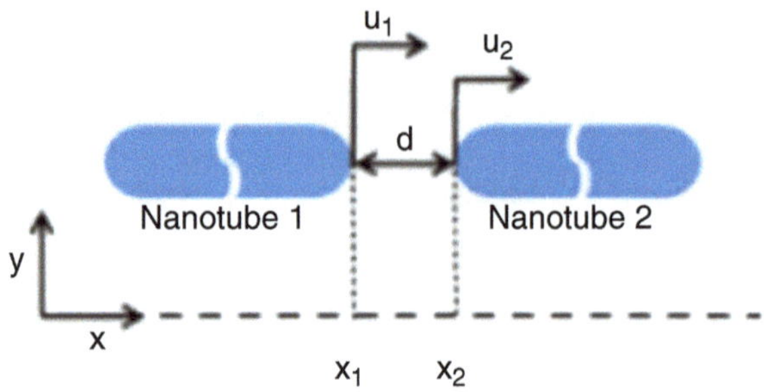

Fig. 7.31 A simple system of separated nanotubes

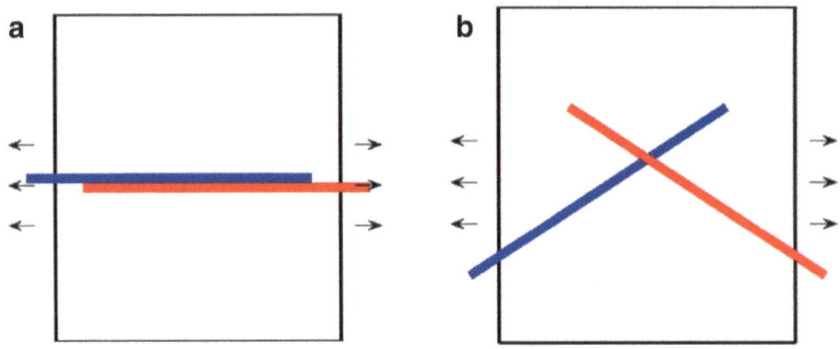

Fig. 7.32 Nanotubes in contact (**a**) perfectly aligned along the loading axis (**b**) randomly oriented

7.3.3.5 Predictions

For the identification of the contribution of each mechanism, various ideal cases have been studied. The most simple case is the one in which two nanotubes are perfectly aligned (Fig. 7.32a). The nanotubes are constantly in contact forming a circuit of resistors in series. Thus, the total resistance varies according to Eq. (7.59), that is linearly with strain as in Fig. 7.28b. The linear response implies that the slope of the diagram can be used as an indication of the impact of strain on the electrical resistance.

For the estimation of this impact, a more complicated system is required; the nanotubes are still constantly in contact, but not aligned along the loading axis (Fig. 7.32b). For simplicity the two nanotubes are assumed to be symmetrically aligned, so that simpler and more "readable" diagrams are obtained. In this case, the total resistance variation depends on the initial nanotube angle, because different orientation leads to different loading.

Figure 7.33 depicts the normalized variation of resistance as a function of the nanotubes orientation. It is clear that the more oriented the nanotubes are, the more sensitive to strain they appear to be. This fact leads to the conclusion that, for the design of strain nanosensors, materials with highly oriented nanotubes should be preferred, such as Carbon Nanofibers (Los Alamos National Laboratory 2006), and Nanocarpets (Laboratoire Francis Perrin CNRS 2004) introduced earlier.

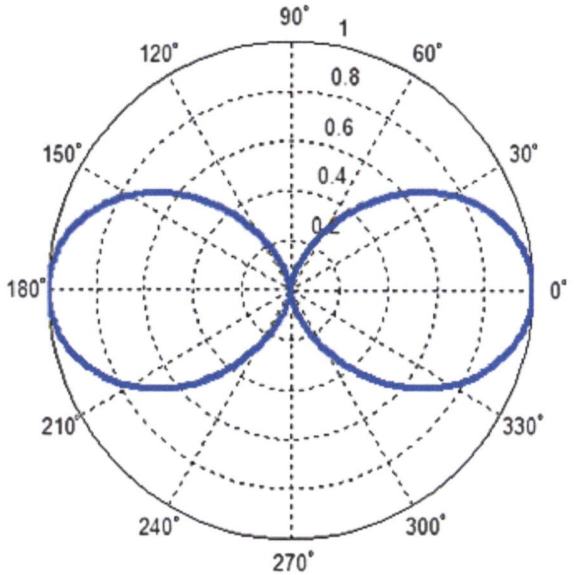

Fig. 7.33 Normalized resistance variation for misaligned nanotubes

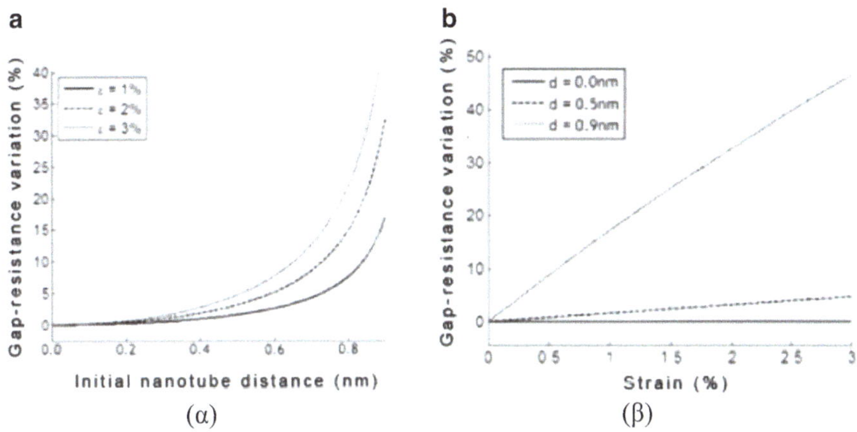

Fig. 7.34 Variation of the Gap Resistance as a function of (**a**) the initial distance, (**b**) strain

Finally, considering nanotubes not directly in contact, the total resistance variation is determined by Eq. (7.64). Depending on the initial nanotube distance, the effects of strain are more or less intense (Fig. 7.34).

An insulating material exhibits conductive behavior as the content of conductive dopants exceeds the critical value of percolation threshold. As the content increases, the average distance between nanotubes decreases, as more nanotubes are dispersed into the same space. This practically means that the more the CNT content the less the impact of strain. Thus, if the percolation threshold is significantly exceeded, the Gap Resistance should be negligible and the response of the materials will be determined only by Eq. (7.59), i.e. the dominant mechanism is the piezoresistive nature of nanotubes.

On the contrary, for materials that undertake large strains and displacements, and knowing that the matrix-nanotube cohesion is preserved only for small strains (Barber et al. 2003), as strain increases, interphase failures are observed and the dominant mechanism seems to be the distortion of conductive networks. However, these phenomena are very different from the ones examined here and exceed by far the scope of this work.

7.3.3.6 Correlation with Experimental Data

The difficulty with the correlation of predictions to experimental data comes from the following factors:

- Usual synthesis methods produce random geometries, and the nanotubes practically always contain impurities and defects. This diverges from the ideal structures assumed during the tight-binding analysis and should lead to the prediction of smaller resistance variations.
- Due to cost, most CNT nanocomposites are fabricated nowadays using multi-wall nanotubes. It has been proved (Tian and Datta 1994; Saito et al. 1993), that the interactions among concentric nanotubes affect only a small percentage of the total response. This means that each multi-wall nanotube can be treated as a circuit of resistors in parallel. Implementation of Kirchhoff's laws means that the resistance variation of a multi-wall nanotube should be the same as that of a single-wall nanotube as studied in this work.
- The nanotubes dispersed into a specimen have various diameters that can be only statistically determined. Depending though on the diameter, a different strain-resistance diagram is obtained.

Correlations with experimental data take place using results of NOESIS (FP6/AEROSPACE), because a very precise description of materials and fabrication process are available to the authors. According to the manufacturer, the nanotubes used have a diameter in the range 1.0–1.5 nm. Following the introduced methodology for the two extreme values, the dashed and dotted lines (Fig. 7.35) are respectively obtained.

Assuming statistically homogeneous distribution of diameters, the solid line represents the effective response of the specimen.

Correlation of predictions with experimental values is shown in Fig. 7.36. It clear that for small strains, the predictions are reasonably good considering the assumptions made.

As strain increases, the theoretical and experimental values seem to diverge. This is due to the following reasons: (a) the structure of nanotubes is not ideal,

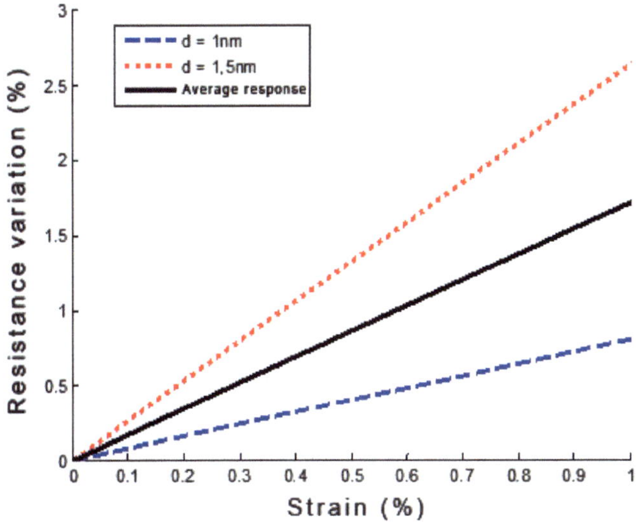

Fig. 7.35 Predictions for the response of a nanocomposite specimen

Fig. 7.36 Correlations with experimental measurements

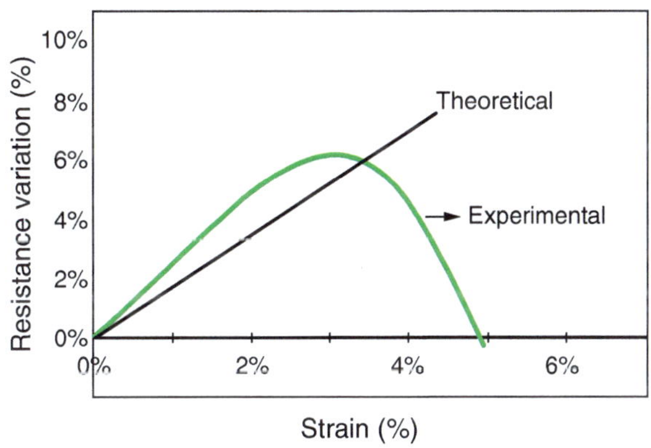

(b) dispersion is practically never perfect, (c) statistical distribution of diameters is not known, (d) local failures at the matrix-dopant interphase are not taken into account, (e) the geometrical change of the specimen is not accounted for etc. However, having in mind that predictions have been made from atomistic level to microscale with no calibrations, the results can be considered as realistic.

The change of the material response after 3% strain and its smooth decrease cannot be explained by the current methodology, nor has it been experimentally explained yet (Vavouliotis 2009); and it would be an interesting extension of this work.

7.4 Summary

In the context of the present research findings, only some nanotube properties have been investigated, but it is clear that they have great potential in modern technology. The predictions presented here cannot, of course, describe every aspect of nanotubes behavior, but they can be used as a basis for an explanation of coupling effects in various material scales, as well as guidelines for the design of innovative materials.

The prediction of mechanical response has been based on the assumption that atomic interactions only within a finite range are necessary to be included in calculations. Predictions are in agreement with theoretical and experimental studies available in the open literature. Future extensions of this work should focus on a detailed description of transient phenomena like progressive damage, buckling etc.

The electrical properties of nanotubes have been successfully correlated to their geometry. The spatial arrangement of atoms determines the electronic interactions and consequently the macroscopically measured electrical resistance. Application of mechanical strain leads to prediction of the coupled electromechanical response. The basic assumptions are: (a) nanotubes are structurally perfect and (b) load is uniformly distributed along the nanotube.

After completion of the previous stage of modeling, nanoscopic models are transferred to microscale. Again it is assumed that nanotubes behave ideally while their mixing with the resin matrix is homogeneous. Although these conditions are usually not met due to various technological limitations, a general conclusion has been obtained: the behavior of materials with good cohesion between the matrix and the dopant is determined mostly by the electromechanical nature of nanotubes, while in materials that can undertake large strains, the dominant mechanism seems to be the modification of microstructure. This fact implies that each material should be treated in a "per case" basis according to its synthesis and microstructure. The two introduced mechanisms are considered to be dominant, however, correlation to experimental data implies that additional phenomena exist. This is quite obvious in Fig. 7.36, where resistance decreases after 3% strain; this behavior has not been explained yet, not even experimentally, thus, it would be a very interesting extension of the present study.

A general conclusion is that the developed numerical and analytical models are not limited in general theoretical conclusions, but they are in reasonably good agreement with experimental reports. Therefore, the modeling framework we have introduced could be used as a guideline for the design of advanced composites at molecular level. Of course, the mathematical formulation does not take into account every possible phenomenon, due to the assumptions made, but the present quantification seems to be adequate for a wide range of engineering applications.

Acknowledgments This research has been supported by the K. Karatheodori program (University of Patras) and the NOESIS project (EU FP6-Aerospace). The authors gratefully acknowledge this support.

References

Aksamija, Z., Ravaioli, U.: Boltzmann transport simulation of single-walled carbon nanotubes. J. Comput. Electron. **7**, 315–318 (2008)

Aksamija, Z., Mohamed, M.Y., Ravaioli, U.: Parallel implementation of Boltamann transport simulation of carbon nanotubes. In: 13th International Workshop on Computational Electronics – IWCE '09, Beijing, pp. 1–4 (2009)

Alexopoulos, N.D., Bartholome, C., Poulin, P., Marioli-Riga, Z.: Structural health monitoring of glass fiber reinforced composites using embedded carbon nanotube (CNT) fibers. Compos. Sci. Technol. **70**(2), 260–271 (2010)

Alig, I., Skipa, T., Lellinger, D., Potschke, P.: Destruction and formation of a carbon nanotube network in polymer melts: rheology and conductivity spectroscopy. Polymer **49**(16), 3524–3532 (2008)

Allen, M.P., Tildesley, D.J.: Computer Simulation of Liquids. Oxford University Press, New York (1989)

Arroyo, M., Belytschko, T.: An atomistic-based finite deformation membrane for single layer crystalline films. J. Mech. Phys. Solid **50**, 1941–1977 (2002)

Ashcroft, N.W., Mermin, N.D.: Solid State Physics. W. B. Saunders Company, Philadelphia (1976a)

Ashcroft, N.W., Mermin, N.D.: Solid State Physics. Holt - Saunders International, Ed. Holt, Rinehart and Winston, New York (1976b)

Avouris, P., Collins, P.G.: Nanotubes for electronics. Sci. Am. **283**, 38–45 (1998)

Avouris, P., et al.: Electrical properties of carbon nanotubes: spectroscopy, localization and electrical breakdown. In: Tomanek, D., Enbody, R.J. (eds.) Science and Applications of Nanotubes. Kluwer Academic Publishers Group, New York (2000)

Avouris, P., Radosavlevic, M., Wind, S.J.: Carbon nanotubes for electronics and optoelectronics. In: Applied Physics of Carbon Nanotubes, pp. 227–251. Springer, Berlin (2005)

Ayuela, A., Chico, L., Jaskolski, W.: Electronic band structure of carbon nanotube superlattices form first-principles calculations. Phys. Rev. B **77**, 085435 (2008)

Bagwell, R.F., Orlando, T.P.: Landauer's conductance formula and its generalization to finite voltages. Phys. Rev. B **40**, 3 (1989)

Balberg, I.: Tunneling and nonuniversal conductivity in composite materials. Phys. Rev. Lett. **59**, 1305–1308 (1987)

Barber, A.H., Cohen, S.R., Wagner, D.H.: Measurement of carbon-polymer interfacial strength. Appl. Phys. Lett. **82**, 4140–4142 (2003)

Bernholc, J., Brenner, D., Nardelli, M.B., Meunier, V., Roland, C.: Mechanical and electrical properties of nanotubes. Annu. Rev. Mater. Res. **32**, 347–375 (2002)

Bloch, F.: Über die Quantenmechanik der Elektronen in Kristallgittern. Zeithschrift für Physik **52**, 555 (1928)

Brenner, D.W.: Empirical potential for hydrocarbons for use in simulating the chemical vapor deposition of diamond films. Phys. Rev. B **42**, 9458–9471 (1990)

Bruus, H.: Introduction to Nanotechnology. Technical University of Denmark, Lyngby (2004)

Burkert, U., Allinger, N.L.: Molecular Mechanics. American Chemical Society, Washington, DC (1982)

Chang, T.-E., Kisluik, A., Rhodes, S.M., Brittain, W.J., Sokolov, A.P.: Conductivity and mechanical properties of well-dispersed single-wall carbon nanotube/polystyrene composites. Polymer **47**, 7740–7746 (2009)

Cox, H.L.: The elasticity and strength of paper and other fibrous materials. Br. J. Appl. Phys. **3**, 72–79 (1952)

Dang, Z.-M., Yao, S.-H., Xu, H.-P.: Effect of tensile strain on morphology and dielectric property in nanotube/polymer nanocomposites. Appl. Phys. Lett. **90**, 012907 (2007)

Datta, S.: Electrical resistance: an atomistic view. Nanotechnology **15**, S433 (2004)

Deng, Z., et al.: Nanotube manipulation with focused ion beam. Appl. Phys. Lett. **88**(2), 023119 (2006)

Dharap, P., Li, Z., Nagarajaiah, S., Barrera, E.V.: Nanotube film based on single-wall carbon nanotubes for strain sensing. Nanotechnology **15**, 379–382 (2004)

Durrant, A.: Quantum Physics of Matter. Taylor & Francis Ltd., Bristol (2000)

Fraysse, J., Minett, A.I., Jaschinski, O., Duesberg, G.S., Roth, S.: Carbon nanotubes acting like actuators. Carbon **40**(10), 1735–1739 (2002)

Gao, X.-L., Li, K.: A Shear-Lag model for carbon nanotube-reinforced polymer composites. Int. J. Solid Struct. **42**, 1649–1667 (2005)

Gartstein, Y.N., Zakhidov, A.A., Baughman, R.H.: Mechanical and electromechanical coupling in carbon nanotube distortions. Phys. Rev. B **68**(11), 115415 (2003)

Gibbs, J.W.: Elements of Vector Analysis, Arranged for the Use of Students in Physics. Yale University, New Haven (1881)

Gibbs J.W., Wilson, E.B.: Vector Analysis. Dover Publications, New York (reprint), 1902 (1960)

Griffiths, D.: Introduction to Quantum Mechanics, 2nd edn. Prentice Hall, Upper Saddle River (2004)

Grimmett, G.: Percolation. Springer, New York (1989)

Grow, R.J., Wang, Q., Cao, J., Wang, D., Dai, H.: Piezoresistance of carbon nanotubes on deformable thin-film membranes. Appl. Phys. Lett. **86**(9), 093104 (2005)

Guo, W., Guo, Y.: Giant electrostrictive deformation in carbon nanotubes. Phys. Rev. Lett. **91**, 115501 (2003)

Hamada, H., Sawada, S., Oshiyama, A.: New one-dimensional conductors: graphitic microtubules. Phys. Rev. Lett. **68**, 1579–1581 (1992)

Haque, A., Ramasetty, A.: Theoretical study of stress transfer in carbon nanotubes reinforced polymer matrix composites. Compos. Struct. **71**, 68–77 (2005)

Huang, K.: Statistical Mechanics. Wiley, New York (1987)

Hutter, K., Johnk, K.: Continuum Methods of Physical Modeling: Continuum Mechanics, Dimensional Analysis, Turbulence. Springer, Berlin/New York (2004)

Javey, A., et al.: High-field quasibalistic transport in short carbon nanotubes. Phys. Rev. Lett. **92**(10), 106804 (2004)

Kang, M.S., et al.: The fabrication of polyaniline/single-walled carbon nanotube fibers containing a highly-oriented filler. Nanotechnology **20**, 085701 (2009)

Kaxiras, E.: Atomic and Electronic Structure of Solids. Cambridge University Press, New York (2003)

Kempel, F., Schlarb, A.K.: Strain sensing with nanoscale carbon fiber – epoxy composites. Appl. Mech. Mater. **13–14**, 239–245 (2008)

Kesten, H.: Percolation Theory for Mathematicians. Birkhauser, Boston (1982)

Koziol, K., et al.: High-performance carbon nanotube fiber. Science **21**, 1892–1895 (2007)

Kumar, S., Murthy, J.Y., Alam, M.A.: Electrical and thermal transport in thin-film nanotube composites with applications to macroelectronics. Int. J. Nanomanuf. **2**, 226–252 (2008)

Laboratoire Francis Perrin CNRS: Aligned Carbon Nanotubes. [Online]. http://www-lfp.cea.fr/ast_visu.php?num=440&lang=ang (2004)

Landauer, R.: Spatial variation of currents and fields due to localized scatterers in metallic conduction. IBM J. Res. Dev. **1**, 223 (1957)

Levine, I.N.: Quantum Chemistry. Prentice Hall, Englewood Cliffs (1991)

Li, C., Chou, T.-W.: A structural mechanics approach for the analysis of carbon nanotubes. Int. J. Solid Struct. **40**(10), 2487–2499 (2003)

Liu, B., Jiang, J., Johnson, H.T., Huang, Y.: The influence of mechanical deformation on the electrical properties of single wall carbon nanotubes. J. Mech. Phys. Solid **52**(1), 1–26 (2004)

Liu, H., Chiashi, S., Ishiguro, M., Homma, Y.: Manipulations of single-walled carbon nanotubes with a tweezers tip. Nanotechnology **19**, 445716 (2008)

Loh, K.J., Lynch, J.P., Shim, B.S., Kotov, N.A.: Tailoring piezoresistive sensitivity of multilayer carbon nanotubes composite strain sensors. J. Intell. Mater. Syst. Struct. **19**, 747–764 (2007)

Los Alamos National Laboratory: NewsBulletin. [Online]. http://www.lanl.gov/news/index.php?fuseaction=nb.story&story_id=8900&nb_date=2006-08-31 (2006)
Lu, C., Mai, Y.W.: Anomalous electrical conductivity and percolations in carbon nanotube composites. J. Mater. Sci. **43**, 6012–6015 (2009)
Luo, X., Qian, G., Fei, W., Wang, E.G., Chen, C.: Systematic study of β-SiC surface structures by molecular-dynamics simulations. Phys. Rev. B **57**, 9234–9240 (1998)
Martin, R.M.: Electronic Structure: Basic Theory and Practical Methods. Cambridge University Press, Cambridge (2004)
Massimi, M.: Pauli's Exclusion Principle. Cambridge University Press, Cambridge (2005)
Minot, E.D.: Tuning the band structure of nanotubes. PhD thesis, Cornell University (2004)
Moniruzzaman, M., Winey, K.I.: Polymer nanocomposites containing carbon nanotubes. Macromolecules **39**, 5194–5205 (2006)
Morse, P.M.: Diatomic molecules according to the wave mechanics. II. Vibrational levels. Phys. Rev. **34**, 57–64 (1929)
Muller, P.: Glossary of terms used in physical organic chemistry (IUPAC recommendations 1994). Pure Appl. Chem. **66**(5), 1077–1184 (1994)
Nairn, J.A.: On the use of Shear-Lag methods for analysis of stress transfer in unidirectional composites. Mech. Mater. **26**, 63–80 (1997)
Nessim, G.D., et al.: Tuning of vertically-aligned carbon nanotube diameter and areal density through catalyst pre-treatment. Nanoletter **8**, 3587–3593 (2008)
Park, M., Kim, H., Youngblood, J.P.: Strain-dependent electrical resistance of multi-walled carbon nanotube/polymer composite films. Nanotechnology **19**, 055705 (2008)
Pham, G.T., Park, Y.-B., Liang, Z., Zhang, C., Wang, B.: Processing and modeling of conductive thermoplastic/carbon nanotube films for strain sensing. Compos. Part B Eng. **39**(1), 209–216 (2008)
Pike, G.E., Seager, C.H.: Percolation and conductivity: a computer study I. Phys. Rev. B **10**, 1421–1434 (1974)
Pozdnyakov, D.V., Galenchik, V.O., Komarov, F.F., Borzdov, V.M.: Electron transport in armchair single-wall carbon nanotubes. Phys. E **33**(2), 336–342 (2006)
Ramasubramaniam, R., Chen, J.: Homogeneous carbon nanotube-polymer composites for electrical applications. Appl. Phys. Lett. **83**, 2928–2930 (2003)
Rapaport, D.C.: The Art of Molecular Dynamics Simulation. Cambridge University Press, New York (1995)
Razavy, M.: Quantum Theory of Tunneling. World Scientific, River Edge (2003)
Reich, S., Thomsen, C., Ordejon, P.: Electronics band structure of isolated and bundled carbon nanotubes. Phys. Rev. B **65**, 155411 (2002)
Saito, R., Fujita, M., Dresselhaus, G., Dressehaus, M.S.: Electronic structure of graphene tubules based on C60. Phys. Rev. B **46**, 1804–1811 (1992)
Saito, R., Dresselhaus, G., Dresselhaus, M.S.: Electronics structure of double-layer graphene tubules. J. Appl. Phys. **73**(2), 494 (1993)
Saito, R., Dresselhaus, G., Dresselhaus, M.S.: Physical Properties of Carbon Nanotubes. Imperial College Press, London (1998)
Sandler, J.K.W., Kirk, J.E., Kinloch, I.A., Shaffer, M.S.P., Windle, A.H.: Ultra-Low electrical percolations threshold in carbon-nanotube-epoxy composites. Polymer **44**(19), 5893–5899 (2003)
Seidel, G.D., Lagoudas, D.C.: Micromechanical analysis of the effective elastic properties of carbon nanotube reinforced composites. Mech. Mater. **38**, 884–907 (2006)
Seidel, G.D., Lagoudas, D.C.: Micromechanical modeling of polymer nanocomposites for use as multifunctional models. In: Collection of Technical Papers: AIAA/ASME/ASCE/AHS/ASC Structures, Structures Dynamics, and Materials Conference, p. 1947 (2008)
Simoes, R., et al.: Low percolation transition in carbon nanotube networks dispersed in a polymeric matrix: dielectric properties and experiments. Nanotechnology **20**, 035703 (2009)

Sinha, N., Ma, J., Yeow, T.W.: Carbon nanotube-based sensors. J. Nanosci. Nanotechnol. **6**, 573–590 (2006)

Slater, J.C.: Wavefunction in a periodic potential. Phys. Rev. **51**, 846–851 (1937)

Slater, J.C.: An augmented plane wave method for the periodic potential problem. Phys. Rev. **92**, 603–608 (1953)

Slater, J.C., Koster, G.F.: Simplified LCAO method for the periodic potential problem. Phys. Rev. **94**, 1498–1524 (1954)

Stankovich, S., et al.: Synthesis of graphene-based nanosheets via chemical reduction of exfoliated graphite oxide. Carbon **45**, 1558–1565 (2007)

Suli, E., Mayers, D.F.: An Introduction to Numerical Analysis. Cambridge University Press, New York (2003)

Tersoff, J.: Empirical interatomic potential for carbon, with applications to amorphous carbon. Phys. Rev. Lett. **61**, 2879–2882 (1998)

Thostenson, E.T., Ziaee, S., Chou, T.-P.: Processing and electrical properties of carbon nanotube/vinyl ester nanocomposites. Compos. Sci. Technol. **69**, 801–804 (2009)

Tian, W., Datta, S.: Aharonov-Bohm-type effect in graphene tubules: a Landauer approach. Phys. Rev. B **49**, 5097–5101 (1994)

Ural, A.: Electronic properties of carbon nanotube percolations films and nanotube film-semiconductor junctions. In: Fullerenes, Nanotubes and Carbon Nanostructures – 215th ESC Meeting, San Francisco, CA, pp. 43–45 (2009)

Vavouliotis, A.I.: New approach in progressive damage monitoring under mechanical loading in fibrous carbon nanotube composites. PhD thesis, University of Patras, Patras (2009)

Wallace, P.R.: The band theory of graphite. Phys. Rev. **71**, 9 (1947)

Xiao, S.P., Belytschko, T.: A bridging domain method for coupling continua with molecular dynamics. Comput. Method Appl. Mech. Eng. **193**, 1645–1669 (2004)

Xiao, K.Q., Zhang, L.C.: The stress transfer efficiency of a single-walled carbon nanotube in epoxy matrix. J. Mater. Sci. **39**, 4481–4486 (2004)

Yang, L., Anantram, M.P., Han, J., Lu, J.P.: Band-gap change of carbon nanotubes: effect of small uniaxial and torsional strain. Phys. Rev. B **60**(19), 13874–13878 (1999)

Ypma, T.J.: Historical development of the Newton-Raphson method. SIAM Rev. **37**(4), 531–535 (1995)

Zhao, X., et al.: Spray deposited fluoropolymer/multi-walled carbon nanotube composite films with high dielectric permittivity at low percolation threshold. Carbon **47**, 561–569 (2009)

Zhou, C., Kong, J., Dai, H.: Intrinsic electrical properties of individual single-walled carbon nanotubes with small band-gaps. Phys. Rev. Lett. **84**, 5604–5607 (2000)

Chapter 8
Improved Damage Tolerance Properties of Aerospace Structures by the Addition of Carbon Nanotubes

Petros Karapappas and Panayota Tsotra

Contents

Abstract The potential use of carbon nanotubes (CNTs) in aerospace structures is considered in this chapter. Various studies are presented on how carbon nanotubes may be the driving force of a new generation of aerospace structures with superior damage tolerance properties, which in turn will lead to novel composite structures for the aerospace industry. This chapter examines the inclusion of CNTs in aerospace grade resins and their reinforcing mechanisms. The conclusion reached is that the main reinforcing mechanisms of carbon nanotubes are: fibre breakage, fibre pull-out, crack bridging and crazing. These are responsible for the improvement of the mechanical properties of composite materials and their structures. In other

P. Karapappas (✉)
Cytec Engineered Materials, LL13 9UZ Wrexham, UK
e-mail: petros.karapappas@cytec.com

P. Tsotra
Huntsman Advanced Materials, Basel 4057, Switzerland

A.S. Paipetis and V. Kostopoulos (eds.), *Carbon Nanotube Enhanced Aerospace Composite Materials*, Solid Mechanics and Its Applications 188,
DOI 10.1007/978-94-007-4246-8_8,

words, the use of carbon nanotubes in aerospace composite structures has been proven to increase fracture toughness, impact strength, post-impact properties and the fatigue life of composites, all these attributes making them more damage tolerant. Finally, a new generation of fibres and fabrics with CNTs grafted or grown on them are presented. They are expected to play a key role in evolution of aerospace composite structures, overcoming any processing issues that have risen due to high CNT-polymer viscosities.

Keywords Damage tolerance • Aerospace structures • Fracture • Fatigue • Impact

8.1 Introduction

In this chapter the potential use of carbon nanotubes in aerospace structures will be examined. First, a simple overview of aircraft structures will be given along with the differences of metallic and composite structures. It is highlighted that they should be treated differently when they are applied to an aircraft structure. Further on, the significance of the damage tolerance design approach and the fail-safe concept of airframe structural design are discussed. Arguments are then presented on how carbon nanotubes may be the driving force of a new generation of aerospace structures with superior damage tolerance properties, which in turn will lead to novel composite structures for the aerospace industry.

All airframes, whatever the aircraft, are designed using the same principles. The smooth exterior provides a streamlined shape, with extra supporting structure underneath to provide the strength and stiffness needed to operate effectively. In many modern aircraft, the covering and part of the framework are made from a single piece of material. The outer skin hides a complex piece of structure that must be strong, stiff and reliable. The modern aeronautical engineering of aircraft design has been an evolutionary process accelerated immensely in recent times from the demanding requirements for safety and the pressures of competitive economics in structural design. The aircraft structures are generally classified as follows:

- Primary-structure critical to the safety of the aircraft.
- Secondary-structure that, if it were to fail, would affect the operation of the aircraft but not lead to its loss.
- Tertiary-structure in which failure would not significantly affect operation of the aircraft.

The structure of most airframe components is made up of four main types of structural member. Ties are members subjected purely to tension. Because tension will not cause the tie to buckle, it does not need to be rigid, although it often is. Ties can be made from rigid items, such as tubes, or simply from wire, like the bracing wires on a biplane. Struts carry compression loads. Because compressive loads can cause the member to buckle, the design of a strut is less simple than a tie. If overloaded, struts will fail in one of two ways: a long, thin strut will buckle;

a short, thick strut will collapse by cracking or crushing, as the material from which it is made is overstressed. A medium strut may do either, or even both, depending on its dimensions and on other factors. Tubes make excellent struts, because the material is evenly loaded, so that the strength-to-weight ratio is high in compression. Beams carry loads at an angle (often at right angles) to their length, and so are loaded primarily in bending. Many of the major parts of an airframe are beams, such as the main spars. The fuselage and wings themselves are structural members, and are beams, because they support the bending loads imposed by weight, inertia and aerodynamic loads. Webs are thin sheets carrying shear loads in the plane of the material. Ribs and the skin itself are shear webs. Thin sheets are ideal for carrying shear, especially if they are supported so that they resist buckling. One may get the impression that each part of an airframe is either a tie or a strut or a beam or a web, but this is not so. Some items, such as wing spars, act almost entirely as one type of member, but others act as different members for different loads. For instance, the fuselage skin may be subjected to tensile and shear loads simultaneously. Pure bending loads almost never exist alone; they are almost always related to a shear load. Consequently a beam will normally carry both bending and shear loads. An aircraft designer nowadays would design for:

- Static ultimate and yield strength
- Fatigue life of the airframe (crack initiation and propagation)
- Static residual strength of damaged structure
- Fatigue life of damaged structure (inspection intervals)
- Thermal stress analysis and design (supersonic aircraft mainly).

The aircraft industry has for the past two decades spent considerable research and development effort to exploit the very attractive structural efficiencies achievable through the use of composite materials and composite structures. Lately, both Boeing and Airbus have focused their efforts on building the first commercial airliners where the usage of composite materials will be more than 50% of the total materials used i.e. the Boeing 787 Dreamliner and the Airbus A350 XWB. Composite materials offer substantial weight savings relative to current metallic structures. Furthermore, the number of parts required to build a composite component may be significantly less than the number of parts needed to construct the same component of metal alloy. In turn, this can lead to considerable labour saving, offsetting the somewhat higher cost of the present composite materials. These features, along with the inherent resistance to corrosion, make composites very attractive candidates for aircraft and aerospace structures. The adoption of composite materials for aircraft structures has been slower than originally foreseen, despite the weight-saving and corrosion and fatigue immunity offered by these materials. The reasons for the restrained use include the high cost of certification and higher materials and production costs for composite components. Composite structures must not be significantly more costly to acquire than those made of aluminium alloys and, to maintain the advantage of weight saving, maintenance costs also, must not be greater. Although a few inroads have been made in terms of reducing certification costs, recently more cost-efficient manufacturing methods have been developed, such as resin-transfer moulding and pultrusion, and improved resin and fibre systems that provide increased toughness are making composites very strong candidates for new designs. However, sensitivity to impact damage and low through-thickness strength are also inhibiting factors. Other issues are poor reliability in estimating development costs and difficulty in accurately predicting structural failure (Niu 1995; Baker 1988).

Initial attempts to certify composite structures simply adopted those requirements already existing for metals without recognizing the inherent differences between the two materials, even though these differences can significantly affect airworthiness considerations. For example, under static loading, composites typically exhibit linear elastic behaviour to failure and are extremely sensitive to stress concentrations. In contrast, metals, with a few rare exceptions, exhibit plastic behaviour above a yield stress and are not notch sensitive under static conditions. Another example of where significant differences exist between composite and metallic structures is in their damage tolerance under compressive loading. Advanced composite structures are much more sensitive to damage, and for this reason there has been an increased requirement on toughness in newly developed composite systems. Typically, certification guidelines deal with the issue of damage-tolerance in composites by requiring new designs to be based on the assumption that damage at the inspection threshold is initially present in the material.

Nevertheless another critical difference includes damage growth due to fatigue. This often represents a critical design condition in metals, whereas composites typically show excellent resistance to such loading. The stress levels associated with design critical load cases in composite materials, such as compression in the presence of impact damage, have traditionally been low enough to ensure that the damage does not grow due to fatigue. Thus, designs in composite materials have typically been determined by static considerations rather than by fatigue. As designers strive to fully use the specific strength and stiffness advantages of composites, the stress levels within components will increase, and fatigue issues must necessarily be given greater consideration in the airworthiness of future aircraft.

Perhaps the most critical difference between composites and metals is in their varying performance under different operational environments. Degradation of composite structures under certain environmental conditions has led to a number of standard certification approaches. Essentially, it is necessary to establish critical material properties after exposure to the extreme thermal and moisture environments to which the structure will be subjected. In addition, it must be demonstrated that there would be no degradation after exposure to chemicals that can be present (e.g. hydraulic fluids, lubricants, fuel, paint strippers and, de-icing fluids) (Baker et al. 2004).

All the above converge to the conclusion that, as the use of advanced composites increases, these materials are exposed to ever harsher environments. Despite their high strength and high stiffness, composites are surprisingly fragile. Damage can come from a number of sources, both during initial processing and in service. Even seemingly minor impact events can have a large effect on thin-walled structures. Since damage can never be entirely avoided, composite structures should be

Table 8.1 Design criteria and failure modes for aircraft structures

Mode of failure	Design criteria	Design input data
Static strength of undamaged structure	Structure must support ultimate loads without failure for 3 s	Static properties
Deformation of undamaged structure	Deformation of the structure at limit loads may not interfere with safe operation	Static properties and creep properties for elevated temperature conditions
Fatigue crack initiation of undamaged structure	(a) Fail-safe structure must meet service life requirements for operational loading conditions	Fatigue properties
	(b) Safe life components must remain crack free in service. Replacement times must be specified for limited life components	
Residual static strength of damaged structure	(a) Fail-safe structure must support 80–100 % limit loads without catastrophic failure	1. Static properties
	(b) A single member failed in redundant structure or partial failure in monolithic structure	2. Fracture Toughness properties
Crack growth life of damaged structure	(a) For fail-safe structure inspection techniques and frequency must be specified to minimise risk of catastrophic failures	1. Crack growth properties
	(b) For safe-life structure must define inspection techniques, frequencies and replacement times such as the probability of failure due to fatigue cracking is extremely remote	2. Fracture Toughness properties

designed to function safely despite the presence of flaws. This concept is called damage tolerance. In other words damage tolerance is the ability to resist fracture from the pre-existent cracks for a given period of time and, is an essential attribute of components whose failure could result in catastrophic failure. Damage tolerance addresses two points concerning an initially cracked structure. First, it is desired to determine fracture load for a specified crack size. Second, it is necessary to predict the length of time required for a 'sub-critical' crack to grow to the size that causes fracture at a given load. It is assumed that the crack can extend in a sub-critical manner by fatigue and/or stress corrosion cracking. Having the composite materials and their structures in mind, we may also define damage tolerance with a simpler description "composite structures should be at least as damage tolerant as the metal structures they replace" (Newaz and Sierakowski 1995). The mode of failure of structures associated with design criteria are shown in Table 8.1.

Designing for damage tolerance includes selecting materials that are inherently damage resistant, identifying sources and types of damage, understanding damage propagation mechanisms, and designing structures to operate with some degree of damage. The damage tolerance design principle comprises two categories: a 'single load path' and a 'multiple load path' structure. A single load path is where the applied loads are eventually distributed through a single member within an assembly, the failure of which could result in loss of the structural integrity of the component involved. A multiple load path is known with redundant structures in which (with the failure of individual elements) the applied loads would be safely distributed to other load-carrying members. Innovative materials research and engineering is essential to achieve the high-strength, heat-resistant, lightweight structures required in modern subsonic and supersonic aircraft. The effects of material selection is to consider the use of materials and stress levels that, after initiation of cracks, provide a controlled slow rate of crack propagation combined with high residual strength. When choosing new materials for airframe applications, it is essential to ensure that there are no compromises in the levels of safety achievable with conventional alloys. Retention of high levels of residual strength in the presence of typical damage for the particular material (damage tolerance) is a critical issue. Durability i.e. the resistance to cyclic stress or environmental degradation and damage, through the service life, is also a major factor in determining through-life support costs. The rate of damage growth and tolerance to damage determine the frequency and cost of inspections and the need for repairs throughout the life of the structure. Damage tolerance as discussed above is a driving factor when designing aerospace structures. Numerous recent studies have demonstrated that carbon nanotubes are able to improve the fracture, fatigue, impact and post-impact properties of composite materials, improving thus the damage tolerance of the composite structures. Three main sections are presented in this chapter related with the damage tolerance of composite materials doped with carbon nanotubes: fracture toughness, fatigue life and impact and post impact properties.

8.2 Fracture Toughness

Fracture control of structures is the strenuous effort by designers, production and maintenance engineers, and inspectors to ensure safe operations without catastrophic fracture failures. Very seldom does a fracture occur due to an unforeseen overload on the undamaged structure. Usually, it is caused by a structural flaw or crack: due to repeated or sustained "normal" service loads a crack may develop (either from a flaw or a stress concentration) and grow slowly in size, due to service loading. Cracks and defects weaken the strength. Therefore, as a crack continues to develop, strength decreases until it becomes so low that service loads cannot be carried any more, and fracture takes place. Fracture control is intended to prevent fracture due to defects and cracks at the loads experienced during operational service. If fracture is to be prevented, the strength should not drop below a certain safe value. In other words, cracks must be prevented from growing to a size, or a number at which the strength would drop below the acceptable limit. In order to determine which size of crack is admissible, one must calculate how the structural strength is affected by cracks (size and number); and in order to determine the safe operational life, one must be able to calculate the time in which a crack grows to the permissible size or number (Broek 1988). In the next pages an elaborate review will be given of the studies that have been made so far on fracture properties of nano-doped polymers and fibre-reinforced composites.

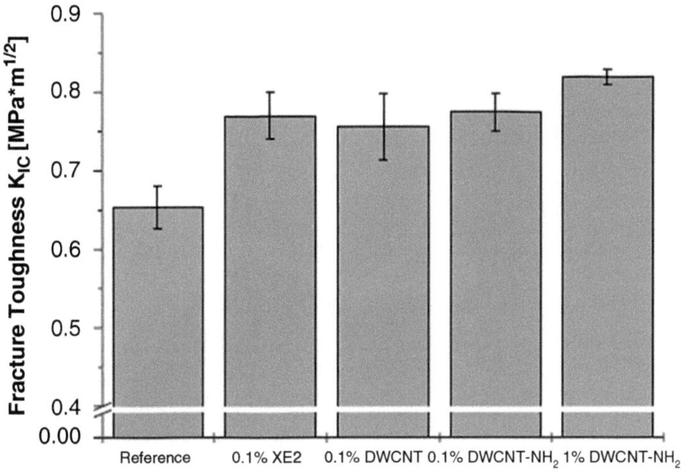

Fig. 8.1 The fracture toughness showed a general increase caused by the nano-scaled reinforcements (Gojny et al. 2004)

8.2.1 Nanopolymers

At this point it should be noted that by the term "nanopolymer" in this chapter, the authors mean polymers with carbon nanotubes added, without any other reinforcing phase. The polymers with both CNTs and carbon, glass or aramid long fibres as the reinforcing phase are called "nanocomposites" i.e. like typical composites but with its matrix doped with CNTs. These and their properties will be presented in the following paragraph. Moreover, in this chapter only epoxy polymers and their composites will be examined since they are the ones that are mainly used in primary and secondary aerospace structures.

When CNTs started drawing the attention of researchers worldwide, as potential fillers in thermosets and thermoplastics, they were compared with other known nanofillers in terms of properties and cost. Likewise, Gonjy et al. (2004) dispersed as received Double-Wall Carbon Nanotubes (DWCNT) and amino-functionalised ($DWCNT\text{-}NH_2$) in bisphenol-A based epoxy resin and directly compared them with Carbon Black (CB) Printex XE2 and the neat epoxy resin. Samples containing 0.1 wt.% DWCNTs, $DWCNT\text{-}NH_2$ and reference samples with same amount of CB were prepared by the use of a 3 roll-mill, also known as calender. The examination of the fracture toughness and the analysis of the obtained data were performed according to ASTM D 5045 (compact tension specimens). All the nanopolymers had a significantly higher fracture toughness compared to the neat epoxy, Fig. 8.1. Researchers could not observe any differences between the nanotubes and the carbon black. This might lead to the conclusion that not "fibre" (nanotube) bridging but crack deflection at small agglomerates seems to be the dominating mechanism for dissipating energy at 0.1% filler content.

However the sample containing 1% wt $DWCNT\text{–}NH_2$ showed higher fracture toughness than all other samples, besides containing numerous voids. Voids are the

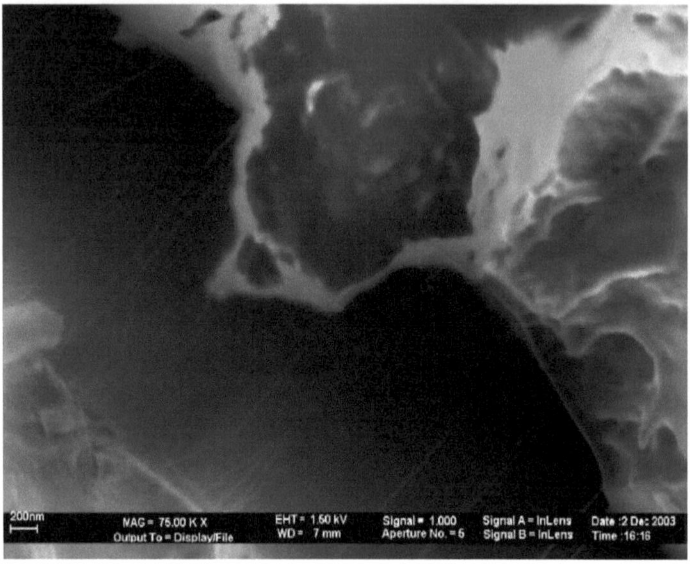

Fig. 8.2 SEM-micrograph of a DWCNT–NH_2/epoxy sample. Crack-bridging is one micro-mechanical mechanism leading to the recorded increase in fracture toughness properties (Gojny et al. 2004)

outcome of high viscosities involved when the amount of CNTs in the epoxy is increased and degassing of the mixture becomes difficult. Therefore the possible fracture toughness, in case of appropriate samples, would have been substantially higher and can be related to the fibre-like structure of the nanotubes and the raising dominance of a crack-bridging mechanism. In order to evaluate these micro-mechanical mechanisms leading to an improvement of the mechanical properties, the researchers investigated the fracture surface under a SEM. The investigations were conducted without any additional surface coating in order to avoid a covering of the DWCNTs. Figure 8.2 presents the DWCNT–NH_2 bridging the micro-cracks having a width of several microns. Remarkably, the bridged length is about 500–1,000 times longer than the average diameter of the nanotubes. This bridging mechanism should reduce the propagation of cracks and contribute positively to the measured enhancement in fracture toughness. The high aspect ratio of nanotubes and the related micro-mechanical properties appear to contribute to an additional increase in fracture toughness properties at higher filler contents.

The same group of researchers carried on the aforementioned study and investigated the influence of different carbon nanotubes on the mechanical properties of epoxy matrix composites (Gojny et al. 2005). In more detail they compared the reinforcing effect of Single-Wall carbon nanotubes (SWCNTs), DWCNTs and Multi-Wall carbon nanotubes (MWCNTs) with CB. The fracture property data of all the above nano-enhanced polymers along with the neat reference epoxy are depicted in Fig. 8.3.

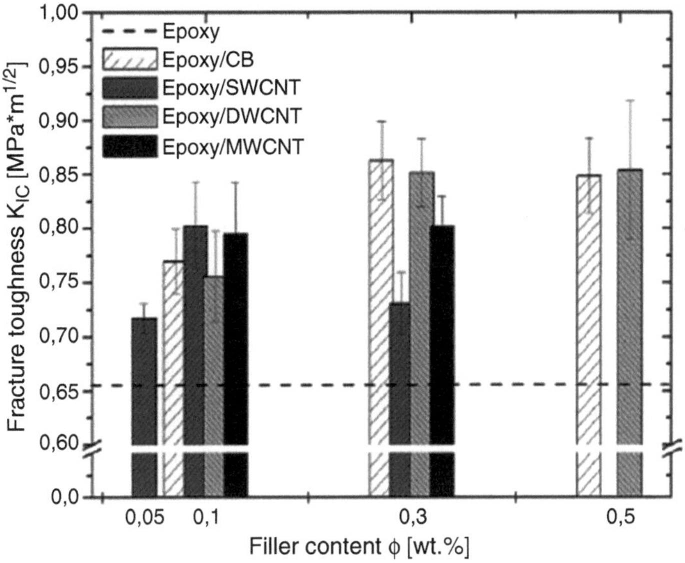

Fig. 8.3 Experimentally obtained $K_{IC,}$ fracture toughness values of epoxy nanopolymers containing CB, SWCNTs, DWCNTs and MWCNTs against the reference epoxy polymer (Gojny et al. 2005)

Nanofillers generally increase fracture toughness of the epoxy matrix significantly at very low filler contents, as shown in Fig. 8.3. The relative improvement of the K_{Ic} value is not dependent on the particle-shape and, therefore, the main fracture mechanism leading to enhanced fracture toughness could be related to the huge surface area of the nanofillers. A partly agglomerated dispersion was observed for all polymer-fillers mixtures, which in turn lead to the conclusion that the localised inelastic matrix deformation, void nucleation and crack deflection at the agglomerates are the dominating toughening mechanisms. A certain dependence of the surface area provided for the nanofillers on the toughening capacity could be found. Large surface areas make possible a more efficient improvement of fracture toughness. The decrease in fracture toughness (e.g., SWCNT at 0.3 wt.%), observed at higher filler contents, is related to the re-agglomeration that takes place due to Van der Waals forces. For all the mechanical characteristics, an exploitation of the theoretical surface area of the nanofillers as interface to the epoxy matrix is related to dispersion and matrix impregnation. Thus, the interface plays a major role in toughening of materials. The next Fig. 8.4 displays the fractured surfaces of a neat epoxy polymer and a nano-reinforced polymer examined under a scanning electron microscope (SEM). The difference of the surface roughness is obvious and is indicative of the toughening effect of the CNTs.

Furthermore, in this study the authors identified possible fracture mechanisms of the CNT that contribute to toughening of the polymers, thus making them more damage tolerant material. The initial situation of the CNT in an ideal case

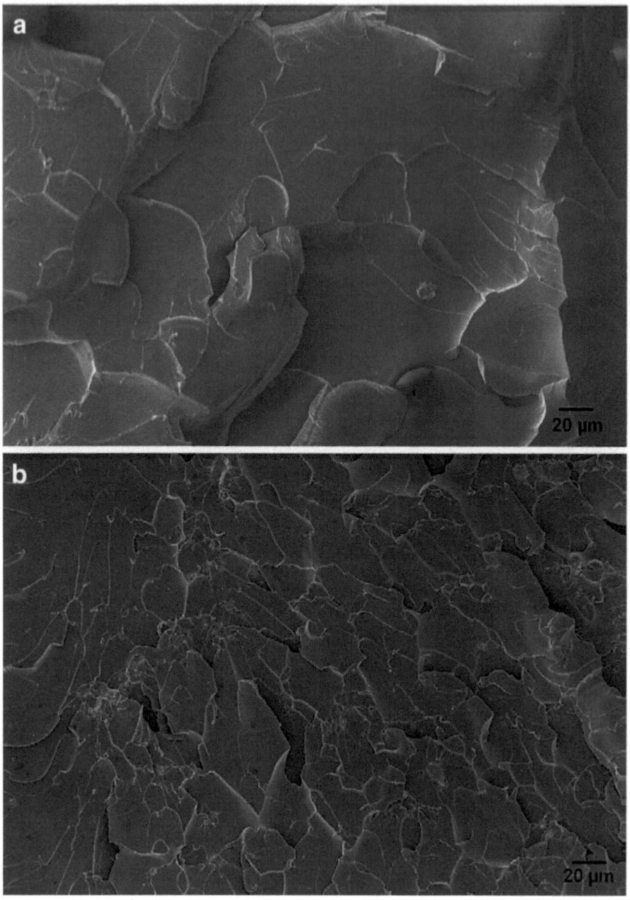

Fig. 8.4 SEM-micrographs of fracture surfaces at 1000X magnification, showing (**a**) the epoxy and (**b**) a DWCNT–NH_2/ nanopolymer. The polymer containing CNTs exhibits a considerably rougher fracture surface compared to the neat epoxy, indicating a toughening effect because of the CNTs (Gojny et al. 2005)

(Fig. 8.5a) is a completely impregnated and isolated one embedded in the matrix. In case of a crack the mechanisms Fig. 8.5b–e can activate, depending on the interfacial adhesion and the mechanical properties of the CNTs. In case of a weak interfacial bonding, plain pull-out of the CNT from the matrix occurs (Fig. 8.5b). On the other hand, a very strong bonding between CNT and matrix will lead to a complete breakage of the CNT (Fig. 8.5c) or to fracture of the outer layer and a telescopic pull-out (sword-sheath) of the inner tube(s) (Fig. 8.5d). If a spatial bonding of the reactive groups at the interface is present, then it enables a partial debonding of the interface, but it would allow for a crack bridging mechanism to activate (Fig. 8.5e). This bridging suppresses a further crack opening. Finally, increasing the applied stresses would lead to failure of the CNT, according to Fig. 8.5c, d.

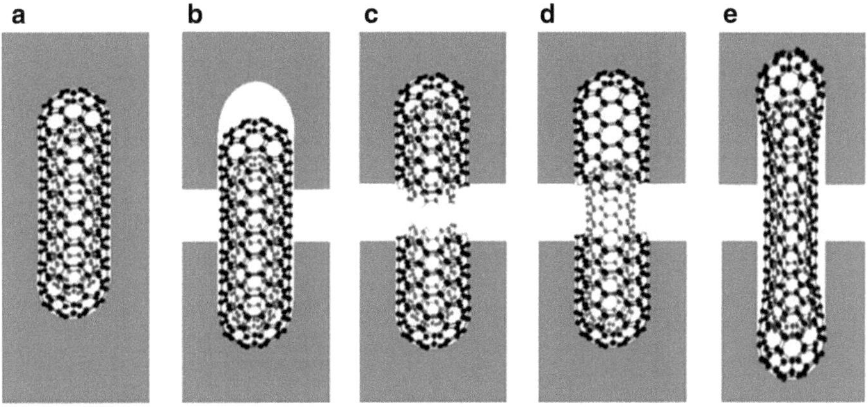

Fig. 8.5 Schematic description of possible fracture mechanisms of CNTs. (**a**) Initial state of the CNT; (**b**) pull-out caused by CNT/matrix debonding in case of weak interfacial adhesion; (**c**) rupture of CNT – strong interfacial adhesion in combination with extensive and fast local deformation; (**d**) telescopic pull-out – fracture of the outer layer due to strong interfacial bonding and pull-out of the inner tube; (**e**) bridging and partial debonding of the interface – local bonding to the matrix enables crack bridging and interfacial failure in the non-bonded regions (Gojny et al. 2005)

Moreover, Ganguli et al. (2008) used MWCNTs and dispersed them in a bi-functional epoxy resin. The MWCNTs were surface modified, both physically and chemically. The physical modifications of the MWCNTs were done by attempting to break the agglomerates with the help of ball milling, while the chemical modifications involved acid functionalisation, washing with purified water, filtering and drying of the CNTs. A contra-rotating mixer was used at high speeds to produce nanopolymers at 0.15 wt.% CNT loading. To evaluate the fracture toughness of the present nanopolymers, single-edged-notch tensile tests were performed on multiple specimens from each material batch. The stress intensity factor increased from 5.6 $MPam^{1/2}$ for the neat to 7.8 $MPam^{1/2}$ for the ball-milled nanopolymers to 10 $MPam^{1/2}$ for the acid-treated MWCNT nanopolymers based on 0.15 wt.% loading.

Thostenson and Chou investigated a scalable calendering approach for achieving dispersion of CVD-grown multi-walled carbon nanotubes through intense shear mixing (Thostenson and Chou 2006). The researchers used a calender to disperse the CNTs in the bisphenol f epichlorohydrin epoxy resin. The nanotube/epoxy suspension was processed at progressively smaller gap settings of 50, 30, and 20 μm then, two different final gap settings were used; 10 and 5 μm, producing thus two different sets of nanopolymers with several CNT weight fractions. Fracture toughness measurements were conducted using the single-edge-notch bending (ASTM D5045) method. The initial crack length was measured directly by taking measurements from the specimen fracture surfaces. Figure 8.6 shows the influence of reinforcement content and processing condition on the fracture toughness of the epoxy nanopolymers. At relatively low nanotube concentrations there were significant enhancements in fracture toughness. For the nanopolymers that were processed to a gap setting of 10 μm, the overall fracture toughness values are higher than for the more highly dispersed structure processed to the most highly dispersed state at 5 μm.

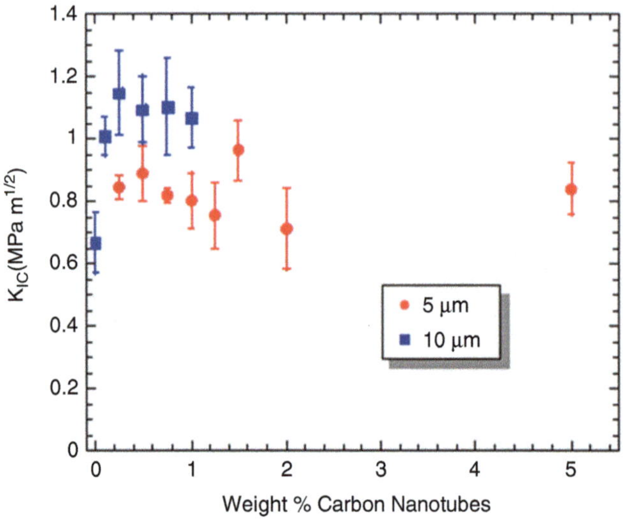

Fig. 8.6 Fracture toughness results showing the influence of processing conditions and reinforcement concentration (Thostenson and Chou 2006)

Figure 8.7 shows the morphologies of the fracture surfaces for the neat epoxy and the nanopolymers. The direction of crack propagation is from the top to bottom of the images. At low magnification the fracture surfaces show river lines which are characteristic of brittle fracture behaviour. Between the river lines in the neat epoxy in Fig. 8.7a the fracture surface is relatively smooth and lacking of any structural features. The nanotube-reinforced fracture surfaces show substantial increases in the micron level surface roughness. This increase in surface roughness is the probable reason that the nanopolymers showed enhanced fracture toughness as also discussed in the previous paragraphs.

The same resin as in the above paragraph was used by Zhou et al., doped with MWCNTs by employing high-intensity ultrasonic processing (Zhou et al. 2008). The weight fraction of the carbon nanotubes ranged from 0 to 0.4 wt.% in an attempt to identify the optimal weight loading for the best mechanical properties. Fracture toughness of neat and nano-doped epoxy was determined from three-point bending tests of the single edge notch specimens. Each of these specimens was cycled 100 times between 4 and 40% of the peak load at 1 Hz and then statically tested. During the static tests, the change in specimen length Δl was measured by recording the ram positions through the displacement transducer of the tensile testing machine. Since non-linearity was seldom observed in load–displacement diagrams, the critical stress intensity factor, K_{Ic} of materials was calculated from the peak load of each load–displacement curve, and was plotted as a function of the CNT weight fraction. Figure 8.8 shows that enhancement reaches a maximum for the critical stress intensity factor at 0.3 wt.%. At the higher contents, fracture toughness decreased as the filler loading was increased.

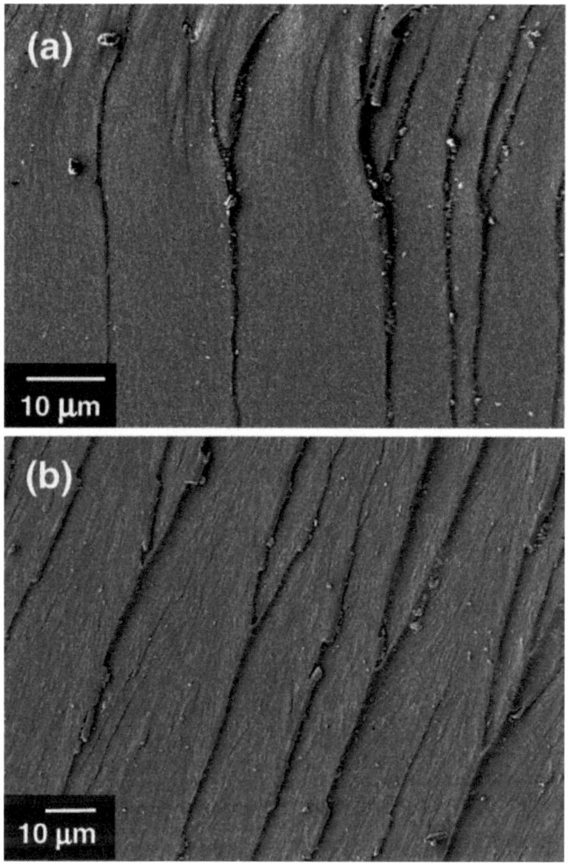

Fig. 8.7 SEM micrographs of (**a**) epoxy and (**b**) nanocomposite fracture surfaces near the region of crack initiation. The difference on the surface roughness is evident (Thostenson and Chou 2006)

Chow and Tan (2010) prepared epoxy/multi-wall carbon nanotube (MWCNT) polymers by using two different mixing techniques i.e., sonication and planetary mixing. Different concentrations (0.25–1.25 wt.%) of MWCNT were incorporated into the epoxy, bisphenol-f. Two different trends in fracture toughness properties were observed as the epoxy/MWCNT were prepared by two different techniques. For the epoxy/MWCNT nanopolymers prepared by sonication, the fracture toughness (K_{Ic}) value of epoxy/MWCNT was first decreased at the beginning at 0.25 wt. % MWCNT and started to increase at 0.75 wt.% MWCNT. At further addition of MWCNT (more than 0.75 wt.%), the K_{Ic} value was considerably reduced. It was commented that at low MWCNT loading, the reduction of fracture toughness of epoxy was due to the amount of filler content, which was inadequate to absorb any fracture energy. When the filler content was increased up to 1.0 and 1.25 wt.%, the percentage of reduction of fracture toughness for epoxy nanopolymers were 30.3 and 24.2%, respectively. At 0.25 wt.% MWCNT, fracture toughness for the epoxy nanopolymers was relatively low for both epoxy nanopolymers prepared using sonication and planetary mixing. At this low weight loading of MWCNT in

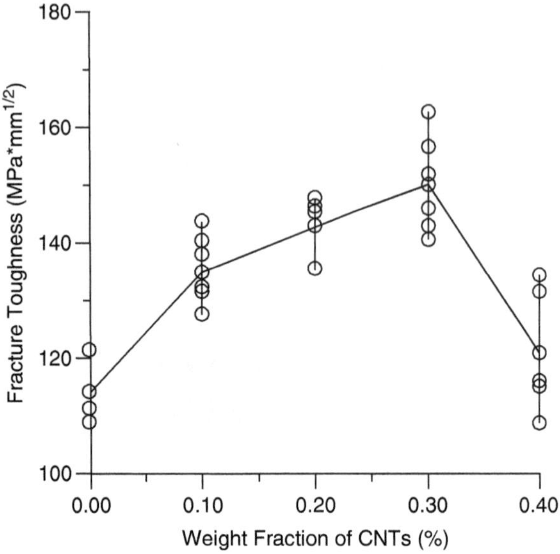

Fig. 8.8 Effects of CNT loading per weight on the fracture toughness of epoxy (Zhou et al. 2008)

epoxy, MWCNT can be quite well dispersed and have good interaction with the epoxy matrix. When the CNT loading is increased in an epoxy matrix, the viscosity of the mixture also increases exponentially, thus making dispersion difficult and also re-agglomeration to occur. A good CNT dispersion will cause the MWCNTs to have strong interfacial bonding with the epoxy matrix and make it difficult to pull them out of the matrix. At this level, the fracture toughness of the epoxy nanopolymers was increased accordingly. This is attributed to the fact that MWCNT can be pulled out of the epoxy matrix and more energy is needed to cause shearing mechanism in between the MWCNT and epoxy. However, it was stated that the fracture toughness of the epoxy would diminish if the loading of the MWCNT is too high. In other words, there is an optimal weight percentage of CNTs for every resin/process above which improvements of properties will not occur and an actual loss of properties will be recorded.

In addition, Yu et al. (2008) introduced as-received MWCNTs in a bisphenol-A epoxy resin by intensive sonication. The resulting epoxy/CNT mixture was then split into two parts and on the one part a degassing agent was introduced to prevent the formation of voids. Compact Tension (CT) specimens were used to evaluate the effect of the CNTs at two different weight fractions on the fracture properties along with effect of the degassing agent. The summary of the above study can be seen in the graph of Fig. 8.9. It was noted that the average fracture toughness ($K_{Ic} = 0.72$ MPam$^{1/2}$) of 3 wt.%-MWCNT/epoxy composite prepared with the aid of a degassing agent is significantly higher than that ($K_{Ic} = 0.55$ MPam$^{1/2}$) of the composite prepared without using a degassing agent. Increasing the MWCNT weight fraction i.e. viscosity increases significantly, makes removing air from the composite mixture difficult. Using a degassing agent to remove the air bubbles from the CNT/epoxy mixture, and thus to decrease the porosity of the composite, is to be given a lot of attention when it comes to nanopolymer processing and

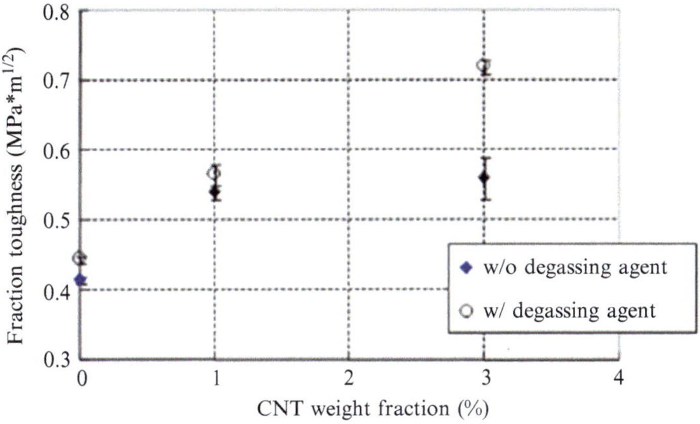

Fig. 8.9 Fracture toughness of MWCNT/epoxy polymer with and without degassing agent (Yu et al. 2008)

manufacturing. One may also note that improvements on the fracture properties took place even when the degassing agent was not added to the epoxy/CNT mixture, strengthening the argument in favour of the use of CNTs as fillers.

Fiedler et al. (2006) in a constructive research highlighted the potential of the CNTs as nanofillers in polymers, but also the limitations and challenges one has to face when dealing with nanoparticles in general. Figure 8.10 shows the surface/volume ratio for a variety of CNTs with different aspect ratios (length/diameter) as a function of their diameter. In addition, the same ratio is shown for spherical nanoparticles like fumed silica (FS) and CB. For comparison, also conventional reinforcements like glass balls (GB), glass and carbon fibres (GF, CF) are given. In the double logarithmic scale of Fig. 8.10 the surface/volume ratio is decreasing linearly with increasing particle diameter. It is obvious that already a small volume content of nanoparticles provides huge surface areas and can enhance the nucleation of polymer crystals in thermoplastic materials or the cross-linking density in thermo-sets, resulting in better mechanical properties by altering the polymer morphology. As a result, the advantage of nanoparticle reinforcement can be a synergistic effect of introducing the reinforcing phase with their high mechanical properties and enhancing the polymer morphology. SWCNTs have dimensions comparable to macromolecules. In suspensions they behave like liquid crystal polymers (LCP). Therefore one may consider nanoparticle-reinforced composites as a polymer blend. The separation between particles becomes very important and depends on the particle size, shape and volume content.

Concluding, one may say that the use of carbon nanotubes as fillers into an epoxy polymer has been proved to be able to improve the damage tolerance of those materials. This is feasible via the reinforcing mechanisms of the CNTs that are capable of absorbing energy while the crack is propagating, increasing thus the fracture toughness of the nano-enhanced polymers in comparison with the neat epoxies. However dispersion and degassing are of key importance in order to fully exploit the potential of these nanofillers.

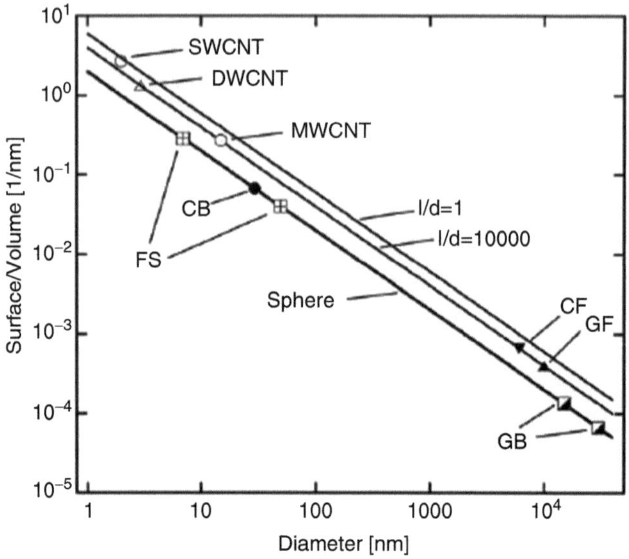

Fig. 8.10 Ratio of particle surface and volume for spherical and fibrous particles as a function of the particle diameter (Fiedler et al. 2006)

8.2.2 Nanocomposites

In this section, the potential of CNTs to enhance fracture toughness and the interlaminar properties of fibre-reinforced epoxy polymers will be presented. All the types of fibre reinforcement i.e. carbon, glass and aramid are considered with an epoxy matrix. Wichmann et al. (2006) used carbon nanotubes and fumed silica nanoparticles to modify the epoxy matrix of glass-fibre-reinforced epoxy composites (GFRPs). A modified DGEBA-based epoxy resin system was used as matrix material, especially suited for resin infusion techniques, because of its very low viscosity of around 250 mPas. The nanotubes used were MWCNTs and DWCNTs. The double-wall carbon nanotubes were also applied with an amino-surface functionalisation (DWCNT-NH_2). In previous works it has been proved that an amino functionalisation enhances the dispersibility in the epoxy matrix and can provide a covalent bonding to the polymer. The potential of CNTs as fillers was examined against reference specimens containing CB and FS. The CB consists of spherical carbon nanoparticles with a diameter of 30 nm while, the FS nanoparticles used had a spherical shape, and a diameter of 7 nm. A surface treatment was also applied to the fumed silica nanoparticles. The GFRPs were produced via a resin transfer moulding (RTM) process. As fibre-reinforcement, glass-fibre non-crimp fabrics were used and two glass-fibre volume contents were chosen for the GFRPs, 37 and 50 vol.%. It must be noted that, due to the high viscosities involved, degassing-time of the mixture prior to injection, as well as the injection time, was significantly extended. Additionally, the RTM-mould was modified to enable

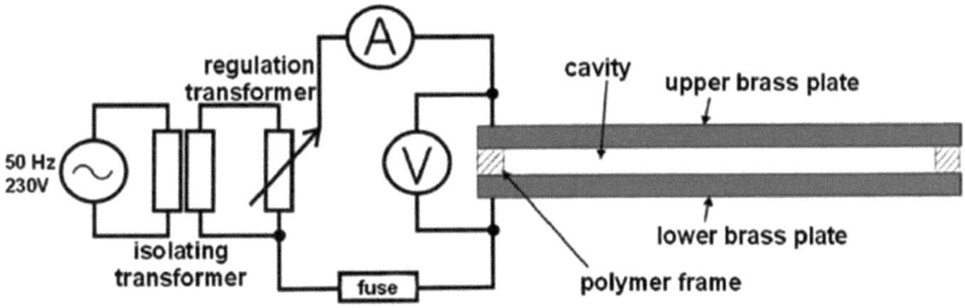

Fig. 8.11 Schematic of the modified RTM-device. The electrical field can be applied between the brass plates (z-direction) (Wichmann et al. 2006)

application of an electrical field during the curing process, in order to induce a preferred orientation of the carbon nanotubes perpendicular to the fibre-plane (z-direction, through thickness, in order to enhance intralaminar properties) and to stimulate the formation of a conductive network of the carbon nanoparticles. The field strength applied during curing was 330 V/cm. A schematic drawing of the modified RTM-device is shown in Fig. 8.11.

The interlaminar shear strength (ILSS), an important design criterion, characterises the ability of a laminate to withstand shear forces and also is an empirical measure of fibre/matrix adhesion. The ILSS was found to be significantly improved by the integration of nanoparticles. The best results were achieved for laminates with a lower volume fraction of glass-fibres. Figure 8.12a shows the ILSS results of the laminates containing 37 vol.% of glass-fibres. It can be seen that all nanoparticles significantly increased the interlaminar shear strength and that the best results were achieved with DWCNTs. The ILSS of the reference laminates was 33.4 MPa. With DWCNT the ILSS increased up to 38.7 MPa. The application of an electrical field in z-direction while curing the composite led in a further minor increase in ILSS value. However, the positive influence on the ILSS seems to decrease with the higher glass-fibre volume content. From Fig. 8.12b it can be seen that all ILSS values measured scatter more or less around the reference value. A slight increase could be observed again for the DWCNT-NH_2 modified laminate, but the DWCNT modified laminates exhibit a lower ILSS than the reference laminate. The application of an electrical field did not lead to any further improvement. An explanation for the different performance of the nanoparticle modification (decreasing with increasing glass-fibre volume content) can be found in the failure mechanism of the composite because of a very weak glass-fibre matrix interface in the composite (also confirmed in the mode I tests presented below), leading to a failure mechanisms clearly dominated by a failure of the glass-fibre/matrix interface. In other words, at low V_f content more matrix material is present, triggering though the reinforcing mechanisms since more CNTs are active during the damage development process. At higher V_f more CNTs are obscured by the presence of fibre reinforcement making it thus more difficult to act as nano-reinforcement.

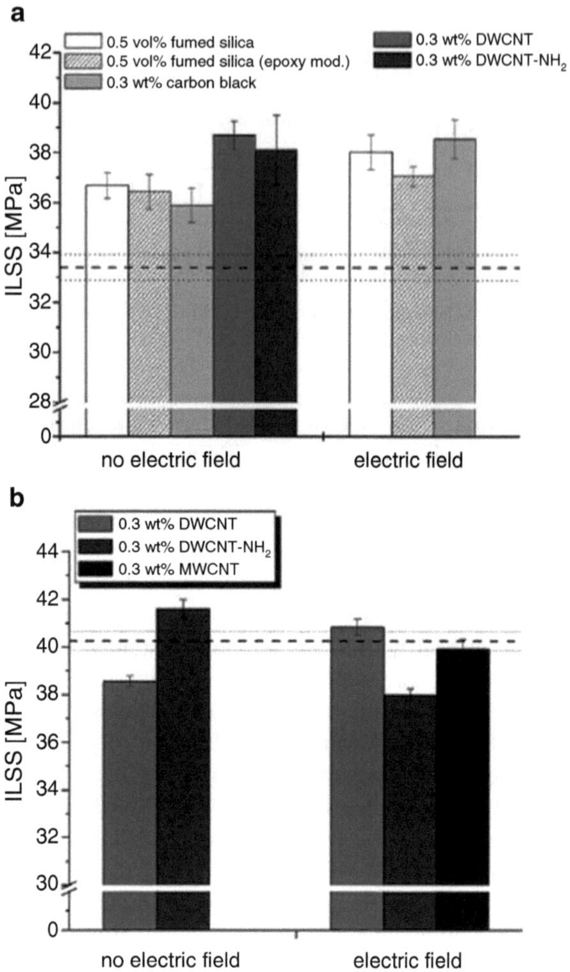

Fig. 8.12 (**a**) ILSS of the glass-fibre/epoxy laminates containing 37 vol.% glass-fibres. (**b**) ILSS of the glass/fibre/epoxy laminates containing 50 vol.% glass fibres. The *black slashed line* represents the reference value and its deviations (Wichmann et al. 2006)

The next graph summarises the results of mode I (opening) and mode II (shear) fracture tests. Mode I tests use a double cantilever beam specimen (DCB) while mode II tests use an end-notch flexure specimen (ENF). A distinct fibre–matrix debonding could be observed, resulting in a considerable contribution of fibre bridging and fibre pull-out. The dominating failure mechanism in the mode I test was glass fibre– matrix interface failure. This mechanism results in relatively high G_{Ic} values. However, the G_{Ic}-values measured all scatter more or less around the reference value, Fig. 8.13a. For the composites containing 50 vol.% of glass-fibres, the G_{Ic} values are slightly reduced in comparison to the reference value. The G_{Ic} value for all the composites containing carbon-based nanoparticles was slightly underestimated, since this material loses its transparency and the actual crack tip can only be detected optically from the edges of the specimen and not inside. This results in a slight underestimation of the crack length and therefore leads to lower

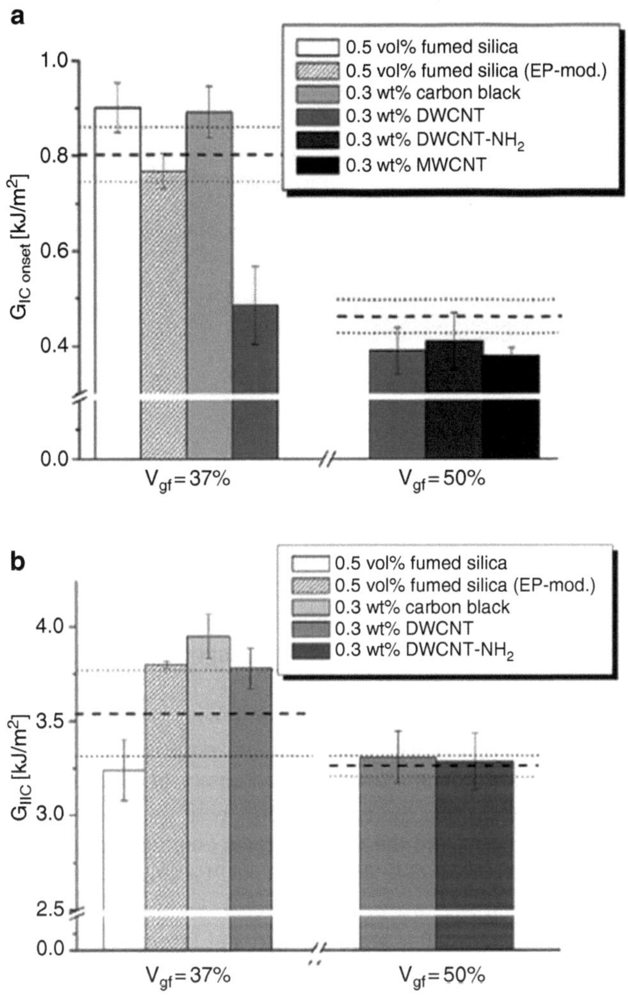

Fig. 8.13 (**a**) GIc-values of the nanoparticle modified FRPs from DCB test. (**b**) GIIc values of the nanoparticle modified laminate from ENF test. The *black slashed line* represents the reference value and its deviations (Wichmann et al. 2006)

G_{Ic} values. In Fig. 8.13b, the results from the ENF tests are shown. Again, no major influence of the nanoparticle modification could be observed. The G_{IIc} values for the laminates containing 37 vol.% of glass-fibres are slightly higher. The G_{IIc} values for the laminates with 50 vol.% of glass-fibres are exactly in the range of the reference value. In contrast to expectations, the application of an electrical field in the z-direction during curing did not result in a further increase of the measured G_{Ic} and G_{IIc} values.

Yokozeki and his colleagues (2007a) used cup-stacked carbon nanotubes (CSCNTs), see Fig. 8.14, and investigated the damage accumulation behaviour in

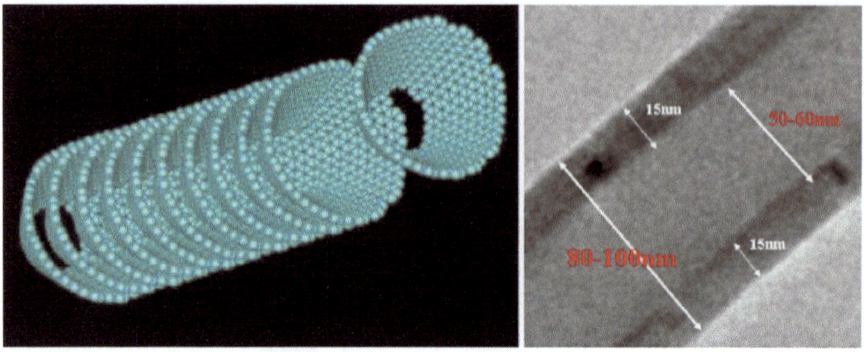

Fig. 8.14 Cup-stacked carbon nanotube by CARBERE®: (*left*) schematic view, (*right*) typical TEM image (Yokozeki et al. 2007a)

CFRP nanocomposite laminates. The resulting nominal aspect ratio of CSCNTs was 10 after being subjected to the dry mill using zirconia beads such as to improve their dispersion in the matrix material. The matrix material used was bisphenol-A based epoxy resin. The mixture of CSCNT/epoxy was diluted with plain epoxy and the curing agent was added to the compounds. Three types of CSCNT-dispersed epoxy with weight fractions of CSCNTs to the compound of 0, 5, and 12 wt.% were prepared.

Then the manufacturing of the nanocomposite with the aforementioned CSCNTs/epoxy blend took place. Unidirectional prepregs were developed using T700SC- 12 K fibres and the above-mentioned epoxy filled with CSCNTs. A wet coater was used to impregnate the carbon fibres with the resin. The prepreg fibre weight was set to 125 g/m^2 and the nominal resin content including CSCNTs was 35 wt.% (the weight percentages of CSCNT in the final three-phase nanocomposites were 0, 1.8, and 4.2 wt.%, respectively). Unidirectional $[0]_{16}$ laminates and cross-ply $[0_2/90_2]$s laminates were stacked and fabricated using an autoclave. The resulting volume fractions of the carbon fibre were 60% for all composites. The mechanical properties of unidirectional carbon fibre- reinforced nanocomposite laminates were evaluated and cross-ply laminates were subjected to tension tests in order to observe the damage accumulation behaviours of matrix cracks. A clear retardation of matrix crack onset and accumulation was found in composite laminates with CSCNT compared to those without CSCNT. Fracture toughness associated with matrix cracking was evaluated based on an analytical model using experimental results. It was then suggested that the dispersion of CSCNT resulted in fracture toughness improvement and residual thermal strain decrease, which is considered to cause the retardation of matrix crack formation improving thus the damage tolerance of the nano-enhanced CFRP in comparison with the neat CFRP.

Moreover, Bekyarova et al. (2007a) used functionalized SWCNTs with carboxylic acid groups as nano-reinforcement for carbon fibre/epoxy composites in diglycidyl ether of bisphenol-f epoxy. Epoxy composites with 0.2 and 0.5 wt%

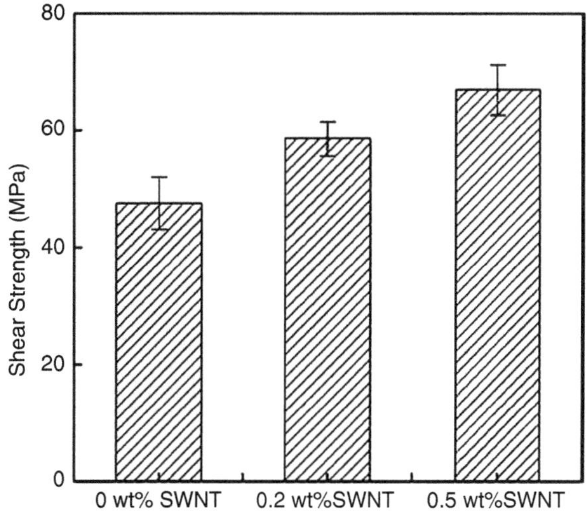

Fig. 8.15 Shear strength of CF/epoxy composites with and without SWNT-COOH (Bekyarova et al. 2007a)

SWNT-COOH loading were mixed with the curing agent and then used for impregnation of eight unidirectional CF layers. The infusion was performed along the fibre direction by the VARTM technique. The shear strength was evaluated by double-notch compression testing (ASTM D3846). Each specimen was notched to half thickness on opposite sides of the laminate at a fixed distance. The specimens were then loaded into an anti-buckling compression fixture and compression loaded by steel platens. The reinforcement effect of the SWNT-COOH is manifested in the shear strength of the composites measured by the double-notch compression method. The shear strength showed an increase with the inclusion of SWNT-COOH; enhancements of 20 and 40% were observed for composites with SWNT-COOH loadings of 0.2 and 0.5 wt %, respectively (Fig. 8.15).

To understand the reinforcement mechanism of SWNT-COOH, the fracture surfaces of the composites that failed in shear were examined by SEM, and typical micrographs are shown in Fig. 8.16. The SEM observation shows fibre/matrix interfacial de-bonding in the composites without SWNTs as indicated by the smooth fibre surfaces and resulting in fibre impressions, Fig. 8.16a. Whereas, the composites with SWNT-COOH showed a reduced level of debonded fibres, see Fig. 8.16b, c. The increase in the interlaminar strength was connected with the reinforcing role of the SWNTs that brings about an improved SWNT-epoxy interface due to cross linking of the carboxylic acid groups of the SWNTs with the epoxy matrix through the formation of an ester bond.

The improved fracture properties of composites with CNTs in the polymer matrix at cryogenic temperatures were demonstrated by Kim et al. (2008). When a composite tank is used to store cryogenic liquids, it undergoes cryogenic aging as well as cycling from room temperature to cryogenic temperature as a load is added. Microcracks can then grow in the composite matrix due to the difference in the coefficients of thermal expansion of the fibre and matrix and between different

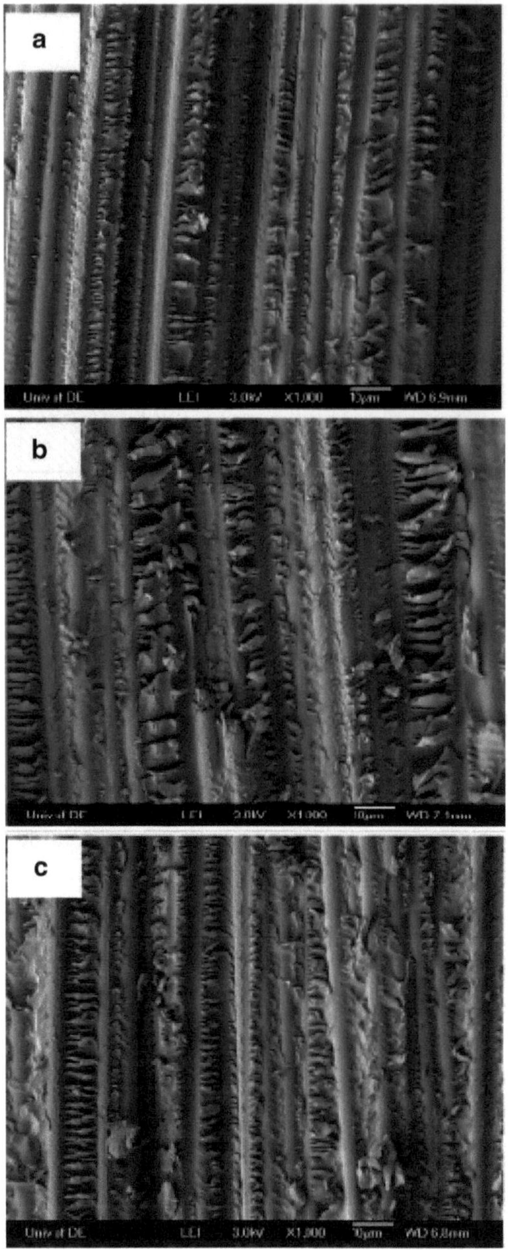

Fig. 8.16 Fracture surface of CFRPS; (**a**) without SWNTs; (**b**) with 0.2 wt % SWNT-COOH; and (**c**) with 0.5 wt % SWNT-COOH (Bekyarova et al. 2007a)

angle layers. Such structural damage gives rise to a degradation of mechanical properties in the structures, such as fibre/matrix interfacial debonding, potholing, and delamination. The outcome of any of these damages would probably be leakage of the contained liquids. A carbon/epoxy prepreg model manufactured by the hot-

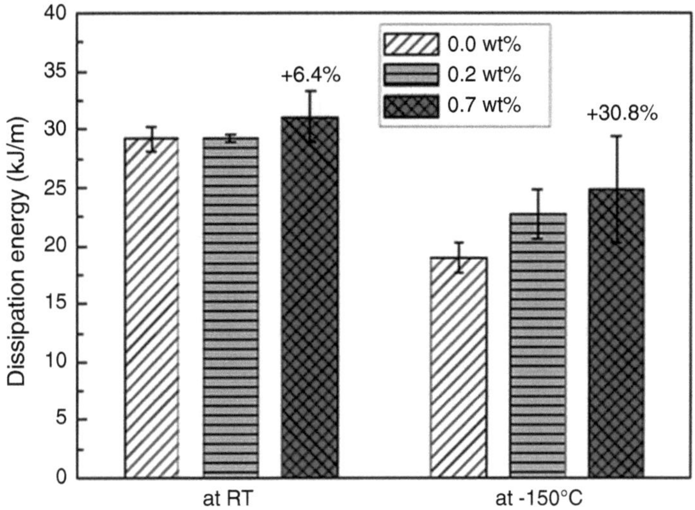

Fig. 8.17 Dissipation energy in the stable crack propagation region for mode I interlaminar fracture (Kim et al. 2008)

melting process can be used as reference material. The nano-doped material was formulated with the same epoxy resin with CNTs added at two weight fractions 0.2 and 0.7%. The toughening effect of the prepreg materials was investigated by comparing their mode I interlaminar fracture toughness using a DCB test at room temperature (RT) and at −150 °C, using an environmental chamber. Unstable crack propagation occurs under a lower crack driving force at cryogenic temperature than at RT. This is due to embrittlement of the epoxy matrix at the cryogenic temperature, which decreases the crack resistance of the composite. In addition, it was found that MWCNTs have little influence on the R-curve behaviour at RT, but a considerable influence at the cryogenic temperature. Therefore, it can be concluded that cracks can propagate stably under higher crack driving force by the addition of MWCNTs at cryogenic temperature. The calculated dissipation energy for all the CFRPs evaluated is graphically illustrated in Fig. 8.17. It was found that all the laminate composites show a decrease in the dissipation energy to the crack propagation at cryogenic temperature in comparison with RT. In other words, lower energy (or driving force) is needed for crack propagation, which is indicative of a reduction of fracture toughness. The MWCNT0.7 specimen shows the highest dissipation energy, both at RT and at −150 °C. At these temperatures, the dissipation energy is 8.4 and 30.8% higher, respectively, than that of the reference specimens. It can be stated that employing MWCNTs as nanofiller in an epoxy matrix, is an effective approach to obtain high fracture toughness with regard to the application of composites in cryogenic conditions.

As known from the literature, Mode I delamination resistance is clearly a matrix-dominated parameter. For this reason one area of great research interest is to produce tougher or more ductile matrix resins. It is anticipated that the enhancement in the matrix fracture toughness can lead to an overall advanced fracture behaviour. The degree of interfacial adhesion between nanotubes and polymers is a key parameter in both production and physical properties of carbon nanotube composites, and is vital in understanding the surface behaviour of nano-composites. Adequate interfacial stress transfer from the matrix to the reinforcement is only possible when the interface has not failed during composite loading. Failure of the interface effectively neutralizes the efficiency of the reinforcement. The study from Karapappas and his colleagues (2009) proved that CNTs can enhance the damage tolerance of composite materials even when a wet lay-up technique was used as the manufacturing method. As received, MWCNTs were dispersed in an epoxy resin with a torus-mill high shear mixer device. Doped resin compounds with three different MWCNT contents of 0.1, 0.5, and 1 wt.%, respectively, were produced. The CF laminas were chosen to be UD with weight of 160 g/m^2. Each panel had 16 plies of carbon fibres and were processed in an autoclave, using the vacuum bag technique. The research team performed mode I and II fracture toughness tests. Figure 8.18 shows typical load–displacement curves during mode I loading for neat epoxy and CNT doped matrix CFRP. The maximum load, P_{max}, is significantly increased in the case where the matrix was doped with 1 wt.% CNTs.

The fracture energy of the CFRPs was calculated using the modified beam theory (MBT) and the areas method. The calculated values are compared in Figure 8.19. The doped CFRPs not only showed an increase in the load bearing capacity but also a significant increase in the fracture energy, G_{Ic}, compared to the reference neat epoxy matrix. The increase in G_{Ic} was of a magnitude of around 60% for the specimens containing 1% MWCNT. The above behaviour was attributed to the significantly large aspect ratio of CNTs which allows them to act as nano-bridges between the notch edges. Extra energy is needed in order to pull them out from the matrix or break them in order to initiate or propagate the crack. SEM pictures of the fractured surfaces of neat epoxy and the nanocomposite with 1% MWCNT can be seen in Fig. 8.20a–c, respectively. In the latter figure the distinctive pull-out of the CNTs during the fracture damage process is obvious, confirming thus the presence of the CNT reinforcing mechanisms. On the other hand, the integration of 0.1% CNTs led to a slight reduction of the G_{Ic} value. This was attributed to the small quantity of MWCNTs. For a small fraction of the nanotubes the viscosity of the resin remains at very low levels leading consecutively to low shear forces during mixing. Therefore, the CNT agglomerations are not subjected to high shear forces and hence it is not feasible to break them and disperse the CNTs, resulting in a mixture with agglomerations Fig. 8.21d. These agglomerations act as defects in the matrix and thus have mechanical properties similar to the reference values.

The mode II results are presented Fig. 8.21. It can be seen that the composites doped with 0.5 and 1% CNT exhibit higher G_{IIc} values than the reference composite. The increase was of about 75 and 45%, respectively. This increase was because of the various energy absorbing mechanisms triggered by the presence of CNTs. The CNTs need extra energy in order to be broken and to be pulled-out of the matrix and, that extra energy contributes to the higher G_{IIc} values. The CNTs also tend to

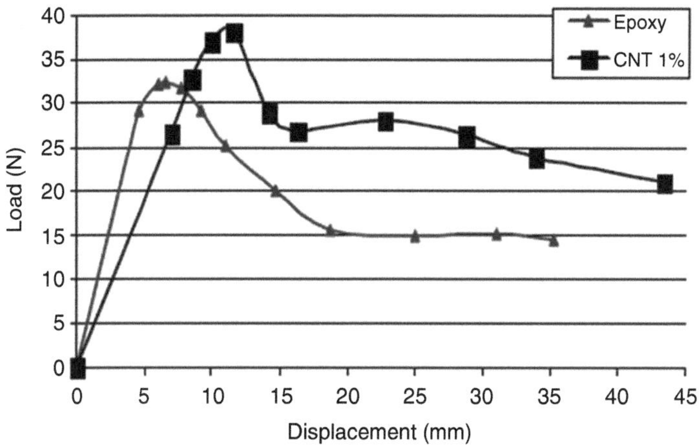

Fig. 8.18 Load vs. Displacement for (**a**) CFRP with neat epoxy matrix and (**b**) CFRP with epoxy matrix doped with 1 wt.% MWCNTs (Karapppas et al. 2009)

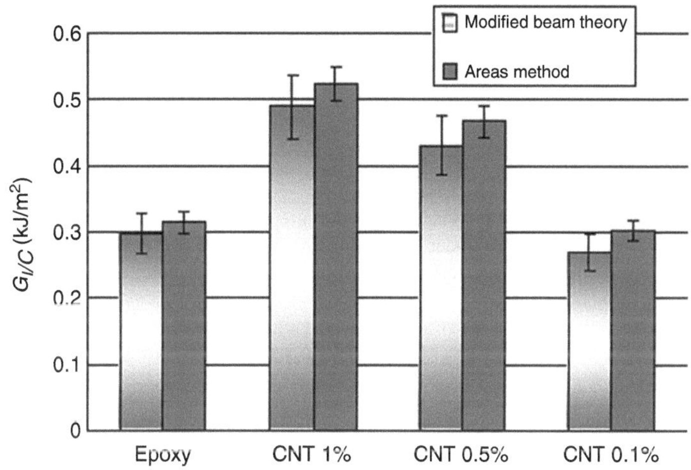

Fig. 8.19 Mode I fracture energy of the different panels, calculated with two different methods (Karapppas et al. 2009)

move away the stress concentration from the crack tip, in this manner helping the CFRP to withstand higher loading before fracture. One may notice that the incorporation of 0.1% CNTs led to a reduction of the G_{IIc} value. The explanation for this phenomenon is the same as in the analysis of the mode I results.

CSCNTs were used by Yokozeki's scientific team (2009a) to ameliorate the mechanical properties of CFRPs. Unidirectional prepregs were developed using T700SC-12 K fibres and the bisphenol-A epoxy filled with CSCNTs (0 and 5 wt.%). The prepreg fibre area weight was set to 125 g/m^2 and the nominal resin content including CSCNTs was 35 wt.%. Unidirectional $[0]_{16}$ laminates were stacked and

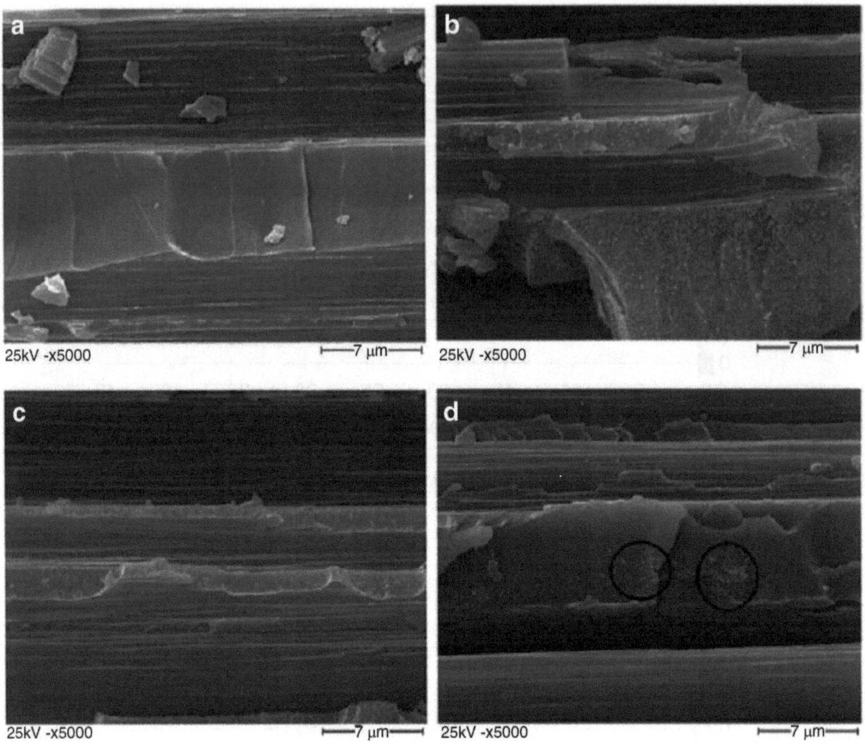

Fig. 8.20 (**a**) SEM picture of neat epoxy sample fractured under Mode I loading, (**b**) SEM picture of 1 % MWCNT fractured sample under Mode I loading. Evidence of good dispersion can be seen. The intensive carbon nanotube pull-out and breakage, which contributed to higher G_{Ic} values, is apparent, (**c**) SEM picture of 0.5 % MWCNT sample fractured under Mode I loading. Evidence of no agglomerations present can be seen. The difference of the amount of CNTs being pulled-out is more than obvious, (**d**) SEM picture of 0.1 % MWCNT sample fractured under Mode I loading. Agglomerations are present and indicated with the black circles. The lack of dispersion can be seen if directly compared with (**b**) and (**c**) (Karapppas et al. 2009)

fabricated using an autoclave. Mode I and mode II interlaminar fracture toughness were measured based on double DCB and ENF tests, respectively. The results are presented in Table 8.2. It can be said that both mode-I and mode-II interlaminar fracture toughness again in this case are improved by the CSCNT inclusion in the matrix. The interlaminar fracture toughness improvement coincides with the trend of matrix cracking evidence of which are related with off-axis loading at 90° where the values of the CNT-doped composites are higher than the neat composite ones.

In continuation of his research work Yokozeki et al. (2009b) examined the effect of CSCNTs of aspect ratio of 10 and 100 designated as AR10 and AR100 respectively. Three weight fractions of CSCNTs were added in the matrix; 0, 5, and 10 wt. %. Unidirectional prepregs were developed using T700SC-12 K fibres and the above-mentioned epoxy filled with 0 and 5 wt.% CSCNTs. The prepreg fibre weight was set to be 125 g/m^2, and the nominal resin content including CSCNTs

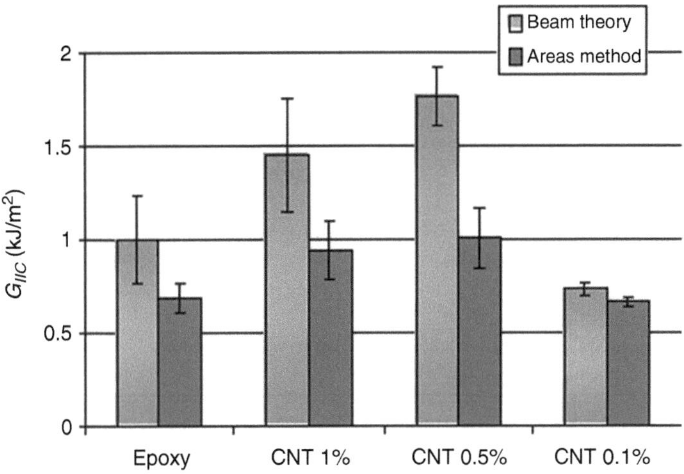

Fig. 8.21 Mode II fracture energy of the different CFRP panels (Karapppas et al. 2009)

Table 8.2 Summary of mechanical properties of CFRPs prepared with and without CSCNTs (Yokozeki et al. 2009a)

CSCNT (%)	90° Stiffness (GPa)	90° Strength (GPa)	G_{Ic} (kJ/m^2)	G_{IIc} (kJ/m^2)
0	8.61 ± 0.02	51.2 ± 2.8	0.086 ± 0.007	0.568 ± 0.115
5	9.11 ± 0.01	57.9 ± 1.7	0.170 ± 0.023	0.732 ± 0.043

was 33 wt.%. In addition, CSCNT-dispersed epoxy films (10 wt.%) were prepared using AR10 and AR100 CSCNTs. Unidirectional $[0]_{36}$ laminates were prepared and processed using an autoclave. Placement of interlayers and CSCNT sprinkle were only performed between the middle layers where the crack would propagate. As a result six types of CFRPs were produced and can be seen in Table 8.3.

The evaluated fracture toughness between the crack growth length of 20 and 60 mm are averaged and summarized in Fig. 8.22 for all samples. All CSCNT-dispersed CFRPs (samples B–F) have high fracture toughness compared to CFRP without CSCNT (sample A). Specifically, sample D exhibits the highest fracture toughness. It must be underlined that just the use of CSCNT-dispersed epoxy (sample B) is capable of contributing to increase the mode-I fracture toughness.

Figure 8.23 displays the mode-II fracture toughness data obtained from the comparison of all the CFRPs. All CSCNT-dispersed CFRPs (B–F) have high fracture toughness compared to CFRP without CSCNT. Specifically, sample D exhibits the highest fracture toughness (about three times higher than reference), which corresponds with the trend in the case of mode-I toughness.

Scientists from the Budapest University of Technology and Economics (Romhany and Szebenyi 2009) prepared carbon fibre/epoxy composites and MWCNTs/carbon fibre/epoxy hybrid composites with 0.1, 0.3, 0.5 and 1 weight % nanotube filling of the matrix and compared their interlaminar properties. Epoxy

Table 8.3 Prepared nanocomposites with and without CSCNTs (Yokozeki et al. 2009b)

Sample	Type	CSCNT in epoxy	CSCNT in interlayers
A	No CNT-reference	–	–
B	CNT in epoxy No CNT in interlayer	AR10, 5 wt.%	–
C	CNT in epoxy CNT sprinkle	AR10, 5 wt.%	AR10, 10 g/m^2
D	CNT in epoxy CNT-dispersed film	AR10, 5 wt.%	AR10, 10 wt.%
E	CNT in epoxy CNT sprinkle	AR10, 5 wt.%	AR100, 10 g/m^2
F	CNT in epoxy CNT-dispersed film	AR10, 5 wt.%	AR100, 10 wt.%

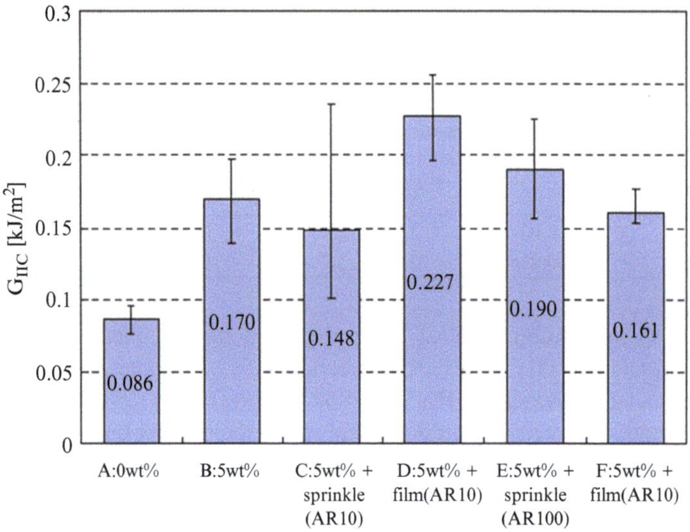

Fig. 8.22 Comparison of mode-I fracture toughness for all the samples (Yokozeki et al. 2009b)

laminating resin was mixed with MWCNTs using a 3-roll mill. Unidirectional carbon fabric was used as fibre reinforcement in the composites. The fabric consisted of 50 k rovings, and had a surface weight of 300 g/m^2. The UD laminates had been produced by hand lamination of 10 plies of carbon fabric impregnated with the resin. The fibre contents were 49.2 ± 1.1, 51.9 ± 2.8, 51.7 ± 3.2, 51.9 ± 2.7, and 53.4 ± 1.9 vol.% in the unfilled and 0.1, 0.3, 0.5 and 1 wt.% MWCNT filled composite respectively. Mode I interlaminar fracture toughness tests have been carried out on the test specimens according to ASTM D 5528–01. The G_{Ic} values increase with nanotube content up to 0.3 wt.% filling, around 13%, after which a decrease in values took place. They also compared the full strain energy rate–crack length increase curves to appreciate the full failure process,

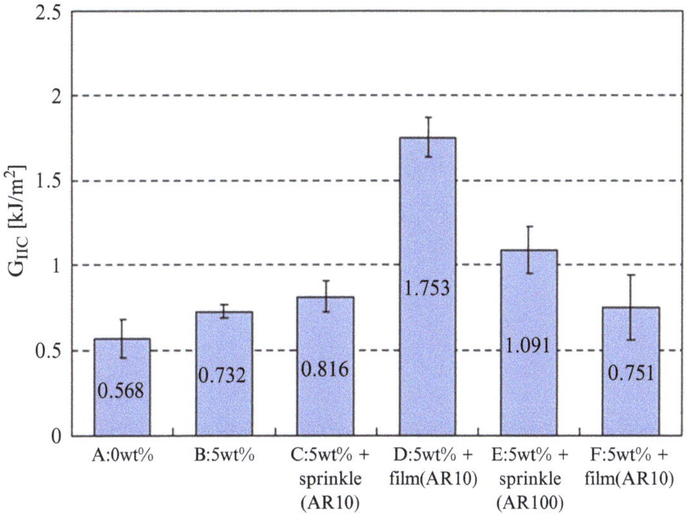

Fig. 8.23 Comparison of mode-II fracture toughness for all the samples (Yokozeki et al. 2009b)

especially after crack initiation. The *R*-curves obtained from averaging of the curves of each specimen of the same CNT content can be seen in Figs. 8.24 and 8.25. It is clear that the nanocomposites containing 0.1 and 0.3 wt.% CNTs significantly outperform the neat matrix composite, while the 0.5 and 1 wt.% CNT filled ones remain around the level of the reference specimens.

Godara et al. (2009) focused their work on increasing understanding of the practical aspects related to processing of three-phase composites with CNTs and their influence on mechanical performance. Carbon nanotubes were incorporated in an epoxy matrix that was then reinforced with carbon fibres. A fixed amount, 0.5 wt. %, of different types of CNTs (functionalized and non-functionalized) were dispersed in the epoxy matrix, and unidirectional prepregs were produced. The prepreg technique was chosen to carry out the work in order to minimise technological difficulties related to dispersion and re-agglomeration of CNTs and resin flow through closely packed fibres. The prepreg technique requires a high viscosity resin that undergoes minimal flow. It reduces the mobility of CNTs during the curing process. The multi-walled carbon nanotube (MWCNT), the thin- WCNT (TMWCNT) and the double-walled CNT (DWCNT) amine group all functionalized. The dispersion of CNTs was performed with high-shear calendaring equipment. As matrix material, an epoxy resin based on diglycidylether of bisphenol A formulated for hot-melt prepreging was used. The effect of the CNTs on the fracture toughness of the UD composite laminates shows that even at relatively low CNT fraction, 0.2–0.25 in the final nanocomposite, there is a noteworthy improvement in the fracture toughness (Fig. 8.26). There is a consistent increase in the G_{Ic} initiation values: ECF < MWCNT < DWCNT < TMWCNT < modified-MWCNT. MWCNTs even though they seem to be the least effective, they still

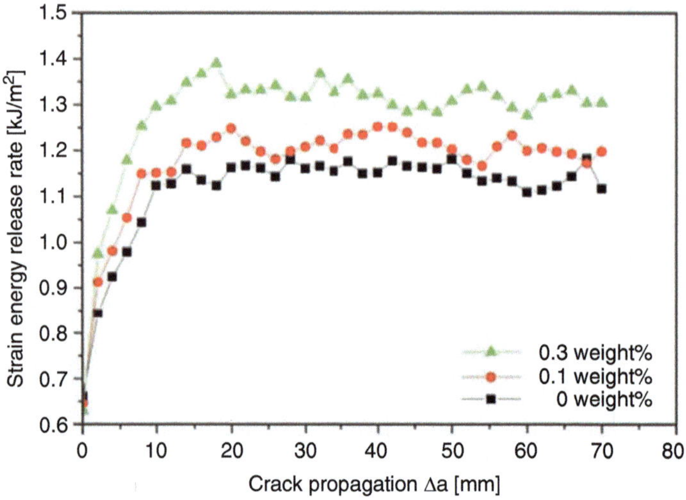

Fig. 8.24 Average *R*-curves for composites containing 0, 0.1 and 0.3 weight% MWCNTs (Romhany and Szebenyi 2009)

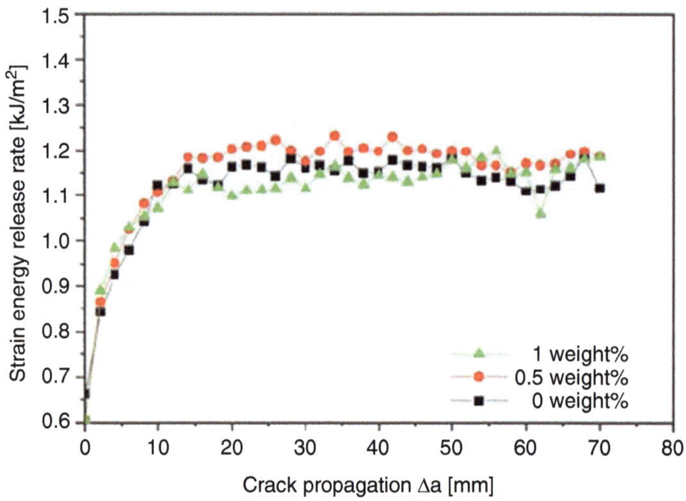

Fig. 8.25 Average *R*-curves for composites containing 0, 0.5 and 1 per weight% MWCNTs (Romhany and Szebenyi 2009)

showed a 21% increase. The increase of properties is even higher for the thinner TMWCNTs nanocomposite. This is because of the greater effectiveness of CNTs in penetrating fibre yarns and minimising the filtering effect. Similar results of increased toughness by functionalized DWCNT are obtained where covalent bonding helped in creating a much more stable and homogenous network of CNT in the matrix, as seen in previous pages. Figure 8.26b represents the average fracture

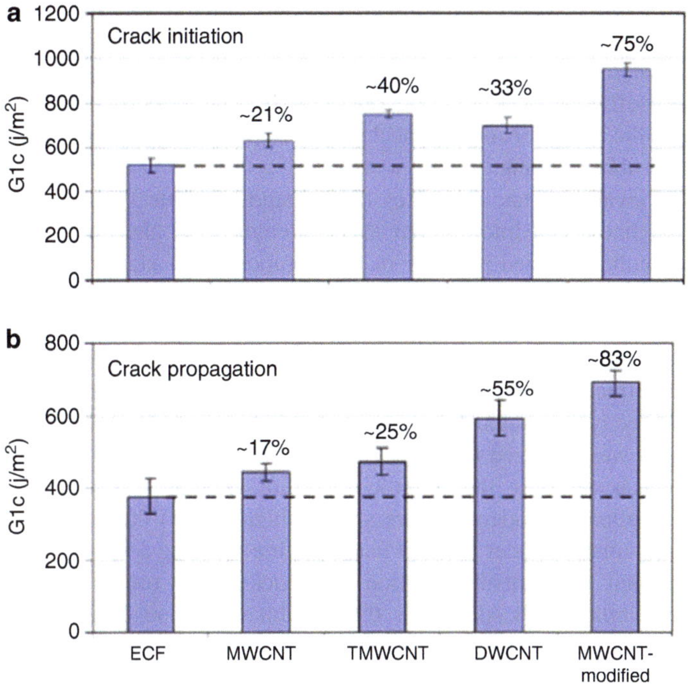

Fig. 8.26 Comparison between fracture energy of (**a**) crack initiation and (**b**) crack propagation between crack lengths 70–90 mm (Godara et al. 2009)

energy required for crack propagation in different composite systems with crack lengths between 70 and 90 mm. The trends of interlaminar fracture toughness are to some extent similar in the crack propagation region. In this region, CNTs generally offer increased resistance to crack propagation: ECF < MWCNT < TMWCNT < DWCNT < modified-MWCNT. Once more, the modified system of the MWCNTs and epoxy shows a significant improvement in the crack propagation resistance indicative of the importance of the optimization of epoxy-CNT interactions. These mechanisms are responsible for the increase in absorption of the applied energy.

In parallel, Warrier et al. (2010) used the same epoxy resin/hardener system as the previous researchers with MWCNTs to enhance the fracture toughness of composites with commercially available E-glass fibres as the reinforcing phase. The MWCNTs were dispersed in the epoxy resin, with a concentration of 0.5 wt.%. A homogenous dispersion was achieved using a calendering machine. This helps very effectively in the exfoliation of CNTs from their pristine bundled microstructure. Then they produced UD prepregs and composite laminates with a thickness of about 3.0 mm were obtained with a final fibre volume fraction ranging between 45 and 50%. When the team evaluated the data from the mode I fracture toughness tests, according to ASTM 5528, they ascertained an increase of 25% for the G_{1c} crack initiation value in comparison with unmodified composite. They also stated that presence of carbon nanotubes in the epoxy matrix had other good side effects, like an increase of glass transition temperature and a decrease of the thermal expansion coefficient by 8 and 12% respectively. The property enhancement was attributed to the energy absorbing mechanisms of the CNTs i.e. fibre breakage, fibre pull-out and fibre bridging.

Finally, Seyan and his associates (2008) studied mode I and II interlaminar fracture toughness and interlaminar shear strength of E-glass non-crimp fabric/carbon nanotube modified polymer matrix composites. A matrix resin containing 0.1 wt.% of amino-functionalized MWCNTs was prepared, utilizing the 3-roll milling technique. Composite laminates were manufactured via VARTM. The CNT-doped composite laminates were found to exhibit 8 and 11% higher mode II interlaminar fracture toughness and interlaminar shear strength values, respectively, as directly compared to the reference laminates. However, no significant improvement was observed for mode I interlaminar fracture toughness values. The ENF test measures only the initiation fracture toughness. The CNT modified composite laminates properties were slightly higher (8%) than those of the base composite laminates. Under mode II loading, fibre bridging does not occur and two other important mechanisms; friction and hackles are responsible for energy absorption. Unlike DCB specimens that exhibit continuous crack growth along the fibre/matrix interface, ENF specimens show discontinuous crack growth by micro-crack coalescence that leads to many hackles occurring at the fracture surface. It is likely that CNTs act as rigid fillers which arrest the crack, preventing or delaying the expansion of micro-cracking within the matrix rich interface area. One would expect a higher amount of hackles to be present at the fracture surface of the CNT-modified composite laminates as compared to that of the base composite laminates. This leads to a relatively high energy absorption by friction in nanotubes modified by composite laminates. In other words, nanotubes may improve the adhesion between the interlayer and the adjacent composite layers at the same time.

8.3 Fatigue

In addition to maintaining static strength in service, structural composites are required to maintain an acceptable level of strength under fluctuating stress conditions, as experienced in the service life of a composite part of an aircraft. The ability to maintain strength under cyclic stresses is called fatigue resistance. In an aircraft wing and empennage, the cyclic stresses are generally highly variable within their design limits; however, in fuselages, where the main stresses result from internal pressurization, a stress cycles to approximately constant peak values. These two types of loading are, respectively, called spectrum and constant amplitude. In testing for fatigue resistance, there are two basic forms of measurement. The first is simply the life-to-failure (or to a certain level of stiffness degradation) at various stress levels; this is the S-N curve, where S is stress and

N is the number of cycles. The second form is the rate-of-growth of damage as a function of cycles at various levels of stress. For metals, the damage is a crack; for composites, it is delamination or a damage zone consisting of localized microcracking and fractured fibres. The ratio between the minimum and maximum stresses in constant amplitude cycling is an important parameter called the R ratio and is the ratio of minimum/maximum stresses. Thus, an R of -1 is a cycle that involves full reversed loading, $R = 0.1$ is tension-tension, and a large positive value, for example $R = 10$ compression/compression. The ratio R generally has a marked influence on fatigue resistance.

8.3.1 Nanopolymers

One of the first attempts to evaluate the effect of the addition of carbon nanotubes on the fatigue properties of epoxy polymers was by Ren et al. (2003) in the early 2000s. The study focused on the fatigue behaviour of unidirectional, aligned SWCNT rope-reinforced epoxy. The SWCNT ropes were synthesized by the hydrogen/argon electric arc discharge method, with lengths up to 100 mm, density of 1.138 g/cm^3 and a volume fraction of 65% in SWCNT bundles. Dog bone specimens with gauge length of about 15 mm were cyclically tested under tension-tension fatigue at 5 Hz, using a sinusoidal wave function at an R ratio of 0.1. Since the SWCNT volume fraction varied from sample to sample, on an S-N plot SWCNT stress could not be contingent upon the applied stress of the composite according to the rule of mixtures. Therefore, the maximum cyclic stress is plotted against the number of cycles to failure of the nanopolymer. The SWCNT stress in the polymer was calculated using $\sigma_{CNT} = (E_{CNT}/E_c)\sigma_c$, where σ_{CNT} and σ_c are the stress of SWCNT and of the polymer, respectively, and E_{CNT} and E_c are the Young's modulus of SWCNT and the polymer respectively. The Young modulus of the SWCNT was estimated around 800 GPa and therefore the maximum cyclic stresses of SWCNT/epoxy were calculated to be between 5.37 and 24 GPa. The S-N data of the nanopolymer is shown in Fig. 8.27. In this figure the tensile data for SWCNT are also included along with unidirection CFRP fatigue data obtained from the literature and presented as the bottom gray region in the S-N graph. A simple linear relation often used for *S–N* curves is $\sigma_a/\sigma_{ult} - 1 - m\log N$, where σ_a and σ_{ult} are the applied and ultimate stress, respectively, N the number of cycles to failure, and m the slope of the normalized *S–N* curve. The *S–N* curve obtained for the SWNT/epoxy composite is very flat, similar to the characteristics of the unidirectional carbon/epoxy composites. Slope m for most unidirectional carbon/epoxy composites ranges from 0.035 to 0.057. For SWNT/epoxy composites, m obtained from linear regression of the quasi-static tensile strength and S-N data was calculated to be 0.042, which is within the range of unidirectional CFRPs. The author underlined that the estimated cyclic stress is at least twice that of unidirectional CFRPs. In other words, the fatigue strength of SWCNT/epoxy is at least twice the carbon fibres.

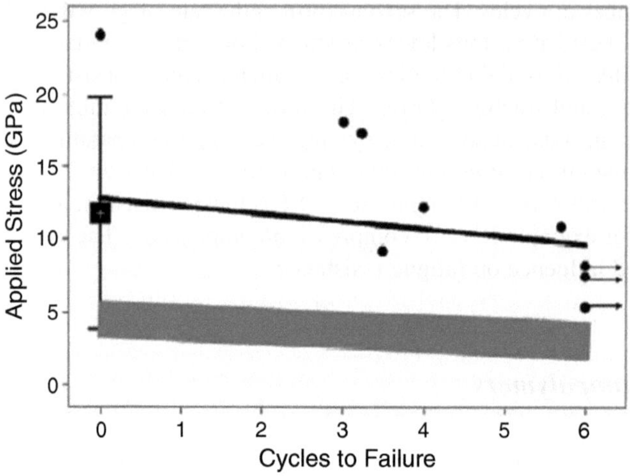

Fig. 8.27 S-N diagram. The *error bar* represents the standard deviation. The *gray rectangular region* represents fatigue data of UD CFRPs from literature (Ren et al. 2003)

Damage and failure modes were examined under a SEM, Fig. 8.28. No SWCNT-bridged transverse matrix cracks were observed on the specimen surface, which was attributed to the fact that the SWCNT ropes were fully aligned in the matrix material. Also, no SWCNT-matrix splitting occurred, a common damage mode of unidirectional CFRPs under fatigue loading. Local failure modes around the SWCNT ropes showed ductile-like failure with plastic deformation of the epoxy and pull-out of SWCNT ropes. Bridging of matrix crack was also clear.

Zhang et al. (2008) performed a detailed study of the effects of nanotube dimensions and dispersion on the fatigue behaviour of CNT-doped polymers. The scientific group systematically varied the tube dimensions by testing nanotubes with different diameters (5–8, 10–20, 20–30, 50–70 nm) and lengths (10–20, 1–2 μm) and characterized its effect on the composite's fatigue crack propagation (FCP) response. Fatigue tests were conducted using the ASTM standard E647-05. An initial sharp notch was created by slicing a fresh razor blade across a v-shape machined slot. Crack length was measured by the compliance method and is confirmed by using a travelling microscope with 30–50X magnification. Results for crack propagation rate versus applied stress intensity are shown in Fig. 8.29 for the neat epoxy and for epoxy filled with various diameters and lengths of MWCNTs. It was noteworthy that the fatigue suppression performance showed a significant improvement with reduction in the tube diameter (from 50–70 to 5–8 nm), while holding the length constant at 10–20 μm. On the other hand, when the length of the MWCNT was decreased (from 10–20 to 1 μm), while holding the diameter constant (10–20 nm), the fatigue crack suppression effect degraded significantly, as shown in Fig. 8.30. The best results were obtained for the MWCNT with the smallest diameter (d = 5–8 nm). Another noteworthy feature of crack growth rate ($\mathrm{d}a/\mathrm{d}N$) versus stress intensity factor amplitude (ΔK) curves is

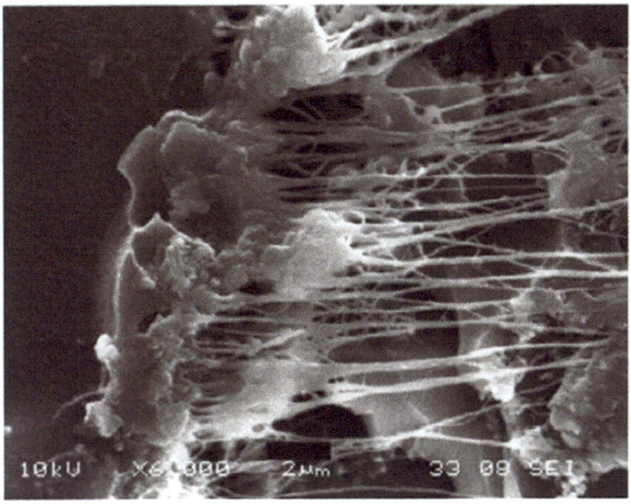

Fig. 8.28 Fatigue fracture surface of SWCNT/epoxy at 6000x. The bridging of matrix crack by the SWCNT ropes is clear. The small contact angles between the nano-reinforcement and the matrix suggest good wetting of the ropes by the epoxy matrix (Ren et al. 2003)

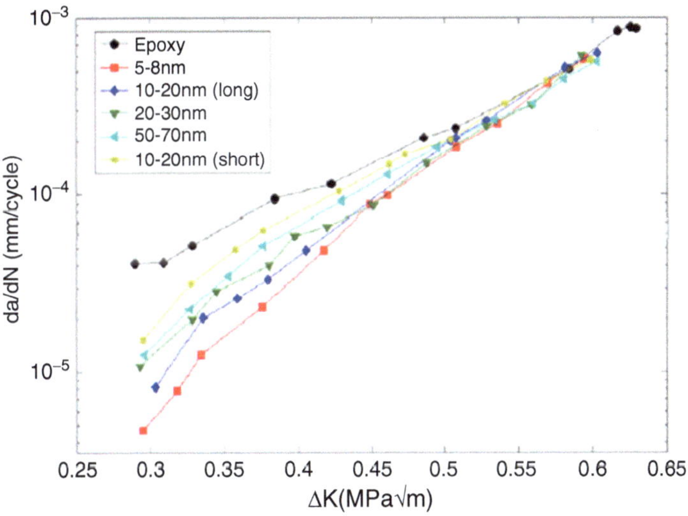

Fig. 8.29 Fatigue crack growth rate plotted as a function of the stress intensity factor amplitude (ΔK) for samples of MWNTs with different diameters. For 5–8 nm MWNT/epoxy samples the crack growth rates are reduced by an order of magnitude at low ΔK (Zhang et al. 2008)

that the performance for all samples rapidly diminishes as the stress intensity is increased. This is caused by shrinking of the fibre-bridging zone in the path of the crack tip as ΔK is increased.

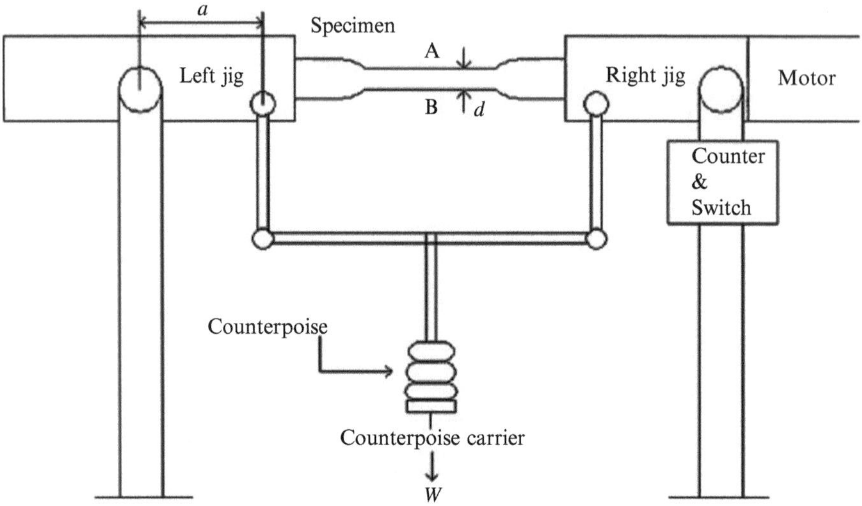

Fig. 8.30 Diagram of the rotary bending fatigue testing machine (Yu et al. 2008)

Figure 8.30 presents the experimental setup for rotary bending fatigue that Yu et al. (2008) used to test the effect of 0.5 wt.% MWCNTs in an epoxy resin. The specimens' diameter *d* was 12 mm and was polished prior to fatigue testing to remove any edge effects.

The stress–life (*S–N*) curves are shown in Fig. 8.31. Under the same loading, nanopolymer specimens do have longer fatigue lives than those of neat epoxy specimen. The fatigue life of composite under stress amplitude of 5.78 MPa is 700,237 cycles, which is 2.7 times the average fatigue life of the neat epoxy specimen. The nanopolymers' fatigue life is 10.5 and 9.3 times the average fatigue life of the neat epoxy, when they are subjected to cyclic loadings with stress amplitudes of 8.67 and 11.56 MPa, respectively.

In this chapter, the main reinforcing mechanisms of the CNTs that are capable of improving the damage tolerance of aerospace grade materials have been identified as fibre pull-out, fibre crack bridging and fibre breakage. A recent research (Zhang et al. 2009a) has also acknowledged *crazing* as a reinforcing mechanism of the CNTs when present in an epoxy polymer. Crazing can be defined as the failure mode of bulk polymers and occurs under prime uni-axial tensile load when the bulk eventually forms denser ligaments (or fibrils) while preserving its continuity. It has been well established that craze phenomena have not been observed in thermosetting polymers such as epoxies due to the high cross-linking density of the epoxy chains, which limits molecular mobility and inhibits craze fibril formation. The thermosetting epoxies typically display a brittle failure as has been demonstrated in the course so far of this chapter. Nevertheless, when a thermosetting epoxy resin was reinforced with amino-functionalized MWCNTs it exhibited crazing. Fatigue crack propagation tests showing crack propagation rate versus stress intensity factor amplitude are shown in Fig. 8.32. For the plain epoxy, epoxy with 0.25 wt.%

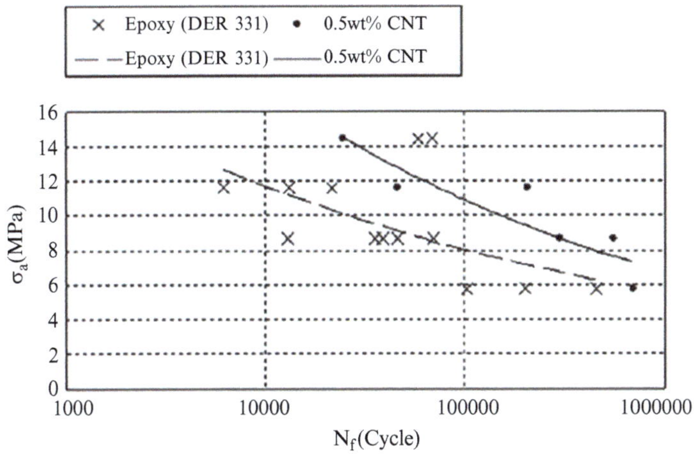

Fig. 8.31 The *S–N* curve of MWCNTs/epoxy and pure epoxy specimens under rotary bending fatigue (Yu et al. 2008)

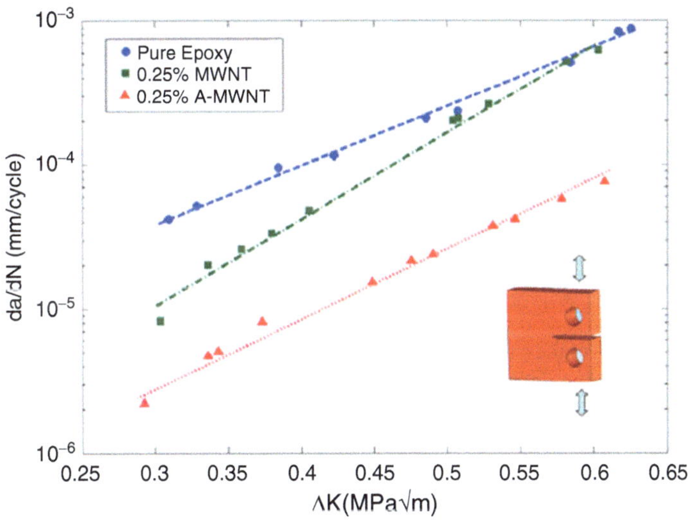

Fig. 8.32 Fatigue crack propagation tests showing crack growth rate (da/dN) plotted as a function of the stress intensity factor amplitude (ΔK). The inset shows a schematic of the compact tension samples used in the testing (Zhang et al. 2009a)

pristine MWNTs, and epoxy with 0.25 wt.% functionalised MWCNTs. The fatigue tests were conducted following the ASTM standard E647-05. From the results one may highlight an over ten times reduction in the crack growth rate for the nanopolymer sample compared to the neat epoxy over the full range of stress intensity factor amplitudes. In contrast, the nanopolymer shows good fatigue

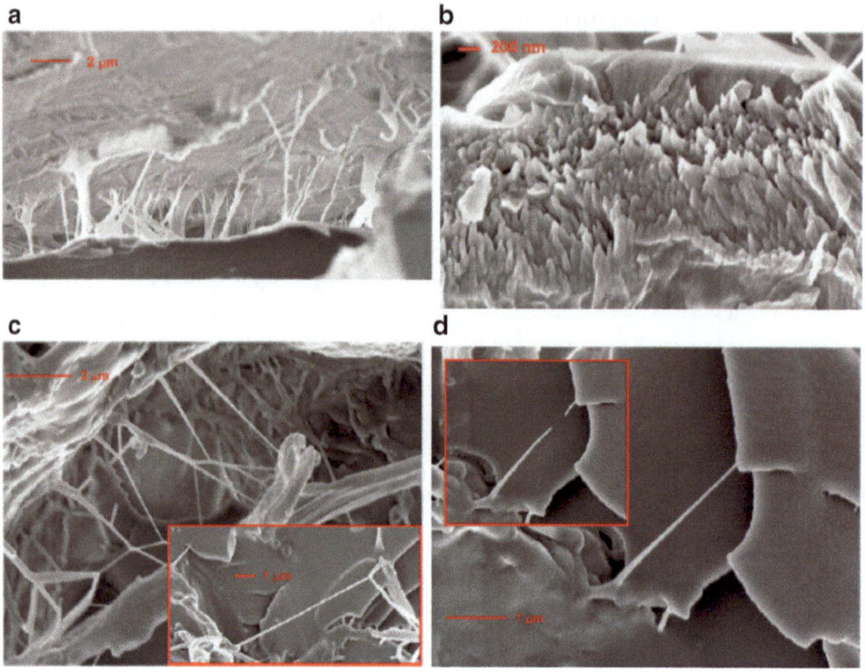

Fig. 8.33 Fractography analysis for the A-MWNT/epoxy nanocomposite. (**a**) SEM micrographs showing crack bridging. The diameter of bridging fibrils is in the 100–1,000-nm range (**b**) SEM micrograph showing dimples that form on the surface after breakage of the bridging fibrils, (**c**) SEM image of stretched out fibrils. The inset shows that the fibrils are up to tenfold longer than A-MWNTs and are much thicker at the ends compared to the middle portion of the fibre. (**d**) SEM image of fibril before heating. The inset shows the same fibril imaged after heating to 100 °C. The fibril breaks under the thermal loading (Zhang et al. 2009a)

suppression performance only at low values of the stress intensity amplitude and its performance rapidly degrades as the amplitude is increased. The fatigue crack propagation results in Fig. 8.32 indicate that a fundamental change in the material response has occurred due to addition of the CNT fillers. SEM analysis of the fracture surface of nanopolymer samples was performed and it was found that the cracks were bridged by fibres as seen in Fig. 8.33a. However, the diameter of the bridging fibres ranges from about 100 to 1 mm. Given that the diameter of an individual MWCNT is around 30 nm, these bridging fibres could not be MWCNTs. Moreover the length of the bridging fibrils is several times larger than the MWCNT length ~1 mm. Also it was observed that for large crack opening displacement the fibrils break and then shrink to form dimples, Fig. 8.33b, suggesting that these bridging fibrils are composed of the epoxy. An interconnected network of stretched out fibrils is shown in Fig. 8.33c. The inset shows an individual fibril that is around 10 mm in length; the diameter of the fibril is significantly thinner in the middle than at the ends. Hence it appears that these bridging fibres are in fact epoxy fibrils that are drawn out in uni-axial tension normal to the crack plane, similar to a craze. When the sample was heated to 100 °C and then re-imaged under SEM, several of the bridging fibres were observed to break due to the elevated temperatures, Fig. 8.33d, which is expected if these were epoxy fibrils. In fact by locally heating individual fibrils using the electron beam of the SEM (power <3 mW) the researchers were able to rapidly break the fibrils, confirming that the bridging fibres were not MCWNTs, but fibrils composed of the epoxy. Therefore the A-MWNTs were initiating the formation of the epoxy craze fibrils and the energy dissipation is due to the pulling and plastic deformation.

Heterogeneous epoxy cross-linking near nanofillers could generate such regions of high molecular mobility and spatial fluctuations (variability) in material properties. Differential scanning calorimetry (DSC) characterization of various samples followed and did not show any exothermal peaks or differences in glass transition temperatures, thus confirming crazing as a reinforcing mechanism to improve damage tolerance when CNTs are present in an epoxy matrix.

8.3.2 Nanocomposites

One of the first studies that established the potential of CNTs as modifiers of epoxy resins that may both increase the fatigue life of carbon fibre-reinforced polymer composites but also as damage sensors was performed by Vavouliotis et al. (2009). This study was a continuation of their previous research (Karapppas et al. 2009) and therefore the CNTs, the epoxy resin and the dispersion method were kept the same as before. The carbon fibre laminas were chosen to be quasi-isotropic $[0, +45, 90, -45]_s$ of 16 plies, with weight of 160 g/m^2. Each panel was wet laid-up and then processed in an autoclave, using the vacuum bag technique. Prior to fatigue tests, standard tensile tests were performed in order to investigate the influence of the CNTs and to collect maximum stress values. The tension-tension fatigue parameters were: frequency $f = 5$ Hz, stress ratio $R = 0.1$, and stress levels of 80, 70 and 60% of maximum stress were chosen and conformed to ASTM D-3479 standard. Using as infinitive fatigue life the 10^6 cycles and forecasting the fatigue limit using a logarithmic fitting curve, it was verified that the nano-doped quasi-isotropic CFRP had an increased fatigue limit (72%) compared with the reference CFRP that had 64%, Fig. 8.34. The enhancement of the mechanical properties was attributed to the failure mechanisms of CNTs. The significant large aspect ratio of CNTs allows them to act as nanobridges between the fibre plies. As a result, more energy is needed so as to pull them out from the matrix or break them in order for the damage to be further developed (Fig. 8.35).

Another study on the effect of CNT on the mechanical properties of CFRPs was recently published and underlined that not only the strength, the stiffness but also the tension-tension fatigue (T-T) and tension-compression fatigue (T-C) properties of the nanocomposites were enhanced by their inclusion in the matrix material (Davis et al. 2010). In more detail, a four-harness satin weave having identical warp and fill yarns of 6,000 filament count, was manufactured using an aerospace grade carbon fibre (tensile strength up 5.5 MPa and elastic modulus near 276 GPa). The nanofillers used were an industrial grade carbon nanotube known as “XD” provided

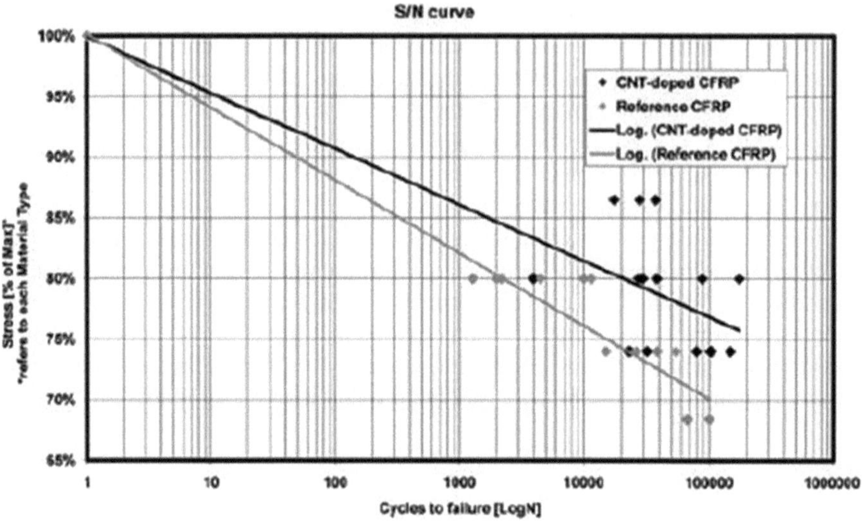

Fig. 8.34 S–N curve, stress S shown as percentage of maximum static tensile stress and fatigue cycles N expressed in logarithmic scale (Vavouliotis et al. 2009)

Fig. 8.35 SEM picture of mid-plane of a fatigued specimen at stress level of 70 %. The good CNT dispersion achieved can be seen along with the extensive CNT pull-out (Vavouliotis et al. 2009)

by Carbon Nanotechnologies, Inc., USA. XD-CNTs are a blend of different carbon nanotubes (CNTs), approximately a third by weight each of single wall, double wall and multi-wall CNTs. To produce the composite laminates, CF were stacked starting from the laminate mid-plane as two halves of an eventual 12- ply symmetric and balance laminate cross-section. For the nanocomposite laminates a solvent spraying technology was used to deposit fluorinated carbon nanotubes (f-XD-CNTs) onto both sides of all the carbon fabric square pieces in the laminate cross-section. The f-CNT solvent solution was sprayed evenly and equally onto both sides of the fabric for total weight percentages of 0.2, 0.3 and 0.5 wt.% defined as the percentage of the ratio of the weight of the deposited CNTs to the weight of each square piece of carbon fabric. The solution eventually evaporates leaving a deposit of CNTs. In the end, a heated vacuum assisted resin transfer molding (H-VARTM®) method was used to fabricate the different CFRPs.

In Fig. 8.36, the straight trendlines through the maximum stress versus cycles to failure (σ_{max}-N) datum points of the 0.3 and 0.5 wt.% f-CNT reinforced materials demonstrate a decreasing slope from that of the neat material. Each datum point represents the cycles to failure results (N) from a single test at the designated σ_{max} and wt.% f-CNT. It was supported that at 0.2 wt.% the CNT fabric–matrix reinforcement was not sufficient to delay the matrix and matrix–fibre interfacial cracking so as to change, on the macroscopic scale, the evolution of damage due to fatigue loadings. At f-XD-CNT reinforcements equal to and greater than 0.3 wt. % it was estimated that a threshold has been reached such that the reinforcement is sufficient to provide resistance to the cyclic or fatigue-like damage process in the CFRPs. The 0.5 wt.% f-XD-CNT reinforced laminate material have a greater durability than the reference CFRP. In other words, these improvements could correlate to a multi-decade increase in cyclic life in the higher cycle regime. All T–T failure modes were via fibre rupture or considered fibre dominant. Figure 8.37a shows for the neat material failures a mostly clean fibre–matrix interface fracture path; whereas, for the f-XD-CNT materials in Fig. 8.37c, the resin material remains attached to fibres. The improved fibre–matrix bonding for the f-XD-CNT materials is believed to account for the improvements in T–T cyclic life and strength as illustrated in Fig. 8.36.

The T–C tests of the f-XD-CNT reinforced material at the higher maximum load cycling levels failed via fibre rupture as the T–T specimens. At lower maximum load cycling and longer life tests, failures were via specimen buckling similar to the neat material. These two different failure modes are illustrated through bi-linear data plots in Fig. 8.38. The difference in the slopes of these two bi-linear data plots is used to measure potential improvements in cyclic life durability, around 70% improvement, of the composite laminates under these T–C loadings when reinforced with CNTs.

Work on how CNTs may improve the fatigue properties of GFRPS has been performed by Grimmer and Dharan (2009). The epoxy resin was blended with 1 wt. % MWCNTs. Then the blended material was used as a matrix material for manufacturing glass fibre- reinforced composites using a woven fabric. The

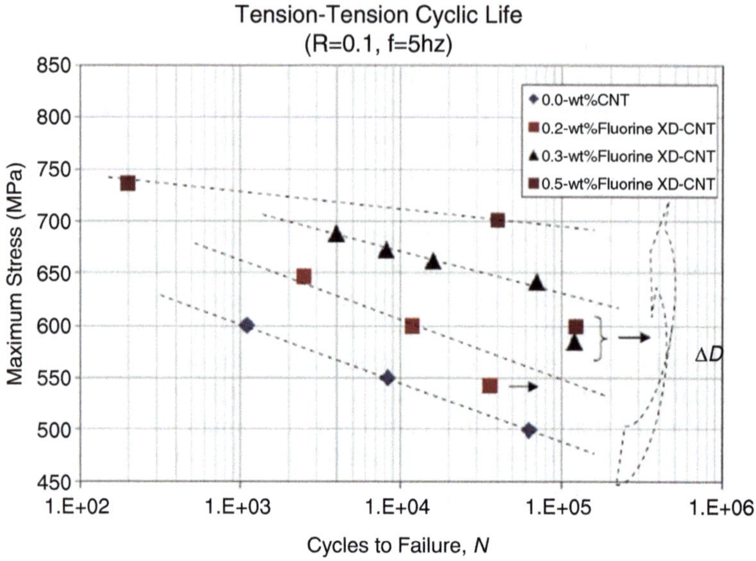

Fig. 8.36 R = 0.1 Tension–tension (*T–T*) cyclic life, maximum stress (*rmax*) vs. cycles to failure (*N*) data from neat (0.0 wt.%) and 0.2, 0.3 and 0.5 wt.% f-XDCNT material specimen tests (Davis et al. 2010)

tension-tension fatigue data obtained from tests at peak stresses of 70, 60, 45 and 30 at f =3 Hz and R = 0.15 are shown in Fig. 8.39.

A significant increase in the number of load cycles to failure for each loading case was observed for the samples that contained the CNTs. The observed increase in life occurs at lifetimes greater than about 10^4 cycles. In this high cycle regime, most of the load cycles are employed for the nucleation and growth of microcracks. The improvement in fatigue life with the addition of CNTs increases as the applied cyclic stress is reduced, making the effect most pronounced at high cycles. At a cyclic stress of 44 MPa, the addition of 1 wt.% of CNTs results in almost a threefold improvement in fatigue life. In composites containing nanotubes, it was alleged that their presence in a very large number prevents initiation and growth of cracks. Furthermore, for a given level of strain energy, a large density of nanoscale cracks will grow more slowly than the lower density of microcracks present in composites not containing CNTs. The result is an increase in the number of cycles required for growth and coalescence which means that high-cycle fatigue life is improved. Nanoscale crack bridging by nanotubes will result in participation of the nanofibres in the fracture process, thereby increasing the fracture energy required for crack propagation, further delaying the failure process. Carrying on the same cyclic mode I delamination crack propagation tests were performed on the GFRPs described above (Grimmer and Dharan 2010). Their findings are depicted in the next Fig. 8.40 below. One may notice that the crack growth rate of the neat GFRPs is bigger which

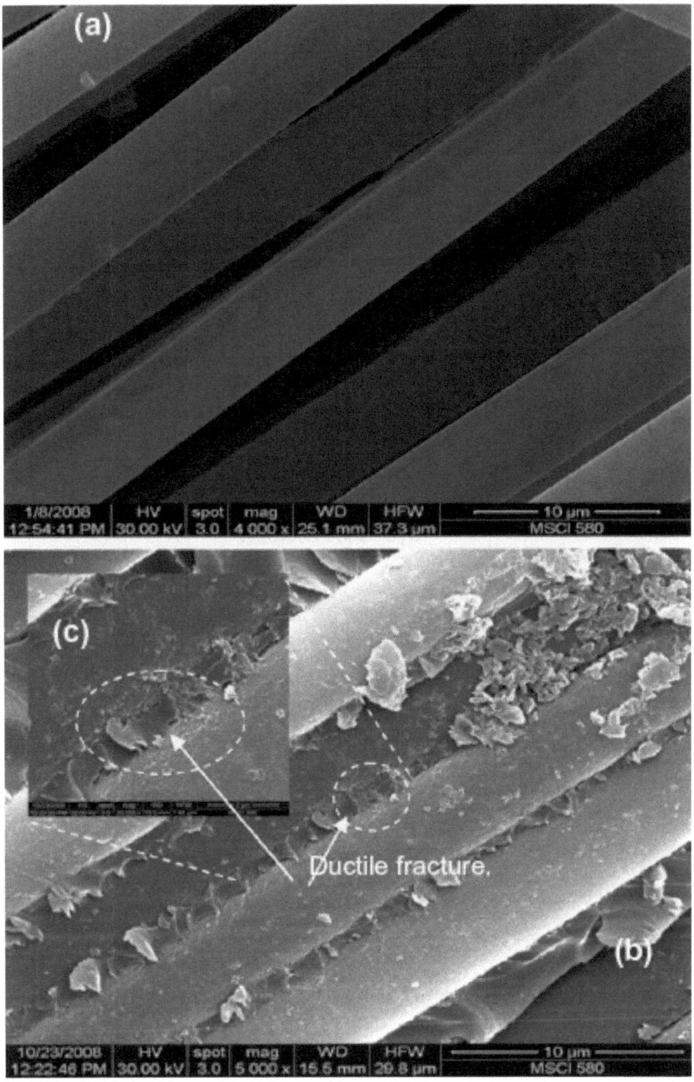

Fig. 8.37 Tension–tension (*T–T*) specimen fracture surfaces. (**a**) Neat specimen brittle fracture surface with no matrix attached, (**b**) and (**c**) 0.3 wt.% f-XD-CNT specimen fracture surface showing ductility, toughened matrix and fibre–matrix interfacial bonding strength (Davis et al. 2010)

actually means that the material is more susceptible to damage than the nanocomposite.

All the above works have established that CNTs can be used to improve the fatigue life of composite aerostuctures because of their reinforcing mechanisms like fibre pull-out, fibre bridging and fibre breakage. However, it is well known in the

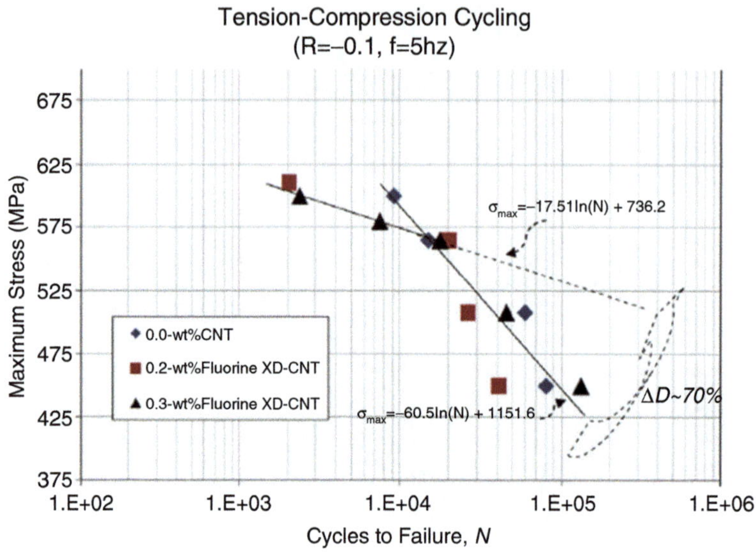

Fig. 8.38 Tension–compression cycling (R = −0.1) of maximum fatigue stress (σ_{rmax}) vs. cycles to failure (*N*) data plots for the neat (0.0 wt.%) and 0.2 and 0.3 wt.% f-XDCNT material specimens. Trendlines are based on only the neat material buckling failure data and the 0.3 wt.% f-XD-CNT fibre rupture failure data (Davis et al. 2010)

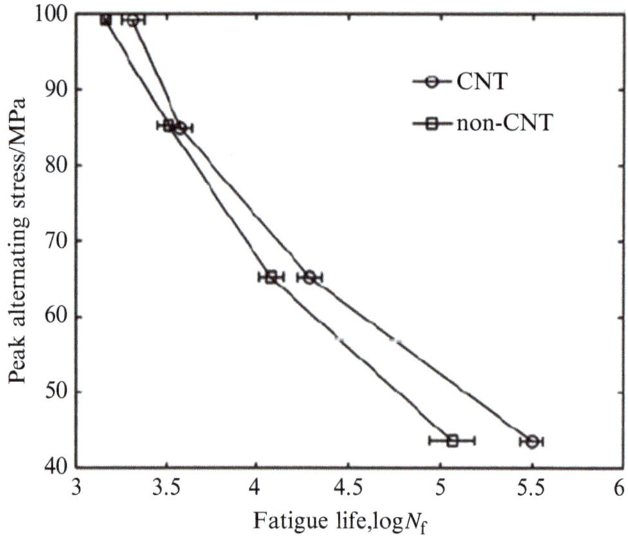

Fig. 8.39 Applied cyclic stress versus the number of cycles to failure of glass fibre-epoxy laminates with and without the addition of 1 wt.% of carbon nanotubes (Grimmer and Dharan 2009)

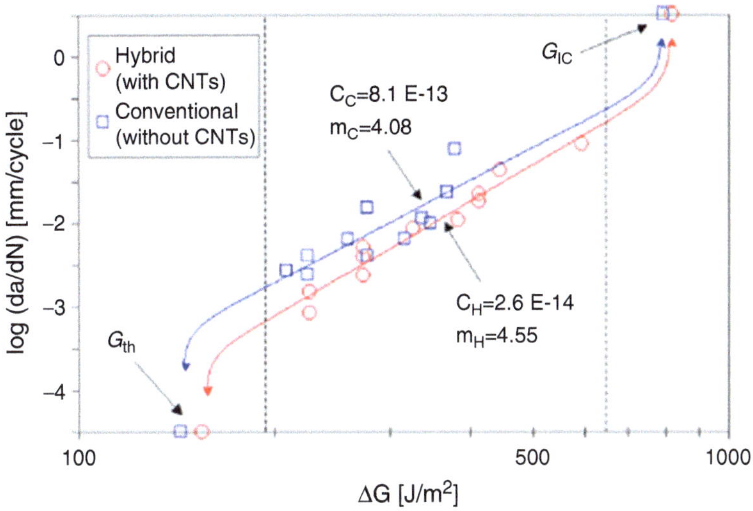

Fig. 8.40 Cyclic mode I delamination crack propagation data for glass fibre-epoxy laminates with and without the addition of 1 % by weight of CNTs. Values shown are the constants *C*, *m* in the equation: da/dN = C(DG)m (Grimmer and Dharan 2010)

composite community that fatigue is a multi-stage phenomenon. According to Talreja (2003), fatigue in composite materials for a typical quasi lay-up has the following sequence; (i) matrix cracking at off-axis plies and some fibre breakage occurs, (ii) matrix cracks are further developed through the thickness of the composite and follow the fibre reinforcement direction. The creation and the propagation of these cracks stabilises at a level, known as critical damage state (CDS), (iii) a consequence of the previous cracks is delamination and fibre breakage, (iv) delamination grows and localised fibre breakage takes place and finally, (v) the axial plies (0°) take up all the applied fatigue loading resulting in fibre breakage and the total fracture of the material (Fig. 8.41).

In the previous pages the contribution of the CNTs to damage tolerance was explained in detail. Matrices with CNTs have enhanced fracture characteristics and therefore they can delay crack appearance and crack propagation at the first stages of fatigue, i.e. delaying the CDS formation. Moreover, fatigue is closely connected with fracture properties of the composite. Once again, by the addition of the CNTs into the polymer matrix it was confirmed that the composites exhibit fracture properties that are superior to those of the neat composites. At the fatigue stages (iii) and (iv) delamination is the main failure mechanism responsible for damage propagation. Since CNT-doped composites outperform normal composites in terms of mode I and II properties, they should also have delayed damage accumulation when subjected to fatigue (Karapappas 2009). The novel process that develops carbon fibres with CNTs grown or grafted on them will also delay fatigue, since they will increase the interfacial properties of the composites and also further reduce delamination.

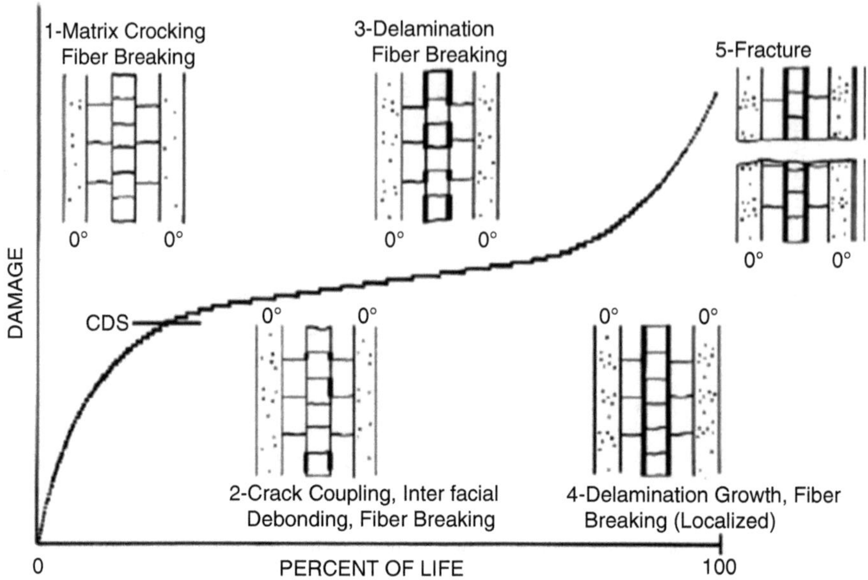

Fig. 8.41 Fatigue damage accumulation stage for quasi-isotropic composites when subjected to axial sinusoidal loading (Talreja 2003)

8.4 Impact and Post Impact

Impact damage in composite airframe components is usually the main preoccupation of designers and airworthiness regulators. This is in part due to the extreme sensitivity of these materials to quite modest levels of impact, even when the damage is almost visually undetectable. Horizontal, upwardly facing surfaces are obviously the most prone to hail damage and should be designed to be at least resistant to impacts of around 1.7 J. The value represents the energy level generally accepted to represent extreme value in (1% probability of being exceeded) hail conditions. Surfaces exposed to maintenance work are generally designed to be tolerant to impacts resulting from tool drops. Monolithic laminates are more damage resistant than honeycomb structures. This is due to their increased compliance. However, if the impact occurs over a hard point such as above a stiffener or frame, the damage may be more severe, and if the joint is bonded, development of disbonding is possible.

8.4.1 Nanopolymers

There are two distinctly different issues in relation to the influence of matrix toughness on impact damage: resistance to damage and residual strength in the presence of damage. Generally, composites with tough matrices are resistant to delamination damage, as measured by delamination size for given impact conditions. However, for a given area of impact damage, both brittle and tough composites suffer about the same degradation in residual strength. Fibre properties significantly influence damage tolerance: the stiffer the fibre, the less damage tolerant it will be. This section will present various works on how CNT have been used so far to enhance the impact properties of epoxy polymers.

The effect of functionalised and non-functionalised MWCNTs on two epoxies with different mechanical properties was studied (Liu and Wagner 2005). The same bisphenol-A epoxy (Epon 828) was used but cured with different curing agents, resulting thus in one being brittle i.e. glassy and the other being rubbery. The toughness of a material is a measure of how much energy it is able to absorb prior to failure. Brittle materials, like a glassy epoxy resin, have high strength but negligible toughness. Conversely, ductile materials have high toughness due to plastic deformation. Among the mechanical properties that there were studied was also tensile impact. Tensile impact tests were performed with a pendulum apparatus (0.5 J) using double-notched specimens (notch depth: 0.4 mm). Figure 8.42 compares the impact resistance of composites using 1 wt.% nanotubes, with pure epoxy resin. The tensile impact strength had improved by 29% for the samples containing 1 wt.% as received CNTs, and an even more considerable 50% improvement in tensile impact strength was observed for the 1 wt.% f-MWNTs based epoxy composites. This was attributed to the toughness increase with higher flexibility and deformability under load of the nanotubes in the matrix. In fact, various studies have shown that carbon nanotubes are able to elastically deform under relatively large stresses, both in tension and compression, leading to a high energy absorbing mechanism.

The following two figures depict the results obtained from the same tensile impact tests as above, on two different epoxy resins (LY564 by Huntsman and Epon 815 by Hexion) used in the general composites industry with and without SWCNTs and MWCNTs (Fidelus et al. 2005). A substantial increase in tensile impact strength was measured (Figs. 8.43 and 8.44), relative to the pure matrices. The improvement ranges between 18% and 35%, respectively, for both resins, for all the CNT fractions. High resolution SEM micrographs of pure epoxy and of SWNT/ epoxy nanopolymers after impact did reveal that pure epoxy fracture surfaces were glassy like whereas the nanocomposite fracture surfaces were rougher in appearance. The increase in impact strength was explained by the presence of cavities bridged by nanotubes, which leads to energy dissipation by CNT pull-out. An additional contribution to energy dissipation arises from the crack deflection at agglomerates.

As with SWCNT systems, a substantial impact resistance arises with respect to the pure matrices for the LY 564 system: up to 70% when using 0.5 wt.% MWCNTs, and up to 50% when using 0.05% MWCNTs. As with SWCNTs nanopolymers, the fracture surfaces generally appeared to be rather rough in the SEM, more so for MWCNT/LY564 than for MWCNT/Epon815.

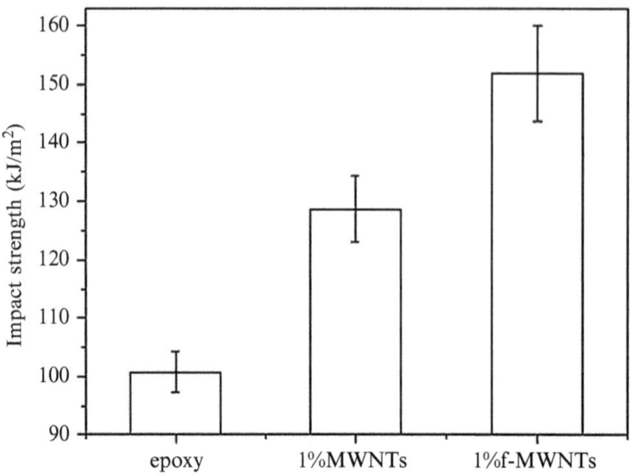

Fig. 8.42 Impact resistance of 1 wt.% MWNTs and f-MWNTs reinforced nanocomposites compared to pure Epon 828 cured with Jeffamine T-403 to produce a glassy system (Liu and Wagner 2005)

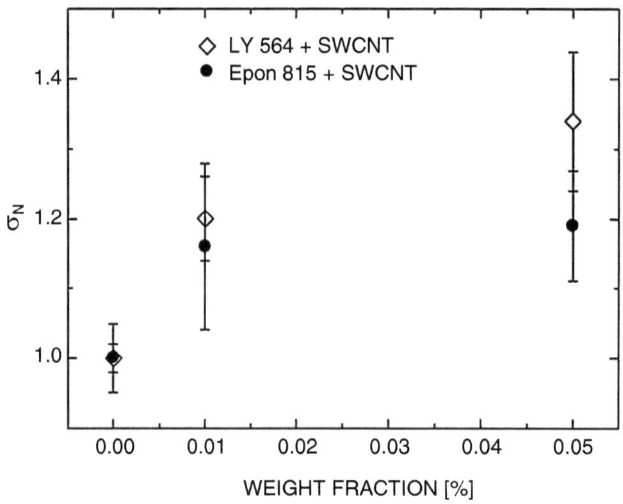

Fig. 8.43 Normalized tensile impact strength as a function of SWCNT weight fraction (Fidelus et al. 2005)

A group of Chinese researchers (Yaping et al. 2006) used the ASTM D-256 standard to perform impact tests on epoxy polymer coupons prepared with and without, as received MWCNTs and amino-functionalised MWCNTs. The data collected from the tests can be seen in a graph-form in the Fig. 8.45 below.

It can be seen that when the content of MWNTs was 0.6%, the impact strength increased by the 8.5 kJ/m^2 of pure substrate to 15.5 kJ/m^2, an increase of about 80%.

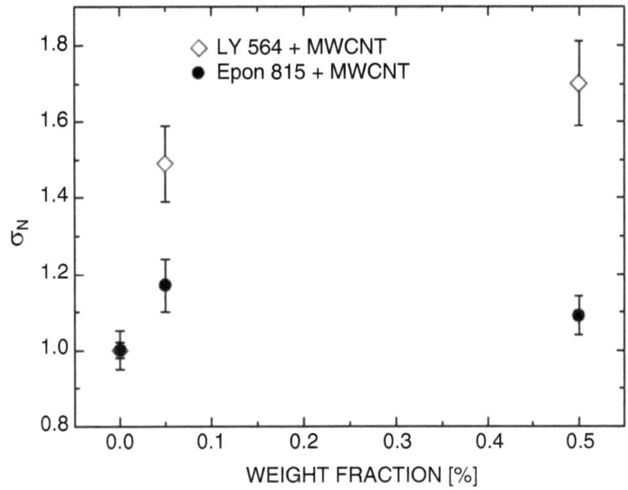

Fig. 8.44 Normalized tensile impact strength as a function of MWCNT weight fraction (Fidelus et al. 2005)

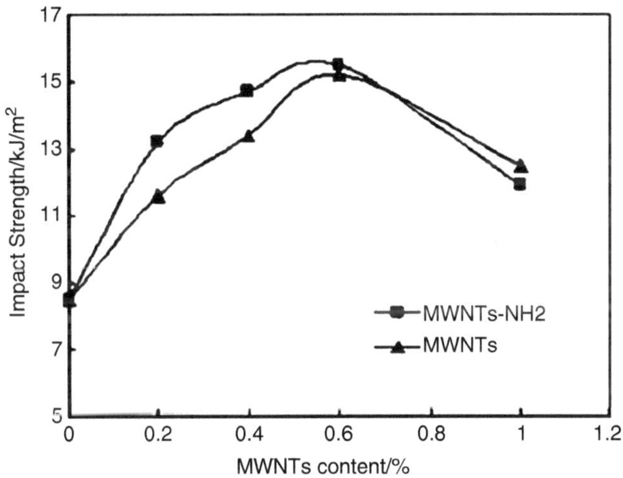

Fig. 8.45 The effect of MWNTs and MWNTs-NH2 content on the impact strength of epoxy polymers (Yaping et al. 2006)

The ductility increase effect was better when the MWNTs-NH_2 was surface treated resulting in higher increase of impact strength. From Fig. 8.46 it is clear that the impact property of the 0.6% content MWNTs-NH_2/epoxy composite has increased by about 100%. The MWNTs-NH_2 with surface treatment is easier to disperse in the epoxy and therefore better toughening effects. The drop of impact properties at 1% wt of MWCNTs is due to the presence of agglomerations acting thus as micro-defects of the material.

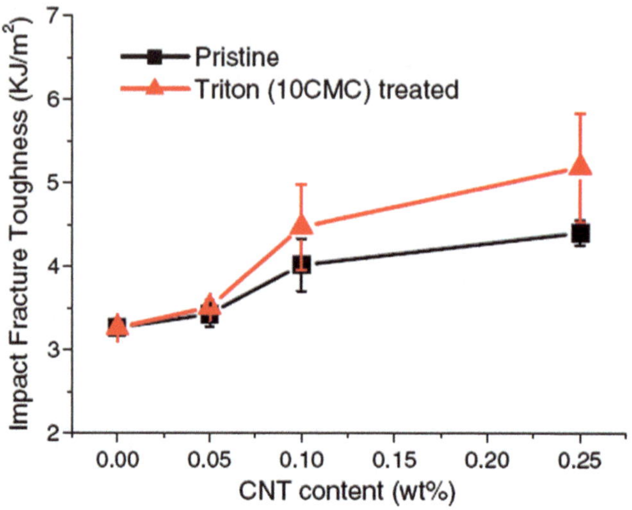

Fig. 8.46 Impact fracture toughness of CNT/epoxy nanocomposites with and without Triton surfactant treatment (Geng et al. 2008)

The same aforesaid ASTM standard was used by another group of scientists from Hong Kong University to assess the effect of the inclusion of MWCNTs in an epoxy matrix with and without the use of a non-ionic surfactant (Triton X-100, VWR International, UK) (Geng et al. 2008). The surfactant was used in order to improve the dispersion by sonication of the CNTs into the epoxy. The enhancement achieved with the addition of CNTs, especially after the surfactant treatment, is shown in Fig. 8.46. It is worth noting that the improvement because of the 0.25 wt. % surfactant-treated CNTs was about 60% compared to the reference neat epoxy.

The morphologies of impact fracture surfaces of the nanopolymers with CNT content of 0.1 wt.% are presented in Fig. 8.47. There were clear differences in morphologies of the as received and surfactant-treated CNT nanopolymer. A rather smooth fracture surface with a small-size, repetitive spatulate pattern was seen for the polymer with as received CNTs, while a rougher and fluctuant morphology with large, elongated radial crack pattern was seen on the composite with surfactant treated CNTs. The fracture morphology with elongated radial crack patterns corresponded to a higher crack growth resistance of the composite. When examined at higher magnifications, large CNT agglomeration with isolated CNT-rich regions was seen for the as received CNTs, Fig. 8.47b, reflecting non-uniform distribution of CNTs in the polymer. In contrast, there were individual CNTs as well as some small CNT bundles, which were distributed more uniformly for the surfactant-treated CNTs, Fig. 8.47d. That difference of the dispersion accounts for the difference effect on the impact properties.

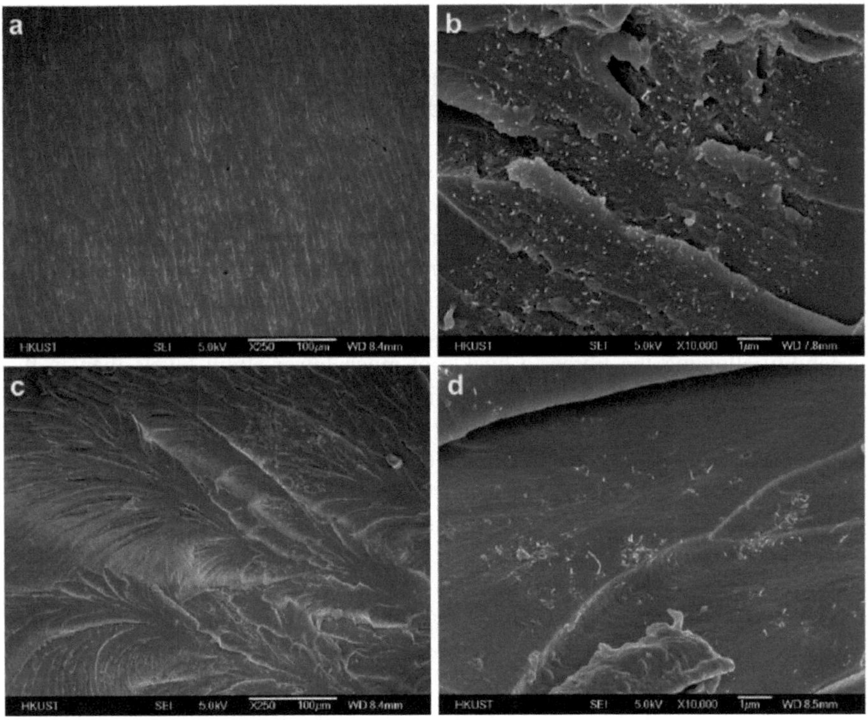

Fig. 8.47 SEM images of impact fracture surfaces of (**a**) and (**b**) pristine; (**c**) and (**d**) Triton surfactant treated CNT/epoxy polymer (Geng et al. 2008)

8.4.2 Nanocomposites

Impact damage can result, for example, from dropped tools, runway stones, or large hailstones. The drastic reduction in residual compression strength and less reduction in tensile strength that can result from impact damage is a major issue in the design and airworthiness certification of these composites. The type of damage resulting from impact on composites depends on the energy level involved in the impact. High-energy impact, such as ballistic damage, results in through-penetration with some minor local delaminations. Lower-energy-level impact, which does not produce penetration, may result in some local damage in the impact zone together with delaminations within the structure and fibre fracture on the back face. Internal delaminations with little, if any, visible surface damage may result from low-energy impact. The actual damage response depends on many intrinsic and extrinsic factors, including the thickness of the laminate, the exact stacking sequence, the shape and kinetic energy of the impactor, and the degree to which the laminate is supported against bending. The strain-to-failure capability of the fibres will determine the degree of back-face damage in a given laminate, and the area of the damage depends on the toughness of the matrix and fibre/matrix bond strength as well as the failure strain and stiffness of the fibres. Also, composites based on woven fibres show less internal damage for a given impact energy than those based on unidirectional material. This is because damage growth between layers is constrained by the weave. High and medium levels of impact energy thus cause surface damage that is relatively easily detected. Low-energy impact produces damage that is difficult to observe visually and is therefore commonly termed barely visible impact damage (BVID). This type of damage is of concern because it may occur at quite low energy levels and is by definition difficult to detect. The effect of BVID on reducing residual compressive strength is well characterized experimentally. However, the actual mechanism has yet to be fully understood. It is clear that in the case of compression loading, the damage constitutes a zone of instability allowing the fibres to buckle at much lower strain levels than in the undamaged region. Unlike glass- and aramid-fibre composites, in which fatigue strength for undamaged structures may be a concern, fatigue of carbon-fibre composites is only a real concern when the laminate also contains low-level impact damage (BVID). Under these circumstances, there is a gradual reduction in residual strength with cycles. In the next pages the role of CNTs in fibre-reinforced epoxy polymers and how they can improve the damage tolerance of the aerospace composite structures will be presented in detail.

Once more prepregs using T700SC-12 K fibres and the bisphenol-A epoxy filled with CSCNTs (0, 5 and 10 wt.%) were developed by Yokozeki and his associates (2007b). The prepreg fibre area weight was set to 125 g/m^2 and the nominal resin content including CSCNTs was 35 wt.%. They produced quasi-isotropic composite panels with stacking sequence; $[0/90/45/-45]_{3S}$ and subjected them among other tests, to low velocity impact and compression after impact (CAI) tests according to the SACMA method (SACMA 1994). The impact test was performed using a weight-drop type machine with a hemispherical impactor of 15.9 mm diameter. The impact energy was set to be 6.67 J/mm according to SACMA the standard.

All load-time curves exhibit peak loads at about 3 ms, and the recorded peak loads are shown in Table 8.4. It is concluded that 0-, 5-, and 10 wt.%-laminates exhibit almost identical time histories, while peak loads of 5- and 10 wt.%-laminates are slightly lower than those of the reference laminates. Almost circular delaminations were recorded in all laminates. Projected delamination areas were measured from C-scan images and are also summarised in Table 8.3. The compressive strengths of the impacted specimens can be found in the same table. An increase around 8% in CAI strength was recorded for 10 wt.%-laminates compared to 0 wt.%-laminates. The effective delamination widths of 10 wt.%-laminates were smaller than those of 0 wt.%-laminates, which in turn can result in a higher CAI strength of 10 wt.%-laminates. However, 0 wt.%- and 5 wt.%-laminates exhibited similar CAI strengths. The scientific group concluded by indicating that the trend of CAI strength increase (or no degradation in CAI strength) of CSCNT-dispersed CFRP laminates was demonstrated, nevertheless, further investigation on CAI strength is necessary to clearly conclude the effect of CSCNT dispersion on CAI strength of CFRP laminates.

Table 8.4 Average values of Impact and CAI tests of quasi-isotropic laminates with and without CSCNTs. The trend of improving the damage resistance of composites with the addition of CNTs is evident (Yokozeki et al. 2007b)

Specimen	Impact peak load (kN)	Delamination area (mm^2)	Delamination width (mm)	CAI strength (MPa)
0 wt.%	7.39	812	30.7	175
5 wt.%	7.14	788	29.8	176
10 wt.%	7.27	800	28.4	188

One of the recent and detailed studies underlines the usage of MWCNTs as fillers in the epoxy matrix of quasi-isotropic laminated prepared by wet lay-up in order to improve not only the CAI properties but also the Compressive Fatigue After Impact (FCAI) properties (Kostopoulos et al. 2010). The researchers prepared an epoxy resin with 0.5 wt.% MWCNTs using the method described in (Karapppas et al. 2009) and then used this as matrix material for the manufacturing of quasi-isotropic $[0, +45, 90, -45]_{2s}$, carbon fibre-reinforced polymers using 16 plies of unidirectional carbon reinforcement of a weight of 160 g/m^2. Such a configuration is often used for structural aerospace composite components. Each panel was liquid resin impregnated and then processed in an autoclave, using the vacuum bag technique. A reference panel was also manufactured with unmodified resin for direct comparison and the volume fraction of both panels was around 58%.

The impact tests were performed according to ASTM D5628- 07. A drop tower equipped with a 3.01 kg hemispherical aluminium impactor with a diameter of 20 mm was used. Following the BVID approach, the impact damage threshold (critical energy) was specified to be at approximately 1 J. The required impact energy was delivered by adjusting the initial height of the impactor. During the tests, the acceleration, force, velocity, deformation and energy versus time were recorded and automatically calculated. Figure 8.48 illustrates the force responses and energies for the five defined impact energy levels for the unmodified (a) and the CNT-modified (b) CFRP laminates. The general behaviour of the two systems does not show any remarkable difference. The time responses of the two materials due to dynamic impact load do not show differences such as delays or changes in the general pattern of the curve. This figure also presents the measured peak impact force recorded during impact for both neat and CNT-modified composite systems versus the impact energies. The measured peak forces were within the same range for both plain and CNT-modified composites.

Impact energy is defined as the total amount of energy introduced to a composite specimen, representing the energy of the impactor. The absorbed energy is the total amount of energy dissipated by the composite specimen during an impact event by the formation of damage inside it. The absorbed energy for every test was calculated as the integral area between the loading and unloading phase of the force-–displacement diagram. In the bar charts, Fig. 8.49, the calculated absorbed energies are displayed as a percentage of the impact energy for the different impact energy levels. The influence of CNT inclusion into the matrix of the composite is distinct with increasing impact energy. At the lower impact energies (2, 8, 12 J), it can be

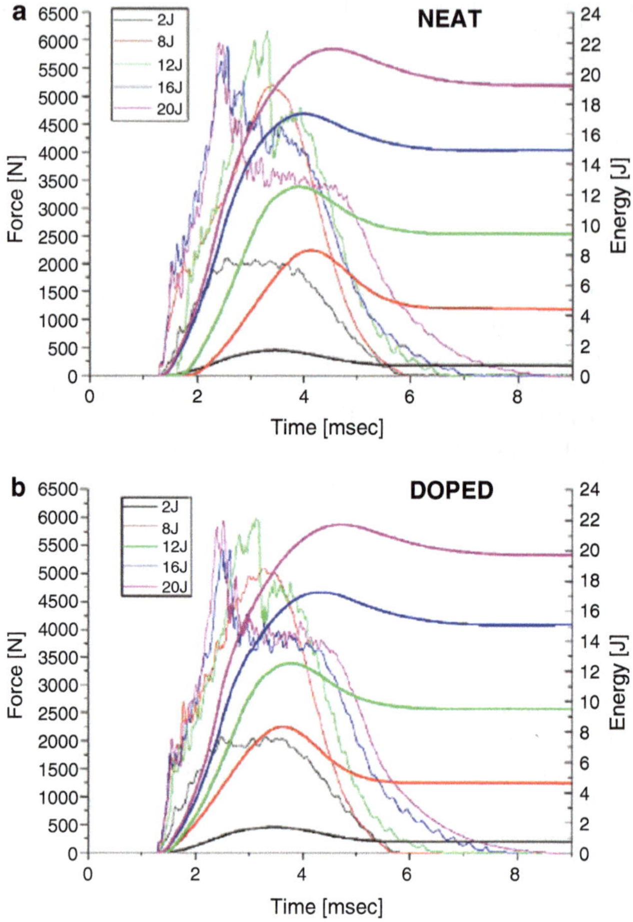

Fig. 8.48 Force and energy histories at different impact energy levels for neat (**a**) and doped (**b**) specimens (Kostopoulos et al. 2010)

noted that no major differences are apparent. However, as the incident energy increases further, up to 16 and 20 J, the modified composites demonstrate a slightly higher absorption performance. However, at this point one can state that the CNT-enhanced CFRP tends to demonstrate a better impact behaviour compared with the neat panel in terms of developed damage. Even though the energy absorbed by the CNT modified composites is slightly higher for the higher impact energy levels, the developed delamination damage is smaller compared against the neat specimens, as is shown in the next C-scan pictures. This accounts for the presence of additional energy absorption mechanisms in the CNT-modified composites.

C-scan measurements were performed in order to evaluate the impact induced delamination damage. A general qualitative note is that the areas of the composites with the CNT modified matrix tend to be smaller compared against the neat composite.

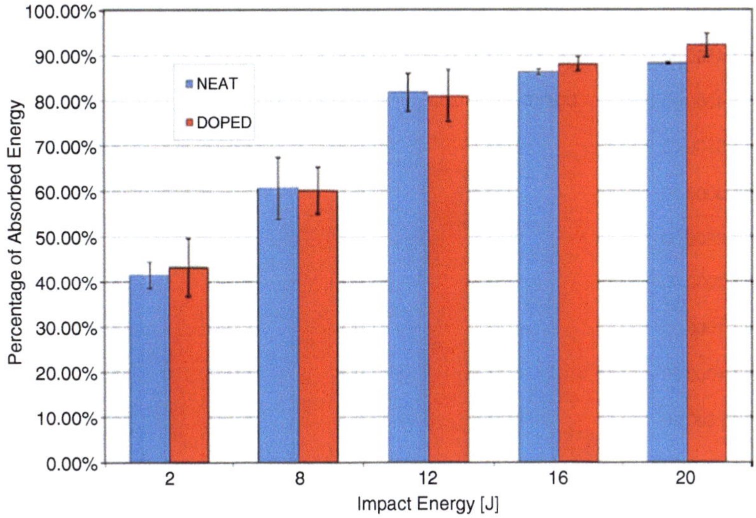

Fig. 8.49 Percentage of absorbed energy versus impact energy levels for both the CNT doped specimens and the neat specimens (Kostopoulos et al. 2010)

This was also verified by the measurement of the delaminated area shown in Fig. 8.50. A closer view of the presented images revealed that the CNT modified specimens tended to be more resistant to delamination in the direction of maximum interlaminar shear that is in the 45° direction where the maximum delamination was expected to occur. This is particularly evident in the case of 12–20 J where the CNT-modified laminates exhibit less delamination particularly in the 45° reinforcement, or the maximum interlaminar shear direction. The inclusion of the CNTs increased the delamination resistance in the maximum interlaminar shear direction, which although only a fraction of the total delamination area may improve the post-impact properties of the CFRP laminate.

CAI tests were performed according to the ASTM D7137 M-07 using the designated anti-buckling jig at a rate of 1.25 mm/min. Once again, inclusion of the CNTs had positive results leading to an increase of approximately 15% for the CAI effective modulus. Moreover, the CNT-modified CFRP laminates were capable of withstanding higher compressive stresses with less deflection. The increase in the CAI strength of the CNT modified CFRPs was around 12–15% for the different impact energy as is clearly shown in Fig. 8.51. Since, the extension of the delamination under mode I loading and the final collapse of the laminate due to buckling is the main failure mechanism during CAI loading, the superior behaviour of CNT-modified CFRP laminates is mainly attributed to their higher mode I fracture toughness properties. The superior mechanical properties of the CNTs, their large surface area and the failure mechanisms are to be responsible for the above enhancement. MWCNTs perform better under compressive load than in tension, i.e. better load transfer between the walls and therefore it is believed the nano-doped

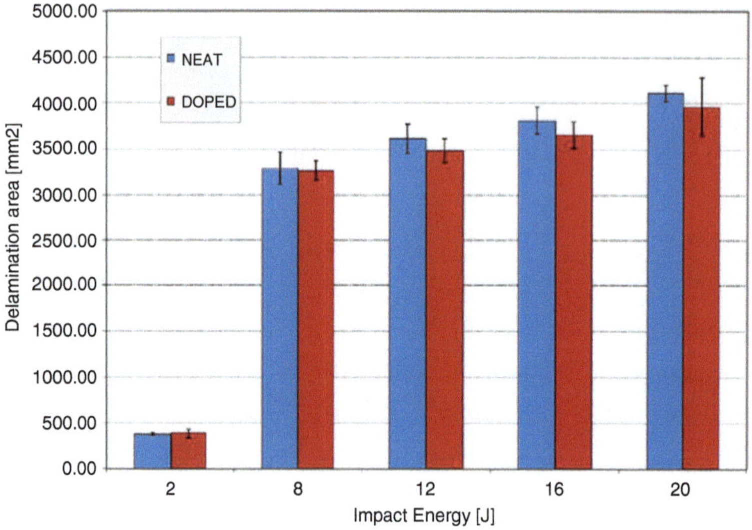

Fig. 8.50 Delamination area versus impact energy levels for both the CNT doped specimens and the neat ones (Kostopoulos et al. 2010)

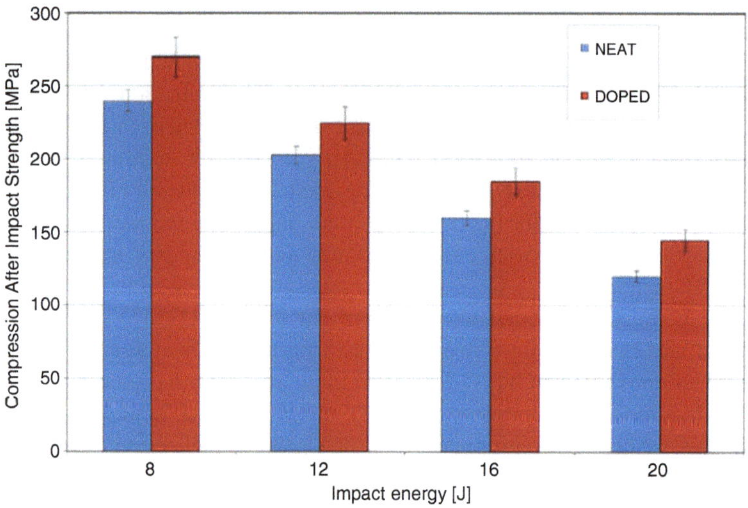

Fig. 8.51 Compressive residual strength versus impact energy levels for both the doped laminates and the neat ones (Kostopoulos et al. 2010)

CFRPs were capable of withstanding higher compressive forces since a part of the applied load is distributed at and within the MWCNTs.

Finally, impacted specimens were subjected to FCAI according to the currently under approval ISO standard (Gower and Shaw 2008). Fatigue was performed at a frequency of 10 Hz for a stress level of 80% of the CAI strength of the

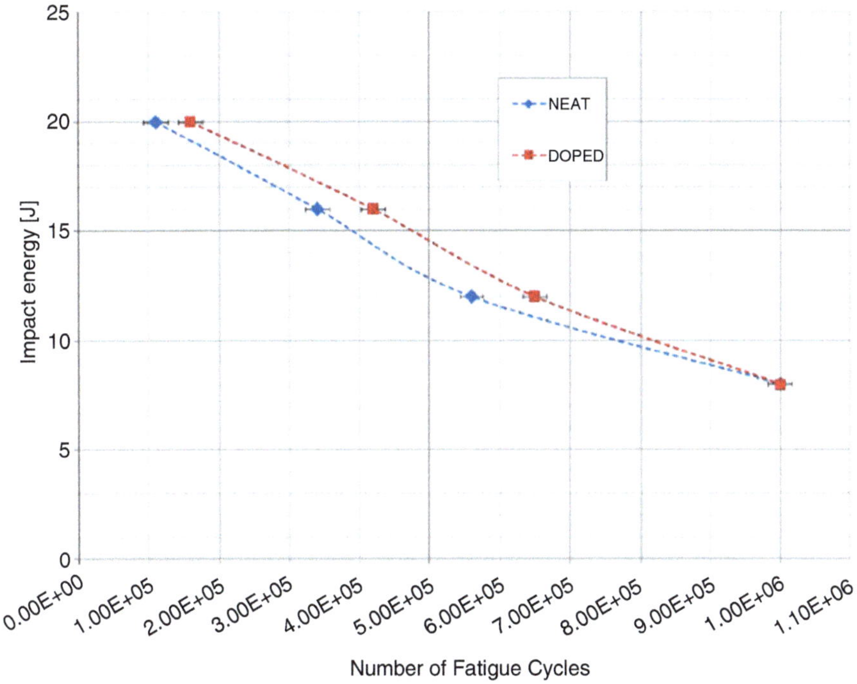

Fig. 8.52 Compressive fatigue after impact (*FCAI*) versus impact energy levels for one stress level (80 % of the CAI strength), R = 10 and frequency of 10 Hz for both doped and neat specimens (Kostopoulos et al. 2010)

impacted panels. The stress ratio was chosen to be R = 10. Figure 8.52 shows that the fatigue life of CFRPs was radically enhanced by the presence of the MWCNTs for all impact energy levels. An extension of fatigue life of at least 20% for all the CNT-modified CFRP laminates compared against the neat laminates is obvious. The CNTs in order to be broken or pulled-out required extra energy and therefore the interlaminar and intralaminar damage was not able to propagate as fast as for the neat composite. Moreover the CNTs were able to bridge the gap and as a result to contribute to the enhanced after-impact properties. Evidence of the above mechanisms can be seen in the SEM micrographs in Fig. 8.53.

Moreover, another scientific team used a standard aerospace grade resin by Cytec, UK and doped it with amino-functionalised DWCNTs to manufacture an orthotropic composite panel (Inam et al. 2010). Plain weave carbon fabrics [0°/90°] were used with density of 0.445 kg/m^2 to manufacture a six-ply panel via vacuum infusion. Four panels were manufactured; one with 0.025 wt.% DWCNT-NH_2, one with 0.05 wt.% DWCNT-NH_2, one with 0.1 wt.% DWCNT-NH_2 and the reference panel without carbon nanotubes.

Energy absorption during impact was measured on flat plates (60 × 60 × 3 mm) using a Ceast instrumented dart impact tester fitted with a data acquisition system. Upon impact the total impact energy can be divided into two parts. The first is the

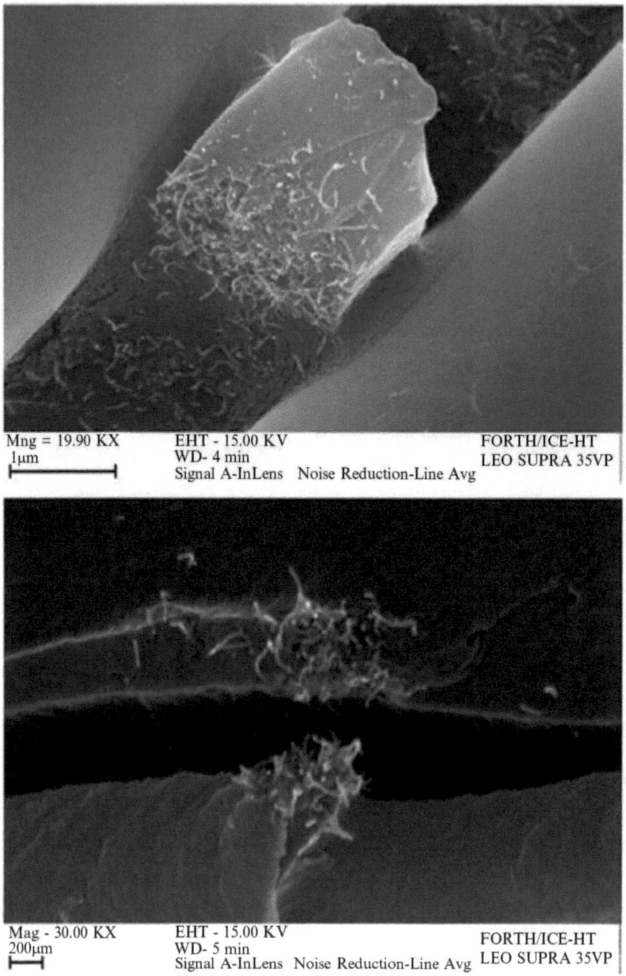

Fig. 8.53 Indicative SEM micrographs of impacted at 16 J specimen with 0.5 wt.% MWCNT. (*Top*) Mid-plane fractured surface picture were the fractured and pulled-out CNTs that contribute to higher impact and CAI properties, are obvious. (*Bottom*) A resin crack being bridged by the MWCNT (Kostopoulos et al. 2010)

elastically stored energy in the composite plate, which is released after maximum deflection by rebouncing of the laminate. This rebouncing energy is successfully transferred back to the impactor. The second is the energy absorbed in the composite laminate available for damage that consequently controls the extent of damage and residual strength. The following bar chart, Fig. 8.54, presents the data for the absorbed energy not only for the nanocomposites but also for the nanopolymers that were then used as matrix material for the manufacturing of the composites with nano-doped matrix. After nano-doping, slightly more energy was absorbed by the nanocomposites. Results in Fig. 8.54 also show a negligible enhancement in the

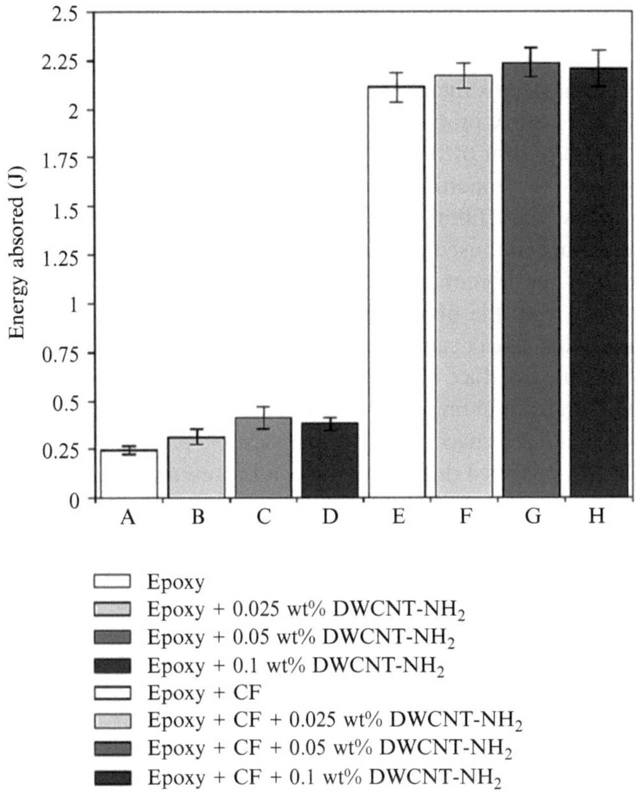

Fig. 8.54 Graph of energy absorbed (area under the curve of force versus displacement) for DWCNT-NH_2 doped polymers and CFRPs (Inam et al. 2010)

energy absorbed by samples F (3% improvement) and G (6% improvement) as a result of the presence of DWCNT-NH_2.

In this section of the book, the positive effect of inclusion of the CNTs in epoxy polymers that in turn can be used as matrix material for fibre-reinforced composites was presented. It was demonstrated that CNTs improve the impact and post-impact properties of composites and on top of that it was highlighted that introduction of the CNTs into the matrix can be done by various methods i.e. prepreging, wet lay-up and infusion providing thus a certain diversity to aerospace composite manufacturers. Nonetheless, as explained in this and in other chapters of this book, the CNT dispersion is a key parameter in order to fully exploit their full potential as epoxy fillers. Moreover, when composite manufacturing is concerned *filtering* is also an issue i.e. CNTs are filtered among fibres and fibre tows resulting thus in uneven CNT distribution. In the next pages a different approach of how to integrate CNTs into aerospace composite structures is explained.

8.5 A Different Approach to Enhance the Damage Tolerance

The ability of composite structures to tolerate impact damage is largely dependent on their fibre and matrix properties. Toughness of composite materials is usually much more than the sum of the toughness of each of the components because it depends also on the properties of the fibre/matrix interface. Therefore, brittle materials such as glass fibres and polyester resin, when combined, produce a tough, strong composite, used in a wide range of structural applications. Control of the strength of the fibre/matrix interface is of vital importance for toughness, particularly when both the fibre and the matrix are brittle. If the interface is too strong, a crack in the matrix can propagate directly through fibres in its path. Thus it is important that the interface be able to disband at a modest stress level, deflecting the crack and thereby avoiding fibre failure. However, if the interface is too weak, the composite will have unacceptably low transverse properties.

In the following pages a different approach is presented for increasing the load transfer properties in composites and in this way enhancing their damage tolerance. Instead of introducing those into the epoxy matrix, CNTs are incorporated into the composite by growing or grafting them on the reinforcement. In this way the problems of dispersion and infiltration of the CNTs, mentioned in the previous pages, can be overcome. However, as shown in the reviewed studies, this route has its own limitations but nevertheless can lead to impressive enhancement of the interfacial properties of composites.

8.5.1 CNT-Modified Fibres

The first attempts started by modifying separate fibres with CNTs and investigating the influence of the processing parameters on the final structure of the CNTs and the fibre properties. In most of the cases the chemical vapour deposition (CVD) method was used for growing CNT on the fibres (Thostenson et al. 2002; Sager et al. 2009; Zhang et al. 2009b; Qian et al. 2010). Although CVD is a promising method due to an easy scale-up and limited equipment needs, it may dramatically decrease the mechanical properties of the fibres by the high temperature and reactive conditions used during the process. Different catalysts and various parameters have been studied in order to achieve an optimum and homogeneous growth of the CNT without scarifying a lot the modulus and strength of the carbon fibres. In 2002 Thostenson et al. (2002) showed the growth of CNT directly on carbon fibres using CVD. The thickness of the nanotube region surrounding the fibre was around 250–500 nm. The effect of the grafting on the properties of the fibres was not studied in this case but the fibre/matrix interface was investigated by the single-fragmentation test on single-fibre composite specimens of epoxy matrix. The Kelly-Tyson model was used in order to calculate the interfacial shear strength assuming a constant interfacial shear stress. The presence of CNT on a carbon fibre's surface

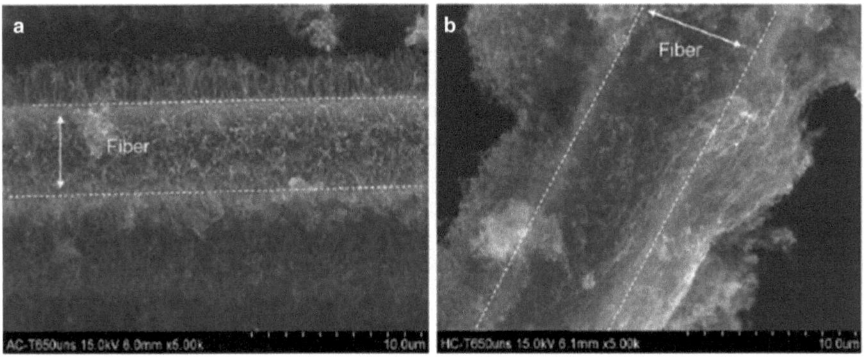

Fig. 8.55 High resolution SEM images of (**a**) carbon fibre with radially aligned MWCNTs and (**b**) carbon fibre with randomly oriented MWCNTs (Sager et al. 2009)

enhanced the interfacial strength by 15%, showing in this way an improved interfacial load transfer via the local reinforcement of the polymer matrix. Various studies have followed using similar techniques, all having the same goal: to increase the interfacial strength between the fibres and the polymer matrix.

Sager et al. (2009) used as well the thermal CVD method for growing MWCNTs on (PAN)-based carbon fibres (T650 from Cytec Industries). Two CVD treatments were used: one produced MWCNTs radially aligned with respect to the fibre surface while the other produced randomly aligned MWCNTs with respect to the fibre surface (Fig. 8.55). In the first step the single fibres were tested showing a negative effect of the surface treatment on the tensile properties of the fibres. Both nanotube coating processes significantly decreased the tensile strength (30–37%) and modulus (9–13%) of the commercial carbon fibres (both sized and unsized). This was attributed to the addition of surface flaws to the fibre via thermal degradation and surface oxidation. However further studies by the same scientific group have eliminated these problems by decreasing the processing temperatures and eliminating the oxygen in the processing chamber (Zhang et al. 2009b). Despite negative results on the fibre properties, the effect of the surface treatment on the interfacial shear strength was studied by single-fibre fragmentation tests. An epoxy matrix was used for fabrication of the test specimen and the interfacial shear strength was calculated via the Kelly-Tyson model. Commercially sized carbon fibres demonstrated the highest shear strength while the unsized fibres had the lowest. Randomly-oriented and aligned MWCNT coated fibres showed an increase of 71 and 11% in interfacial shear strength over untreated unsized fibres. The increase was attributed to an increase in both the adhesion of the matrix to the fibre and interface shear yield strength due to their presence on nanotubes.

Quian et al. (2010) followed a similar growing method and testing of the interfacial properties as the above but studied as well the influence of CNT grafting on the wetting behaviour between carbon fibres and the poly-methylmethacrlylate (PMMA) matrix via direct contact angle measurements. An iron catalyst was used to grow CNTs on (PAN)-based unsized carbon fibres (IM7 from Hexcel

Composites) using a CVD method. The homogeneous distribution of the iron catalyst particles on the surface of the carbon fibres resulted in a homogeneous growth of randomly-oriented, curly MWCNTs. The grafted carbon fibres showed an increased BET surface area and decreased tensile strength compared to the as-received fibres. However the decrease was lower than in the previous studies (around 15–17%). The contact angles were determined via a drop-on-fibre test and demonstrated a good wetability of the CNT grafting by the polymer. The fragmentation tests showed that the CNT grafting led to an increase of the interfacial shear strength by 26% compared to the as-received fibres.

8.5.2 CNT-Modified Fabrics

The successful effort to graft/grow CNTs on different fibres became the base of intensive studies for applying the same or similar techniques for modifying fibres' laminas with CNTs. We may consider the above studies as an attempt to transform the one-dimensional (1D) structure of the single fibre specimens into 2D. The following studies focus on enhancing the out-of-plane as well as, the through thickness properties of the traditional composites. CNT grafting is introduced on the fibres' laminas as a mechanism for improving the polymer/fibre interface and thus the out-of-plane mechanical properties. Veedu et al. (2006) presented this concept on SiC woven cloth. Well-aligned MWCNTs (CNT forests) were grown perpendicular to the fabric using a CVD process. The fabrics were then impregnated by a high-temperature epoxy resin and subsequently stacked to form multilayer composites. The final CNT content in the composite was 2% by weight while the SiC weight content was 63%. The interlaminar properties were studied by Mode I and Mode II fracture tests. The CNT modified composites showed an improvement of 348 and 54% in fracture toughness, G_{IC} and G_{IIC}, respectively, compared to the composite without CNTs. This effect was investigated by SEM and explained by the interlocking of the SiC fibres with the epoxy matrix via the CNTs. In the case of the Mode II loading the shearing effect of the matrix at the routes of the CNTs resulted in a lower improvement of the fracture toughness, while under the Mode I loading the pull-out effect of the CNT forests led to an impressive improvement of the fracture performance of the composites. It was also demonstrated that the presence of CNTs did not affect negatively the in-plane properties of the composites. Flexural modulus and strength were increased compared to the composites without CNTs. Apart from the improvement of mechanical properties the presence of CNTs enhanced by 514% the damping of the composites which can be encountered as a sign of improved fatigue properties for this kind of structures. Moreover the through-thickness thermal and electrical conductivity of the 3D composites were significantly increased via the presence of vertical arrays of the CNTs in the thickness direction.

Following another approach, Bekyarova et al. (2007b) used electrophoresis for deposition of MWCNT and SWCNT on woven carbon fabrics. The CNT modified fabrics were infiltrated with an epoxy resin via vacuum-assisted resin transfer moulding (VARTM). This was a quite demanding step because the presence of CNTs on the surface of the carbon fibres resulted in a huge increase of the surface area of the reinforcement. However investigation of the cross-sections of the composites via SEM showed that most of the CNTs remained bonded to the surface of the carbon fibres after the injection and curing process. The CNT modified composites showed an increase of about 30% in interlaminar shear strength (ILSS) compared to the composites without CNTs and moreover enhanced in-plane tensile properties. The enhanced performance becomes more impressive when we take into account that in this case only ~0.25% of MWCNT was present in the composites. The studies of the out-of-plane electrical conductivity of the CNT modified composites showed that both MWCNTs and SWCNTs resulted in an enhancement of this property compared to the composites without CNTs. The advantage of the specific approach is that the electrophoretic deposition process can be easily scaled-up for industrial applications and together with the VARTM method which is widely nowadays used for the production of aerospace parts can result in composites with enhanced mechanical and electrical performance.

Other studies within the last 2 years have concentrated on the CVD process for growing CNTs on fabrics. Mathur et al. (2008) have applied the method on three different types of carbon fibre substrates: unidirectional carbon fibre tow, plain weave carbon fibre cloth and carbon fibre felt. After the CVD process an amount of up to 9.1, 8 and 18.4% by weight of CNT were grown for each type of carbon substrate, respectively, after 90 min of deposition time. Composite specimens were prepared by impregnating the carbon fibre substrates with phenolic resin via a solvent-based prepreg process. In this study only the flexural properties of the different composites were tested for various content of CNT. Increase in the modulus and strength was demonstrated for all types of carbon fibre substrates when a CNT content of around 9% was introduced to the fibre tow and weave cloth and an 18% wt.% CNT was grown on the fibre felt. Kepple at al. (2008) have as well used the CVD process to grow CNT on commercial woven carbon fibre fabrics (T-300 6 K from Cytec). A special technique was used, called the 'napkin ring' technique, which allowed the manufacturing of up to 620 cm^2 of carbon fibre laminas coated at once with CNT. A CNT layer of about 20 μm thickness was grown on the laminas as it was observed by SEM. Four-layers of laminas were impregnated with epoxy resin by wet-lay up, stacked and cured at room temperature. The CNT modified composites showed an enhanced fracture toughness of 50% compared to the composites without CNT. This improvement in toughness occurred without sacrificing the stiffness of the composites: the flexural modulus increased by 5%. The comparison of the fracture surfaces showed that the presence of the CNTs engaged more fibres to break during the crack growth and in this way increased the total required energy.

Systematic studies of in-site growth of CNTs on alumina woven cloth were conducted by Garcia et al. (2008a). The CVD process used for this purpose resulted in the growth of extremely long, dense and aligned CNT on the surface of the alumina fibres (Fig. 8.56). The composite specimens were fabricated with the

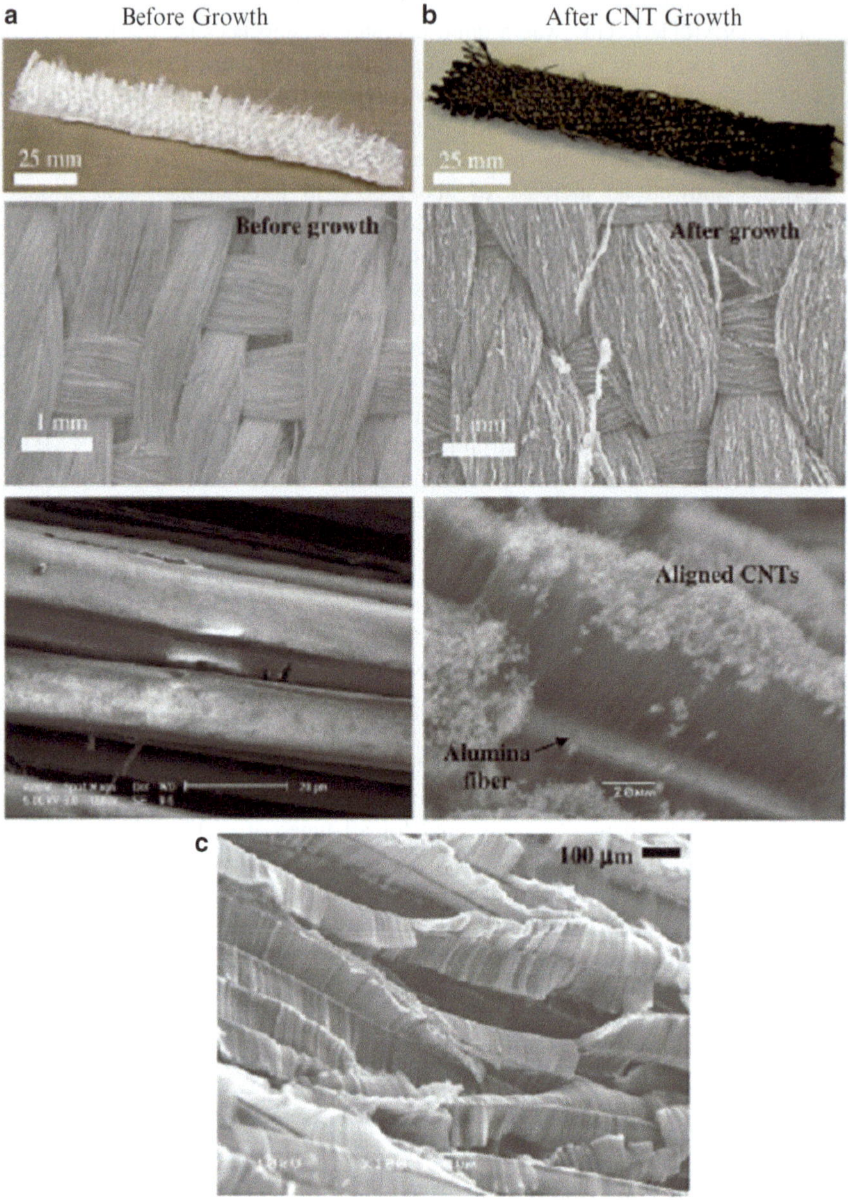

Fig. 8.56 Woven alumina cloth used in laminate fabrication: (**a**) As-received without CNT, (**b**) With aligned CNTs grown radially from the fibre surfaces, (**c**) Aligned CNT coverage over multiple fibre in a tow (Kepple et al. 2008)

wet-lay up process. Each ply was impregnated with a room-temperature curable epoxy resin. One to four plies were stacked together and cured under vacuum. The CNT volume content in the final composite laminates varied between 1 and 3% while the alumina fibre content was 60%. The CNT content was controlled by changing the CNT length via increasing the growth time during the CVD process. Investigation of the composite laminates via SEM showed that the CNTs grew not only on the surface of the alumina fabric but on the fibres in the interior of the cloth as well. Moreover it was observed that the well-aligned and organised CNTs around the fibres were easily impregnated by the epoxy resin, mainly via capillary action along the axis of the CNTs. Improvement in the in-plane and through-thickness electrical conductivity was monitored as the content of CNTs grown on the alumina fibres increased. A conductive network of the aligned CNTs was formed at about 0.5% CNT volume fraction leading to a drop of electrical resistivity in the range of 10^2 Ohm mm. The mechanical performance of the composites was studied via an interlaminar shear strength test on specimens with 4-plies of woven cloth. The CNT modified laminates showed an increase of 69% in the interlaminar shear strength compared to the reference laminate without CNTs. This significant enhancement of the interlaminar properties was attributed to the well-aligned CNT forests compared to the CNT structures grown by electrophoresis which resulted in a maximum increase of 30% in ILSS (Bekyarova et al. 2007b). The aligned CNT forests were additionally expected to give positive results regarding the Mode I toughness and other interlaminar properties. This was demonstrated in a later work by Wicks et al (2010). CNT were grown via a CVD process on woven SiC cloth, creating so-called fuzzy-fibre plies (FFRP). The effect of CNT on Mode I interlaminar toughness and tension-bearing strength was investigated. The specimen for the Mode I fracture test is shown in Fig. 8.57. They consisted of two fuzzy-fibre plies between the standard plies as the behaviour of interest concentrates into the mid-plane section. Fibre glass plates were bonded externally for reinforcing the specimens. The mode I test results showed an increase by 60 and 76% in the initiation and steady-state toughness, respectively, for the CNT modified laminates compared to the laminates without CNT. The steady-state toughness of the CNT containing specimens was determined to be around 4 kJ/m^2. These values are in the range of toughness of laminates with z-spin or stitched reinforcement. Microscopical evaluation of the fracture surfaces showed that the CNT are present on both sides of the crack as shown in Fig. 8.58. The thickness of the interlaminar region was the same for the specimens with and without CNT. The bridging of the CNT during the crack opening resulted in additional fracture mechanisms via their breaking or pulling out. The tension-bearing test was used to evaluate the stress transfer behaviour of the CNT reinforced laminates. It was demonstrated that the "nano-stiched" interlayers led to an increase in the bearing stiffness and strength.

Further studies of Garcia et al. (2008b) used the above described growing of aligned CNT for joining the interfaces of commercial prepregs. CVD was used for growing on a silica substrate CNTs in a volume content of around 1%, having 8 nm and 60–150 μm diameter and length, respectively. The so-called vertical aligned CNT (VACNT) forests were then applied on the surface of prepregs via the process shown in Fig. 8.59. The prepreg is fixed on a cylinder which is rolled under pressure

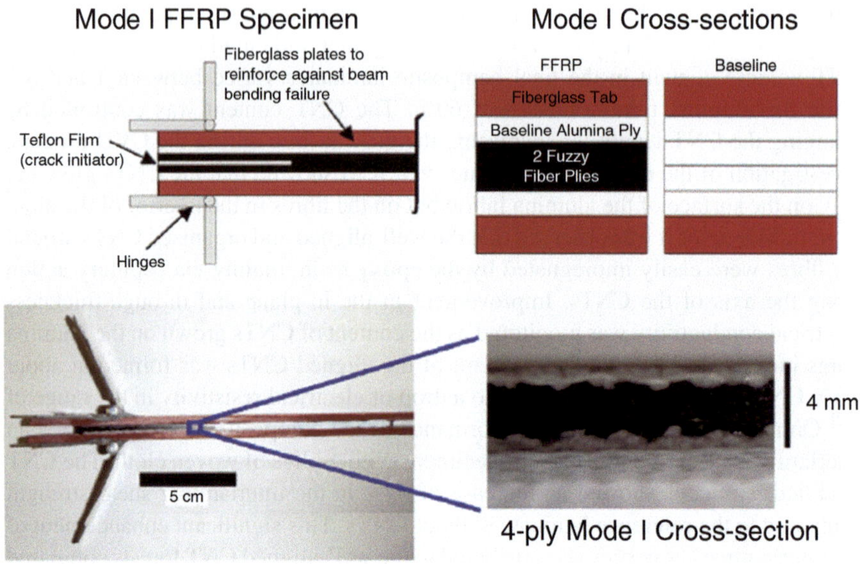

Fig. 8.57 Illustration and images of Mode I fracture specimens (Wicks et al. 2010)

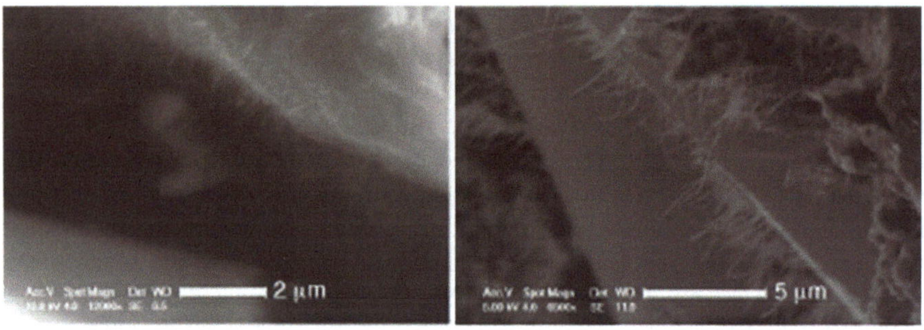

Fig. 8.58 SEM images of the fracture surface showing the CNT pull-out (Wicks et al. 2010)

on the SiC substrate with the VACNTs. Due to the tackiness of the prepregs the CNTs are sticky on their surfaces and become detached from the SiC cloth. The SEM investigation of the prepreg surfaces showed that the CNT retained their alignment and they were not broken during the transplantation process (Fig. 8.59c, d). Two different types of commercial aerospace grade prepreg were used (IM7/977-3 from Cytec and AS4/8552 from Hexcel). The specimens for the fracture testing consisted of 24-layers (140 × 20 mm) from each prepreg type. A 90 × 20 mm CNT forest was applied on the surface of one of the mid-plane prepregs as explained above. The height of the CNT forest varied between 60, 120 and 150 µm. A pre-crack was also inserted via a 50 mm Teflon layer in the mid-plane. The prepregs were then cured in an autoclave according to the manufacturer's instructions. SEM investigation of the final composites showed

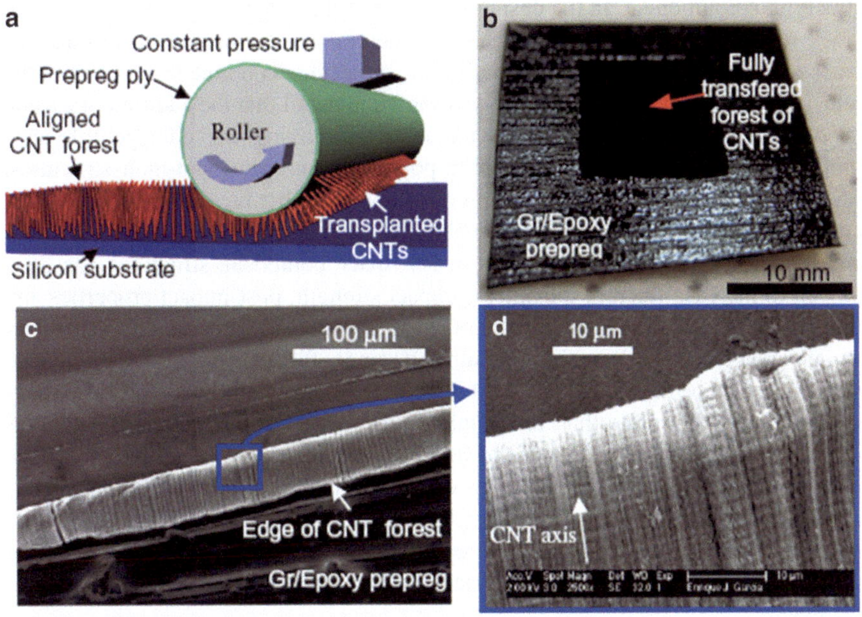

Fig. 8.59 Transplantation process of CNTs on prepreg: (**a**) Illustration of the process; (**b**) CNT forest fully transferred on the surface of the prepreg; (**c** and **d**) SEM images of the CNT forests showing the CNT alignment after transplantation (Garcia et al. 2008b)

that the CNT were wetted by the epoxy matrix of the prepregs forming an interlayer. It was also observed that the CNT penetrated the ply structure, forming in this way "interlaminar nano-stitches". However it was not possible to observe the alignment of the CNT after the curing of the prepregs. The results of preliminary mode I and mode II fracture tests showed increased fracture toughness for the CNT containing prepregs compared to the reference prepregs. This enhancement was attributed to bridging effects of the nanotubes and the extra fracture energy needed for pulling them out. As the number of specimens tested was quite small, further testing is required in order to verify these findings. The concept is surely very interesting as it can be applied, apart from the prepreg/composite technology, to the joining and repair of composites which are quite critical processes for the aerospace industry.

8.6 Conclusions

Damage tolerance is the property of a material or a structure to sustain defects or cracks safely, until such time that action is, or can be, taken to eliminate the cracks. Elimination can be affected by repair or by replacing the cracked structure or component. In the design stage one still has the option to select a more crack resistant material or, improve the structural design such as to ensure that cracks will not become dangerous during the projected economic service life of the relevant structure. In this chapter, it was shown in detail that the addition of carbon nanotubes in small quantities is capable of improving the damage tolerance properties of polymers, fibre-reinforced polymer composites and their structures. The reinforcing mechanisms of carbon nanotubes i.e. fibre breakage, fibre pull-out, crack bridging and crazing are responsible for the aforesaid improvement. In other words, the use of carbon nanotubes in aerospace composite structures has been proven to increase fracture toughness, impact strength, post-impact properties and fatigue life of composites, making them less susceptible to damage. This is actually an advantage when designing an aircraft since full advantage of composite structures can be taken by their extensive use in both primary and secondary structures. In addition to that, fewer joints can be used in a structure, reducing as a consequence the total weight of the structure and the cost, while at the same time increasing the flexibility of a design concept. Finally, it is evident that a new generation of fibres and fabrics with CNTs grafted or, grown on them are to play an important role in revolutionising aerospace composite structures, overcoming any processing issues that have risen due to high CNT-polymer viscosities involved.

References

Baker, A.: Development and potential of advanced fibre composites for aerospace applications. Mater. Forum **11**, 217–231 (1988)

Baker, A., Dutton, S., Kelly, D. (eds.): Composite Materials for Aircraft Structures. AIAA Education Series, 2nd edn. American Institute of Aeronautics and Astronautics, Reston (2004)

Bekyarova, E., et al.: Functionalized single-walled carbon nanotubes for carbon fibre-epoxy composites. J. Phys. Chem. C **111**, 17865–17871 (2007a)

Bekyarova, E., et al.: Multiscale carbon nanotube – carbon fibre reinforcement for advanced epoxy composites. Langmuir **23**, 3970–3974 (2007b)

Broek, D.: The Practical Use of Fracture Mechanics. Kluwer Academic Publishers, Dordrecht (1988)

Chow, W.S., Tan, P.L.: Epoxy/multiwall carbon nanotube nanocomposites prepared by sonication and planetary mixing technique. J. Reinf. Plast. Compos. **29**, 2331–2342 (2010)

Davis, D., et al.: Improvements in mechanical properties of a carbon fibre epoxy composite using nanotube science and technology. Compos. Struct. **92**, 2653–2662 (2010)

Fidelus, J.D., et al.: Thermo-mechanical properties of randomly oriented carbon/epoxy nanocomposites. Compos. Part A **36**, 1555–1561 (2005)

Fiedler, B., et al.: Fundamental aspects of nano-reinforced composites. Compos. Sci. Technol. **66**, 3115–3125 (2006)

Ganguli, S., et al.: Effect of loading and surface modification of MWCNTs on the fracture behaviour of epoxy nanocomposites. J. Reinf. Plast Compos. **25**, 175–188 (2008)

Garcia, E.J., et al.: Fabrication and multifuctional properties of a hybrid laminate with aligned carbon nanotubes grown in situ. Compos. Sci. Technol. **68**, 2034–2041 (2008a)

Garcia, E.J., et al.: Joining prepreg composite interfaces with aligned carbon nanotubes. Compos. Part A **39**, 1065–1070 (2008b)

Geng, Y., et al.: Effects of surfactant treatment on mechanical and electrical properties of CNT/ epoxy nanocomposites. Compos. Part A **39**, 1876–1883 (2008)

Godara, A., et al.: Influence of carbon nanotube reinforcement on the processing and the mechanical behaviour of carbon fibre/epoxy composites. Carbon **47**, 2914–2923 (2009)

Gojny, F.H., et al.: Carbon nanotube-reinforced epoxy-composites:enhanced stiffness and fracture toughness at low nanotube content. Compos. Sci. Technol. **64**, 2363–2371 (2004)

Gojny, F.H., et al.: Influence of different carbon nanotubes on the mechanical properties of epoxy matrix composites – a comparative study. Compos. Sci. Technol. **65**, 2300–2313 (2005)

Gower, M.R.L., Shaw, R.M.: Assessment of the applicability of compression after impact (CAI) and open hole tension (OHT) methods for use under fatigue loading. In: ECCM 13 Conference Proceedings, Stockholm, Sweden, 2–5 June, 2008

Grimmer, G.S., Dharan, C.K.H.: High-cycle fatigue life extension of glass fibre/ polymer composites with carbon nanotubes. J.Wuham Univ. Technol. Mater. Sci. **24**, 167–173 (2009)

Grimmer, G.S., Dharan, C.K.H.: Enhancement of delamination fatigue resistance in carbon nanotube reinforced glass fibre/polymer composites. Compos. Sci. Technol. **70**, 901–908 (2010)

Inam, F., et al.: Multiscale hybrid micro-nanocomposites based on carbon nanotubes and carbon fibres. J. Nanomater. **2010**, 12 (2010)

Karapappas, P.: Improvement of mechanical, fracture and fatigue properties of FRPs with the addition of CNTs in the matrix. Ph.D. thesis, University of Patras, Greece (2009)

Karapppas, P., et al.: Enhanced fracture properties of carbon reinforced composites by the addition of multi-wall carbon nanotubes. J. Compos. Mater. **43**, 977–985 (2009)

Kepple, K.L., et al.: Improved fracture toughness of carbon fibre composite functionalized with multi walled carbon nanotubes. Carbon **46**, 2026–2033 (2008)

Kim, M.G., Hong, J.S., Kim, C.G.: Enhancement of the crack growth resistance of a carbon/epoxy composite by adding multi-walled carbon nanotubes at a cryogenic temperature. Compos. Part A **39**, 647–654 (2008)

Kostopoulos, V., et al.: Impact and after-impact properties of carbon fibre reinforced composites enhanced with multi-wall carbon nanotubes. Compos. Sci. Technol. **70**, 553–563 (2010)

Liu, L., Wagner, H.D.: Rubbery and glassy epoxy resins reinforced with carbon nanotubes. Compos. Sci. Technol. **65**, 1861–1868 (2005)

Mathur, R.B., et al.: Growth of carbon nanotubes on carbon fibre substrates to produce hybrid/ phenolic composites with improved mechanical properties. Compos. Sci. Technol. **68**, 1608–1615 (2008)

Newaz, G., Sierakowski, R.L.: Damage Tolerance in Advanced Composites. Technomic Publishing AG, Basel (1995)

Niu, M.C.-Y.: Airframe Structural Design. Conlimit Press Ltd, Hong Kong (1995)

Qian, H., et al.: Carbon nanotube grafted carbon fibres: a study of wetting and fibre fragmentation. Compos. Part A **41**, 1107–1114 (2010)

Ren, Y., et al.: Tension-tension fatigue behaviour of undirectional single-walled carbon nanotube reinforced epoxy composite. Carbon **41**, 2177–2179 (2003)

Romhany, G., Szebenyi, G.: Interlaminar crack propagation in MWCNT/fibre reinforced hybrid composites. Express Polym. Lett. **3**, 145–151 (2009)

SACMA: Recommended Method 2R-94. The Suppliers of Advanced Composite Materials Association, Arlington (1994)

Sager, R.J., et al.: Effect of carbon nanotubes on the interfacial shear strength of T650 carbon fibre in an epoxy matrix. Compos. Sci. Technol. **69**, 898–904 (2009)

Seyhan, A.T., Tanoglu, M., Schulte, K.: Mode I and mode II fracture toughness of E-glass non-crimp fabric/carbon nanotube (CNT) modified polymer based composites. Eng. Fract. Mech. **75**, 5151–5162 (2008)

Talreja, R.: Comprehensive Composite Materials. Volume 2, Fatigue of Polymer Matrix Composites. Elsevier, Oxford (2003)

Thostenson, E.T., Chou, W.-T.: Processing-structure-multi-functional property relationship in carbon nanotube/epoxy composites. Carbon **44**, 3022–3029 (2006)

Thostenson, E.T., et al.: Carbon nanotube/carbon fibre hybrid multiscale composites. J. Appl. Phys. **69**, 6034–6037 (2002)

Vavouliotis, A., et al.: Multistage fatigue life monitoring on carbon fibre reinforced polymers enhanced with multiwall carbon nanotubes. Plast. Rubbers Compos. **38**, 124–130 (2009)

Veedu, V.P., et al.: Multifunctional composites using reinforced laminae with carbon-nantube forests. Nat. Mater. **5**, 457–462 (2006)

Warrier, A., et al.: The effect of adding carbon nanotubes to glass/epoxy composites in the fibre sizing and/or the matrix. Compos. Part A **41**, 532–538 (2010)

Wichmann, M.H.G., et al.: Glass-fibre-reinforced composites with enhanced mechanical and electrical properties – benefits and limitations of a nanoparticle modified matrix. Eng. Fract. Mech. **73**, 2346–2359 (2006)

Wicks, S.S., et al.: Interlaminar and intralaminar reinforcement of composite laminates with aligned carbon nanotubes. Compos. Sci. Technol. **70**, 20–28 (2010)

Yaping, Z., et al.: Functionalized effect on carbon nanotube/epoxy nano-composites. Mater. Sci. Eng. A **435**, 145–149 (2006)

Yokozeki, Y., Iwahori, Y., Ishiwata, S.: Matrix cracking behaviours in carbon fibre/epoxy laminates filled with cup-stacked carbon nanotubes (CSCNTs). Compos. Part A **38**, 917 (2007a)

Yokozeki, T., et al.: Mechanical properties of CFRP laminates manufactured from unidirectional prepregs using CSCNT-dispersed epoxy. Compos. Part A **38**, 2121–2130 (2007b)

Yokozeki, T., et al.: Fabrication of CNT-dispersed CFRP using length-controlled CNTs: measurement of CNT length and characterization of mechanical properties. Tsinghua Sci. Technol. **14**, 100–104 (2009a)

Yokozeki, T., et al.: Fracture toughness improvement of CFRP laminates by dispersion of cup-stacked carbon nanotubes. Compos. Sci. Technol. **69**, 2268–2273 (2009b)

Yu, N., Zhang, Z.H., He, S.Y.: Fracture toughness and fatigue life of MWCNT/epoxy composites. Mater. Sci. Eng. A **494**, 380–384 (2008)

Zhang, W., Picu, R.C., Koratkar, N.: The effect of carbon nanotube dimensions and dispersion on the fatigue behaviour of epoxy nanocomposites. Nanotechnology **19**, 285709 (2008)

Zhang, W., et al.: Heterogeneity in epoxy nanocomposites initiates crazing: significant improvements in fatigue resistance and toughening. Small **5**, 1403–1407 (2009a)

Zhang, Q., et al.: Hierarchical composites of carbon nanotubes: influence of growth condition on fibre tensile properties. Compos. Sci. Technol. **69**, 594–601 (2009b)

Zhou, Y., et al.: Fabrication and characterization of carbon/epoxy composites mixed with multiwall carbon nanotubes. Mater. Sci. Eng. A **475**, 157–165 (2008)

Chapter 9
Environmental Degradation of Carbon Nanotube Hybrid Aerospace Composites

Nektaria-Marianthi Barkoula

Contents

Abstract This chapter focuscs on the environmental response of carbon fibre-reinforced epoxy composites, where the matrix has been modified with carbon nanotubes. These newly developed hybrid aerospace systems have been recently introduced as alternatives to conventional high performance polymer composites due to their improved mechanical properties, toughness and damage sensing abilities as discussed in detail in previous chapters. First an attempt is made to outline the conditions that lead to environmental degradation specifically in aerospace environments. Next to that the response of typical aerospace composites to

N.-M. Barkoula (✉)
Department of Materials Science and Engineering, University of Ioannina,
PO Box 1186, GR-45 110 Ioannina, Greece
e-mail: nbarkoul@cc.uoi.gr

A.S. Paipetis and V. Kostopoulos (eds.), *Carbon Nanotube Enhanced Aerospace Composite Materials*, Solid Mechanics and Its Applications 188,
DOI 10.1007/978-94-007-4246-8_9,

these environments is discussed. Following, the benefits and challenges in using hybrid aerospace composites in in-service conditions is presented. The degradation of hybrid composites due to exposure on hydro/hygrothermal loadings and galvanic corrosion is presented based on preliminary results. In this section, the focus is on epoxy based composites reinforced with carbon fibres. Matrix modification of these systems is provided by the addition of carbon nanotubes.

Keywords Environmental degradation • Carbon fibre-reinforced epoxies • Carbon nanotubes • Hygrothermal • Hydrothermal • UV radiation • Galvanic corrosion • Aerospace patch • Hybrid composites

9.1 Introduction

As indicated by the title of this book and chapter respectively, the scope of this chapter is to highlight the research performed in the area of hybrid aerospace composites and specifically to discuss their response to environmental loading and in turn their degradation due to this exposure. It is outside the scope of this chapter to compile a full list of papers published in aerospace composites, where environmental studies are performed. The idea behind this chapter is instead to focus on newly developed hybrid aerospace systems that have been successfully introduced, by modifying the most common aerospace matrices (i.e. epoxies (EPs)) with carbon nanotubes (CNT). In order to do so, this chapter will first introduce the conditions that lead to environmental degradation specifically in aerospace environments, and will briefly present the response of typical aerospace composites to these environments. Following that a brief definition of hybrid aerospace composites will be given, highlighting the challenges in using them in in-service conditions. This will allow the reader to understand the differences between conventional and hybrid aerospace composites in terms of their environmental response and elucidate why environmental degradations studies are important for this class of materials. As previously mentioned, the focus of this chapter is to review potential issues that will arise under environmental loadings of aerospace hybrid composites due to the introduction of CNTs. The certification process in both civil and military aircraft involves rigorous testing in order to qualify a new material for aerospace structures with very strict requirements in terms of performance and durability. Based on that it becomes clear why the environmental response of these newly developed hybrid composites is of substantial interest. As the previously mentioned CNT hybrid aerospace composites are fairly new systems, the literature regarding their environmental response is very limited. This chapter will discuss the most common degradation routes in aerospace composites, which are exposure to hydro/hygrothermal loadings and galvanic corrosion when metallic parts are in contact with hybrid composite systems. In this section, the focus will be on carbon fibre-reinforced composites (CFRPs). Matrix modification of these systems will be provided by the addition of CNTs. The way matrix modification influences the water uptake of these systems as well as their thermomechanical properties will be discussed in detail. Finally, the effect of matrix modification with nanotubes on the galvanic corrosion of aerospace patches will be discussed. The synergy between the modification of the matrix and the substrate will be discussed using two main methods, electrical potential measurements and adhesion measurements. Most of the results presented in this chapter come from preliminary studies (Barkoula et al. 2009, 2010; Gkikas et al. 2010) and are still in progress.

9.2 The Nature of Environmental Degradation in Composite Materials with Focus on Aerospace Applications

9.2.1 Conditions That Promote Environmental Degradation

Usually when speaking about environmental degradation in composite materials we refer to deterioration of the material properties due to exposure in specific environments. Depending on the environment, different conditions prevail that may include one of the following, alone or in combination: high and low temperatures, UV radiations, humidity, liquid and gas exposure, electrical fields, wear due to debris/sand and enzyme attack. In this chapter we will focus on those conditions that are present in typical aerospace applications, where polymer composites are employed. Polymer matrix composites (PMCs) find use in both civil and military aircraft, where the effect of the environment on their structural performance is part of the certification process (Dao et al. 2006a). PMCs, such as GFRPs, are being used in spacecraft. The space environment is highly complex, and all its constituents can degrade the properties of spacecraft materials to some extent. It is not however within the scope of the current chapter to discuss the conditions that prevail in that environment.

The most obvious environmental exposure is related to temperature changes. Mechanical, electrical and optical properties of PMCs are highly influenced by temperature. The magnitude of property changes is tightly linked to the operation and storage temperatures of the composite parts (Mahieux 2006). Civil aircrafts during their service are exposed to thermal cycles with temperatures varying between ambient and very low temperatures. Many studies have been devoted in the past to determining how to simulate temperature/moisture profiles. The best describe the environmental exposure of composite structures and especially of aircraft during their lifetime (Bank et al. 1995; Reynolds and Mc Manus 2000; Shin et al. 2000a, b; Jedidi et al. 2005, 2006; Youssef et al. 2008). Figure 9.1 shows a representative cycle for such accelerated temperature/moisture profiles (Reynolds and Mc Manus 2000).

Military aircraft on the other hand experience different temperature conditions compared to civil ones, varying between desert, tropical and even arctic environments. Typical heat damage can result also from fires, lightning strikes,

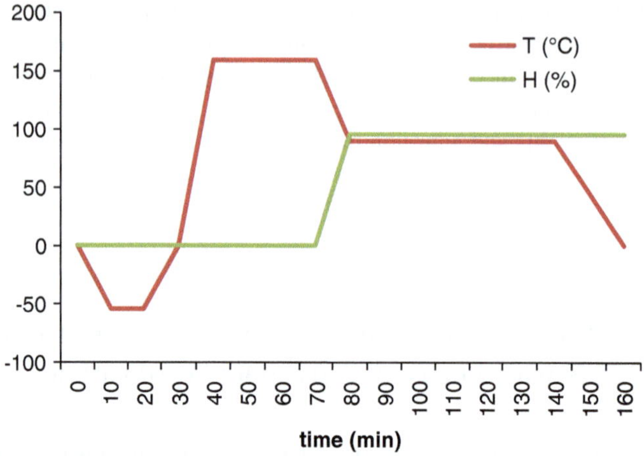

Fig. 9.1 Temperature and relative humidity representative cycles reproduced after Reynolds and Mc Manus (2000)

supersonic dashes, jet engine exhaust, heating blankets, or curing ovens/autoclaves (Matzkanin and Hansen 1998). Depending on the application, composite components may be nonuniformly heated to temperatures in excess of recommended maximum values (either short term or prolonged exposure); these components, although not visibly blistered or delaminated, may have been seriously degraded and may be no longer flight worthy (Matzkanin and Hansen 1998). Also, the damage may result as a combination of thermally cycling the composite above the glass transition temperature (T_g) of the polymer, and oxidative degradation of the polymer or the polymer-fibre interface (Matzkanin and Hansen 1998).

Although loading at different temperatures is very common, most studies available in the literature deal with composite exposure to water under its various forms, during storage or operation. It is well known that moisture can degrade mechanical properties of PMCs, especially at elevated temperature. However, composites can also be exposed to more aggressive liquid and gaseous environments. Simple addition of salt to water can also induce high-rate corrosion damage (Mahieux 2006). Rain erosion on the other hand can occur in high speed vehicles. In this case, erosion mechanisms and rates are a function of speed, droplet size and time of exposure (Mahieux 2006).

The level of damage resulting from radiation exposure depends on the strength of radiation flux, distance from source, time of exposure and temperature (Zunjarrao et al. 2006). Amorphous structures are more strongly affected by radiation than are crystalline structures (weakness of bonds). Radiations can have beneficial or detrimental effects on PMCs. Low intensity radiations, representing 96% of sunlight radiation at the earth's surface (Kojima et al. 1993), have no detrimental effects on the PMCs. The energy provided is much below the level required to break molecular bonds. Degradation is reported to occur for higher energy levels (wavelengths ranging from 280 to 400 nm). Since most polymers have bond dissociation energy in the range of the wavelength of UV radiation (290–400 nm), they are affected greatly by exposure to the solar spectrum (Singh et al. 2010).

9.2.2 Aging Mechanisms

The aging mechanisms that prevail in PMCs depend on the loading conditions, as described in Sect. 9.2.1. PMCs will experience different degradation mechanisms if the exposure is related to single exposure to temperature, moisture or radiation or their combined application. The aging mechanisms that have been indentified to degrade the performance of PMCs are grouped into three major groups, i.e. physical, chemical and mechanical – stress induced aging (Schoeppner et al. 2008).

Physical aging occurs at temperatures well below the polymer's T_g, where the material is in a nonequilibrium state and undergoes changes towards thermodynamic equilibrium. This leads to changes in stiffness, yield stress, density, viscosity, diffusivity, and fracture energy (toughness) as well as embrittlement in some polymeric materials (Schoeppner et al. 2008). The rate that these changes occur depends on the distance of the aging temperature from the material's T_g (Schoeppner et al. 2008). It is a reversible process that is influence by stress and temperature. Physical aging is thermo-reversible for all amorphous polymers by heating the polymer above its T_g and subsequently rapidly quenching the material. It is assumed that this thermo-reversible behavior does not occur in thermoset materials due to the tendency for elevated temperature to affect their extent of cross-linking and/or influence chain scission. Operational mean temperature and lifetime thermal history have a strong influence on the rate of physical aging. For PMCs that are used at temperatures near the material's T_g, physical aging may dramatically affect the time-dependent mechanical properties (creep and stress relaxation) and rate-dependent failure processes (Schoeppner et al. 2008).

Chemical aging on the other hand is a nonreversible process that includes chain scission reactions and/or additional crosslinking, hydrolysis, deploymerization, and plasticization. Hydrolysis and oxidation are the primary forms of chemical degradation in high-temperature PMCs. Oxidation leads to chemical bond breaks, i.e. reduction in molecular weight, mechanical response changes, and mass loss due to outgassing of oxidation byproducts (Schoeppner et al. 2008). At typical PMC operating temperatures, cross-linking and oxidation are the dominant chemical aging mechanisms. Thermo-oxidative degradation becomes increasingly important as the exposure temperature and time increase. Frequently, such aging results in an increase in cross-linking density that can severely affect mechanical properties by densification and increasing the T_g.

Mechanical degradation mechanisms are irreversible processes that can be observed on the macroscopic scale. If this is combined with physical and/or chemical aging the resulting degradation is more severe. Creep-relaxation and thermomechanical cycling tests are most often used to evaluate the effects of long-term mechanical loading on PMCs at elevated temperatures. Although the aforementioned aging mechanisms lead to changes primarily in the polymer-dominated properties, namely the transverse properties (perpendicular to the fibre direction) and the shear properties, the fibre-dominated properties may be affected by degradation of the fibre and deterioration of the fibre–matrix interface (Schoeppner et al. 2008). The degradation mechanisms include matrix cracking, delamination, interface degradation, fibre breaks, and inelastic deformation. In some cases, mechanical degradation mechanisms dominate only after chemical or physical aging mechanisms have altered the polymer properties.

Exposure to liquids and water under its various forms involves diffusion phenomena. The exact effects of diffusion on PMCs depend on the nature of constituent materials and that of the solvent. Diffusion can be reversible and irreversible depending on the degradation mechanisms that prevail and the type of polymer matrix (polar vs. apolar) and can have positive or negative effects depending on the temperature (Mahieux 2006). Molecular degradation by hydrolysis or micro-cracking is irreversible. Swelling or plasticization are reversible and disappear with desorption. Swelling translates into localized time-dependent stresses on the fibres. The combined effect of water exposure with mechanical stresses leads to changes in the failure mechanisms of PMCs due to modification of the stress transfer around the fibres, stress corrosion of the fibres and/or debonding at the fibre-matrix interface (Mahieux 2006). Considerable discussion and disagreement has occurred on the types of molecular environment of adsorbed water (bonded and nonbonded) and the types of absorption kinetics (Fickian or non-Fickian) (Jelinski et al. 1985; Xiao et al. 1997; Ngono et al. 1999; Zhou and Lucas 1999a; Buehler and Seferis 2000; Musto et al. 2000, 2002; Liu et al. 2002; Patel and Case 2002; Bockenheimer et al. 2004).

Several studies on the environmental degradation of CFRPs (Nam and Seferis 1992; Bowles et al. 1993, 1998; Colin et al. 2001; Morgan et al. 2002; Fox et al. 2004; Dao et al. 2007a) reveal the variation of properties from the surface inwards because of the diffusion of oxygen through the material. Moisture enters the material at a speed determined by the material's moisture diffusivity. The moisture content in a thin layer next to the edge or surface of the material is highly affected. The interior of the specimen, on the other hand, slowly approaches an equilibrium moisture concentration determined by the ambient relative humidity. If a Fickian response is assumed, then the typical response of polymers or PMCs in terms of moisture absorption vs. time is illustrated in Fig. 9.2. As can be seen the moisture content (M) is plotted as a function of the square root of time. The slope of the linear part of the curve can be used to calculate the diffusivity coefficient D, while the asymptote to the curve gives the $M\infty$ value. $M\infty$ value is the moisture content of a material at saturation. In real experiments, it happens that after the $M\infty$ value is reached, the mass of the specimen starts decreasing, a phenomenon that is related to material loss (degradation) due to prolonged exposure to the liquid or gaseous environment (Mahieux 2006). If a material is exposed to moisture on all six sides, then according to Shen and Springer (1976) Eq. (9.1) can be used to connect the moisture content (M) with the diffusivity coefficient D:

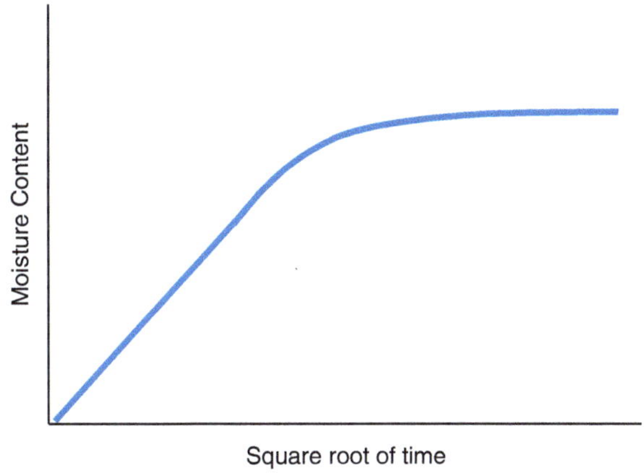

Fig. 9.2 Moisture absorption curve for typical polymers or PMCs

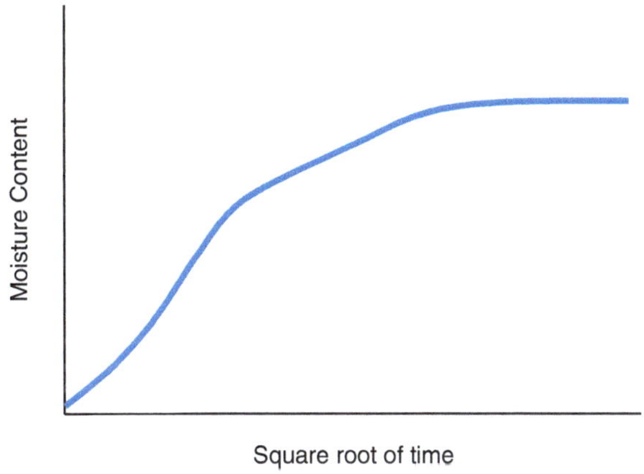

Fig. 9.3 Moisture absorption curve for partially cured polymers or PMCs

$$M = \frac{4M\infty}{h}\sqrt{\frac{tD}{\pi}}, \tag{9.1}$$

where:

M: is the moisture content at any given time,
M∞: is the moisture content at saturation,
h: is the thickness of the specimen,
t: is the time of exposure.

This behavior is typical in the many neat-resin systems (Xiao and Shanahan 1997; Ivanova et al. 2001; Merdas et al. 2002; Lin 2006; Apicella and Nicolais 1984; Zhou and Lucas 1999a; Musto et al. 2000, 2002; Liu et al. 2002; Bockenheimer et al. 2004) as well as on some fully cured EP composites (McKague et al. 1975; Buehler and Seferis 2000; Patel and Case 2002). However, when the resin systems are not fully reacted, the response in terms of moisture uptake might involve an additional stage. The three stages include an induction period with a slow weight increase followed by the middle stage with a high absorption rate and finally a plateau with a very low weight increase rate as illustrated in Fig. 9.3 (Dao et al. 2007b) The duration and intensity of each stage depends mainly on the aging conditions (temperature, relative humidity) as well as the nature of the resin and the level of reaction reached at the time of exposure. The induction period is linked to the various effects of material extraction at the surface of the composite and the defect group (unreacted monomer and oligomer end groups) reactions with water (Dao et al. 2007b). The intermediate, high-absorption rate may relate to the osmotic effects of unreacted monomers (Dao et al. 2007b).

Heat on the other hand enters the material governed by Fourier's law. Temperature affects the composite both physically (e.g. shrinkage and thermal mismatch in plies leading to microcracking) and chemically (e.g. thermal reactions and oxidation coupled with oxygen diffusion into the material). Temperature conductivity for the materials under consideration are typically orders of magnitude higher than the moisture diffusivities. Hence, temperature gradients can generally be ignored when studying moisture effects (Reynolds and Mc Manus 2000). Temperature and moisture can affect the properties of the material, either reversibly or irreversibly via chemical reactions. Fibre-matrix debonding (due to stresses created by all of the above combined with material property changes), and ply cracking (also caused by stresses and possibly aided by microdamage) are observed as a result of the combined application of temperature and moisture, therefore most discussion here is concentrated on the thermal and moisture states, as these are believed to be the drivers of the observed damage.

Finally, the main effects of radiation on composites include curing as well as degradation, embrittlement and gas emission. Ultra-violet exposure can lead to the formation of three-dimensional networks in the polymer (cross-linking). This method can be used effectively during manufacturing to promote curing. If the material is only partially cross-linked at the end of the manufacturing process, care should be taken that exposure to UV radiation might create property changes due to the additional crosslinks. Those property changes can be beneficial (increased strength) or adverse (increased brittleness) depending on the application (Mahieux 2006).

9.2.3 Experimental Techniques for Environmental Performance Evaluation

Many studies of the hot/wet ageing of aerospace composites and neat resin materials have been carried out (McKague et al. 1975; Morgan and O'neal 1978; Springer 1982; Apicella and Nicolais 1984, 1987; Collings and Stone 1985; Luoma and Rowland 1986; Clark et al. 1990; Xiao and Shanahan 1997; Hancox 1998; Hough et al. 1998; Ivanova et al. 2001; Merdas et al. 2002; Lin 2006). The academic literature has mainly relied on spectroscopic and dynamic mechanical studies to determine the chemistry and physics of moisture interaction with simple (noncommercial) formulations of neat resin.

The key experimental techniques used to identify changes due to environmental exposure involve spectroscopic, thermal, thermomechanical analysis mainly reported in the academic literature (McKague et al. 1975; Morgan and O'neal 1978; Springer 1982; Collings and Stone 1985; Luoma and Rowland 1986; Apicella and Nicolais 1987; Xiao and Shanahan 1997; Ivanova et al. 2001; Merdas et al. 2002) and mechanical and fatigue tests mainly reported by the aerospace industry (Apicella and Nicolais 1984; Clark et al. 1990; Hancox 1998; Hough et al. 1998; Lin 2006). A brief summary of both areas was presented in the review by Hancox (1998) which included both literature references and standard test methods. FTIR spectroscopy has been used extensively to study the thermal and photochemical oxidation of EP resin systems over many years (Bellenger and Verdu 1983, 1985; Luoma and Rowland 1986; Morgan and Mones 1987; Garton 1989; Grenier-Loustalot et al. 1990; Musto et al. 2001; Bondzic et al. 2006).

Many of these systems have been aerospace type formulations (although generally highly simplified) and a very good library of peak positions for each molecular structure has been built up (Dao et al. 2006a). Differential scanning calorimetry (DSC) on the other hand is useful since it measures the exothermic energy of any residual reactions present in a composite material. This has been used in the past by the aerospace industry to estimate the remaining cure percentage of the PMCs (Dao et al. 2006a). Dynamic mechanical analysis (DMA) of PMCs provides information related to molecular motions and hence chemical/mechanical changes over a large temperature and/or frequency range. The aerospace industry is focusing on the temperature of loss of modulus (T_g, E' onset) and the shape and position of the loss factor (tanδ) peak (Dao et al. 2007a). It is important to keep in mind that, as with most mechanical test methods, DMA measures an average result over a relatively thick sample and the chemical ageing changes in a composite generally occur very selectively from the surface.

It is important to note that most environmental studies for the prediction of long-term aging changes make use of high aging temperatures and extreme moisture conditions for relatively short times. These are assumed to be equivalent to much longer times under realistic conditions, based on the use of Arrhenius-type extrapolation methods. It is not straightforward to use this type of extrapolation in the case of composites, where the interactions of the multiple phases can make such extrapolations unreliable (Tian and Hodgkin 2010).

9.3 Typical Aerospace Composites and Their Environmental Performance

The prediction of the service life of composite structures made out of PMCs subjected to environmental loading is challenging, mainly due to the complex physical, chemical, and thermomechanical mechanisms involved. The scope of the current paragraph is to review the effect of environmental degradation on typical aerospace polymer matrix composites, with focus on CFRPs.

One of the challenges for the applicability of CFRPs is related to their environmental durability (Joshi 1983; Kenig et al. 1989; Frassine and Pavan 1994; Lai and Young 1995; Selzer and Friedrich 1995; Ogi and Takeda 1997; Wood and Bradley 1997; Asp 1998; Chou and Ding 2000; Wang et al. 2002; Wang and Chung 2002; Botelho et al. 2006a, b; Ray 2006). It is well known that EP matrix composites are susceptible to heat and moisture particularly when they operate in varying environments. The aging mechanisms in CFRPs can be very complex depending on the moisture absorption levels, moisture reactions as well as cure chemistry, of the resin system. When the EP resin in the composite is incompletely cured, the ageing chemistry is deferent to that of a complete cured system. This is due to the fact that material extraction effects can remove unreacted monomers and oligomers from the surface areas and moisture reactions can deactivate monomers, (Xiao and Shanahan 1997; Xiao et al. 1997) but not break completed chains, further in the body of the composite. This means that highly accelerated ageing conditions used to test any partly cured composite could produce material with very different properties from those seen in "inservice" conditions (Dao et al. 2007b). The accelerated thermal aging of fully cured EP resins has been extensively reported (Kerr and Haskins 1987; Tsotsis and Lee 1998; Tsotsis et al. 1999). These studies have shown that amide and acid groups are initially formed by resin breakdown (Pearce et al. 1981; Bellenger and Verdu 1983, 1985; Luoma and Rowland 1986; Morgan and Mones 1987; Garton 1989; Grenier-Loustalot et al. 1990; St John and George 1994; Dyakonov et al. 1996; Bondzic et al. 2006). Due to the fibre/matrix interaction the degradation pathways are very complex and temperature and humidity dependent (Dao et al. 2006b).

Thermal oxidation is one of the irreversible degradation mechanisms observed in CFRPs that use organic resin systems as matrix material. It has been reported that sufficiently high temperatures cause resin degradation. Carbon fibres are more resistant to oxidation than the polymer matrix; however, the properties of the fibre–matrix interface are influenced and a reduction in the room-temperature mechanical strength properties of the composite is expected (Matzkanin and Hansen 1998; Schoeppner et al. 2008). The effect of oxidation on the T_g is polymer dependent. Some polymers initially have a decrease and then an increase in the T_g, others may have only a decrease, and still others may only have an increase in the T_g. This may be due to competing chemical and physical aging phenomenon or differences in the oxidation reaction mechanisms (Schoeppner et al. 2008). Since aerospace composites operate at temperatures near their initial design T_g, changes in the T_g can have detrimental effects on their performance (Schoeppner et al. 2008). Although these composites can appear visually and microscopically to be undamaged, there is a significant loss in properties (60% of their original strength (Matzkanin and Hansen 1998)). The surface of the material may also present embrittlement and cracking, which in turn results in loss of the impact strength of the material. Therefore CFRPs exposed to overheating conditions can suffer irreversible and catastrophic damage in a very short time (Matzkanin and Hansen

1998). It has been emphasized in the past that the exact temperature of ageing can have a great effect on the chemistry of degradation of the composite matrix, mainly at the surface. Therefore accelerated, mechanical property and ageing studies could be misleading without knowledge of the chemical changes taking place (Dao et al. 2006a).

On the other hand exposure to moisture leads to both reversible and irreversible changes of the material properties in CFPRs. The amount of moisture absorbed by the matrix is significantly different from that absorbed by the reinforcing phase. The EP resins used in aerospace applications absorb approximately 5–6% by weight at full saturation. This leads to about 1.5–1.8% moisture weight gain in CFRPs with the usual 60% fibre volume fraction (Mangalgiri 1999). The presence of moisture and stresses associated with moisture-induced expansion may deteriorate the matrix related properties of the composite and as a result, have an adverse effect on damage tolerance and structural stability. It has been concluded (Joshi 1983; Kenig et al. 1989; Frassine and Pavan 1994; Lai and Young 1995; Selzer and Friedrich 1995; Ogi and Takeda 1997; Wood and Bradley 1997; Asp 1998; Chou and Ding 2000; Wang et al. 2002; Wang and Chung 2002; Botelho et al. 2006a, b; Ray 2006) that the higher the temperature the higher the moisture uptake rate of the composites and the delamination nucleation. Furthermore the interfacial adhesion degradation is dependent on the conditioning temperature and exposure time. Some of the mechanisms occurring during moisture absorption include weakening of the fibre-matrix interface (Joshi 1983; Selzer and Friedrich 1995; Botelho et al. 2006a; Ray 2006), plasticization and swelling of the matrix and in some cases even softening of the matrix (Selzer and Friedrich 1995). Among the properties of polymer matrix composites that are negatively affected by moisture uptake is the stiffness (Ogi and Takeda 1997; Chou and Ding 2000), the interfacial strength (Wood and Bradley 1997), the interlaminar interface (Joshi 1983; Kenig et al. 1989; Frassine and Pavan 1994; Selzer and Friedrich 1995; Asp 1998; Wang et al. 2002; Wang and Chung 2002; Botelho et al. 2006a, b; Ray 2006), the damping ratio (Lai and Young 1995).

Different trends have been reported on the effect of moisture in the T_g with most of the studies reporting a reduction of the T_g due to exposure to moisture (Weitsman 1991; Maggana and Pissis 1999; Li et al. 2001; Nogueira et al. 2001; Mohd Ishak et al. 2001). The opposite is observed by Zhou and Lucas (1999a, b) and Papanicolaou et al. (2006). According to their findings, water molecules bind with EP resins through hydrogen bonding. Two types of bound water were found in EP resins. The binding types are classified as Type I or Type II bonding, depending on differences in the bond complex and activation energy. They revealed that the change of the T_g does not depend solely on the water content absorbed in EP resins, that the T_g depends on the hygrothermal history of the materials. They also proposed that for a given EP system, higher values of the T_g resulted for longer immersion time and higher exposure temperature and the water/resin interaction characteristics (Type I and Type II bound water) have a quite different influence on the T_g variation. Type I bound water disrupts the initial interchain Van der Waals force and hydrogen bonds, resulting in increased chain segment mobility acting as a plasticizer and decreasing the T_g. In contrast, Type II bound water contributes, comparatively, to an increase in the T_g in water saturated EP resin by forming a secondary crosslink network. Another unexpected finding is the detection of a significant increase in the composite T_g value at the surface of the material versus the center, where normally a drop in the T_g due to moisture plasticization would be expected. However, molecular chain stiffening caused by surface oxidation is likely the reason (Tian and Hodgkin 2010).

Apart from aging due to heat and moisture, UV degradation and radiation becomes significant especially for space structures. Photooxidative reactions take place during the exposure of polymers to UV radiation. This alters the chemical structure by molecular chain scission or chain crosslinking and results in material deterioration (Ranby and Rabek 1975). For prolonged exposure to UV radiation, the matrix dominated properties, such as interlaminar shear strength, flexural strength, and flexural stiffness can suffer severe deterioration (Chin et al. 1997; Liau and Tseng 1998; Shin et al. 2000a, b; Kumar et al. 2002; Signor et al. 2003). Furthermore, degradation phenomena due to UV radiation and moisture when acting together can significantly accelerate the degradation process of the matrix. Kumar et al. (2002) studied the combined effect of UV and water vapor condensation and found that cyclic exposure leads to a synergistic degradation mechanism causing extensive matrix erosion and resulting loss of mechanical properties (Singh et al. 2010). The combined exposure of UV radiation and condensation resulted in the loss of mass of both materials due to the erosion of EP by a synergistic physico-chemical process that was previously identified and characterized by Kumar et al. (2002). They suggested the formation of photo-oxidative byproducts that underwent dissolution by water vapor condensation and run-off results in the removal of surface layers degraded by UV radiation. Therefore, cyclic exposure to both UV radiation and water vapor condensation results in a continual material degradation and erosion process (Singh et al. 2010). Recently, Woo et al. also suggested that the presence of moisture can enhance the mobility of free radicals and ions and, thereby, enhance the photo-oxidative effects of UV radiation (Woo et al. 2007, 2008).

9.4 Environmental Performance of Hybrid Aerospace Nanocomposites

9.4.1 Benefits and Challenges of Aerospace Nanocomposites Related to In-Service Conditions

It has been extensively discussed in the previous chapters why novel hybrid nanocomposite materials have been proposed as aerospace composites and as candidates in the aircraft repair. As previously mentioned, the objective is to use a nano-sized phase, which in small weight fractions may significantly alter the macroscopic properties of the material. In short, the nanophase enhances the damage tolerance characteristics of a composite, by improving their fracture toughness and fatigue performance. At the same time, the introduction of specific nanodopants such as CNTs renders the material conductive. Finally, the network of the nano-sized filler follows the macroscopic changes of the structure exhibiting both real time change in its conductivity with applied strain, and monotonic conductivity decrease with the initiation and propagation of damage within the composite volume. Mapping the electric resistance changes in the region of appalled repair will monitor the damage/debonding initiation and propagation. In parallel, the presence of CNTs into the adhesive, increases the bonding performance of the adhesive.

If CNTs can give a large interfacial bonding strength with matrix materials, great load transfer ability can be achieved, because a strong bonding allows shear stress to build up without causing interfacial failure. The presence of moisture is expected to alter the interfacial stress transfer characteristics of CNT-reinforced composites and the hot/wet aging response of the CNT modified composites. It is well discussed in the previous chapters that the mechanical properties of CNT-reinforced composite systems are critically dependent on the integrity of the interface. Any change due to the environmental degradation in the matrix and/or the interface is expected to have a clear effect on the overall macroscopic response of the composite.

At the same time, CNTs are well known to exhibit extremely hydrophobic behaviour, which is expected to inhibit the electrochemical degradation of the parent material if this is aluminium. Alternative approaches consist of reducing the redox potential in metal/CNT galvanic cells, offering a challenging new technology for their use in corrosive environments, such as the locus of the repair. Therefore, the assessment of the environmental response as well as that of the effects of galvanic corrosion in metal/CFRP interfaces is critical for the applicability of CNT-modified composites in aerospace structures, as structural components and/or repair systems.

Although the environmental response of CFRPs has received a lot of focus as shown in previous paragraphs, very few papers have been published on the environmental degradation of CNT-reinforced composites (Zhang and Wang 2006a, b; Windle 2007; Yip and Wu 2007). From the analysis presented in Sects. 9.2 and 9.3 one can conclude that in the case of CFRPs, the properties dominated by the matrix or the fibre-matrix interface are degraded by moisture absorption, whereas the properties that are dominated by the fibres are less influenced.

Zhang and Wang (2006a) provided an analytical method for the investigation of the hygrothermal effects on the interfacial stress transfer characteristics of CNT-reinforced composites. This study omits the van der Waals force interaction and considers transverse isotropic characteristic of thermal expansion coefficients of CNTs. Furthermore, the thermoelastic theory and conventional fibre pullout models are being considered for the analytical model. The thermal expansion coefficient of CNTs is considered as a nonlinear function of temperature change, the thermal expansion coefficient of polymer matrix is isotropic and a linear function of temperature changes, and the moisture concentration change in CNTs is neglected.

This study concludes that the mismatch of the thermal and moisture expansion coefficients between the CNTs and polymer matrix may be more important in governing interfacial stress transfer characteristics of CNT-reinforced composite systems (Zhang and Wang 2006a). It was found that the interfacial maximum shear stress decreases linearly with the increase of moisture concentration change in polymer matrix because the coefficient of moisture expansion is independent of the moisture concentration change (Zhang and Wang 2006a). This study also highlights that increasing of temperature change or moisture concentration change results in the decrease of the interfacial maximum shear stress at the ends of the interface which helps the structural stability of CNT–polymer composites, with the thermal effect being more dominant (Zhang and Wang 2006a). Finally, the architecture of the CNTs (armchair vs. chiral and zigzag) is critical for the hygrothermal response of the CNT-polymer composites (Zhang and Wang 2006a).

The initial frictional pull-out force has been used in the past to evaluate the interface integrity and structural stability of CNT-reinforced composite systems. The work by Zhang and Wang (2006a, b) is an analytical approach on the hygrothermal effects on the pull-out force and on the interfacial stress transfer of CNT-reinforced composites that takes into consideration the mismatch of the thermal and moisture expansion coefficients of CNTs and polymer. It was concluded that the initial frictional pull-out force of CNT-reinforced composite systems increased with the increase of temperature and moisture concentration variations. The magnitude of the initial frictional pull-out force was significantly dependent on the temperature variation, the degree of moisture concentration, the chiral vectors of the CNTs, as well as the number of layers of the CNTs (Zhang and Wang 2006b).

Most experimental studies on the effect of hygro/hydrothermal exposure on nanomodified composites focus on systems modified with nano-clays. One interesting study is that of Singh et al. (2010), which investigates the response of nanoclay modified EPs due to the combined effect of temperature-humidity and UV radiation. It was found that the presence of nanoscale clay inhibited moisture uptake, which was however lower compared to that of the unmodified system. It was therefore concluded that, nanoclay acted as a barrier and significantly hindered the moisture absorption of the EP matrix. The exposure to moisture did not result in significant changes of the flexure modulus, however, it led to degradation in flexural strength of both modified and unmodified systems with the modified ones being more resistant (Singh et al. 2010). The combined UV radiation and condensation resulted in reduction of weight due to the erosion of EP with the EP-clay specimens showing higher deterioration. The variation of the mass was governed by two competing mechanisms, namely, decrease in mass due to loss of EP and increase in mass due to moisture absorption. For exposure to UV radiation and condensation the flexure modulus decreased for both materials with increasing exposure duration. The decrease in modulus was greater for the unmodified EP specimens as compared to the EP-clay nanocomposite. As in the case of exposure to moisture, the combined effect of UV radiation and condensation led to degradation in strength. The decrease in flexural strength was lower for the EP-clay nanocomposite. Based on these results it can be concluded that the clay particles provide resistance to moisture transport. Nevertheless, the moisture absorption process in polymer clay nanocomposites is governed by numerous factors, including the cross-linking density around the clay layers, degree of net cure, and the total exposed surface area of the clay platelets (Kim et al. 2005; Hwang et al. 2009). In addition, the interaction between UV exposure, moisture, and EP-clay chemistry is driven by complex physicochemical mechanisms. Therefore, further investigation is warranted to analyze the chemical and microstructural characteristics of the degradation process and establish the process kinetics in polymer nanocomposites subjected to varied environments (Singh et al. 2010).

Further to the aforementioned study on the combined effect of UV radiation on the properties of nano-modified composites, some more studies have been published in the case of CNT-modified composites. All available studies focus on spacecraft applications and demonstrate a radiation-hardening effect, observed down to nanotube loadings of less than 1 wt.% (both SWNTs and MWNTs) (Nielsen et al. 2008). CNTs have been introduced to polyimide films (Delozier et al. 2004; Qu et al. 2004; Smith et al. 2004a, b, 2005; Watson et al. 2005) in order to dissipate any electrostatic charge accumulated during handling or in the charged orbital environment. They have been also dispersed in poly(methyl methacrylate) (PMMA) matrices (Harmon et al. 2002; Muisener et al. 2002). It was found that they reduce the degradation of mechanical properties from exposure to g-radiation. It was suggested that the p-conjugated CNTs acted as radiation sinks, able to effectively dissipate the energy deposited by ionizing radiation (Harmon et al. 2002; Muisener et al. 2002; Tatro et al. 2004). Najafi et al. (Najafi and Shin 2005; Najafi et al. 2005) observed a strong protective effect against UV treatment and EB-irradiation, due to the effective dispersion and dissipation of the deposited energy by the conductive nanotube network and a strong interaction of the CNTs with the radical species produced in the degradation processes. CNT-modified PE (Pulikkathara et al. 2003, 2005; Wilkins et al. 2005), poly(vinyl alcohol) (Minus et al. 2006) and poly(acrylonitrile) (Chae et al. 2006) composites have been also studied for use in structural components and radiation shielding application. All studies reveal an increase in resistance compared to unmodified polymers.

9.4.2 *Hygrothermal Response of Carbon Nanotube Hybrid Aerospace Composites*

As already discussed the specific surface area of nano-sized particles is huge (Windle 2007), indicating that a large proportion of the surrounding matrix will be in contact with the interface or even a separate phase – the interphase – will be developed with properties different from those of the bulk matrix. One important point is that in cross-linking resins, the ability of CNTs to absorb or donate electrons may well affect the cross-linking density. Matrix and reinforcement-matrix interface are more prone to absorb water and alter their properties.

The fact that CNT-reinforced composites possess increased interfacial area may be beneficial for their fracture toughness; however, this could prove to be their weakest point in terms of in-service durability. It is expected that the hygrothermal conditioning of the CNT-reinforced composites will alter their macroscopic response, especially the properties controlled by the matrix and the interface, as well as their viscoelastic response and more specifically their damping properties, the T_g and dynamic modulus. Since the focus of the current chapter is on the CNT hybrid aerospace composites, this paragraph will present some preliminary results obtained by the research group of the authors. These results have been published in (Barkoula et al. 2009, 2010), and present the effect of hygrothermal loading on the water uptake, the interlaminar shear strength (ILSS), and the thermomechanical response of CNT-modified EPs and CFRPs. An attempt is also made to relate the hygrothermally induced changes of the material to electrical resistivity changes (Barkoula et al. 2009).

To this end, multi-wall CNTs were incorporated in a commercial EP system via high shear mechanical mixing which was subsequently used for the manufacturing of quasi-isotropic laminates CFRPs, using the wet layup method. Modified matrices with CNTs content varying from 0.1 to 1% were manufactured. All modified resins were used to manufacture un-reinforced rectangular cast specimens. The resin with the 0.5% CNT content was subsequently used for manufacturing of the modified CFRPs. All systems were subjected to hygrothermal loading. During the environmental conditioning, the composites were weighted in specified intervals and the water absorption vs. time was recorded for both the modified and a reference system (Barkoula et al. 2010). At the same time intervals the electrical resistance was recorded for the modified and unmodified systems. After maximum exposure the conditioned composite systems as well as the reference materials were tested in interlaminar shear. The conditioned composite systems were subsequently tested in dynamic three-point bending in order to study their viscoelastic behaviour. The properties of the modified systems were compared to the properties of unmodified composites that were subjected to identical conditioning (Barkoula et al. 2010).

Details on the materials used and the testing procedures can be found in Barkoula et al. (2009, 2010). In short, multiwalled CNTs were used in combination with Araldite LY564/Aradur HY2954 from Huntsman Advanced Materials, Switzerland. A shear mixing device was used for the dispersion of the CNTs. Modified matrices with CNT content varying from 0.1 to 1% were manufactured. All modified resins were used to manufacture un-reinforced rectangular cast specimens. The resin with the 0.5% CNT content was subsequently used for manufacturing of the modified CFRPs. Sixteen plies of quasi-isotropic CF laminas [(0/+45/−45/90)2]s, were used for the manufacturing of CFRPs. Each panel was hand laid-up and then processed in an autoclave, using the vacuum bag technique. A reference panel was also manufactured with unmodified resin for direct comparison. Two laminates of CFRP materials were tested in total; one having a modified matrix with the addition of CNTs, and the other having an unmodified (neat) matrix. The specimens were conditioned at 80 °C, and were exposed up to about 1,200 h.

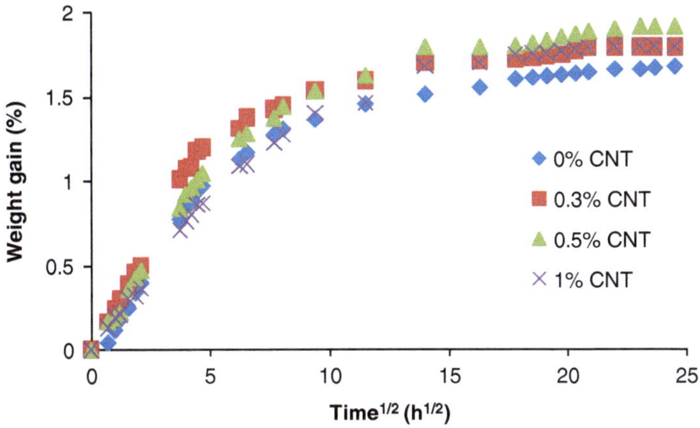

Fig. 9.4 Weight gain versus square root of time for the unmodified and CNT-modified EP matrices reproduced after Barkoula et al. (2009)

The moisture uptake kinetics was measured at different intervals of the conditioning time. The weight gain was calculated according to the equation

$$M(t)(\%) = \frac{m_w - m_d}{m_d} * 100, \tag{9.2}$$

where:

m_d: is the dry weight of the specimen,
m_w: is the wet weight of the specimen.

Three-point bend tests were made for the determination of the ILSS according to the BS EN ISO 14130 [44]. The ILSS tests were performed using a 5 kN load for increased accuracy during the load measurements at a crosshead speed of 1 mm/min.

DMA measurements prior and after exposure were performed on a DMA Netzsch 242 device in flexular configuration. Thermal scans from 35 to 200 °C were conducted at a heating rate of 1 °C/min at 1 Hz and constant amplitude (30 μm).

Figure 9.4 presents the weight gain of the unmodified and CNT-modified EP matrices and Fig. 9.5 the weight gain of the 0% and the 0.5% CNT-modified CFRP laminates. As can be seen in both figures, all systems reached saturation within the time frame of the hygrothermal exposure. The unmodified EP exhibited the least weight gain at saturation compared to all modified systems (Fig. 9.4). The CNT-modified EPs exhibited increased water uptake ranging without clear trend between the CNT content and the relative weight gain for the modified matrix systems (Fig. 9.4). The highest increase is however observed at 0.5% CNT content which could be linked to an optimum dispersion of the CNTs and the increased interfacial area at this CNT level. The pronounced difference in the water uptake between the matrices and the laminates, depicted in Fig. 9.5, was due to the presence of carbon fibres, which did not exhibit any water absorption. There were no visible

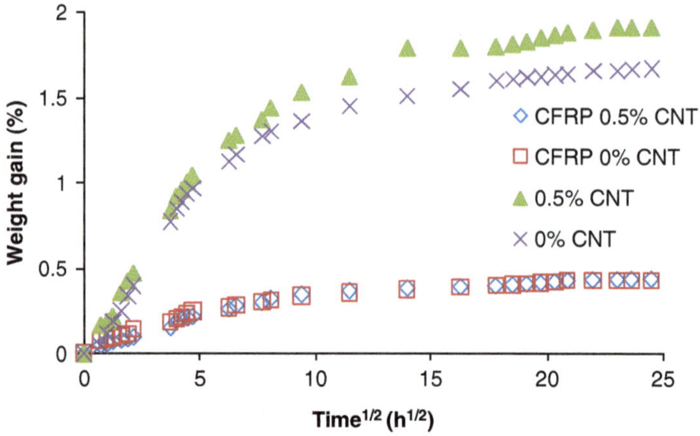

Fig. 9.5 Weight gain versus square root of time for the 0 and 0.5% CNT-modified EP matrices and the 0 and 0.5% CNT-modified CFRP specimens, reproduced after Barkoula et al. (2009)

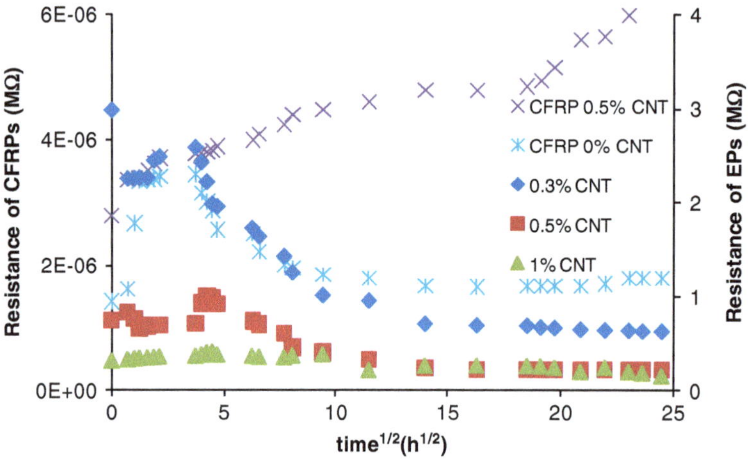

Fig. 9.6 Resistance versus square root of time for the unmodified and CNT-modified EP matrices and CFRP specimens, reproduced after Barkoula et al. (2009)

differences between the two laminates (Fig. 9.5) similarly to the case of a glass fibre/EP system modified with a montmorillonite system (Chow 2007). It can therefore be concluded that the presence of the fibre reinforcement was masking any increase in the water uptake created by the CNTs.

The modification of the EP resins rendered them conductive enabling the monitoring of the electrical resistance throughout the exposure (Barkoula et al. 2009). As can be seen in Fig. 9.6 for all studied EP systems the resistance reached a peak value after approximately 20 h of exposure or at 1% weight gain. After that, the resistance decreased monotonically until the end of the exposure, i.e. at 600 h.

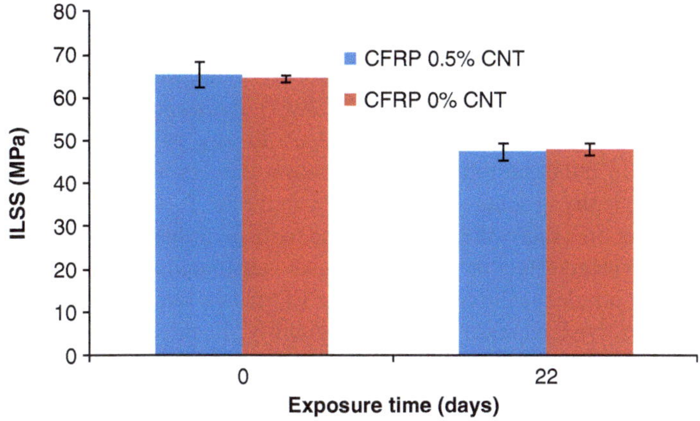

Fig. 9.7 ILSS before and after exposure for the unmodified and CNT-modified CFRP specimens reproduced after Barkoula et al. (2009)

Similar behaviour was observed for the 0% CNT CFRPs with the peak value being reached at almost the same exposure time, i.e. 20 h and at 0.2% weight gain. On the contrary in the case of the 0.5% CNT modified CFRP laminates the resistance increased monotonically throughout the experiment. One would expect that the inclusion of both carbon fibres and CNTs in the EP, would affect the resistance of the composite in the same way regarding its response to hygrothermal exposure. However, in the case of the CNT-modified CFRP laminates a paradox was observed. This paradox was manifested by the fact that the resistance was monotonically increasing with weight gain. The inclusion of a small weight fraction of a conductive phase (CNTs) to an otherwise conductive material (due to presence of carbon fibres), although it was hardly affecting the initial resistance of the system, was totally altering its electrical behaviour. This was attributed to a synergistic effect between the two conductive phases, i.e. the carbon fibres and the CNTs (Barkoula et al. 2009).

Finally, in Fig. 9.7 the ILSS of all composite laminates prior to and after exposure is depicted. The inclusion of 0.5% CNT in the composite matrix did not affect the interlaminar performance of the composite systems. This is consistent with the fact that there was no obvious difference in the water uptake between the modified and the unmodified laminated specimen (Fig. 9.5). Although the inclusion of an additional interface was expected to cause deterioration of the ILSS, this was not verified in the experimental campaign.

The effect of CNT addition on the thermomechanical properties of EP based composites has been investigated in the last decade with contradicting trends. According to the dispersion state of the CNTs, different E′ and loss modulus (E″) behavior are expected (Hatakeyama and Quinn 1999; Li et al. 2000; Mitchell et al. 2002). Some studies report strong effect in the E′ on the glassy state, due to improved interaction between the nanotubes and the EP matrix, which reduces the mobility of the EP matrix around the nanotubes and leads to an observed increase in thermal stability (Valentini et al. 2003; Gojny and Schulte 2004; Fidelusa et al. 2005; Chen et al. 2008; Montazeri and Montazeri 2011; Montazeri et al. 2011; Prolongo et al. 2011). A slight effect was measured on the rubbery state (Montazeri and Montazeri 2011; Montazeri et al. 2011; Prolongo et al. 2011). It was stated that the nanotube content was not sufficient (0.5 wt.%) to lead to any reinforcement since the molecular motion and the amplitude of this motion are very high and the macromolecule is not practically in contact with particles above the T_g. Gojny et al. (Gojny and Schulte 2004) report exactly the opposite trend, with no influence on the glassy state and more pronounced effect on the rubbery state, attributed again to the interfacial interaction which reduces the mobility of the EP matrix around the nanotubes and leads to the observed increase in thermal stability. In this study this effect was expected to appear around and above the T_g, due to the limited potential movement of the polymeric matrix below (Gojny and Schulte 2004).

In terms of loss modulus (E″), the dispersed nanotubes dissipate energy due to resistance against viscoelastic deformation of the surrounding EP matrix [Montazeri and Montazeri 2011 (Prolongo et al. 2011; Gojny and Schulte 2004; Fidelusa et al. 2005). Other possible mechanisms are rearrangements of molecules and nanotubes as well as internal friction between the nanotubes and the polymer matrix (Li et al. 2004). The loss modulus increased at 0.5 and 1 wt.%, followed by a continuous decrease in the peak height at higher CNT values (Montazeri and Montazeri 2011)]. The increase in the E′ was limited for nanotube contents up to 0.5 wt.%. When the nanotube value reached 1 and 2 wt.%, the E′ decreased. This was attributed to agglomeration of the nanotubes at high weight content (Gojny and Schulte 2004; Seyhan et al. 2007; Prolongo et al. 2011; Montazeri and Montazeri 2011).

Contradicting results have been reported on the effect of CNTs addition to the T_g. The addition of nanotubes to the EP results in a shift of the T_g. Some studies report shift of the T_g towards higher values, which was attributed to restricted mobility of the polymer chains in the matrix due to the presence of the nanotubes (Gojny and Schulte 2004; Ramanathan et al. 2005; Wang et al. 2006; Shen et al. 2007a; Chen et al. 2008; Montazeri et al. 2011; Prolongo et al. 2011). It was also observed that the increase of the T_g is more pronounced when the curing time is not optimized and less pronounced when sufficient curing of the resin occurs (Montazeri et al. 2011). This gain in thermostability was again interpreted as a reduction of the mobility of the EP matrix around the nanotubes by interfacial interactions (Gojny and Schulte 2004). Other studies report either a very small shift of 1–3 °C or even a decrease as more nanotubes are present (Fidelusa et al. 2005; Miyagawa et al. 2006; Shen et al. 2007b; Chen et al. 2008; Montazeri and Montazeri 2011). This was linked to reduced cross-linking tendency of the resin (Fidelusa et al. 2005; Miyagawa et al. 2006). The penetration of nanotubes into a free volume of polymer decreases the cross-link density but the rigidity and the tensile modulus of polymer increases (Won et al. 1990; Montazeri and Montazeri 2011). On the other hand, the tanδ peaks associated to the T_g were broader and shoulders with both, CNT content and pre-curing treatment. These phenomena can be attributed to the covalent bonds between EP and amino-functionalized CNTs,

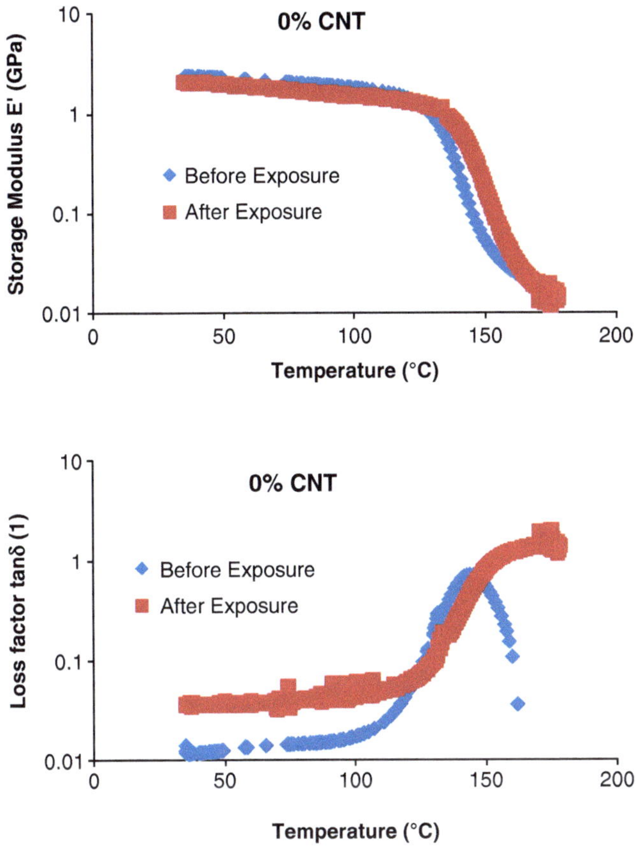

Fig. 9.8 Storage Modulus (E′) and loss factor (tanδ) of the EP with 0% CNTs as a function of temperature before and after hydrothermal exposure

inducing different cross-linking regions into the EP matrix (Prolongo et al. 2011). Figures 9.8, 9.9, 9.10 and 9.11 present the DMA results of the EPs as a function of the % CNT content, before and after exposure. In these graphs the variation of the E′ and the tanδ as a function of temperature is presented. From these figures it can be observed that the water absorption led into degradation of the E′ value, which shifted to lower values. In the case of the tanδ, it can be seen that the water exposure introduced a broadening of the peak around the T_g which is slightly more pronounced at increased CNT contents. Next to that no considerable trend of the T_g can be observed due to the water exposure. The variation of the T_g was small and non-monotonic both as a function of CNT content, as well as a function of the water exposure. This analysis refers to the effect of the CNT addition on thermomechanical properties. The preliminary results below are presented on the combined effect of the CNT addition with hygrothermal exposure on thermomechanical properties (Barkoula et al. 2010).

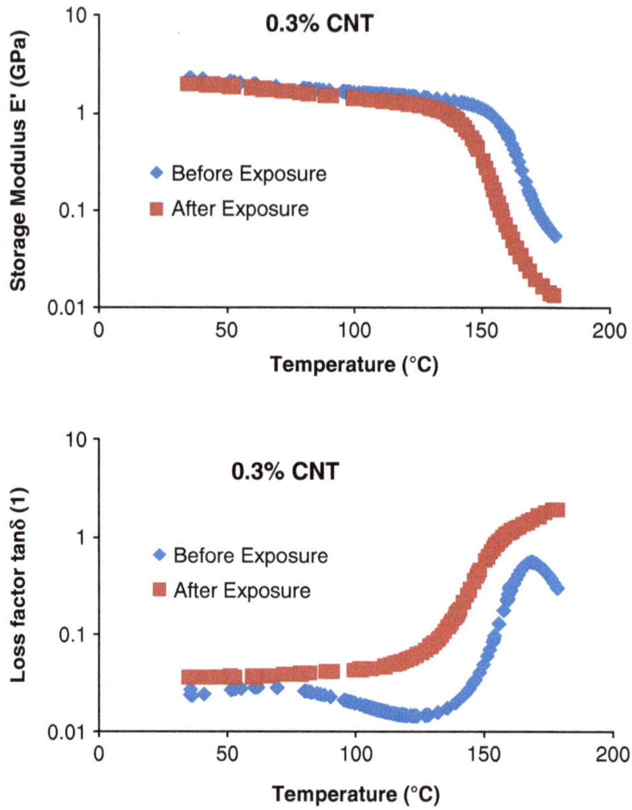

Fig. 9.9 Storage Modulus (E′) and loss factor (tanδ) of the EP with 0.3% CNTs as a function of temperature before and after hygrothermal exposure

Figures 9.12 and 9.13 presents the DMA results of the unmodified and CNT-modified CFRPs before and after exposure. As in the case of unmodified and CNT-modified EPs the variation of the E′ and the tanδ as a function of temperature is presented. In all cases it can be observed that there was no influence of the exposure on the stiffness and the T_g of the unmodified and CNT-modified composites. This can be attributed to the water absorption results presented in Fig. 9.5, where no significant difference was observed in the water uptake of the CFRP specimens, due to the masking effect provided by the presence of the carbon fibres as explained above.

Figure 9.14 summarizes the results presented in Figs. 9.8, 9.9, 9.10 and 9.11 for the unmodified and CNT-modified EPs, before and after exposure. The data for the E′ for all specimens are taken from the glassy region (35 °C) and the rubbery region (130 °C). From these data it can be seen that for low CNT content the E′ remains almost constant at the glassy region before and after exposure, while a small increase can be observed at the rubbery region. This is in line with previous data

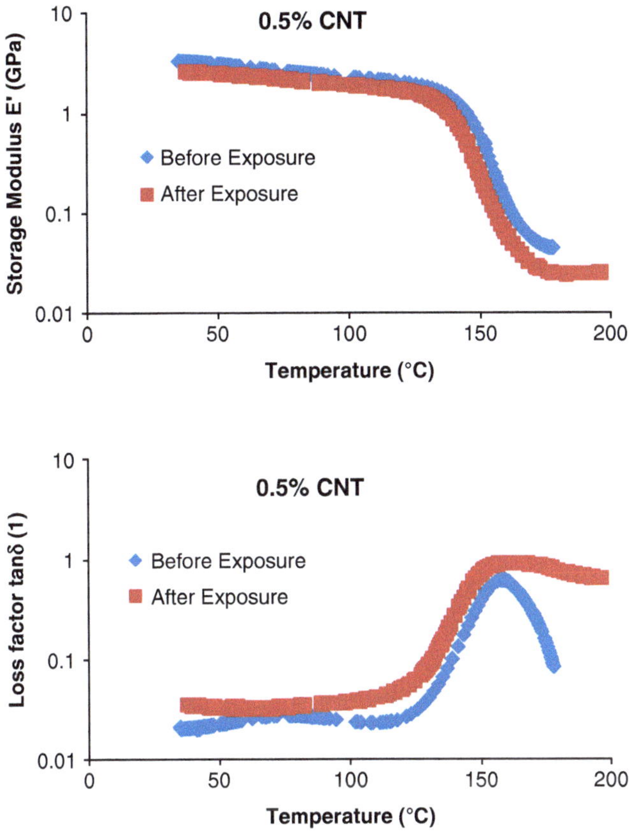

Fig. 9.10 Storage Modulus (E′) and loss factor (tanδ) of the EP with 0.5% CNTs as a function of temperature before and after hygrothermal exposure

where the addition of CNTs had a more pronounced effect on the properties above the T_g (Gojny and Schulte 2004). Further CNT addition results in pronounced enhancement of the E′ in both regions, while CNT contents as high as 1% lead to again a lowering of the E′ which is more pronounced in the rubbery state. This holds for both exposed and non-exposed specimens. The incorporation of low weight fractions of CNTs (up to 0.3%) into the EP matrix caused small changes in the T_g also (shift to higher values). Above a certain fraction the opposite effect can be observed. This behavior can be explained in terms of the interaction of the CNTs with the EP at the CNT/EP interface. Due to the higher surface area and the interfacial interactions, a reduced mobility of the EP is obtained, which leads to increased stiffness and increased thermal stability. The dispersed nanotubes dissipate energy due to resistance against viscoelastic deformation of the surrounding EP matrix. Above a certain CNT, which here was estimated at about 0.5%, the nanotubes tend to agglomerate, leading to less energy dissipating in the system

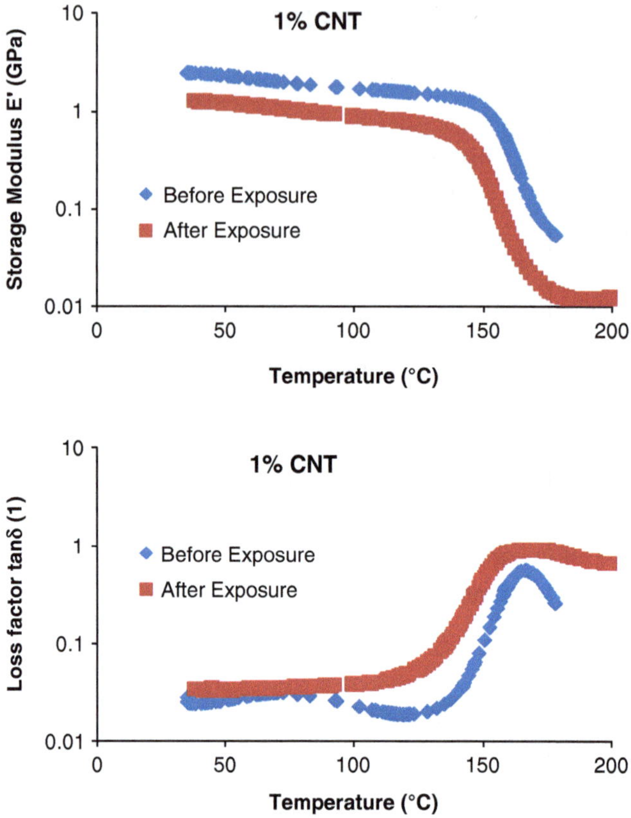

Fig. 9.11 Storage Modulus (E′) and loss factor (tanδ) of the EP with 1% CNTs as a function of temperature before and after hygrothermal exposure

under visco-elastic deformation. The decrease of E′ due to water absorption can be explained by increased mobility around the CNT/EP interface due to the presence of water and lowering of the stress transfer efficiency of the modified systems. It is interesting to note there was no monotonic change in the T_g before and after hygrothermal exposure. From Fig. 9.14 it can be seen that the T_g increased after exposure in most cases compared to the un-exposed condition. In the past a decrease of the T_g has been observed due to water absorption, in EP based systems. This has been attributed mainly to plasticization of the EP by moisture. Another explanation could be less cross-linking of the interface due to the presence of water. Though in most of the published literature (Weitsman 1991; Maggana and Pissis 1999; Li et al. 2001; Nogueira et al. 2001; Mohd Ishak et al. 2001) an increase in the peak of tanδ and a shift to lower temperatures of the T_g region with an increase of the water content is observed, in our case, the opposite is observed as in case of Zhou and Lucas (1999a, b) and Papanicolaou et al. (2006). This kind of behavior

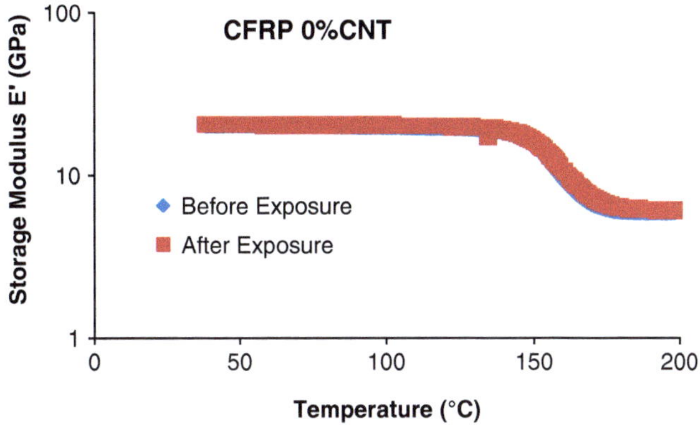

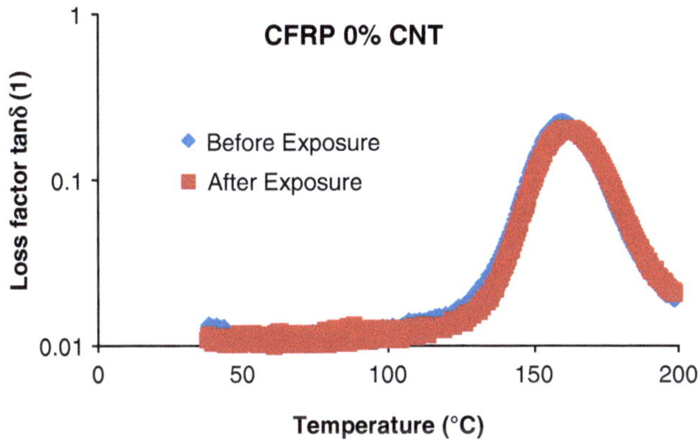

Fig. 9.12 Storage Modulus (E′) and loss factor (tanδ) of the CFRP with 0% CNTs as a function of temperature before and after hygrothermal exposure

can only be explained on the basis of respective recent findings reported (Papanicolaou et al. 2006; Zhou and Lucas 1999a, b). As aforementioned, water molecules bind with EP resins through hydrogen bonding. Two types of bound water were found in EP resins. The binding types are classified as Type I or Type II bonding, depending on differences in the bond complex and activation energy. They revealed that the change of the T_g does not depend solely on the water content absorbed in EP resins, that the T_g depends on the hygrothermal history of the materials. They also proposed that for a given EP system, higher values of the T_g resulted from longer immersion time and higher exposure temperature and the water/resin interaction characteristics (Type I and Type II bound water) have quite different influences on the T_g variation. Type I bound water disrupts the initial interchain Van der Waals force and hydrogen bonds, resulting in increased chain segment mobility acting as a plasticizer and decreasing the T_g. In contrast, Type II bound water contributes, comparatively, to an increase of the T_g in water saturated EP resin by forming a secondary cross-link network.

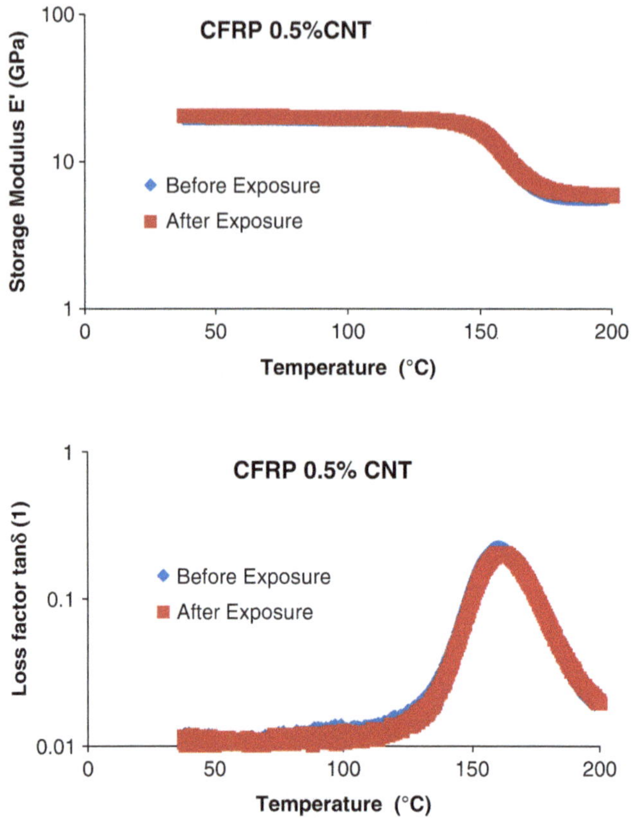

Fig. 9.13 Storage Modulus (E′) and loss factor (tanδ) of the CFRP with 0.5% CNTs as a function of temperature before and after hygrothermal exposure

9.4.3 Response of Carbon Nanotube Hybrid Aerospace Composites in Galvanic Corrosion

The application of bonded composite patches as doublers to repair or reinforce defective metallic structures is becoming recognized as a very effective and versatile repair procedure for many types of damage. Although mechanically fastened patches are usually endorsed by aircraft manufacturers, adhesively bonded patches have been reported to perform better than bolted patches (Baker 1999). Various applications of this technology include the repair of cracking, localized reinforcement after removal of corrosion damage and reduction of fatigue strain (Baker 1997). The bonded repair on the cracked metallic structure allows for the restoration of strength and stiffness of the structure, as well as hindering further crack growth by reducing the stress intensity factor. Aircraft alloys are specially designed alloys that impart high strength and light weight to aircraft, but are often susceptible

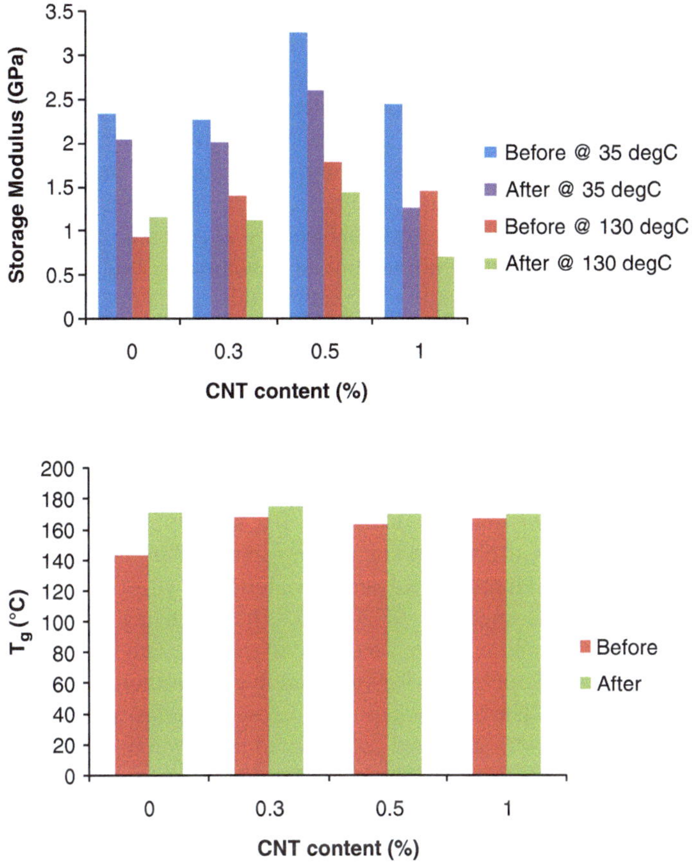

Fig. 9.14 Storage Modulus (E′) at the glassy (35 °C) and rubbery region (130 °C) and T_g of the unmodified and CNT-modified EPs as a function of CNT content before and after hygrothermal exposure

to corrosion, especially the two most commonly used Al (aluminum) alloys AA 2024T-3 and AA 7075T-6 (Buchheit 1995; Buchheit et al. 1997; Ilevbare et al. 2000). These are phase separated alloys that are in themselves highly complex metal-in-metal composites, but tend to have weakness towards local galvanic corrosion because of this structure. Also, these two alloys are among the most difficult to protect of all Al alloys (Reynolds et al. 1997; Bierwagen and Tallman 2001; Bierwagen et al. 2007). The requirements for an effective repair start from the structural enhancement of the repair system and its interface with the parent structure. It is expected that the addition of CNTs as discussed in previous chapters will enhance the damage tolerance of the repair systems as well as allow the tailoring of the thermal properties of the patch system in order to minimize the thermal stresses that are present due to the thermal coefficient mismatch between the patch and the parent material. As far as the electrode potential difference is concerned,

CNTs inclusions in bimetallic systems are reported to alter the REDOX (reduction oxidation) potential of the system, reducing the presence of localized corrosion. This paragraph will discuss some preliminary results (Gkikas et al. 2010) on the effect of the introduction of CNTs in aerospace adhesive repair systems in order to enhance the adhesion and control the galvanic corrosion between the patch and the substrate. This is expected to have a big impact on enabling the usage of CFRP patches in ageing Al aircrafts, which is hindered by galvanic effects between the Al substrate and the graphite fibres. The CNT-enhanced adhesive can be tailored to mediate the effects of galvanic corrosion in Al structures by bridging the galvanic potential between the substrate and the patch. In this study, the adhesion and the galvanic corrosion properties were assessed for both a reference and a modified adhesive. Both the mechanical and the corrosion properties of the modified systems were studied.

Details on the materials used and the testing procedures can be found in (Barkoula et al. 2009; Gkikas et al. 2010). In short, multiwalled CNTs were incorporated in a commercial EP system via high shear mechanical mixing. Modified EPs with CNT content of 0.5 and 1% were manufactured. The resin system was composed from two-component liquid shim adhesive, Epibond 1590 – 3 mm A/B from Huntsman Advanced Materials, Switzerland. The substrate used for this study was anodized Al 2024T3. The anodizing process – surface preparation was performed in-house according to the Standard Guide for Preparation of Al Surfaces for Structural Adhesives Bonding (Phosphoric Acid Anodizing) (ASTM: D 3933 – 98). The electrochemical corrosion studies were performed on: (a) anodized Al (Al), (b) anodized Al covered by unmodified adhesive film (Al_EP_0% CNTs), (c) anodized Al covered by doped adhesive film 0.5% CNTs (Al_EP_0.5% CNTs) and (d) anodized Al covered by doped adhesive film 1% CNTs (Al_EP_1% CNTs). The effect of the CNTs on the adhesion efficiency was studied using the lap shear test, according to the Standard Test Method for Lap Shear Adhesion for Fibre Reinforced Plastic (FRP) Bonding (ASTM: D 5868 – 95). The materials used were Al substrate, unmodified adhesive film and doped adhesive films with 0.5% CNTs and 1% CNTs. Two sets of specimens were tested, i.e. one without surface treatment and one surface treated. Testing was performed at a displacement rate of 13 mm/min.

In Fig. 9.15 the rest potential of the Al covered with unmodified and CNT-modified adhesive films as well as of the neat Al relative to the reference electrode is depicted. The Al_EP_0.5% CNTs is the least prone to corrosion. However it exhibits significant variation in rest potential values. As was expected, Al is slightly lower in the electrochemical series than Al covered with adhesive film, since the adhesive film acts as insulation. Finally doping the adhesive film with CNTs bridges or even reverses (when increasing the CNT content) the rest potential closer between the mixture and Al. The last one can be seen from the rest potential of the Al_EP_1% which approaches or becomes lower than the rest potential of Al after 70,000 s (20 h) immersion in the conductive solution.

In Fig. 9.16, typical current density/time curves are depicted. The current density was measured in pairs of anodized Al with the aforementioned substrates a–d.

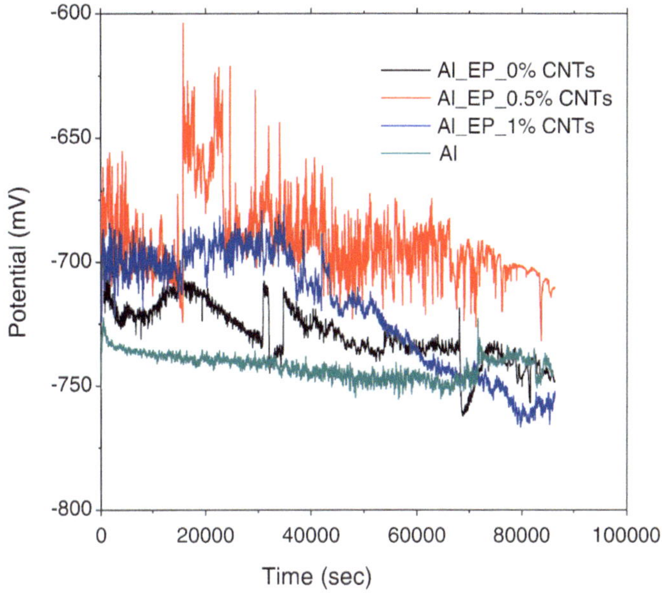

Fig. 9.15 The rest potential versus time curves for Al, Al_EP_0% CNTs, Al_EP_0.5% CNTs and Al_EP_1% CNTs substrate systems

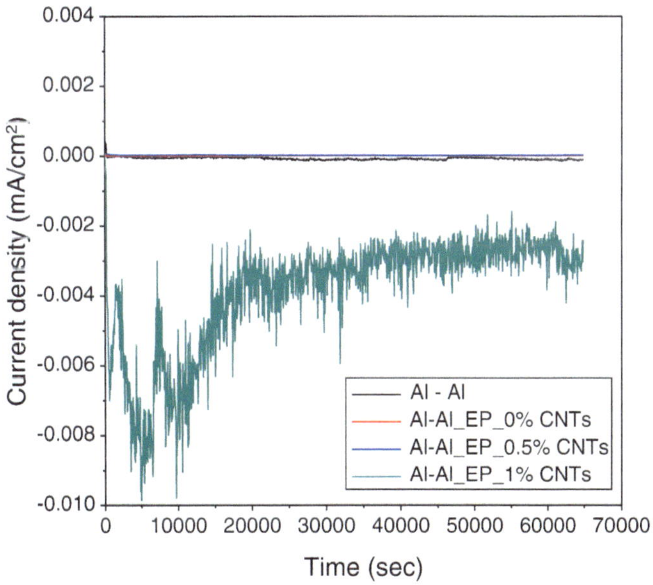

Fig. 9.16 Current density versus time for the pairs: Al-Al, Al-Al_EP_0% CNTs, Al-Al_EP_0.5% CNTs and Al-Al_EP_1% CNTs reproduced after Gkikas et al. (2010)

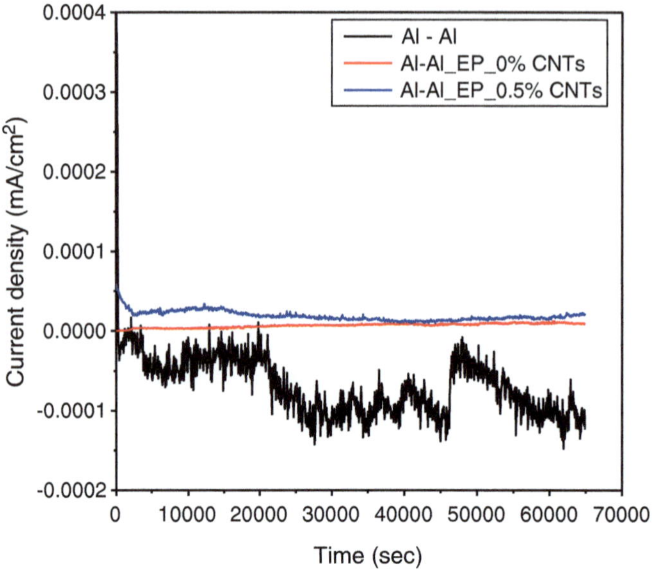

Fig. 9.17 Current density vs. time for Al vs. Al with adhesive films (doped and undoped). Current density versus time for the pairs: Al-Al, Al-Al_EP_0% CNTs and Al-Al_EP_0.5% CNTs Note that the variation in current density is two orders of magnitude smaller than the variation in Fig. 9.16, reproduced after Gkikas et al. (2010)

As can be seen the current density was minimum when the Al_EP_1% CNTs was used as the second electrode. This means that this electrode was corroded and corrosion was transferred to the Al with the adhesive film.

In Fig. 9.17 the same results as in the case of Fig. 9.16 are presented with the only difference that the scale of the y-axis is lower so that the systems that are not visible in Fig. 9.16 can be seen. The need to change scale in the case of the CNT-modified systems is directly attributed to the conductive nature of the CNTs that enhances ionic exchange and therefore promotes galvanic corrosion. It can be argued that the presence of the CNTs will be beneficial as it will mediate the effects of the localized corrosion that leads to stress failure and premature failure. More specifically:

1. In pairs of Al-Al, Al-Al_EP_0% CNTs and Al-Al_EP_0.5% CNTs there is minimum or no galvanic corrosion because these three materials have similar rest potentials. As can be seen in Fig. 9.17 where the afore-mentioned pairs are depicted separately, the current density between these three pairs is two orders of magnitude lower than the current density measured between Al-Al_EP_1% CNTs.
2. Al-Al_EP_1% CNTs exhibit higher current density which is consistent with the higher rest potential compared to the plain Al substrate. In this pair Al acts as cathode as the current density is negative, which proves that the CNT incorporation may result in reversal of the galvanic potential.

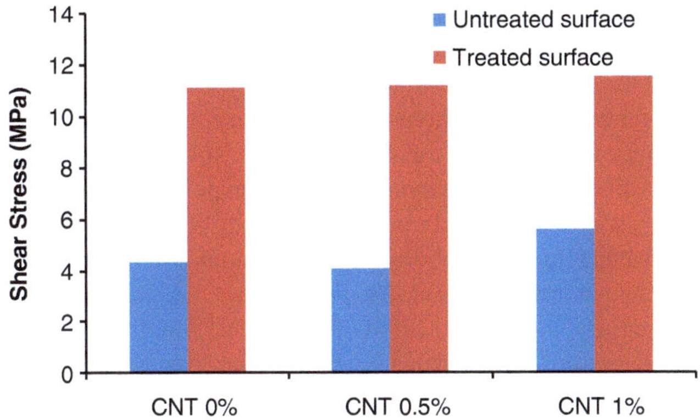

Fig. 9.18 Effect of surface treatment and CNT content on the adhesion of Al with unmodified and CNT EP substrates reproduced after Gkikas et al. (2010)

As can be seen from the lap shear tests (Fig. 9.18) CNT doping enhances adhesion in all but one cases. The enhancement of the adhesion is more pronounced for 1% CNT content and the untreated substrate. However, the adhesion enhancement is within the experimental scatter of the studied system. This enhancement may be attributed to the CNT/EP interface which activated mechanisms at the nanoscale such as crack bifurcation and arrest, delaying thus the global shear failure (Kostopoulos et al. 2007).

9.5 Summary

The scope of the current chapter was to review all available data related to the environmental degradation of carbon nanotube hybrid aerospace composites. These newly developed hybrid aerospace systems have been recently introduced as alternatives to conventional high performance polymer composites due to their improved mechanical properties, toughness and damage sensing abilities as discussed in detail in previous chapters. In order to be qualified for the aerospace industry their environmental response was of key interest as explained in details at the introductions of this chapter. Due to the lack of extensive literature on such systems, in this chapter an attempt was made to highlight possible issues due to environmental exposure based on previous experience on CFRPs. The degradation of hybrid composites due to exposure on hydro/hygrothermal loadings and the galvanic corrosion response of CNT-modified patches were discussed based on preliminary results.

The current work focussed on the effect of hygrothermal exposure of EP matrices and CFRPs with and without CNT modification. The weight gain as well as the electrical resistance of the exposed systems was measured as a function of exposure time. There were very little differences between the unmodified and the modified EP resins in term of weight gain. The unmodified EP system exhibited slightly less water uptake than the modified systems. In the case of the composite laminates, there was practically no observed difference in terms of weight gain versus time. This was consistent with the fact that there was no notable difference for the interlaminar shear strength of the composite laminates prior to and after exposure. The modification of the EP resins rendered them conductive enabling the monitoring of the electrical resistance throughout the exposure. For all studied systems the resistance reached a peak value after approximately 20 h of exposure or at 1% weight gain. After that, the resistance decreased monotonically until the end of the exposure, i.e. at 600 h. Similar behaviour was observed for the unmodified CFRPs with the peak value being reached at almost the same exposure time, i.e. 20 h and at 0.2% weight gain. On the contrary in the case of the 0.5% CNT modified CFRP laminates the resistance increased monotonically throughout the experiment. This was attributed to a synergistic effect between the two conductive phases, i.e. the carbon fibres and the CNTs. Furthermore, the inclusion of CNTs in the matrix of otherwise conventional CFRPs is promising as far as the monitoring of hygrothermal degradation by means of electrical resistance measurements is concerned.

The thermomechanical results reveal that the exposure of the CNT-modified EPs into hygrothermal loading influenced slightly their viscoelastic properties. In all cases it was observed that the water absorption led into degradation of the Storage Modulus (E′), which shifted to lower values. In the case of the loss factor, it was seen that the water exposure introduced a broadening of the peak around the glass transition temperature (T_g) which was slightly more pronounced at increased CNT contents. Next to that no considerable trend of the T_g was observed due to water exposure. The variation of the T_g was small and non-monotonic both as a function of CNT content, as well as a function of water exposure. The CNT-modified CFRPs did not show any deterioration due to exposure into hydrothermal loading, which was explained due to the lack of any kind of difference in the water up-take curves.

Finally CNTs were introduced in EP adhesives in order to tailor the galvanic behaviour of the composite patches with Al substrate and enhance their adhesion. As was shown, CNTs may alter the galvanic behavior of the adhesive leading even to the reversal of the rest potential with Al. This is very promising regarding the use of CFRP bonded patches on Al substrates in corrosive environments. On the other hand the adhesion enhancement which stemmed from the CNT doping was present but within the experimental scatter of the system.

Acknowledgements The author would like to acknowledge the EU (IAPETUS PROJECT, Grant Agreement Number: ACP8-GA-2009-234333) for financial support. Part of the presented experimental work is performed within the framework of the PhD study of PhD candidate Giorgos Gkikas, supervised by Prof. A. Paipetis.

References

Apicella, A., Nicolais, L.: Role of processing on the durability of epoxy composites in humid environments. Ind. Eng. Chem. Prod. Res. Dev. **23**, 288–297 (1984). doi:10.1021/i300014a024

Apicella, A., Nicolais, L.: Network structure and plasticization of epoxy-based resins. Makromol. Chem. Macromol. Symp. **7**, 97–113 (1987). doi:10.1002/masy.19870070111

Asp, L.E.: The effects of moisture and temperature on the interlaminar delamination toughness of a carbon/epoxy composite. Compos. Sci. Technol. **58**, 967–977 (1998). doi:10.1016/S0266-3538(97)00222-4

Baker, A.A.: Joining and repair of aircraft composite structures. In: Mallick, P.K. (ed.) Composite Engineering Handbook. Marcel Dekker, New York (1997)

Baker, A.A.: Bonded composite repair of fatigue-cracked primary aircraft structure. Compos. Struct. **47**, 431–443 (1999). doi:10.1016/S0263-8223(00)00011-8

Bank, L.C., Gentry, T.R., Barkatt, A.: Accelerated test methods to determine the long-term behavior of FRP composite structures: environmental effects. J. Reinf. Plast. Compos. **14**, 559–587 (1995). doi:10.1177/073168449501400602

Barkoula, N.M., Paipetis, A., Matikas, T., Vavouliotis, A., Karapappas, P., Kostopoulos, V.: Environmental degradation of carbon nanotube modified carbon fibre reinforced laminates: an electrical resistivity study. Mech. Compos. Mater. **45**, 21–32 (2009). doi:10.1007/s11029-009-9059-8

Barkoula, N.M., Gkikas, G., Makri, A., Matikas, T.E., Paipetis, A.: Effect of the environmental degradation on the viscoelastic response of nano modified epoxies and CTRPs. In: Proceedings of the 14th European Conference on Composite Materials (ECCM14), Budapest, Hungary. Paper ID: 045-ECCM14 (2010)

Bellenger, V., Verdu, J.: Photooxidation of amine crosslinked epoxies II. Influence of structure. J. Appl. Polym. Sci. **28**, 2677–2688 (1983). doi:10.1002/app. 1983.070280901

Bellenger, V., Verdu, J.: Oxidative skeleton breaking in epoxy–amine networks. J. Appl. Polym. Sci. **30**, 363–374 (1985). doi:10.1002/app. 1985.070300132

Bierwagen, G.P., Tallman, D.E.: Choice and measurement of crucial aerospace coating system properties. Prog. Org. Coat. **41**, 201–217 (2001). doi:10.1016/S0300-9440(01)00131-X

Bierwagen, G., Battocchi, D., Simoes, A., Stamness, A., Tallman, D.: The use of multiple electrochemical techniques to characterize Mg-rich primers for Al alloys. Prog. Org. Coat. **59**, 172–178 (2007). doi:10.1016/j.porgcoat.2007.01.022

Bockenheimer, C., Fata, D., Possart, W.: New aspects of aging in epoxy networks. II. Hydrothermal aging. J. Appl. Polym. Sci. **91**, 369–377 (2004). doi:10.1002/app. 13093

Bondzic, S., Hodgkin, J., Krstina, J., Mardel, J.: Chemistry of thermal ageing in aerospace epoxy composites. J. Appl. Polym. Sci. **100**, 2210–2219 (2006). doi:10.1002/app. 23692

Botelho, E.C., Pardini, L.C., Rezende, M.C.: Hygrothermal effects on the shear properties of carbon fibre/epoxy composites. J. Mater. Sci. **41**, 7111–7118 (2006a). doi:10.1007/s10853-006-0933-7

Botelho, E.C., Pardini, L.C., Rezende, M.C.: Damping behavior of hygrothermally conditioned carbon fibre/epoxy laminates. J. Appl. Polym. Sci. **106**, 3143–3148 (2006b)

Bowles, K.J., Jayne, D., Leonhardt, T.A.: Isothermal aging effects on PMR-15 resin. SAMPE Q. **24**, 2–9 (1993)

Bowles, K.J., McCorkle, L., Ingram, L.: Comparison of graphite fabric reinforced PMR-15 and Avimid N composites after long-term isothermal aging at various temperatures. J. Adv. Mater. **30**, 27–35 (1998)

Buchheit, R.G.: Compilation of corrosion potentials reported for intermetallicphases in aluminum alloys. J. Electrochem. Soc. **142**, 3994–3996 (1995)

Buchheit, R.G., Grant, R.P., Hlava, P.F., McKenzie, B., Zender, G.L.: Local dissolution phenomena associated with S phase (Al2CuMg) particles in aluminum alloy 2024-T3. J. Electrochem. Soc. **144**, 2621–2628 (1997)

Buehler, F.U., Seferis, J.C.: Effect of reinforcement and solvent content on moisture absorption in epoxy composite materials. Compos. Part A-Appl. Sci. Manuf. **31**, 741–748 (2000). doi:10.1016/S1359-835X(00)00036-1

Chae, H.G., Minus, M.L., Kumar, S.: Oriented and exfoliated single wall carbon nanotubes in polyacrylonitrile. Polymer **47**, 3494–3504 (2006). doi:10.1016/j.polymer.2006.03.050

Chen, W., Lu, H., Nutt, S.R.: The influence of functionalized MWCNT reinforcement on the thermomechanical properties and morphology of epoxy nanocomposites. Compos. Sci. Technol. **68**, 2535–2542 (2008). doi:10.1016/j.compscitech.2008.05.011

Chin, J.W., Nguyen, T., Aouadi, K.: Effects of environmental exposure on Fibre-Reinforced Plastic (FRP) materials used in construction. JCTR **19**, 205–213 (1997)

Chou, P.J.C., Ding, D.J.: Characterization of moisture absorption and its Influence on composite structures. J. Thermoplast. Compos. **13**, 207–225 (2000). doi:10.1177/089270570001300303

Chow, W.S.: Water absorption of epoxy/glass fibre/organo-montmorillonite nanocomposites. Express Polym. Lett. **1**, 104–108 (2007). doi:10.3144/expresspolymlett.2007.18

Clark, G., Saunders, D.S., van Blaricum, T.J., Richmond, M.: Moisture absorption in graphite/epoxy laminates. Compos. Sci. Technol. **39**, 355–375 (1990). doi:10.1016/0266-3538(90)90081-F

Colin, X., Marais, C., Verdu, J.: Thermal oxidation kinetics for a poly(bismaleimide). J. Appl. Polym. Sci. **82**, 3418–3430 (2001). doi:10.1002/app. 2203

Collings, T.A., Stone, D.E.W.: Hygrothermal effects in CFC laminates: damaging effects of temperature, moisture and thermal spiking. Compos. Struct. **3**, 341–378 (1985). doi:10.1016/0263-8223(85)90061-3

Dao, B., Hodgkin, J., Krstina, J., Mardel, J., Tian, W.: Accelerated aging versus realistic aging in aerospace composite materials. J. Appl. Polym. Sci. **102**, 4291–4303 (2006a). doi:10.1002/app.24862

Dao, B., Hodgkin, J., Krstina, J., Mardel, J., Tian, W.: Accelerated aging versus realistic aging in aerospace composite materials. II. Chemistry of thermal aging in a structural composite. J. Appl. Polym. Sci. **102**, 3221–3232 (2006b). doi:10.1002/app. 24573

Dao, B., Hodgkin, J., Krstina, J., Mardel, J., Tian, W.: Accelerated ageing versus realistic ageing in aerospace composite materials. III. The chemistry of thermal ageing in bismaleimide based composites. J. Appl. Polym. Sci. **105**, 2062–2072 (2007a). doi:10.1002/app. 26320

Dao, B., Hodgkin, J.H., Krstina, J., Mardel, J., Tian, W.: Accelerated ageing versus realistic ageing in aerospace composite materials. IV. Hot/wet ageing effects in a low temperature cure epoxy composite. J. Appl. Polym. Sci. **106**, 4264–4276 (2007b). doi:10.1002/app. 27104

Delozier, D.M., Watson, K.A., Smith, J.G., Connell, J.W.: Preparation and characterization of space durable polymer nanocomposite films. Compos. Sci. Technol. **65**, 749–755 (2004). doi:10.1016/j.compscitech.2004.10.026

Dyakonov, T., Mann, P.J., Chen, Y., Stevenson, W.T.K.: Thermal analysis of some aromatic amine cured model epoxy resin systems–II: residues of degradation. Polym. Degrad. Stabil. **54**, 67–83 (1996). doi:10.1016/0141-3910(96)00096-1

Fidelusa, J.D., Wiesela, E., Gojny, F.H., Schulte, K., Wagnera, H.D.: Thermo-mechanical properties of randomly oriented carbon/epoxy nanocomposites. Compos. Part A **36**, 1555–1561 (2005). doi:10.1016/j.compositesa.2005.02.006

Fox, B.L., Lowe, A., Hodgkin, J.H.: Investigation of failure mechanisms in aged aerospace composites. Eng. Fail. Anal. **11**, 235–241 (2004). doi:10.1016/j.engfailanal.2003.05.010

Frassine, R., Pavan, A.: The combined effects of curing and environmental exposure on fracture properties of woven carbon/epoxy laminates. Compos. Sci. Technol. **51**, 495–503 (1994). doi:10.1016/0266-3538(94)90082-5

Garton, A.: Infrared spectroscopic characterization of epoxy matrix composites. J. Macromol. Sci. Part A-Pure Appl. Chem. **26**, 17–41 (1989)

Gkikas, G., Paipetis, A., Barkoula, N.M., Lekatou, A., Sioulas, D.: Enhanced bonded aircraft repair using nano-modified adhesives. In: Proceedings of the 14th European Conference on Composite Materials (ECCM14), Budapest, Hungary. Paper ID: 660-ECCM14 (2010)

Gojny, F.H., Schulte, K.: Functionalisation effect on the thermo-mechanical behavior of multiwall carbon nanotube/epoxy composite. Compos. Sci. Technol. **64**, 2303–2308 (2004). doi:10.1016/j.compscitech.2004.01.024

Grenier-Loustalot, M.F., Metras, F., Grenier, P., Chenard, J.Y., Horny, P.: Mecanismes reactionnels, cinetique et structure du reseau de systemes thermodurcissables type TGDDM-DDS. Eur. Polym. J. **26**, 83–100 (1990). doi:10.1016/0014-3057(90)90102-A

Hancox, N.L.: Overview of effects of temperature and environment on performance of polymer matrix composite properties. Plast. Rubber Compos. Process Appl. **27**, 97–105 (1998)

Harmon, J.P., Muisener, P.A.O., Clayton, L., D'Angelo, J., Sikder, A.K., Kumar, A., et al.: Ionizing radiation effects on interfaces in carbon nanotube-polymer composites. Mater. Res. Soc. Symp. Proc. **697**, 425–435 (2002)

Hatakeyama, T., Quinn, F.: Thermal Analysis: Fundamentals and Application to Polymer Science, 2nd edn. Wiley, Chichester (1999)

Hough, J.A., Xiang, Z.D., Jones, F.R.: The effect of thermal spiking and resultant enhanced moisture absorption on the mechanical and viscoelastic properties of carbon fibre reinforced epoxy laminates. Key Eng. Mater. **144**, 27–42 (1998). doi:10.4028/www.scientific.net/KEM.144.27

Hwang, T., Pu, L., Kim, S.W., Oh, Y.S., Nam, J.D.: Synthesis and barrier properties of poly (vinylidene chloride-co-acrylonitrile)/SiO2 hybrid composites by sol-gel process. J. Membr. Sci. **345**, 90–96 (2009). doi:10.1016/j.memsci.2009.08.029

Ilevbare, G.O., Scully, J.R., Yuan, J., Kelly, R.G.: Inhibition of pitting corrosion on aluminum alloy 2024-T3: effect of soluble chromate additions vs. chromate conversion coating. Corrosion **56**, 227–242 (2000)

Ivanova, K.I., Pethrick, R.A., Affrossman, S.: Hygrothermal aging of rubber-modified and mineral-filled dicyandiamide-cured DGEBA epoxy resin. II. Dynamic mechanical thermal analysis. J. Appl. Polym. Sci. **82**, 3477–3485 (2001). doi:10.1002/app. 2209

Jedidi, J., Jacquemin, F., Vautrin, A.: Design of accelerated hygrothermal cycles on polymer matrix composites in the case of a supersonic aircraft. Compos. Struct. **68**, 429–437 (2005). doi:10.1016/j.compstruct.2004.04.009

Jedidi, J., Jacquemin, F., Vautrin, A.: Accelerated hygrothermal cyclical tests for carbon/epoxy laminates. Compos. Part A-Appl. Sci. Manuf. **37**, 636–645 (2006). doi:10.1016/j.compositesa.2005.05.007

Jelinski, L.W., Dumais, J.J., Cholli, A.L., Ellis, T.S., Karasz, F.E.: Nature of the water-epoxy interaction. Macromolecules **18**, 1091–1095 (1985). doi:10.1021/ma00148a008

Joshi, K.M.: The effect of moisture on the shear properties of carbon fibre composites. Composites **14**, 196–200 (1983). doi:10.1016/0010-4361(83)90005-8

Kenig, S., Moshonov, A., Shucrun, A., Maron, G.: Environmental effects on shear delamination of fabric-reinforced epoxy composites. Int. J. Adhes. Adhes. **9**, 38–45 (1989). doi:10.1016/0143-7496(89)90145-0

Kerr, J.R., Haskins, J.F.: Time-Temperature-Stress Capabilities of Composite Materials for Advanced Supersonic Technology Application. NASA CR-178272 (1987)

Kim, J.K., Hu, C., Woo, R.S.C., Sham, M.L.: Moisture barrier characteristics of organoclay-epoxy nanocomposites. Compos. Sci. Technol. **65**, 805–813 (2005). doi:10.1016/j.compscitech.2004.10.014

Kojima, Y., Usuki, A., Kawasumi, M., Okada, A., Kurauchi, T., Kamigaito, O.: Sorption of water in nylon 6-clay hybrid. J. Appl. Polym. Sci. **49**, 1259–1264 (1993). doi:10.1002/app.1993.070490715

Kostopoulos, V., Tsotra, P., Karapappas, P., Tsantzalis, S., Vavouliotis, A., Loutas, T.H., Paipetis, A., Friedrich, K., Tanimoto, T.: Mode I interlaminar fracture of CNF or/and PZT doped CFRPs via acoustic emission monitoring. Compos. Sci. Technol. **67**, 822–828 (2007). doi:10.1016/j.compscitech.2006.02.038

Kumar, B.G., Singh, R.P., Nakamura, T.: Degradation of carbon fibre-reinforced epoxy composites by ultraviolet radiation and condensation. J. Compos. Mater. **36**, 2713–2733 (2002). doi:10.1177/002199802761675511

Lai, J.Y., Young, K.F.: Dynamics of graphite/epoxy composite under delamination fracture and environmental effects. Compos. Struct. **30**, 25–32 (1995). doi:10.1016/0263-8223(94)00017-4

Li, G., Lee-Sullivan, P., Thring, R.: Determination of activation energy for glass transition of an epoxy adhesive using dynamic mechanical analysis. J. Therm. Anal. Calorim. **60**, 377–390 (2000). doi:10.1023/A:1010120921582

Li, Y., Miranda, J., Sue, H.J.: Hygrothermal diffusion behavior in bismaleimide resin. Polymer **42**, 7791–7799 (2001). doi:10.1016/S0032-3861(01)00241-5

Li, X., Gao, H., Scrivens, W.A., Fei, D., Xu, X., Sutton, M.A., Reynolds, A.P., Myrick, M.L.: Nanomechanical characterization of single-walled carbon nanotube reinforced epoxy composites. Nanotechnology **15**, 1416–1423 (2004). doi:10.1088/0957-4484/15/11/005

Liau, W.B., Tseng, F.P.: The effect of long-term ultraviolet light irradiation on polymer matrix composites. Polym. Compos. **19**, 440–445 (1998). doi:10.1002/pc.10118

Lin, Y.C.J.: Investigation of the moisture-desorption characteristics of epoxy resin. Polym. Res. **13**, 369–374 (2006). doi:10.1007/s10965-006-9053-y

Liu, M., Wu, P., Ding, Y., Chen, G., Li, S.: Two-dimensional (2D) ATR−FTIR spectroscopic study on water diffusion in cured epoxy resins. Macromolecules **35**, 5500–5507 (2002). doi:10.1021/ma011819f

Luoma, G.A., Rowland, R.D.: Environmental degradation of an epoxy resin matrix. J. Appl. Polym. Sci. **32**, 5777–5790 (1986). doi:10.1002/app. 1986.070320710

Maggana, C., Pissis, P.: Water sorption and diffusion studies in an epoxy resin system. J. Polym. Sci. Part B-Polym. Phys. **37**, 1165–1182 (1999). doi:10.1002/(SICI)1099-0488(19990601) 37:11<1165::AID-POLB11>3.0.CO;2-E

Mahieux, C.A.: Environmental Degradation in Industrial Composites. Elsevier, Oxford (2006)

Mangalgiri, P.D.: Composite materials for aerospace applications. Bull. Mater. Sci. **22**, 657–664 (1999). doi:10.1007/BF02749982

Matzkanin, G.A., Hansen, G.P.: Heat Damage in Graphite Epoxy Composites: Degradation, Measurement and Detection: A State-of-the-Art Report. NTIAC-SR-98-02 (1998)

McKague, E.L., Halkias, J.E., Reynolds, J.D.: Moisture in composites: the effect of supersonic service on diffusion. J. Compos. Mater. **9**, 2–9 (1975). doi:10.1177/002199837500900101

Merdas, I., Thominette, F., Tcharkhtchi, A., Verdu, J.: Factors governing water absorption by composite matrices. J. Compos. Sci. Technol. **62**, 487–492 (2002). doi:10.1016/S0266-3538 (01)00138-5

Minus, M.L., Chae, H.G., Kumar, S.: Single wall carbon nanotube templated oriented crystallization of poly(vinyl alcohol). Polymer **47**, 3705–3710 (2006). doi:10.1016/j. polymer.2006.03.076

Mitchell, C.A., Bahar, J.L., Arepalli, S.: Dispersion of functionalized carbon nanotubes in polystyrene. Macromolecules **35**, 8825–8830 (2002). doi:10.1021/ma020890y

Miyagawa, H., Rich, M.J., Drzal, L.T.: Thermo-physical properties of epoxy nanocomposites reinforced by carbon nanotubes and vapor grown carbon fibres. Thermochim. Acta **442**, 67–73 (2006). doi:10.1016/j.tca.2006.01.016

Mohd Ishak, Z., Ariffin, A., Senawi, R.: Effects of hygrothermal aging and a silane coupling agent on the tensile properties of injection molded short glass fibre reinforced poly(butylene terephthalate) composites. Eur. Polym. J. **37**, 1635–1647 (2001). doi:10.1016/S0014-3057(01) 00033-7

Montazeri, A., Montazeri, N.: Viscoelastic and mechanical properties of multi walled carbon nanotube/epoxy composites with different nanotube content. Mater. Des. **32**, 2301–2307 (2011). doi:10.1016/j.matdes.2010.11.003

Montazeri, A., Khavandi, A., Javadpour, J., Tcharkhtchi, A.: Viscoelastic properties of multi-walled carbon nanotube/epoxy composites using two different curing cycles. J. Nano Res. **13**, 33–39 (2011). doi:10.4028/www.scientific.net/JNanoR.13.33

Morgan, R.J., Mones, E.T.: The cure reactions, network structure, and mechanical response of diaminodiphenyl sulfone-cured tetraglycidyl 4,4′ diaminodiphenyl methane epoxies. J. Appl. Polym. Sci. **33**, 999–1020 (1987). doi:10.1002/app. 1987.070330401

Morgan, R.J., O'neal, J.E.: The durability of epoxies. Polym. Plast. Tech. Eng. **10**, 49–116 (1978). doi:10.1080/03602557808055845

Morgan, R.J., Shin, E.E., Lincoln, J.: Thermal properties of high temperature polymer matrix fibrous composites. In: Cheng, S.Z.D. (ed.) Handbook of Thermal Analysis and Calorimetry, Elsevier Science BV, Amsterdam (2002). doi: 10.1016/S1573-4374(02)80015-2

Muisener, P.A.O., Clayton, L., D'Angelo, J., Harmon, J.P., Sikder, A.K., Kumar, A., et al.: Effects of gamma radiation on poly(methyl methacrylate)/single-wall nanotube composites. J. Mater. Res. **17**, 2507–2513 (2002). doi:10.1017/S0884291400061914

Musto, P., Ragosta, G., Mascia, L.: Vibrational spectroscopy evidence for the dual nature of water sorbed into epoxy resins. Chem. Mater. **12**, 1331–1341 (2000). doi:10.1021/cm9906809

Musto, P., Ragosta, G., Russo, P., Mascia, L.: Thermal-oxidative degradation of epoxy and epoxy-bismaleimide networks: kinetics and mechanism. Macromol. Chem. Phys. **202**, 3445–3458 (2001). doi:10.1002/1521-3935(20011201)202:18<3445::AID-MACP3445>3.0.CO;2-N

Musto, P., Ragosta, G., Scarinzi, G., Mascia, L.: Probing the molecular interactions in the diffusion of water through epoxy and epoxy–bismaleimide networks. Polym. Sci. Part B: Polym. Phys. **40**, 922–938 (2002). doi:10.1002/polb.10147

Najafi, E., Shin, K.: Radiation resistant polymer-carbon nanotube nanocomposite thin films. Colloid Surf. A **257**, 333–337 (2005). doi:10.1016/j.colsurfa.2004.10.076

Najafi, E., Lee, J.O., Shin, K.: Materials for space applications. Mater. Res. Soc. Symp. Proc. **851**, 279–285 (2005)

Nam, J.D., Seferis, J.C.: Anisotropic thermo-oxidative stability of carbon fibre reinforced polymeric composites. J. C. SAMPE. Q. **24**, 10–18 (1992)

Ngono, Y., Marechal, Y., Mermilliod, N.: Epoxy–amine reticulates observed by infrared spectrometry. I: hydration process and interaction configurations of embedded H2O molecules. J. Phys. Chem. B **103**, 4979–4985 (1999). doi:10.1021/jp984809y

Nielsen, K.L.C., Hill, D.J.T., Watson, K.A., Connell, J.W., Ikeda, S., Kudo, H., Whittaker, A.K.: The radiation degradation of a nanotube-polimide nanocomposite. Polym. Degrad. Stabil. **93**, 169–175 (2008). doi:10.1016/j.polymdegradstab.2007.10.010

Nogueira, P., Ramírez, C., Torres, A., Abad, M.J., Cano, J., López, J., López-Bueno, I., Barral, L.: Effect of water sorption on the structure and mechanical properties of an epoxy resin system. J. Appl. Polym. Sci. **80**, 71–80 (2001). doi:10.1002/1097-4628(20010404)80:1<71::AID-APP1077>3.0.CO;2-H

Ogi, K., Takeda, N.: Effect of moisture content on non-linear deformation behaviour of CF/epoxy composites. J. Compos. Mater. **31**, 530–551 (1997). doi:10.1177/002199839703100601

Papanicolaou, G.C., Kosmidou, T., Vatalis, A.S., Delides, C.G.: Water absorption mechanism and some anomalous effects on the mechanical and viscoelastic behavior of an epoxy system. J. Appl. Polym. Sci. **99**, 1328–1339 (2006). doi:10.1002/app. 22095

Patel, S.R., Case, S.W.: Durability of hygrothermally aged graphite/epoxy woven composite under combined hygrothermal conditions. Int. J. Fatig. **24**, 1295–1301 (2002). doi:10.1016/S0142-1123(02)00044-0

Pearce, P.J., Davidson, R.G., Morris, C.E.M.: Hydrolytic stability of some uncured epoxy resins. J. Appl. Polym. Sci. **26**, 2363–2372 (1981). doi:10.1002/app. 1981.070260722

Prolongo, S.G., Gude, M.R., Urena, A.: Improving the flexural and thermomechanical properties of amino functionalized carbon nanotube/epoxy composites by using a pre-curing treatment. Compos. Sci. Technol. **71**, 765–771 (2011). doi:10.1016/j.compscitech.2011.01.028

Pulikkathara, M.X., Shofner, M.L., Wilkins, R., Vera, J.G., Barrera, E.V., Rodriguez-Macias, F.J., Vaidyanathan, R.K., Green, C.E., Condon, C.G.: Fluorinated single wall nanotube/polyethylene composites for multifunctional radiation protection. Nanomater. Struct. Appl. **740**, 365–370 (2003)

Pulikkathara, M.X., Pena-Paras, L., McIntosh, D., Chipara, M., Wilkins, R., Barrera, E.V., Dye, D., Zaleski, J.M.: Proton beam induced modifications in multi-functional polyethylene-based carbon nanotube composites. Mater. Res. Soc. Symp. Proc. **851**, 261–266 (2005)

Qu, L.W., Lin, Y., Hill, D.E., Zhou, B., Wang, W., Sun, X.F., et al.: Polyimide-functionalized carbon nanotubes: synthesis and dispersion in nanocomposite films. Macromolecules **376055–6060** (2004). doi:10.1021/ma0491006

Ramanathan, T., Liu, H., Brinson, L.C.: Functionalized SWNT/polymer nanocomposites for dramatic property improvement. J. Polym. Sci. Part B: Polym. Phys. **43**, 2269–2279 (2005). doi:10.1002/polb.20510

Ranby, B., Rabek, J.: Photodegradation, Photo-Oxidation and Photostabilization of Polymers. Wiley, London (1975)

Ray, B.C.: Temperature effect during humid ageing on interfaces of glass and carbon fibres reinforced epoxy composites. J. Colloid Interface Sci. **298**, 111–117 (2006). doi:10.1016/j.jcis.2005.12.023

Reynolds, T.G., Mc Manus, H.L.: Accelerated tests of environmental degradation in composite materials. In: Grant, P., Rousseau, C.Q. (eds.) Composite Structures: Theory and Practice, ASTM STP 1383:513–525. American Society for Testing and Materials, West Conshohocken (2000)

Reynolds, L.B., Twite, R., Donley, M.S., Bierwagen, G.P., Khobaib, M.: Preliminary evaluation of the anticorrosive properties for aircraft coatings by electrochemical methods. Prog. Org. Coat. **32**, 31–34 (1997). doi:10.1016/S0300-9440(97)00098-2

Schoeppner, G., Tandon, G., Pochiraju, K.: Predicting thermooxidative degradation and performance of high-temperature polymer matrix composites. In: Talreja, R. (ed.) Multiscale Modeling and Simulation of Composite Materials and Structures, Springer, New York (2008). doi: 10.1007/978-0-387-68556-4_9

Selzer, R., Friedrich, K.: Influence of water up-take on interlaminar fracture properties of carbon fibre-reinforced polymer composites. J. Mater. Sci. **30**, 334–338 (1995). doi:10.1007/BF00354392

Seyhan, A.T., Gojny, F.H., Tanoglu, M., Schulte, K.: Rheological and dynamic-mechanical behavior of carbon nanotube/vinyl ester–polyester suspensions and their nanocomposites. Eur. Polym. J. **43**, 2836–2847 (2007). doi:10.1016/j.eurpolymj.2007.04.022

Shen, C., Springer, G.S.: Moisture absorption and desorption of composite materials. J. Compos. Mater. **10**, 2–20 (1976). doi:10.1177/002199837601000101

Shen, J.F., Huang, W.S., Wu, L.P., Hu, Y.Z., Ye, M.X.: The reinforcement role of different amino-functionalized multi-walled carbon nanotubes in epoxy nanocomposites. Compos. Sci. Technol. **67**, 3041–3050 (2007a). doi:10.1016/j.compscitech.2007.04.025

Shen, J.F., Huang, W.S., Wu, L.P., Hu, Y.Z., Ye, M.X.: Thermo-physical properties of epoxy nanocomposites reinforced with amino-fucntionalized multi-walled carbon nanotubes. Compos. Part A **38**, 1331–1336 (2007b). doi:10.1016/j.compositesa.2006.10.012

Shin, K.B., Kim, C.G., Hong, C.S., Lee, H.H.: Prediction of failure thermal cycles in graphite/epoxy composite materials under simulated low earth orbit environments. Compos. Part B-Eng. **31**, 223–235 (2000a). doi:10.1016/S1359-8368(99)00073-6

Shin, E.E., Morgan, R.J., Zhou, J., Lincoln, J., Jurek, R., Curliss, D.B.: Hygrothermal durability and thermal aging behavior prediction of high-temperature polymer-matrix composites and their resins. J. Thermoplast. Compos. **1340–57** (2000b). doi:10.1177/089270570001300104

Signor, A.W., VanLandingham, M.R., Chin, J.W.: Effects of ultraviolet radiation exposure on vinyl ester resins: characterization of chemical, physical and mechanical damage. Polym. Degrad. Stabil. **79**, 359–368 (2003). doi:10.1016/S0141-3910(02)00300-2

Singh, R.P., Khait, M., Zunjarrao, S.C., Korach, C.S., Pandey, G.: Environmental degradation and durability of epoxy-clay nanocomposites. J. Nanomater. **352746** (2010). doi:10.1155/2010/352746

Smith, J.G., Connell, J.W., Delozier, D.M., Lillehei, P.T., Watson, K.A., Lin, Y., et al.: Space durable polymer/carbon nanotube films for electrostatic charge mitigation. Polymer **45**, 825–836 (2004a). doi:10.1016/j.polymer.2003.11.024

Smith, J.G., Delozier, D.M., Connell, J.W., Watson, K.A.: Carbon nanotube-conductive additive-space durable polymer nanocomposite films for electrostatic charge dissipation. Polymer **45**, 6133–6142 (2004b). doi:10.1016/j.polymer.2004.07.004

Smith, J.G., Connell, J.W., Watson, K.A., Danehy, P.M.: Optical and thermo-optical properties of space durable polymer/carbon nanotube films: experimental results and empirical equations. Polymer **46**, 2276–2284 (2005). doi:10.1016/j.polymer.2005.01.022

Springer, G.S.: Chapter 3: Moisture absorption in fibre-resin composites. In: Pritchard, G. (ed.) Developments in Reinforced Plastics. Applied Science, London (1982)

St John, N.A., George, G.A.: Diglycidyl amine – epoxy resin networks: kinetics and mechanisms of cure. Prog. Polym. Sci. **19**, 755–795 (1994). doi:10.1016/0079-6700(94)90032-9

Tatro, S.R., Clayton, L.M., Muisener, P.A.O., Rao, A.M., Harmon, J.P.: Probing multi-walled nanotube/poly(methyl methacrylate) composites with ionizing radiation. Polymer **45**, 1971–1979 (2004). doi:10.1016/j.polymer.2004.01.012

Tian, W., Hodgkin, J.: Long-term aging in a commercial aerospace composite sample: chemical and physical changes. J. Appl. Polym. Sci. **115**, 2981–2985 (2010). doi:10.1002/app. 31394

Tsotsis, T.K., Lee, S.M.: Long-term thermo-oxidative aging in composite materials: failure mechanisms. Compos. Sci. Technol. **58**, 355–368 (1998). doi:10.1016/S0266-3538(97)00123-1

Tsotsis, T.K., Keller, S., Bardis, J., Bish, J.: Preliminary evaluation of the use of elevated pressure to accelerate thermo-oxidative aging in composites. Polym. Degrad. Stabil. **64**, 207–212 (1999). doi:10.1016/S0141-3910(98)00190-6

Valentini, L., Biagiotti, J., Kenny, J.M., Manchado, M.A.L.: Physical and mechanical behavior of single-walled carbon nanotube/polypropylene/ethylene–propylene–diene rubber nanocomposites. J. Appl. Polym. Sci. **89**, 2657–2663 (2003). doi:10.1002/app. 12319

Wang, S., Chung, D.D.L.: Effect of moisture on the interlaminar interface of a carbon fibre polymer-matrix composites, studied by contact electrical resistivity measurement. Compos. Interface. **9**(5), 453–458 (2002). doi:10.1163/15685540260256546

Wang, S., Kowalik, D.P., Chung, D.D.L.: Effects of the temperature, humidity, and stress on the interlaminar interface of carbon fibre polymer-matrix composites, studied by contact electrical resistivity measurements. J. Adhes. **78**, 189–200 (2002). doi:10.1080/00218460210384

Wang, J.G., Fang, Z.P., Gu, A.J., Xu, L.H., Liu, F.: Effect of amino-fucntionalization of multi-walled carbon nanotubes on the dispersion with epoxy resin matrix. J. Appl. Polym. Sci. **100**, 97–104 (2006). doi:10.1002/app. 22647

Watson, K.A., Ghose, S., Delozier, D.M., Smith, J.G., Connell, J.W.: Transparent, flexible, conductive carbon nanotube coatings for electrostatic charge mitigation. Polymer **46**, 2076–2085 (2005). doi:10.1016/j.polymer.2004.12.057

Weitsman, Y.: Moisture in composites: sorption and damage. In: Reifsnider, K.L. (ed.) Fatigue of Composite Materials, pp. 385–429. Elsevier Science Publishers, Amsterdam (1991)

Wilkins, R., Pulikkathara, M.X., Khabashesku, V.N., Barrera, E.V., Vaidyanathan, R.K., Thibeault, S.A.: Materials for space applications. Mater. Res. Soc. Symp. Proc. **851**, 267–278 (2005)

Windle, A.: Two defining moments: a personal view. Compos. Sci. Technol. **67**, 929–930 (2007). doi:10.1016/j.compscitech.2006.07.037

Won, Y.G., Galy, J., Ge'rard, J.F., Pascault, J.P., Vellenger, V., Verdu, J.: Influence of sequence of chemical reactions on kinetics and solid state behavior for bisphenol A diglycidyl ether–bisphenol A–sulfanilamide ternary blends. Polymer **31**, 179–186 (1990). doi:10.1007/PL00010783

Woo, R.S.C., Chen, Y., Zhu, H., Li, J., Kim, J.K., Leung, C.K.Y.: Environmental degradation of epoxy-organoclay nanocomposites due to UV exposure. Part I: photo-degradation. Compos. Sci. Technol. **67**, 3448–3456 (2007). doi:10.1016/j.compscitech.2007.03.004

Woo, R.S.C., Zhu, H., Leung, C.K.Y., Kim, J.K.: Environmental degradation of epoxy-organoclay nanocomposites due to UV exposure. Part II: residual mechanical properties. Compos. Sci. Technol. **68**, 2149–2155 (2008). doi:10.1016/j.compscitech.2008.03.020

Wood, C.A., Bradley, W.L.: Determination of the effect of seawater on the interfacial strength of an interlay E-glass/graphite/epoxy composite by In-Situ Observation of Transverse Cracking in an Environmental SEM. Compos. Sci. Technol. **57**(8), 1033–1043 (1997). doi:10.1016/S0266-3538(96)00170-4

Xiao, G.Z., Shanahan, M.E.R.: Water absorption and desorption in an epoxy resin with degradation. J. Polym. Sci. Part B: Polym. Phys. **35**, 2659–2670 (1997). doi:10.1002/(SICI)1099-0488(19971130)35:16<2659::AID-POLB9>3.0.CO;2-K

Xiao, G.Z., Delamar, M., Shanahan, M.: Irreversible interactions between water and DGEBA/DDA epoxy resin during hygrothermal aging. J. Appl. Polym. Sci. **65**, 449–458 (1997). doi:10.1002/(SICI)1097-4628(19970718)65:3<449::AID-APP4>3.0.CO;2-H

Yip, M.C., Wu, H.Y.: Fatigue and electrical properties of CNT/phenolic composites under moisture-temperature effects. Key Eng. Mater. **334–335**, 769–772 (2007). doi:10.4028/www.scientific.net/KEM.334-335.769

Youssef, Z., Jacquemin, F., Gloaguen, D., Guillen, R.: A multi-scale analysis of composite structures: application to the design of accelerated hygrothermal cycles. Compos. Struct. **82**, 302–309 (2008). doi:10.1016/j.compstruct.2007.01.008

Zhang, Y.C., Wang, X.: Hygrothermal effects on interfacial stress transfer characteristics of carbon nanotubes-reinforced composites system. J. Reinf. Plast. Compos. **25**, 71–88 (2006a). doi:10.1177/0731684406055456

Zhang, Y.C., Wang, X.: Hygrothermal effects on initial frictional pull-out force of CNTs-reinforced composites. J. Therm. Stress. **29**, 67–92 (2006b). doi:10.1080/01495730500257425

Zhou, J., Lucas, J.P.: Hygrothermal effects of epoxy resin. Part I: the nature of water in epoxy. Polymer **40**, 5505–5512 (1999a). doi:10.1016/S0032-3861(98)00790-3

Zhou, J., Lucas, J.P.: Hygrothermal effects of epoxy resin. Part II: variations of glass transition temperature. Polymer **40**, 5513–5522 (1999b). doi:10.1016/S0032-3861(98)00791-5

Zunjarrao, S.C., Sriraman, R., Singh, R.P.: Effect of processing parameters and clay volume fraction on the mechanical properties of epoxy-clay nanocomposites. J. Mater. Sci. **41**, 2219–2228 (2006). doi:10.1007/s10853-006-7179-2

MIX
Papier aus verantwortungsvollen Quellen
Paper from responsible sources
FSC® C105338

If you have any concerns about our products,
you can contact us on
ProductSafety@springernature.com

In case Publisher is established outside the EU,
the EU authorized representative is:
Springer Nature Customer Service Center GmbH
Europaplatz 3, 69115 Heidelberg, Germany

Printed by Libri Plureos GmbH
in Hamburg, Germany